Mannava V. K. Sivakumar · Raymond P. Motha (Eds.)

Managing Weather and Climate Risks in Agriculture

World Meteorological Organization

India Meteorological Department

Government of India,
Ministries of Science and Technology
and Earth Sciences

Mannava V. K. Sivakumar
Raymond P. Motha
Editors

Managing Weather and Climate Risks in Agriculture

With 134 Figures

Springer

Dr. Mannava V.K. Sivakumar
Agricultural Meteorology Division
World Meteorological Organization
7bis, Avenue de la Paix
1211 Geneva 2
Switzerland

Dr. Raymond P. Motha
USDA/OCE/WAOB
1400 Independence Ave. SW
Room 5133
Washington D.C. 20250
USA

Library of Congress Control Number: 2007928832

ISBN   978-3-540-72744-6 Springer Berlin Heidelberg New York

This work is subject to copyright. All rights are reserved, whether the whole or part of the material is concerned, specifically the rights of translation, reprinting, reuse of illustrations, recitation, broadcasting, reproduction on microfilm or in any other way, and storage in data banks. Duplication of this publication or parts thereof is permitted only under the provisions of the German Copyright Law of September 9, 1965, in its current version, and permission for use must always be obtained from Springer-Verlag. Violations are liable to prosecution under the German Copyright Law.

Springer is a part of Springer Science+Business Media
springeronline.com
© Springer-Verlag Berlin Heidelberg 2007

The use of general descriptive names, registered names, trademarks, etc. in this publication does not imply, even in the absence of a specific statement, that such names are exempt from the relevant protective laws and regulations and therefore free for general use.

Cover design: deblik, Berlin
Production: Almas Schimmel
Typesetting: Satz-Druck-Service (SDS), Leimen
Printed on acid-free paper   30/3180/as   5 4 3 2 1 0

Foreword

Decision making in agricultural production is a complex process in which many risks need to be considered for an informed decision to be made. Farmers face many types of risks related to production, marketing, legal, social and human aspects. In many parts of the world, weather and climate are one of the biggest production risk and uncertainty factors impacting on agricultural systems performance and management. Extreme climatic events such as severe droughts, floods, cyclonic systems or temperature and wind disturbances strongly impede sustainable agricultural development. Hence weather and climate variability is considered in evaluating all environmental risk factors and coping decisions.

Coping with agrometeorological risk and uncertainties is the process of measuring or otherwise assessing agrometeorological risks and uncertainties and then developing strategies to cope with these risks. There are many challenges. In many developing countries technology generation, innovation and adoption are too slow to sufficiently counteract the increasingly negative effects of degrading environmental conditions. Even in the high rainfall areas, increased probability of extreme events can for example cause increased nutrient losses due to excessive leaching, runoff and water logging. Lack of attention to preparedness and response strategies is a major challenge.

Currently there are many opportunities that can assist in coping effectively with agrometeorological risks and uncertainties. One of the most important strategies is improved use of climate knowledge and technology, which includes the development of monitoring and response mechanisms to current weather. By providing new, quantitative information about he environment within which the farmers operate or about the likely outcome of alternative or relief management options, uncertainties in crop productivity can be reduced. Quantification is essential and computer simulations can assist such information and may be particularly useful to quantitatively compare alternative management and relief options in areas where seasonal climatic variability is high and/or that are prone to extremes. Given the current recognition of the importance of preparedness to cope with risks and uncertainties as compared to the practice of reactive responses, it is necessary to take stock of the opportunities that exist in coping with agrometeorological risks, to develop suitable practices/strategies and to disseminate them widely.

It is with this background that WMO had organized the International Workshop on Agrometeorological Risk Management: Challenges and Opportunities in conjunction with the 14th Session of the Commission for Agricultural Meteorology of WMO held in New Delhi, India. The workshop was co-sponsored by the Asia-Pacific Network for Global Change Research (APN), the Bureau of Meteorology,

Australia; the Centre Technique de Coopération Agricole et Rurale – Technical Centre for Agricultural and Rural Co-operation (CTA); the Food and Agriculture Organization of the United Nations (FAO); the India Meteorological Department, Météo-France; the Ministries of Science and Technology and Earth Sciences, Government of India; the UK Met Office; and the United States Department of Agriculture (USDA).

The workshop reviewed the components of farmers' agrometeorological coping strategies with risks and uncertainties in different regions of the world and discussed the major challenges to these coping strategies, such as reducing the vulnerability of different agro-ecosystems to weather and climate related risks and uncertainties, access to technological advances, particularly in developing countries, and attention to preparedness and response strategies. Structural measures such as irrigation, water harvesting etc., and non-structural measures such as use of seasonal to inter-annual climate forecasts and improved application of medium-range weather forecasts for strategic and tactical management of agriculture were addressed. A special evening symposium on weather risk insurance for agriculture reviewed the use of crop insurance strategies and schemes to reduce the vulnerability of the farming communities to agrometeorological risks.

I hope that the papers presented in this book will serve as a significant source of information to all agencies and organizations involved with designing and implementing appropriate strategies and related services to farmers in their efforts cope with weather and climate risks.

M. Jarraud
Secretay-General
World Meteorological Organization

Preface

In many parts of the world climate change and extreme climatic events such as severe droughts, floods, storms, tropical cyclones, heat-waves, freezes and extreme winds are one of the biggest production risk and uncertainty factors impacting agricultural systems performance and management. These events direct influence on the quantity and quality of agricultural production, and in many cases adversely affect it. Although agrometeorology particularly deals with production risks and evaluation of possible production decisions, to solve local problems of farming systems the other risk factors have to be taken into account. Inappropriate management of agroecosystems, compounded by severe climatic events such as recurrent droughts, from West Africa to northern Sudan, have tended to make the drylands increasingly vulnerable and prone to rapid degradation and hence desertification.

In the context of the need for increased agricultural productivity to meet the food and nutritional needs of the growing populations in the world, coping with agrometeorological risk and uncertainties is a very important issue and there are many challenges as well as opportunities as explained in the foreword by Mr M. Jarraud, the Secretary-General of WMO. Accordingly, the Management Group of the Commission for Agricultural Meteorology (CAgM) of WMO recommended the organization of the International Workshop on Agrometeorological Risk Management: Challenges and Opportunities from 25 to 27 October 2006 in New Delhi, India in conjunction with the 14th Session of the Commission for Agricultural Meteorology of WMO. The workshop, hosted by the India Meteorological Department (IMD) and the Ministry of Science and Technology and Earth Sciences of the Government of India, was attended by 188 participants from 78 countries. The specific objectives of the workshop were:
- To identify and assess the components of farmers' agrometeorological coping strategies with risks and uncertainties in different regions of the world, e.g. extreme climatic events (droughts, floods, cyclonic systems, temperature and wind disturbances etc.), inadequate attention to agroclimatic characteristics of a location, lack of timely information on weather and climate risks and uncertainties, lack of crop diversification etc;
- To discuss the major challenges to these coping strategies with agrometeorological risks, such as reducing the vulnerability of different agro-ecosystems to weather and climate related risks and uncertainties, access to technological advances -- particularly in developing countries --, attention to preparedness and response strategies, to agrometeorological services, to training of intermediaries between NMHSs and farmers etc;

- To review the opportunities for farmers to cope with agrometeorological risks and uncertainties in different parts of the world, e.g. with structural measures (irrigation, water harvesting, microclimate management and manipulation and other preparedness strategies) and non-structural measures (use of seasonal to inter-annual climate forecasts, improved application of medium-range weather forecasts) for strategic and tactical management of agriculture;
- To provide on-farm examples of appropriate coping strategies for minimizing agrometeorological risks and uncertainties and of sustainable agriculture;
- To review, through appropriate case studies, the use of crop insurance strategies and schemes to reduce the vulnerability of the farming communities to agrometeorological risks;
- To discuss and recommend suitable policy options, such as agrometeorological services for coping with agrometeorological risks and uncertainties in different parts of the world.

Altogether there were 8 sessions (including opening and closing session) in the workshop during which 25 invited papers were presented. In the workshop sessions, firstly weather and climate events and risks to farming from droughts, floods, cyclones and high winds, and extreme temperatures were identified through risk and risk characterization. Papers on approaches to dealing with risks highlighted preparedness planning, risk assessments and improved early warning systems which can lessen the vulnerability of society to weather and climate risks. Enterprise diversification, contract hedging, crop insurance, weather derivatives and weather index insurance play a key role in developing agricultural risk management strategies. A special session examined the use of crop insurance strategies and schemes to reduce the vulnerability of the farming communities to risks posed by weather and climate extremes.

A number of strategies were identified to cope with risks. These include the use of seasonal forecasts in agriculture, forestry and land management to assist alleviation of food shortages, drought and desertification. The use of integrated agricultural management and crop simulation models with climate forecasting systems give the highest benefits. Strategies to improve water management and increase the efficient use of water included crop diversification and better irrigation. Especially important was the application of local indigenous knowledge. A combination of locally adapted traditional farming technologies, seasonal weather forecasts and warning methods were important for improving yields and incomes. Challenges to coping strategies were many and identified in several papers. Particularly important was the impact of different sources of climate variability and change on the frequency and magnitude of extreme events. Lack of systematic data collected from disasters impeded future preparedness, as did the need for effective communication services for the timely delivery of weather and climate information to enable effective decision making. Finally a range of policy options to cope with such risks were presented. These included contingency planning, use of crop simulation modelling, and use of agrometeorological services.

All the participants in the workshop were engaged in discussions on these papers and developed several useful recommendations for all organizations involved in agrometeorological risk management, particularly the National Meteorological and Hydrological Services. These have been presented in the final paper in this book.

As Editors of this volume, we would like to thank all the authors for their efforts and for their cooperation in bringing out this volume in time. We are most grateful to the India Meteorological Department (IMD) and the Ministry of Science and Technology and Earth Sciences of the Government of India for hosting this meeting and to the Secretary-General of WMO for his continuous support and encouragement.

M.V.K. Sivakumar
R.P. Motha
Editors

Contents

1 Extreme Weather and Climate Events, and Farming Risks
John Hay

1.1 Introduction 1
1.2 Risk and Risk Management – Some Basic Concepts 3
    1.2.1 Step A – Risk Scoping. 4
    1.2.2 Step B – Risk Characterization and Evaluation 4
    1.2.3 Step C – Risk Management. 4
    1.2.4 Step D – Monitoring and Review 6
1.3 Farming Risks 6
1.4 Risk Characterization 7
    1.4.1 Weather Extremes. 9
    1.4.2 Climate Anomalies 11
1.5 Changing Risk 12
1.6 Risk Management 16
1.7 Conclusions. 18
Acknowledgements. 18
References. 19

2 Preparedness and Coping Strategies for Agricultural Drought Risk Management: Recent Progress and Trends
Donald A. Wilhite

2.1 Introduction 21
2.2 Agricultural Drought Risk Management: Understanding the Hazard and Societal Vulnerability. 22
2.3 Drought as Hazard. 22
2.4 Drought: Understanding Vulnerability 23
2.5 Drought Types and Vulnerability 24
2.6 Decision-Support Tools for Drought Risk Management. 26
    2.6.1 U.S. Drought Monitor 26
    2.6.2 Drought Monitor: Decision Support System. 29
    2.6.3 The Drought Impact Reporter: A Web-based Impact Assessment Tool and Database 31
    2.6.4 Drought Risk Atlas 35
    2.6.5 Vegetation Drought Response Index (VegDRI) 35

	2.6.6	Ranching Drought Plan: A Drought Planning Tool for Livestock and Forage Producers.	36
2.7	Summary		37
	References.		37

3 Challenges to Coping Strategies with Agrometeorological Risks and Uncertainties in Africa
Elijah Mukhala, Adams Chavula

3.1	Introduction	39
3.2	Farmers' coping strategies	41
3.3	Provision of climatic information	42
	3.3.1 Intra-seasonal distribution.	43
	3.3.2 Language/terminology and communication.	43
	3.3.3 Capacity of key institutions	43
	3.3.4 Stakeholder awareness/training.	44
	3.3.5 Tailored forecasts	44
	3.3.6 Key relationships	44
	3.3.7 Timely issuance	45
3.4	Key recommendations to address identified weaknesses	45
	3.4.1 Forecast improvement	45
	3.4.2 Access to technological advances	47
	3.4.3 Structural and non-structural measures to mitigate risks and uncertainties	47
	3.4.4 Crop insurance activities.	48
3.5	Conclusions.	49
	References.	50

4 Challenges to Coping Strategies with Agrometeorological Risks and Uncertainties in Asian Regions
L.S. Rathore, C.J. Stigter

4.1	Introduction	53
4.2	Challenges to disaster risk mainstreaming	54
	4.2.1 Adaptation strategies.	54
	4.2.2 High intensity rainfall and floods.	55
	4.2.3 Tropical storms, tornadoes and strong winds	55
	4.2.4 Extreme temperatures including heat waves and cold waves . . .	56
	4.2.5 Droughts.	56
	4.2.6 Wildfires and bushfires.	57
4.3	Challenges to contingency planning and responses	58
4.4	Challenges to mitigation practices as a coping strategy	58
	4.4.1 Impact reductions.	58
	4.4.2 High intensity rainfall and floods.	59
	4.4.3 Tropical storms, tornadoes and strong winds	60
	4.4.4 Extreme temperatures including heat waves and cold waves. . . .	60

		4.4.5 Droughts	61
		4.4.6 Wildfires and bushfires	61
4.5	Challenges to preparedness as a coping strategy		62
4.6	Challenges to methodologies in disaster science to support preparedness		63
		4.6.1 Early warning systems for assessing agrometeorological risks	63
		4.6.2 Remote sensing for spatial information	64
		4.6.3 Data analysis in research	65
4.7	Agrometeorological Advisory Service (AAS)		65
4.8	Conclusion and recommendation		66
	References		66

5 Challenges and Strategies to face Agrometeorological Risks and Uncertainties – Regional Perspective in South America
Constantino Alarcón Velazco

5.1	Introduction	71
5.2	Climatology and Geography of South America	72
5.3	Natural Phenomena that affect Agriculture in South America	73
	5.3.1 Landslides	73
	5.3.2 Volcanic Eruptions	73
	5.3.3 Floods	73
	5.3.4 Droughts	74
	5.3.5 Extreme temperatures including heat waves and cold waves	74
	5.3.6 The Climate Change	74
	5.3.7 El Niño – La Niña Phenomenon	75
	5.3.8 Strong winds	76
5.4	Vulnerability of the Region	76
5.5	Capacities and Resources in the region to face Agrometeorological Risks and Uncertainties	77
	5.5.1 Economic status of the countries	77
	5.5.2 Government policies	77
	5.5.3 Creation and strengthening of specialized institutions	77
	5.5.4 Risk identification and analysis	78
	5.5.5 Monitoring networks and early warning	78
	5.5.6 Information on the risks for decision making	79
5.6	Defining Policies and Tools to face Agrometeorological Risks and Uncertainties	79
5.7	Strategies to cope with Agrometeorological Risks and Uncertainties	80
5.8	Conclusions	81
	References	82

6 Agrometeorological Risk and Coping Strategies – Perspective from Indian Subcontinent
N. Chattopadhyay, B. Lal

6.1 Introduction 83
6.2 Extreme weather events and its impacts on Indian agriculture. 84
    6.2.1 Cold wave 84
    6.2.2 Drought.. 86
    6.2.3 Fog 88
    6.2.4 Thunderstorm, Hailstorm and Dust storm................. 88
    6.2.5 Heat waves 89
    6.2.6 Tropical Cyclones 90
    6.2.7 Floods... 93
6.3 Crop Insurance... 94
6.4 Strategies adopted in areas with high weather risk................ 95
6.5 Conclusions.. 97
References.. 97

7 Challenges to coping strategies in Agrometeorology: The Southwest Pacific
James Salinger

7.1 Introduction 99
7.2 El Niño-Southern Oscillation (ENSO) 99
7.3 Decadal Variability.. 103
7.4 Regional Warming.. 103
7.5 Challenges to Agriculture and Forestry 105
    7.5.1 ENSO.. 105
    7.5.2 Tropical Cyclones 107
    7.5.3 Regional Warming 107
7.6 Discussion and Conclusions 108
References... 110

8 Challenges to agrometeorological risk management – regional perspectives: Europe
Lučka Kajfež Bogataj, Andreja Sušnik

8.1 Introduction 113
8.2 Seasonal weather forecasts for crop yield modeling in Europe 114
8.3 Climate change as a challenge to agrometeorological risk managementin Europe.. 115
    8.3.1 The use of different high resolution climate models in Europe .. 115
    8.3.2 Expected impacts of climate change in Europe during this century 117
    8.3.3 Increasing drought Risk with Global Warming in Europe..... 119
    8.3.4 Options for future adaptation strategies 120
    8.3.5 European agrometeorological research needs 121

8.4.	Conclusions	122
	References	123

9 Methods of Evaluating Agrometeorological Risks and Uncertainties for Estimating Global Agricultural Supply and Demand
Keith Menzie

9.1	Introduction	125
9.2	Sources of Risk	125
9.3	Risk, Uncertainty, and the Agricultural Marketing System	127
9.4	Information - the Key to Efficient Market Function	129
9.5	Global Crop Assessment Methods and Risk Reduction – Tools and Analysis	130
9.6	Global Crop Assessment Methods and Risk Reduction – the Case of Brazilian Soybeans	136
9.7	Conclusions	139
	References	140

10 Weather and climate and optimization of farm technologies at different input levels
Josef Eitzinger, Angel Utset, Miroslav Trnka, Zdenek Zalud, Mikhail Nikolaev, Igor Uskov

10.1	Introduction	141
10.2	Strategies for optimizing farm technologies in various agricultural systems	144
	10.2.1 Optimization of farm technologies and water resources	145
	10.2.2 Optimization of farm technologies and soil resources	149
	10.2.3 Optimization of farm technologies and crop resources	152
	10.2.4 Optimization of farm technologies and the microclimate of crop stands	162
10.3	Conclusions	164
	Acknowledgement	165
	References	165

11 Complying with farmers' conditions and needs using new weather and climate information approaches and technologies
C.J. Stigter, Tan Ying, H.P. Das, Zheng Dawei, R.E. Rivero Vega, Nguyen Van Viet, N.I. Bakheit, Y.M. Abdullahi

11.1	Introduction	171
11.2	Complying with conditions and needs	172
11.3	Differentiated information needs and channels for various farmers	174
	11.3.1 Information demands of different income levels in poor areas of China	174
	11.3.2 Differentiation between income levels in poor areas of China	176

		11.3.3	Information channels for different income levels in poor areas of China 178

11.3.3 Information channels for different income levels
in poor areas of China 178
11.3.4 Demand and supply of information for different income levels
in poor areas of China 179
11.3.5 General implications of the findings for different income levels
in poor areas of China 179
11.4 Implications for information approaches and technologies 180
11.4.1 Poor farmers 180
11.4.2 Low-income farmers 181
11.4.3 Middle-income farmers 181
11.4.4 Richer farmers 182
11.4.5 Other developing countries 183
11.5 What WMO/CAgM should realize as implications of the above 186
References 187

12 Information Technology and Decision Support System for On-Farm Applications to cope effectively with Agrometeorological Risks and Uncertainties
Byong-Lyol Lee

12.1 Introduction 191
12.1.1 On-Farm Applications Against Risks 191
12.2 Risk & Uncertainty in Agriculture 193
12.2.1 Agrometeorological risks 193
12.2.2 Risk Management in Agrometeorology 193
12.3 Decision-making Support Against Risks 194
12.3.1 Emergency Response System 194
12.4 Information Technology Required 199
12.4.1 Requirements for Agrometeorological Products 199
12.4.2 Requirements for DMSS Infrastructure 200
12.5 Resource Sharing System: Case of WAMIS 200
12.5.1 WAMIS as a Web Portal 201
12.6 Discussion & Conclusions 206
References 207

13 Coping Strategies with Agrometeorological Risks and Uncertainties for Crop Yield
Lourdes V. Tibig, Felino P. Lansigan

13.1 Challenges and opportunities 209
13.2 Types of coping strategies with agrometeorological risks
and uncertainties for crop yield 210
13.2.1 Optimal and sustainable utilization of resources 210
13.2.2 Change in cultural practices or improved farming practices . . .213
13.2.3 Modifications of resource potential including controlled
micro-climates 214
13.2.4 Local indigenous knowledge systems/networks 214

		13.2.5 Access to extension services 214
		13.2.6 Technological innovations 215
		13.2.7 Others, including resilience and divestment of natural capital . . 216
13.3	Some examples of coping strategies 216	
	13.3.1 Canada 216	
	13.3.2 The United States 218	
	13.3.3 Latin America 219	
	13.3.4 Africa. 219	
13.4	Case studies: Regional/national coping strategies 220	
	13.4.1 The Philippine experience: Empowering farmers for rural development. 220	
	13.4.2 The West African semi-arid tropics (WASAT) 227	
	13.4.3 Improving rice-based cropping systems in the Indo-Gangetic Plains and in north-west Bangladesh. 229	
	13.4.4 Special Case: Small-holder rubber production in South Sumatra, Indonesia 232	
13.5	Conclusions. 233	
	References 233	

14 Water management in a semi-arid region: an analogue algorithm approach for rainfall seasonal forecasting
Giampiero Maracchi, Massimiliano Pasqui and Francesco Piani

14.1	Introduction 237
14.2	Methods and Dataset 238
14.3	Skill evaluation 241
14.4	Conclusions. 242
	References. 243

15 Water Management – Water Use in Rainfed Regions of India
YS Ramakrishna, GGSN Rao, VUM Rao, AVMS Rao and KV Rao

	Abstract 245
15.1	Introduction 245
15.2	Water Resources of the Country 247
15.3	Rainwater Management. 248
15.4	Issues and Perspective in Water Management 248
15.5	Strategies for Improving the Water Management and Water Use Efficiencies 249
15.6	Water Management through Watershed Program 251
	Conclusions. 262
	References 262

16 Examples of coping strategies with agrometeorological risks and uncertainties for Integrated Pest Management
A.K.S. Huda, T. Hind-Lanoiselet, C. Derry, G. Murray and R.N. Spooner-Hart

16.1 Introduction 265
    16.1.1 Crop Diseases - Stripe rust in wheat and Sclerotinia rot in canola 267
    16.1.2 Implications for technology transfer 269
    16.1.3 Resource allocation for risks 270
    16.1.4 Supportive Decision-Making Tools 271
    16.1.5 Effectiveness of decision-making tools 271
    16.1.6 Importance of Experimental Observation 272
    16.1.7 Desirable level of complexity 272
    16.1.8 Economic balance in control 273
    16.1.9 Towards the Future 273
16.2 Conclusions 277
Acknowledgements 278
References 278

17 Coping Strategies with Agrometeorological Risks and Uncertainties for Drought Examples in Brasil
O. Brunini, Y. M. T. da Anunciação, L. T.G. Fortes, P. L. Abramides, G. C. Blain, A. P. C. Brunini, J. P. de Carvalho

17.1 Introduction 281
17.2 Methodologies to Assess Precipitation Anomaly and Drought 286
    17.2.1 Meteorological Indices 286
    17.2.2 Agrometeorological Indices 290
17.3 Results and Analysis 296
    17.3.1 Meteorological Aspects of Drought Monitoring and Prediction .. 296
    17.3.2 Agrometeorological Aspects of Drought 304
    17.3.3 Drought Monitoring and Mitigation Center 306
    17.3.4 Climatic Risk Zoning 309
17.4 Conclusions 313
References 313

18 Coping Strategies with Desertification in China
Wang Shili, Ma Yuping, HouQiong, Wang Yinshun

18.1 Introduction 317
18.2 Status of Desertification in China 318
    18.2.1 Status of Desertified Land 318
    18.2.2 Status of land most vulnerable to sand encroachment 319
    18.2.3 Dynamic Changes of Desertification 319

18.3 Development and Causes of Desertification in North China 319
    18.3.1 Development and Cause Analysis 319
    18.3.2 Possible influence of climate change on desertification. 321
18.4 Desertification Monitoring in China. 321
    18.4.1 Indicator system for desertification monitoring and evaluation . . 321
    18.4.2 Desertification monitoring in China 322
18.5 China's Key Forestry Programs on Combating Desertification. 324
    18.5.1 Program for converting cropland to forest/shrubbery 324
    18.5.2 Programme of Combating desertification in the wind
           sand sources areas affecting Beijing and Tianjin city 325
    18.5.3 Three-North Shelterbelt Programme and Shelterbelt Programme
           in upper and middle reaches of the Yangtze River 326
18.6 Practical Strategies and Countermeasures to Combat Desertification. . . 326
    18.6.1 Stabilizing sands techniques system 327
    18.6.2 Shelterbelt techniques system 329
    18.6.3 Typical models in combating desertification in China 330
18.7 Services for combating desertificationin Chinese Meteorological Offices. . 334
    18.7.1 Research on desertification development and combating
           in terms of meteorological conditions 334
    18.7.2 Monitoring and assessing services to combating desertification
           of grassland 336
    18.7.3 Monitoring and predicting of dust storms in China 338
18.8 Conclusions and Discussion 339
    Acknowledgements. 340
    References. 340

19 Coping strategies with agrometeorological risksand uncertainties for water erosion, runoff and soil loss
P.C. Doraiswamy, E.R. Hunt, Jr., V.R.K. Murthy

19.1 Introduction 343
19.2 Agrometeorological coping strategies 344
19.3 Soil Management Strategies. 346
    19.3.1 Organic Matter 346
    19.3.2 Tillage Practices. 346
    19.3.3 Crop Management Strategies 349
    19.3.4 Mechanical Control Strategies. 350
19.4 Conclusions. 351
    References 352

20 Developing a global early warning system for wildland fire
Michael A. Brady, William J. de Groot, Johann G. Goldammer, Tom Keenan, Tim J. Lynham, Christopher O. Justice, Ivan A. Csiszar, Kevin O'Loughlin

20.1 Introduction	355
20.1.1 EWS-Fire Proposal	356
20.2 Objectives and Expected Impact of EWS-Fire	357
20.3 Planned System Development	358
20.3.1 Warning System Design	358
20.3.2 Operational Implementation	359
20.3.3 Technology Transfer	360
20.4 Implementing Organizations and Division of Tasks	360
20.4.1 Natural Resources Canada - Canadian Forest Service (CFS)	361
20.4.2 Bureau of Meteorology Research Centre (BMRC)	361
20.4.3 Bushfire Cooperative Research Centre, Australia	362
20.4.4 University of Maryland (UMD), USA, acting on behalf of GOFC-GOLD	362
20.4.5 Global Fire Monitoring Center (GFMC), Germany on behalf of the UNISDR Wildland Fire Advisory Group / Global Wildland Fire Network and the United Nations University (UNU)	362
20.4.6 Global Observation of Forestand Land Cover Dynamics (GOFC-GOLD) Secretariat, Edmonton, Canada	362
20.5 Sustainability	362
20.6 Case Study in EWS-Fire Development	363
20.6.1 System Development	363
20.6.2 Operational FDRS	364
20.6.3 Lessons Learned	364
References	366

21 Scientific and Economic Rationale for Weather Risk Insurance for Agriculture
Peter Höppe

21.1 Introduction	367
21.2 Natural Disasters and Losses	367
21.3 Climate Change and Natural Disasters	370
21.4 Agricultural Risk Insurance	371
21.4.1 Crop Insurance Products	373
21.4.2 Crop Insurance in Developing Countries	373
21.5 Conclusions	374
References	375

22 Weather index insurance for coping with risks in agricultural production
Ulrich Hess

22.1 Introduction 377
    22.1.1 Are there any effective precedents for agricultural insurance mechanismsin developing countries? 377
    22.1.2 Is this kind of insurance only suitablefor large-scale commercial farmers? 378
    22.1.3 Is India's insurance program sustainable? 378
22.2 Risk and Risk Management in Agriculture 379
    22.2.1 Informal risk management mechanisms 380
22.3 Crop Insurance Programs in Developed Countries 383
    22.3.1 The United States 384
    22.3.2 Canada 385
    22.3.3 Spain 386
    22.3.4 Experiences of developed countries provide inadequate models for developing countries 387
22.4 Weather index insurance alternatives 388
    22.4.1 Basic characteristics of an index. 388
    22.4.2 Structure of index insurance contracts 389
    22.4.3 Relative advantages and disadvantages of index insurance 391
    22.4.4 The trade-off between basis risk and transaction costs 391
    22.4.5 Where index insurance is inappropriate 392
22.5 Application of weather index insurance in developing countries: The role of government 393
    22.5.1 Premise: The concept of risk layering. 393
    22.5.2 Policy instruments 395
22.6 Overview of ongoing agricultural risk pilot programs. 400
    22.6.1 India 400
    22.6.2 Malawi 401
22.7 Conclusions 403
    References 404

23 Weather Risk Insurance for Coping with Risks to Agricultural Production
Pranav Prashad

23.1 Weather and Indian Agriculture 407
23.2 Introduction to Weather Insurance 407
    23.2.1 Process of making an index based product 407
23.3 Advantages of Index based Insurance products like Weather Insurance 409
23.4 Initiatives in Weather Insurance 409
23.5 Innovative ways to reach to the hinterland – reduction of basis risk . . . 410
23.6 Designing Crop and situation specific products 410
    23.6.1 Wheat 410
    23.6.2 Apples 411

	23.6.3 Salt manufacturing 412
23.7 Snapshot of 2005-2006. 412
23.8 Distribution: a key challenge 412
23.9 Conclusions. 414

24 Contingency planning for drought – a case study in coping with agrometeorological risks and uncertainties
Roger C Stone, Holger Meinke

24.1 Introduction 415
24.2 The basis of drought contingency planning. 416
24.3 Preparedness strategies 422
24.4 Risk management systems and tools. 426
24.5 Issues associated with contingency planning for drought under climate change 428
24.6 Conclusions. 430
References. 430

25 Agrometeorological services to cope with risks and uncertainties
Raymond P. Motha, V.R.K. Murthy

25.1. Introduction 435
25.2 Weather, Natural Disasters, and Agriculture 435
    25.2.1 Fundamental importance of weather in agriculture 436
    25.2.2 Impact of natural disasters in agriculture, rangeland, forestry, and environment 436
    25.2.3 The role of Indigenous Technical Knowledge (ITK) in agrometeorological services 439
    25.2.4 The role of contemporary technological advances in agrometeorological services 441
25.3. Operational Agrometeorological Services to Cope with Risks and Uncertainties of Natural Disasters 444
    25.3.1 United States (U.S.A.). 444
    25.3.2 India 447
25.4 Strategies to Improve the Agrometeorological Services to Cope with Risks and Uncertainties. 449
    25.4.1 Improving the agrometeorological services 450
    25.4.2 Improving the support systems of agrometeorological services. . .456
    25.4.3 A comprehensive agrometeorological service strategy to cope with risks and uncertainties 458
25.5 Conclusions. 459
References. 460

26 Using Simulation Modelling as a Policy Option in Coping with Agrometeorological Risks and Uncertainties
Simone Orlandini, A. Dalla Marta, L. Martinelli

26.1 Introduction 463
26.2 Conditions of model implementation and application. 464
26.3 Examples of Using Agrometeorological Models 466
    26.3.1 Models for soil erosion 466
26.4 Water balance and irrigation 468
26.5 Crop protection. 471
26.6. Early Warning Systems (EWS) 474
26.7 Conclusions 475
    References 476

27 Managing Weather and Climate Risks in Agriculture Summary and Recommendations
Mannava V.K. Sivakumar, Raymond P. Motha

27.1 Introduction 477
27.2 Risk and Risk Management in Agriculture 477
27.3 Addressing Agrometeorological Risk Management during the Workshop 478
27.4 Workshop Summary. 479
    27.4.1 Risk in Agriculture 479
    27.4.2 Risk and Risk Characterization 480
    27.4.3 Approaches to Dealing with Risks 481
    27.4.4 Risk Coping Strategies 483
    27.4.5 Perspectives for Farm Applications 484
    27.4.6 Challenges to Coping Strategies 485
27.5 Recommendations 486
    27.5.1 Risk Management 486
    27.5.2 Risk Management Tools 486
    27.5.3 Research Needs 487
    27.5.4 Emphasis on User Needs 488
    27.5.5 Communication 488
    27.5.6 Marketing 489
    References 489

Subject Index 493

List of Contributors

Y.M. Abdullahi
Ahmadu Bello University
National Agricultural Extension
and Rural Living Services
Zaria, Nigeria
E-mail: ymabdullahi@yahoo.com

P.L. Abramides
Instituto Agronômico
Av- Barão de Itapura, 1481
13.020-902 ,Campinas
Sao Paulo, Brazil
E-mail: pedro@apta.sp.gov.br

N.I. Bakheit
Sinnar University
Faculty of Agriculture,
Abu Naama, Sinnar, Sudan
E-mail: Nagibrahim@hotmail.com

G.C. Blain
Instituto Agronômico -Ciiagro
Av- Barão de Itapura, 1481
13.020-902 ,Campinas
Sao Paulo, Brazil
E-mail: gabriel@iac.sp.gov.br

Lučka Kajfež Bogataj
University of Ljubljana
Agronomy Department
Jamnikarjeva 101, 000
Ljubljana, Slovenia
E-mail: lucka.kajfez.bogataj@bf.uni-lj.si

Michael A. Brady
Global Observation of Forest
and Land Cover Dynamics
(GOFC-GOLD) Project Office
c/o Canadian Forest Service
5320-122 St.,
Edmonton AB, Canada T6H 3S5
E-mail: mbrady@nrcan.gc.ca

A.P.C. Brunini
Instituto Agronômico –Ciiagro-Cepa
Av- Barão de Itapura, 1481
13.020-902, Campinas
Sao Paulo, Brazil
E-mail: andrew@cepaempresa.com.br

O. Brunini
Instituto Agronomico
R. Fernao de Magalhaes 1080
Sao Paulo, Brazil
E-mail: brunini@iac.sp.gov.br

J.P de Carvalho
Instituto Agronômico
Av- Barão de Itapura, 1481
13.020-902, Campinas
Sao Paulo, Brazil
E-mail: jotape@iac.sp.gov.br

N. Chattopadhyay
India Meteorological Department
Agrimet Division
Shivajinagar
Pune, India
E-mail: agrimet_pune@yahoo.com

Adams Chavula
Agricultural Meteorologist
Malawi Meteorological Services
Blantyre, Malawi
Email: adamschavula@metmalawi.com

Ivan A. Csiszar
University of Maryland
Department of Geography
2181 LeFrak Hall
College Park, MD 20742, U.S.A.
E-mail: icsiszar@hermes.geog.umd.edu

Y.M.T. Da Anunciaçao
Instituto Nacional de Meteorologia
Eixo Monumental Via S1
70680-900, Brasília
DF, Brazil
E-mail: marina.tanaka@inmet.gov.br

H.P. Das
India Meteorological Department
Agrimet Division
Shivajinagar
Pune 5, India
E-mail: hpd_ag@rediffmail.com

C. Derry
University of Western Sydney
Hawkesbury Campus
Locked Bag 1797
Penrith South D.C. NSW1797, Australia
E-mail: c.derry@uws.edu.au

P.C. Doraiswamy
United States Department
of Agriculture (USDA)
Agricultural Research Service
1400 Independence Avenue, S.W.
Room 114, Hydrology and Remote Sensing Laboratory
Washington, D.C. 20705, U.S.A..
E-mail: pdoraiswamy@hydrolab.arsusda.gov

J. Eitzinger
Institute of Meteorology
Univ. of Natural Resources
and Applied Life Sciences (BOKU)
Peter Jordan Str. 82
A-1190 Wien, Austria
E-mail: josef.eitzinger@boku.ac.at

L.T.G. Fortes
Instituto Nacional de Meteorologia
Eixo Monumental Via S1
70680-900, Brasília
DF- Brasil
E-mail : lfortes@inmet.gov.br

William J. de Groot
Natural Resources Canada
Canadian Forest Service
5320-122 St., Edmonton, AB
Canada T6H 3S5
E-mail: bdegroot@nrcan.gc.ca

Johann G. Goldammer
The Global Fire Monitoring Center
Max Planck Institute for Chemistry
c/o Freiburg University
Georges-Koehler-Allee 75
D - 79110 Freiburg, Germany
E-mail: johann.goldammer@fire.uni-freiburg.de

John Hay
Institute for Global Change
Adaptation Science
Ibaraki University
Mito City, Japan
E-mail: johnhay@mx.ibaraki.ac.jp

Ulrich Hess
World Food Programme (WFP)
Via C.G.Viola 68
Parco dei Medici
Rome, Italy
E-mail: ulrich.hess@wfp.org

Peter Höppe
Department of Geo Risks Research
Munich Reinsurance Company AG
D-80791 Munich, Germany
E-mail: phoeppe@munichre.com

A.K.S. Huda
University Western Sydney
Hawkesbury Campus
Locked Bag 1797
Penrith South D.C. NSW1797, Australia
E-mail: s.huda@uws.edu.au

Christopher O. Justice
University of Maryland
Department of Geography
2181 LeFrak Hall
College Park, MD 20742, U.S.A.
E-mail: justice@hermes.geog.umd.edu

Tom Keenan
Weather Forecasting Group
Bureau of Meteorology Research Centre
GPO Box 1289K
Melbourne, VIC, Australia 3001
E-mail: T.Keenan@bom.gov.au

B. Lal
India Meteorological Department
New Delhi, India
E-mail: lalrp@yahoo.com

T. Hind-Lanoiselet
New South Wales Department
of Primary Industries
Wagga Wagga Agricultural Institute
PMB Wagga Wagga
NSW 2650 Australia
E-mail: tamrika.hind@dpi.nsw.gov.au

Felino P. Lansigan
INSTAT and SESAM
University of the Philippines Los Banos (UPLB)
4031 Laguna,Philippines
E-mail: fplansigan@yahoo.com/fpl@instat.uplb.edu.ph

Tim J. Lynham
Natural Resources Canada
Canadian Forest Service
1219 Queen St. East,
Sault Ste. Marie, ON, Canada P6A 2E5
E-mail: tlynham@nrcan.gc.ca

Byong Lyol Lee
Korea Meteorological Administration
208-16 Seodun-dong, Gwonson-gu
Suwon 441-856, Republic of Korea
E-mail: bllee@kma.go.kr

G. Maracchi
I.A.T.A. - C.N.R.
National Research Council
Institute of Agrometeorology & Environmental
Analysis for Agriculture
P.le delle Cascine, 18
I-50144 Florence, Italy
E-mail: g.maracchi@ibimet.cnr.it

A. Dalla Marta
Department of Agronomy
and Land Management
University of Florence
Piazzale delle Cascine, 18 50144
Florence, Italy
E-mail: anna.dallamarta@unifi.it

L. Martinelli
Department of Agronomy
and Land Management
University of Florence
Piazzale delle Cascine, 18 50144
Florence, Italy
E-mail: luca.martinelli@unifi.it

Holger Meinke
Department of Plant Sciences
Wageningen University
P.O. Box 430
NL 6700 AK Wageningen, The Netherlands
E-mail: holger.meinke@wur.nl

Keith Menzie
United States Department
of Agriculture (USDA)
World Agricultural Outlook Board
Office of the Chief Economist
1441 Independence Avenue, S.W.
Room 4438 South Building
Washington, D.C. 20250, U.S.A.
E-mail: kmenzie@oce.usda.gov

Elijah Mukhala
SADC Secretariat
Food Agriculture and Natural
Resources Direcotorate
P/B 0095
Gaborone, Botswana
Email:emukhala@yahoo.com

Raymond P. Motha
United States Department
of Agriculture (USDA)
Office of the Chief Economist
World Agricultural Outlook Board
1441 Independence Avenue, S.W.
Room 4419 South Building
Washington, D.C. 20250, U.S.A.
E-mail: rmotha@oce.usda.gov

G. Murray
New South Wales Department
of Primary Industries
Wagga Wagga Agricultural Institute
PMB Wagga Wagga NSW 2650 Australia
E-mail: gordon.murray@dpi.nsw.gov.au

V.R.K. Murthy
Acharya N.G.Ranga Agricultural University
College of Agriculture,
Department of Agronomy
Rajendranagar, Hyderabad-500 030
Andhra Pradesh, India
E-mail: vrkmurthy11@hotmail.com

M. Nikolaev
Agrophysical Research Institute (ARI)
Grazhdansky pr. 14
195220 St. Petersburg, Russia
E-mail: clenrusa@mail.ru

Kevin O'Loughlin
Bushfire Cooperative Research Centre
Level 5, 340 Albert St. East
Melbourne, VIC, Australia 3002
E-mail: kevin.oloughlin@bushfirecrc.com

S. Orlandini
Department of Agronomy
and Land Management
University of Florence
Piazzale delle Cascine, 18
I - 50144 Florence, Italy
E-mail address: simone.orlandini@unifi.it

M. Pasqui
Institute of Biometeorology –
National Research Council
Laboratory for Meteorology
and Environmental Modelling
Via Caproni, 8
I – 50145 Florence, Italy
E-mail: m.pasqui@ibimet.cnr.it

F. Piani
Institute of Biometeorology –
National Research Council
Laboratory for Meteorology
and Environmental Modelling
Via Madonna del Piano, 10
I – 50019 Sesto Fiorentino (FI), Italy
E-mail: f.piani@ibimet.cnr.it

P. Prashad
ICICI Lombard Bank
Zenith House, Keshavrao Khade Marg
Mahalaxmi
Mumbai 400 034, India
E-mail: pranav.prashad@icicilombard.com

Hou Qiong
Inner Mongolia Meteorological Institute
No.49 Hailar Street, Hohhot,
Inner Mongolia, China, 010051
E-mail: Qiong_hou@sina.com

Y. Ramakrishna
Central Research Institute
for Dryland Agriculture (CRIDA)
Santoshnagar
Hyderabad 500059, India
E-mail: ramakrishna.ys@crida.ernet.in

G.G.S.N. Rao
Central Research Institute
for Dryland Agriculture (CRIDA)
Santoshnagar
Hyderabad 500059, India
E-mail: ggsnrao@crida.ernet.in

V.U.M. Rao
Central Research Institute
for Dryland Agriculture (CRIDA)
Santoshnagar
Hyderabad 500059, India
E-mail: vumrao@crida.ernet.in

A.V.M.S. Rao
Central Research Institute
for Dryland Agriculture (CRIDA)
Santoshnagar
Hyderabad 500059, India
E-mail: vumrao@crida.ernet.in

K.V. Rao
Central Research Institute
for Dryland Agriculture (CRIDA)
Santoshnagar
Hyderabad 500059, India
E-mail: vumrao@crida.ernet.in

L.S. Rathore
National Centre for Medium Range
Weather Forecasting
A-50, Institutional Area, Phase II, Sector-62
NOIDA (UP), 201 307, India
E-mail: lsrathore@ncmrwf.gov.in

James Salinger
National Institute of Water
and Atmospheric Research (NIWA)
P.O. Box 109-695, New Market
Auckland, New Zealand
E-mail: j.salinger@niwa.co.nz

Wang Shili
Institute of Eco-environment
and Agrometeorology
Chinese Academy of Meteorological Sciences
No. 46 Zhongguancun, Nandajie
Beijing, China,100081
E-mail: wangsl@cams.cma.gov.cn

R.N. Spooner-Hart
University Western Sydney
Hawkesbury Campus
Locked Bag 1797
Penrith South D.C. NSW1797,Australia
E-mail: r.spooner-hart@uws.edu.au

C.J. Stigter
Agromet Vision and INSAM
Groenestraat 13, 5314 AJ, Bruchem,
The Netherlands & Jl. Diponegoro 166,
68214 Bondowoso, Indonesia
E-mail: cjstigter@usa.net

Roger C. Stone
Australian Centre for Sustainable Catchments
Faculty of Sciences, University
of Southern Queensland,
Darling Heights, Toowoomba, Australia, 4350
E-mail: stone@usq.edu.au

Andreja Sušnik
Environmental Agency
of the Republic of Slovenia
Agrometeorological Department
Vojkova 1b, 1000
Ljubljana, Slovenia
E-mail: andreja.susnik@rzs-hm.si

Lourdes V. Tibig
Philippine Atmospheric, Geophysical
and Astronomical Services Administration
(PAGASA)
PAGASA Science Garden Complex
Agham Road, Quezon City, Philippines
E-mail: lvtibig@yahoo.com

M. Trnka
Institute of Agriculture Systems
and Bioclimatology
Mendel University of Agriculture and Forestry
Zemedelska 1
61300 Brno, Czech Republic
E-mail: mirek_trnka@yahoo.com

I. Uskov
Agrophysical Research Institute (ARI)
Grazhdansky pr. 14
195220 St. Petersburg, Russia
E-mail: office@agrophys.ru

A. Utset
Instituto Tecnologico Agrario de Castilla y
Leon (ITACYL)
Ctra. Burgos km 119
47071 Valladolid, Spain
E-mail: utssuaan@jcyl.es

R.E. Rivero Vega
Meteorological Centre
of Camagüey Province,
Camagüey, Cuba
E-mail: roger@cmw.insmet.cu

Constantino Alarcón Velazco
Servicio Nacional de Meteorología
e Hidrología (SENAMHI)
Jr. Cahuide N° 785 Jesus María
Lima 11. Perú
E-mail: calarcon@senamhi.gob.pe

Nguyen van Viet
Agrometeorological Research Centre
Institute of Meteorology and Hydrology
Ministry of Natural Resources
and Environment
5/62 Nguyen Chi Thanh Street
Dong Da District
Hanoi, Viet Nam
E-mail: agromviet@hn.vnn.vn

Donald A. Wilhite
National Drought Mitigation Center
University of Nebraska-Lincoln
819 Hardin Hall
Lincoln, NE 68583-0988, U.S.A.
E-mail: dwilhite@unlnotes.unl.edu

Tan Ying
China Agricultural University
College of Humanity and Development,
Department of Media and Communication
Beijing, China
E-mail: tanying9966@sohu.com

Wang Yingshun
Xilinhot National Climate Observatory
of Inner Mongolia
No.10 of Group 11 in Eerdemuteng Street,
Xilinhot City
Inner Mongolia, China, 026000
E-mail:Wys5959@yahoo.com.cn

Zheng Dawei
China Agricultural University
Department of Agricultural Meteorology
College of Resources and Environment
Beijing, China
E-mail: zhengdawei44@263.net

Z. Zalud
Institute of Agriculture Systems
and Bioclimatology
Mendel University of Agriculture
and Forestry
Zemedelska 1
61300 Brno, Czech Republic
E-mail: zalud@mendelu.cz

CHAPTER 1

Extreme Weather and Climate Events, and Farming Risks

John Hay

1.1 Introduction

Extreme weather events, and climatic anomalies, have major impacts on agriculture. Of the total annual crop losses in world agriculture, many are due to direct weather and climatic effects such as drought, flash floods, untimely rains, frost, hail, and storms. High preparedness, prior knowledge of the timing and magnitude of weather events and climatic anomalies and effective recovery plans will do much to reduce their impact on production levels, on land resources and on other assets such as structures and infrastructure and natural ecosystems that are integral to agricultural operations. Aspects of crop and livestock production, as well as agriculture's natural resource base, that are influenced by weather and climatic conditions include air and water pollution; soil erosion from wind or water; the incidence and effects of drought; crop growth; animal production; the incidence and extent of pests and diseases; the incidence, frequency, and extent of frost; the dangers of forest and bush fires; losses during storage and transport; and the safety and effectiveness of all on-farm operations (Mavi and Tupper 2004).

Figure 1.1 illustrates how the climate influences agricultural production – specific climatic conditions, including absence of extremes, are required for optimum production. There are major gaps between the actual and attainable yields of crops, largely attributable to the pests, diseases and weeds, as well as to losses in harvest and storage.

When user-focused weather and climate information are readily available, and used wisely by farmers and others in the agriculture sector, losses resulting from adverse weather and climatic conditions can be minimized, thereby improving the yield and quality of agricultural products. While most emphasis should be placed on preparedness and timely management interventions, there will always be a need for the capacity to recover quickly and minimize the residual damages of adverse events and conditions (Stigter et al. 2003).

This paper focuses on a risk-based approach to managing the detrimental consequences of extreme weather events and climatic anomalies such as those described above. Basic concepts related to risk and to risk management are explained, followed by a discussion of farming risks. Details of risk characterization procedures are provided, along with some practical examples. Given the important consequences of climate change for agriculture, attention is given to projection of risk levels into the future. Again some practical examples are provided. Finally, relevant aspects of risk management are discussed. Overall conclusions are also presented.

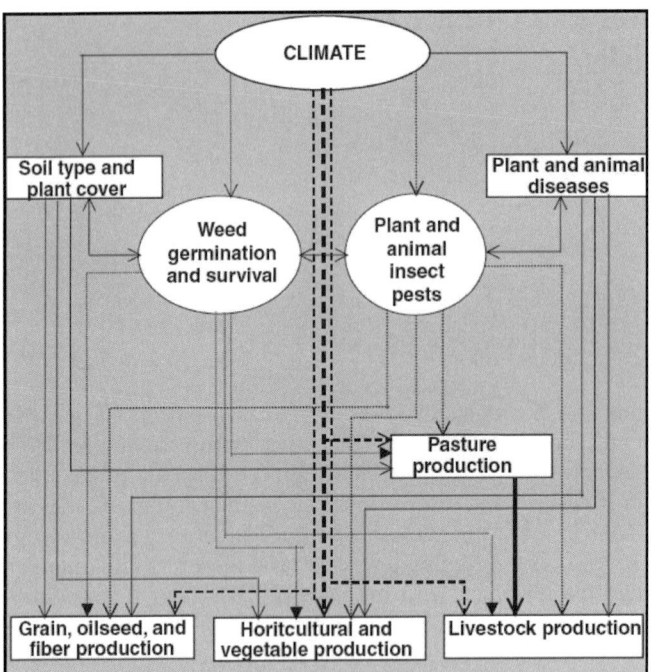

Fig. 1.1. The role of climate in agricultural production (from Mavi and Tupper 2004).

Why a risk-based approach? In recent decades there have been major advances in short-term and seasonal weather forecasting, as well as in long-term climate modelling. These have yielded major improvements in early warnings and advisories as well as in longer-term planning. This is resulting in increasing emphasis on proactive rather than reactive management of the adverse consequences of extreme weather events and anomalous climatic conditions on agriculture. It is also increasing the diversity of options available to farmers and others in the agriculture sector to manage those impacts. Increasingly, farm managers and other practitioners are seeking more rational and quantitative guidance for decision making, including cost benefit analyses. As will be demonstrated in the following sections, a risk-based approach to managing the adverse consequences of weather extremes and climate anomalies for agriculture goes a long way towards meeting these requirements. It also provides a direct functional link between, on the one hand, assessing exposure to the adverse consequences of extreme weather and anomalous climatic conditions and, on the other, the identification, prioritization and retrospective evaluation of management interventions designed to reduce anticipated consequences to tolerable levels.

Finally, risk assessment and management procedures have already been embraced by many sectors in addition to agriculture – e.g. health, financial, transport, energy, and water resources. As will be shown in the following section, a risk-based approach provides a common framework that facilitates coordination and cooperation amongst various players and stakeholders, including the sharing of information that might otherwise be retained by information "gate keepers".

1.2
Risk and Risk Management – Some Basic Concepts

Risk considers not only the potential level of harm arising from an event or condition, but also the likelihood that such harm will occur. In the present context, risk events include weather-related hazards such as extreme daily rainfall and frost. Risk conditions are climate-related and include hazards such as droughts and heat waves. Risk levels can change, including as a result of potentially detrimental changes in the climate (e.g. warming, decreasing rainfall). Changes in levels of exposure, due to altering levels of investment, can also influence risk levels. As defined above, risk combines both the likelihood of a harm occurring and the consequences of it doing so. Thus, in risk terms, an unlikely hazard or condition causing considerable harm (e.g. a category 5 hurricane, such the cyclone in the state of Orissa that devastated parts of India in 1999), may be compared to a hazard or condition which causes less harm but has a higher probability of occurring (e.g. a seasonal drought). By way of illustration, Figure 1.2 shows the likelihood of given extreme daily rainfall amounts for Delhi, India. A relatively common daily rainfall of, say, 30 mm will obviously result in far less devastation than the maximum observed daily rainfall of 192 mm.

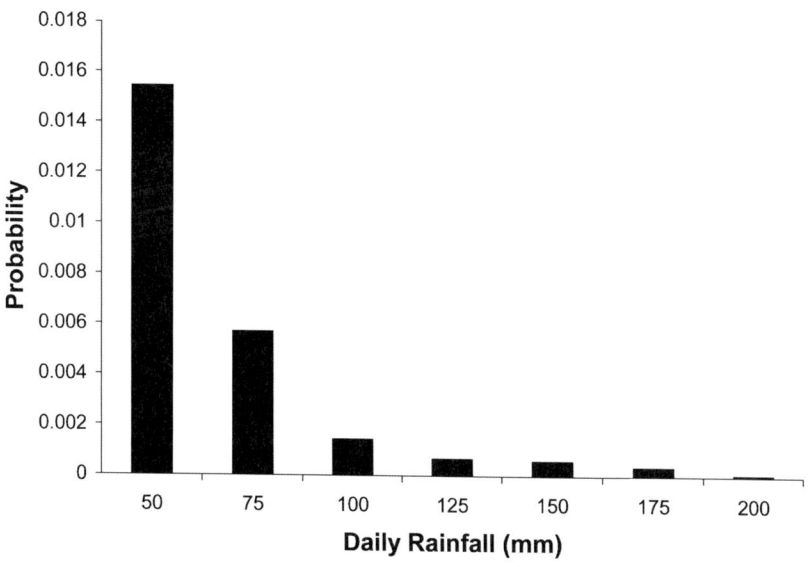

Fig. 1.2. Probability of a daily rainfall (mm) in 25 mm bands up to the given amount. Based on 1969 to 2004 data for Delhi, India. Data courtesy of India Meteorological Department.Ebunte

Harm may be expressed in many ways, such as loss of production in tonnes or number of livestock fatalities. Where the harm can be due to several different causes, use of the same units to describe the harm makes it possible to combine the different categories of risk. The result is the total risk. Thus:

$Total\ risk = \Sigma_i\ (Likelihood_i * Harm_i)$

There is a well established approach to characterizing and managing risks (Figure 1.3). As noted above, the risk-based methodology makes explicit the link between weather- and climate-related risks and the actions required to reduce them to acceptable levels. The widely-used procedures for characterizing and managing risk provide the basis for procedures which relate more specifically to characterizing and managing weather- and climate-related risks of relevance to the agriculture sector (Figure 1.4).

1.2.1
Step A – Risk Scoping

Through a consultative process, involving stakeholders as well as relevant experts, as required, risk reduction targets and criteria are established. These are based on identifying acceptable levels of risk. Existing information sources, experience and expert judgment are used, as appropriate, to identify possible weather- and climate-related risk events and conditions. These in turn lead to identification of the associated sources of stress and the components ("receptors") of the agricultural system on which the stresses act. The pathways for these interactions are also identified.

1.2.2
Step B – Risk Characterization and Evaluation

For each of the risk events identified in 1.2.1 above, scenarios are developed in order to provide a basis for estimating the likelihood of each risk event, for present conditions and into the future if change is anticipated, for example as a consequence of climate change. The extent to which the climate changes into the future will influence the probability of the risk event occurring. The consequences of a given risk event are quantified in terms of individual and annualized costs. The overall findings are compiled into a risk profile.

1.2.3
Step C – Risk Management

In this step a number of questions are asked – all are with reference to the targets agreed in 1.2.1 above. Actions taken depend on the responses to a series of questions.

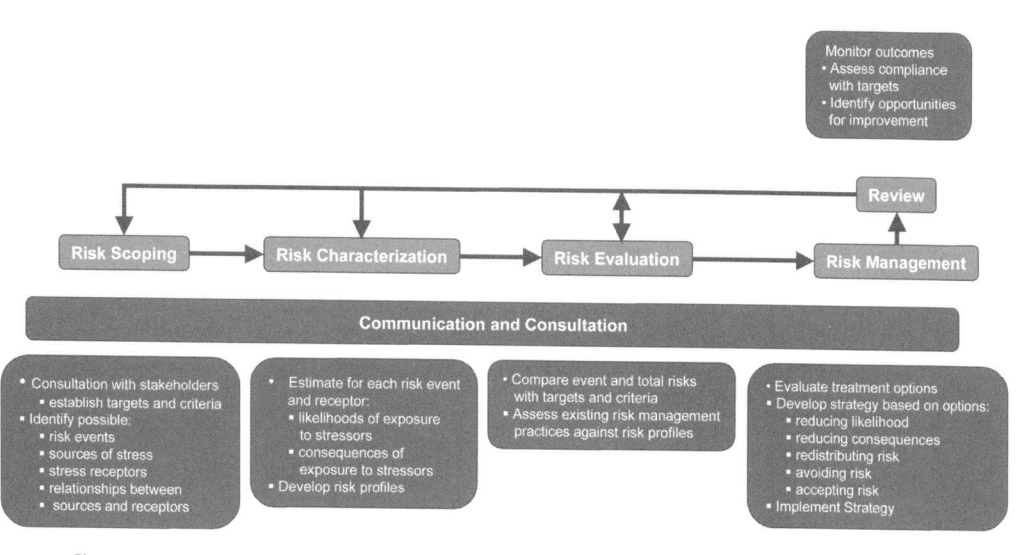

Chapter 1: Extreme Weather and Climate Events, and Farming Risks 5

Fig. 1.3. Generic methodology for characterizing and managing risks.

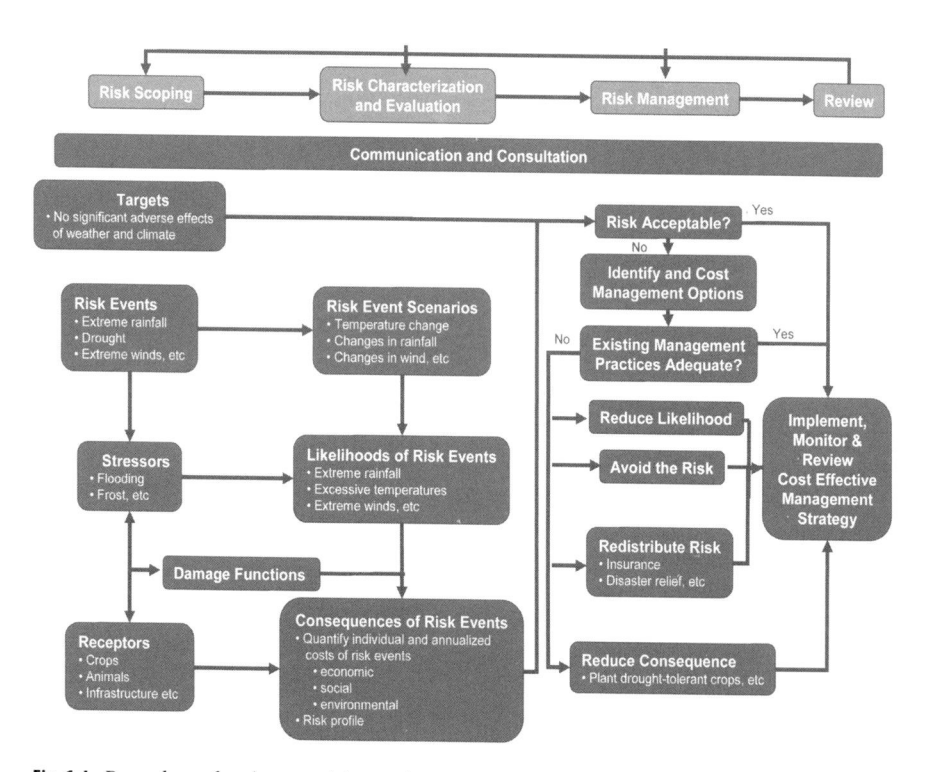

Fig. 1.4. Procedures for characterizing and managing risks, and their application to the agriculture sector.

Is the risk acceptable? – If "yes", it is appropriate to continue with current management approaches. These should include monitoring and reviewing as the acceptability of the risk may change over time into the future. If "no", risk management options are identified, and assessed in terms of costs and benefits.

Are the current risk management options adequate? – If "yes", it is appropriate to continue with the current approaches. Again, these include monitoring and reviewing, in part due to the possibility that the acceptability of the risk may change over time. If "no", one or more of the following risk management strategies should be implemented:

- Take actions to reduce the likelihood of the risk event occurring. For example, reduce greenhouse gas emissions by the agriculture sector and thereby reduce the rate of climate change and the resulting increased frequency of risk events such as drought and frosts.
- Avoid the risk. For example, avoid planting crops in areas that are exposed to risk events of concern.
- Redistribute the risk. For example, provide access to crop insurance cover or ensure disaster relief programmes are in place.
- Reduce the consequences. For example, plant drought tolerant crops if drought is a risk event of concern.

1.2.4
Step D – Monitoring and Review

The next step is to implement the risk management programme, and monitor and review the risk management outcomes in relation to the agreed targets. If the targets are not met it will be necessary to repeat one or more of the following steps: i) identify the problem and formulate a response plan; ii) enhance the quality of the risk characterization procedures and findings; and iii) enhance the quality of the risk management procedures and outcomes.

The process of risk characterization and management is iterative, to ensure that the quality of the outcomes are always consistent with the risk reduction targets that are established, reviewed, revised and reaffirmed through consultative processes.

1.3
Farming Risks

In both the developing and developed worlds risk exposure and management are important aspects of farming. Variations in the weather, climate, yields, prices, government policies, global markets and other factors can cause wide swings in farm production and, in the case of commercial agriculture, in farm income. Risk management involves choosing among strategies that reduce the social and financial consequences of these variations in production and income.

Five general types of risk in the agriculture sector are recognized (USDA 2006a):

- **Production risk** derives from the uncertain natural growth processes of crops and livestock. Weather, disease, pests, and other factors affect both the quantity and quality of commodities produced;
- **Price or market risk** refers to uncertainty about the prices producers will receive for commodities or the prices they must pay for inputs. The nature of price risk varies significantly from commodity to commodity;
- **Financial risk** results when the farm business borrows money and creates an obligation to repay debt. Rising interest rates, the prospect of loans being called by lenders, and restricted credit availability are also aspects of financial risk;
- **Institutional risk** results from uncertainties surrounding government actions. Tax laws, regulations for chemical use, rules for animal waste disposal, and the level of price or income support payments are examples of government decisions that can have a major impact on the farm business; and
- **Human or personal risk** refers to factors such as problems with human health or personal relationships that can affect the farm business. Accidents, illness, death, and divorce are examples of personal crises that can threaten a farm business.

This paper focuses on production risks, and specifically the way extreme weather events and anomalous climate conditions contribute to production risk. In this context, production risk is the risk associated with undesirable and often unanticipated weather and climatic conditions that affect the performance of crops and livestock. The relationships between weather, climate and production risk are well recognised (George et al. 2005). Some examples should suffice to illustrate the strength and importance of these relationships.

Climate-based models have been used to predict the potential for soybean rust spore production in the southern USA. This makes it possible to define regions where the climate is more favourable for rust to develop, expressed as the frequency of years a higher production of spores would be likely (Del Ponte and Yang 2006).

Figure 1.5 shows the likelihood that soybean rust in Texas, USA, will reach a severity of over 20 percent by late June if the rust is found in late May.

Figure 1.6 shows the strong influence of rainfall on cereal production in Niger.

Figure 1.7 shows drought risk for Gujarat, which is situated on the western coast of India. The drought risk map was obtained by integrating risk maps for both agricultural and meteorological drought. High drought risk prevails in nearly 30% of the area. This comprises districts that are major producers of food grains as well as oilseeds, emphasizing a critical need for drought management plans in these districts (Chopra 2006).

Figure 1.8 shows how anomalous climatic conditions in India influence food production.

1.4
Risk Characterization

As shown in Figure 1.3, risk characterization is an important step in the overall process of risk management. This section describes the methodology and pro-

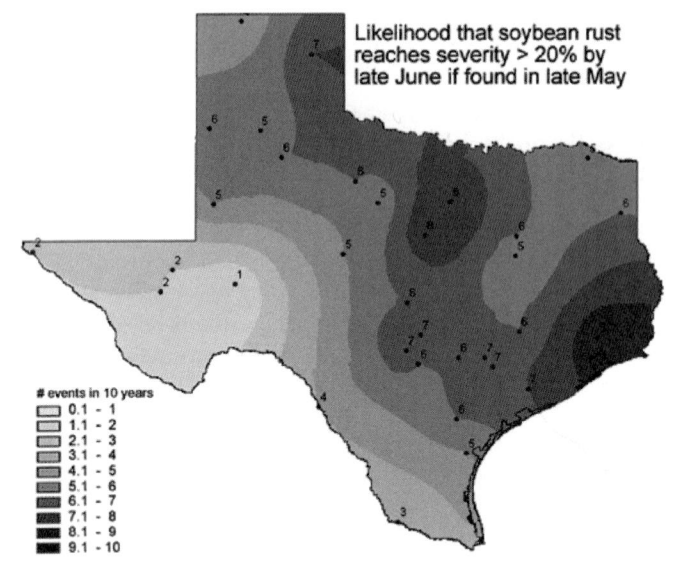

Fig. 1.5. Likelihood that soybean rust will reach a severity of over 20 percent by late June if found in late May (from Del Ponte and Yang 2006).

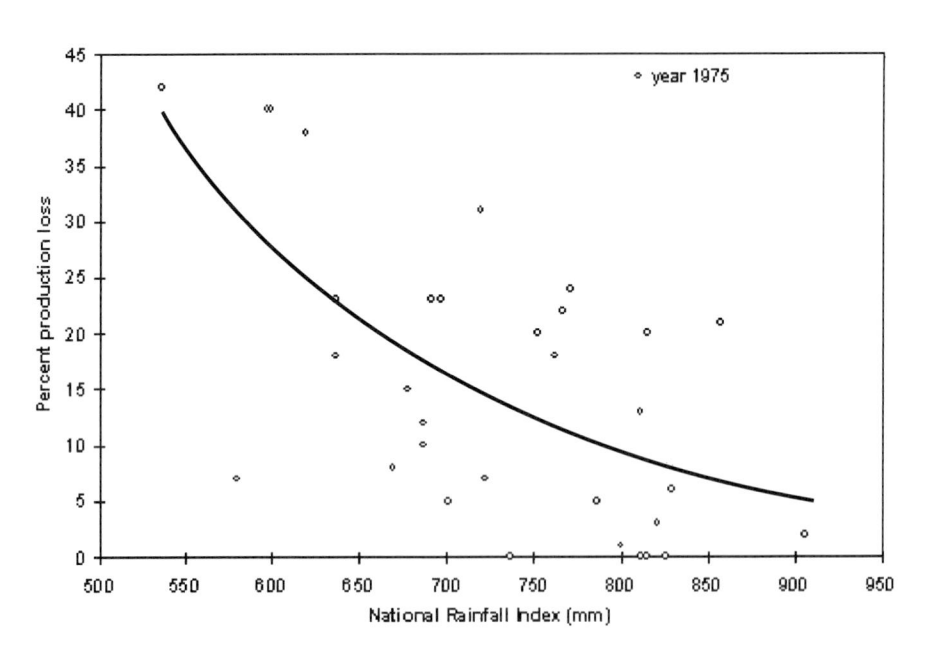

Fig. 1.6. Percent decrease in total cereal production for Niger as a function of the National Rainfall Index (Gommes 1998).

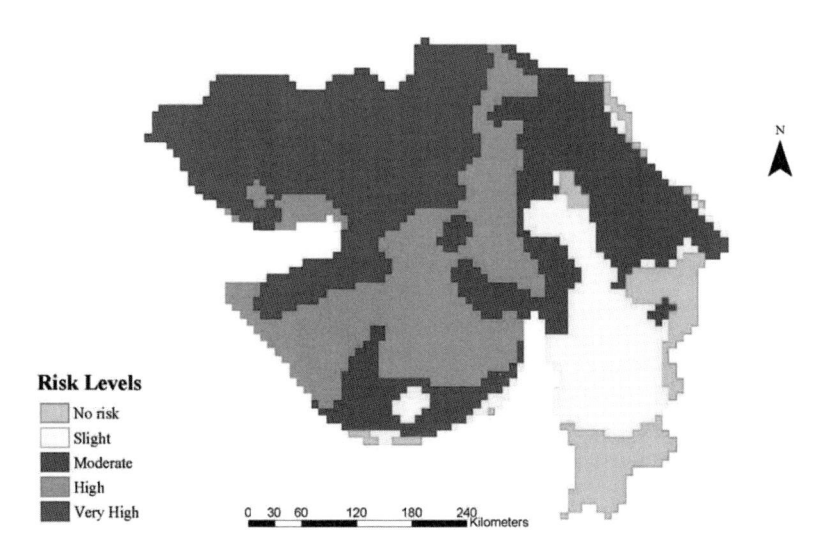

Fig. 1.7. Drought risk for Gujarat, India, determined by integrating risk maps for both agricultural and meteorological drought (Chopra 2006).

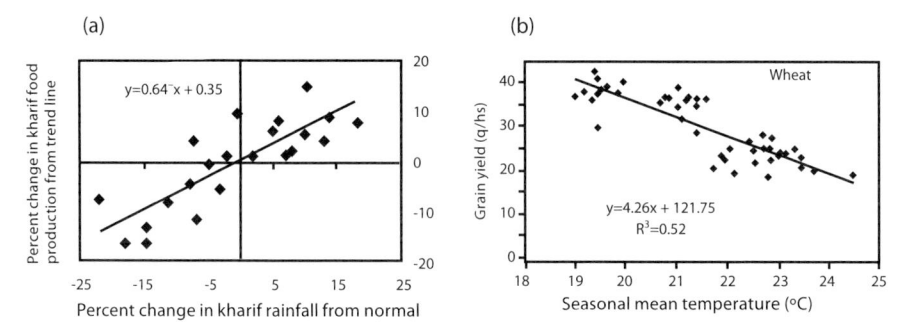

Fig. 1.8. Relationship between (a) monsoon season food production and seasonal rainfall and (b) regional wheat yields with seasonal temperature (Government of India 2004).

vides some illustrative results for characterizing levels of risk associated with both weather extremes and climate anomalies.

1.4.1
Weather Extremes

The *return period* (also know as the *recurrence interval* of an event) is a statistical measure of how often an extreme event of a given magnitude is likely to be equalled or exceeded, within a given time frame. For example, a "fifty-year rainfall event" is one which will, on the average, be equalled or exceeded once in any fifty-year period. Note: it does not mean that the event occurs every fifty years.

The likelihood or probability that an event of specific magnitude will be equalled or exceeded in any given year is the inverse of the return period, that is, 1 / Return Period

A one in fifty-year event has one chance in fifty of occurring in any specified year, that is, its probability is 1/50. Thus the probability equals 0.02, or 2%.

In some cases it is useful to know the probability that an event of at least a given magnitude will occur within a specified number of years, say five years. This probability can be calculated using the following equation:

probability of occurrence in n years = $1 - (1 - \text{probability of occurrence in any year})^n$

For example, the probability that an event with a probability of 0.2 will occur in the next five years is:

$1 - (1-0.2)^5 = 0.67$

Note again that this probability applies only on average, and cannot be considered a forecast.

Table 1.1 provides return periods and probabilities for given extremes in daily rainfall, based on observed data for Delhi and Pune, India. It is clear that extreme rainfall events of a given magnitude at Delhi are substantially more frequent than those observed in Pune.

Similarly, return periods and probabilities for specified values of maximum air temperature and extreme wind speeds are given in Tables 1.2 and 1.3, respectively. The results show that both extreme high temperatures and extreme wind gusts are much more common at Delhi, relative to Pune.

Table 1.1. Return Periods for Daily Rainfalls of Given Amounts, for Delhi and Pune, India. Based on Data for 1969 to 2004, inclusive. [Data courtesy of India Meteorological Department]

Daily Rainfall of at Least (mm)	Delhi		Pune	
	Return Period (y)	Probability	Return Period (y)	Probability
50	1.1	0.94	1.2	0.80
75	1.3	0.75	2.3	0.40
100	2.0	0.49	5.8	0.20
125	3.6	0.28	16	0.06
150	6.9	0.14	48	0.02
175	14	0.07	140	0.01
200	28	0.04	>400	0.00

Chapter 1: Extreme Weather and Climate Events, and Farming Risks   11

Table 1.2. Return Periods for Maximum Temperatures of Given Amounts, for Delhi and Pune, India. Based on Data for 1969 to 2004, inclusive. [Data courtesy of India Meteorological Department]

Maximum Temperature of at Least (°C)	Delhi		Pune	
	Return Period (y)	Probability	Return Period (y)	Probability
41	1	1	1.6	0.61
42	1	0.99	6.2	0.16
43	1.2	0.86	31	0.03
44	1.9	0.53	160	0.01
45	4.0	0.25	>800	0.0
46	9.7	0.10		
47	25	0.04		

Table 1.3. Return Periods for Maximum Annual Wind Gusts of Given Amounts, for Delhi and Pune, India. Based on Data for 1969 to 2004, inclusive. [Data courtesy of India Meteorological Department]

Daily Annual Wind Gust of at Least (km h^{-1})	Delhi		Pune	
	Return Period (y)	Probability	Return Period (y)	Probability
50	1	1	1.1	0.90
75	1.1	0.92	2.2	0.46
100	2.4	0.42	6.6	0.15
125	9.1	0.11	23	0.04
150	41	0.02	83	0.01

1.4.2
Climate Anomalies

Drought has a major impact on agricultural production, making it an important risk condition. Figure 1.9 shows the frequency of drought for Delhi, where in this instance drought is defined as months when the rainfall is at or below the ten-percentile for that month. It is clear that, based on this indicator, there is a high risk of at least a brief drought occurring in any given year. The risk of a prolonged drought is also very real.

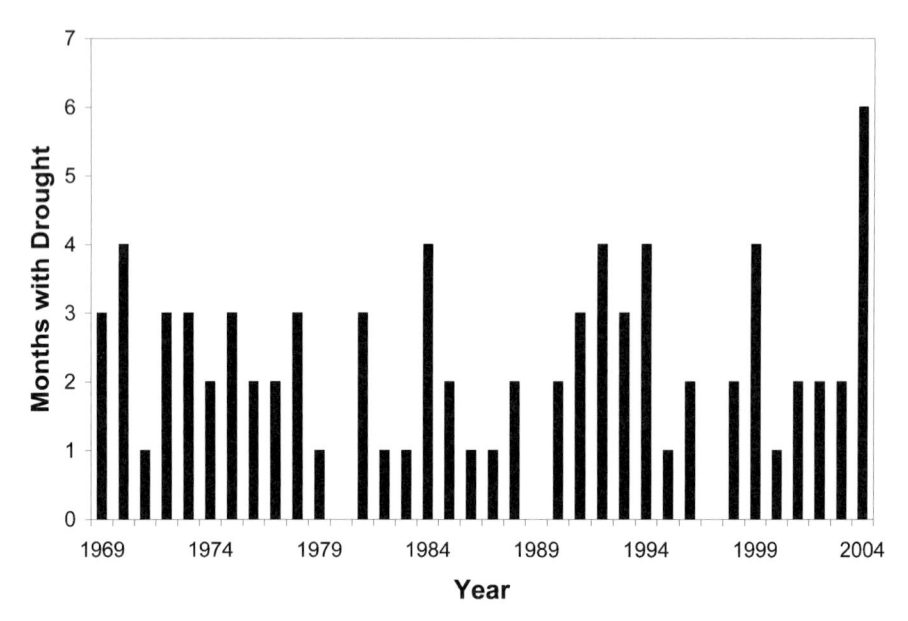

Fig. 1.9. The frequency of drought for Delhi, India, for 1969 – 2004. Drought is defined here as a month when the rainfall is at or below the ten-percentile for that month. Data courtesy of India Meteorological Department.

1.5
Changing Risk

Agriculture is one of the main sectors likely to be impacted by climate change. This section presents the results of analyses designed to show how risk levels for rainfall and temperature extremes, and drought, are projected to change over the remainder of the current century. One such change of importance to India is illustrated in Figure 1.10. A substantial increase in drought risk is expected during the current century. One consequence is a major decline in irrigated wheat yields in northern India during the coming decades (Figure 1.11). Clearly, and as would be expected, as the time horizon increases the confidence in the projections decreases.

Future changes in risk are estimated using the outputs of selected global climate models (GCMs)[1] run for a range of greenhouse gas emission scenarios (Figure 1.12). Table 1.4 lists the combination of models and emission scenarios on which the risk projections are based.

Differences in climate projections give rise to uncertainties in the estimated values of future climate risks. There are numerous sources of uncertainty in projections of the likelihood components of climate-related risks. These include uncertainties in greenhouse gas emissions as well as in modelling the complex interactions and responses of the atmospheric and ocean systems. Policy and decision makers need to be cognizant of uncertainties in projections of the likelihood components of extreme events.

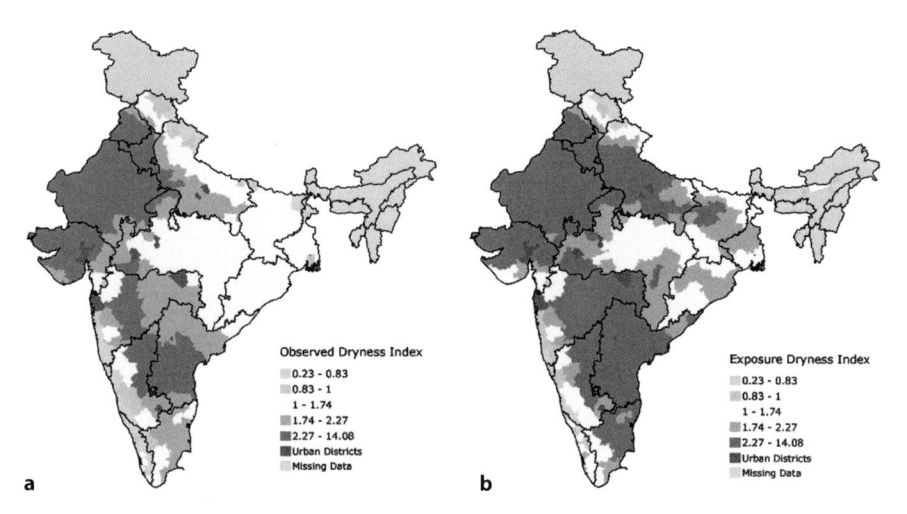

Fig. 1.10. Areas in India prone to drought (a) today and (b) in the mid 21st century, determined using a dryness index. The light shading indicates areas where rain exceeds evaporation. The darker shading identifies regions where evaporation is greater than precipitation – the darker the shading the drier the region, except that urban areas have the darkest shading (Schreiner 2004).

Fig. 1.11. Simulated impact of global climate change on irrigated wheat yields in North India (Aggarwal 2002).

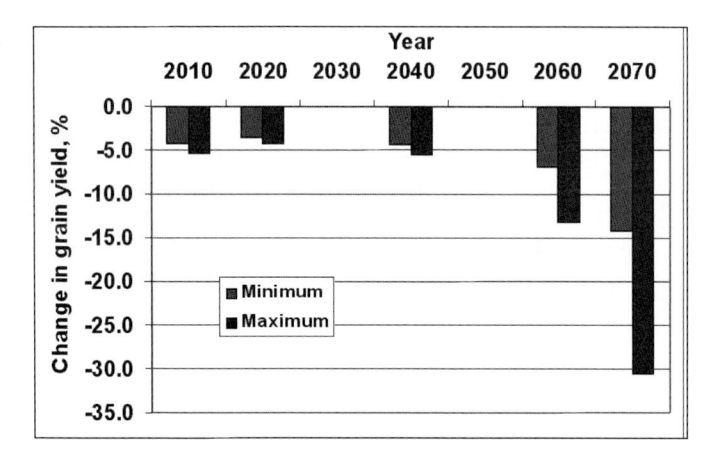

Best estimates of future risk levels are based on an average of the estimates using a multi model and emission scenario ensemble. The range in uncertainty is determined using a model and emission scenario combination that produces the maximum and minimum rate of change in future risk levels.

Projected changes in the return periods of extreme daily rainfall events (Figure 1.13) are based on estimates using a multi model and emission scenario ensemble (see Table 1.4). It is anticipated that global warming will reduce the return periods

Table 1.4. Available Combinations of Global Climate Models and Emission Scenarios[1]

	CGCM	CSIRO	Hadley	NIES	GFDL	See Text
A1B	T, P[1]	T, P	T, P	T, P	S	W[1]
A1F	T, P	T, P	T, P	T, P	S	W
A1T	T, P, S	T, P, S	T, P, S	T, P	S	W
A2	T, P, S	T, P, S	T, P, S	T, P	S	W
B1	T, P, S	T, P, S	T, P, S	T, P	S	W
B2	T, P, S	T, P, S	T, P, S	T, P	S	W

[1] T = temperature, P = precipitation, W = wind

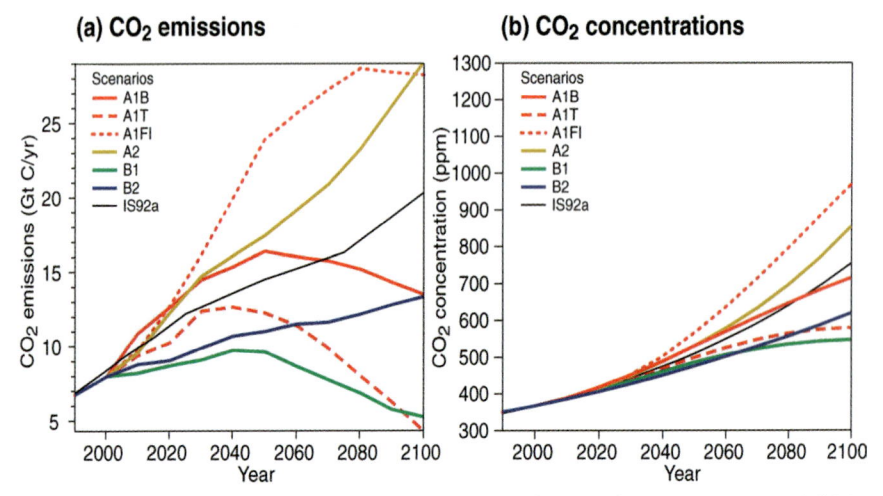

Fig. 1.12. Scenarios of CO_2 gas emissions and consequential atmospheric concentrations of CO_2 (from IPCC 2001).

of extreme daily rainfall events for Delhi – that is, the likelihood of such extreme events will increase in the future.

Projected changes in the return periods of extreme maximum temperature (Figure 1.14) are based on estimates using a multi model and emission scenario ensemble (see Table 1.4). It is anticipated that global warming will also reduce the return periods of extreme maximum temperatures for Delhi.

Estimates of changes in maximum wind gusts are based on the assumption that such wind gusts will increase by 2.5, 5 and 10 per cent per degree of global warming. Thus the emission scenarios listed in Table 1.4 are explicitly included in the estimates. The best estimate of the increase in maximum wind gusts is determined

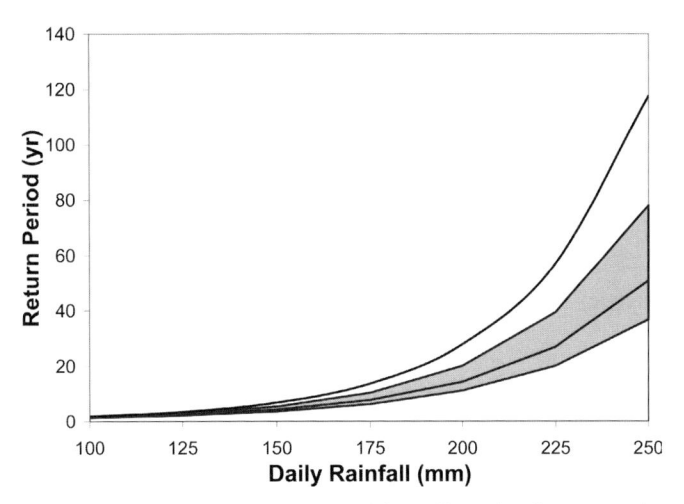

Chapter 1: Extreme Weather and Climate Events, and Farming Risks

Fig. 1.13. Relationship between daily rainfall and return period for Delhi, India, for present day (black line) and 2050 (blue lines). The uncertainty envelope shows the maximum and minimum estimates of return periods for 2050, based on all possible combinations of the available global climate models and emission scenarios.

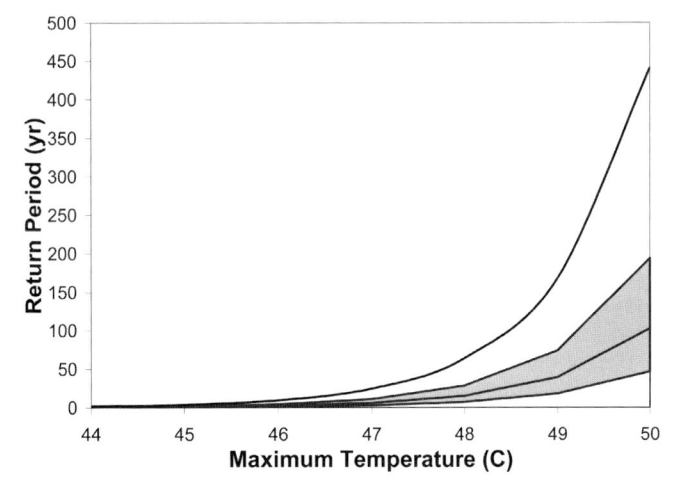

Fig. 1.14. Relationship between maximum temperature and return period for Delhi, India, for present day (black line) and 2050 (blue lines). The uncertainty envelope shows the maximum and minimum estimates of return periods for 2050, based on all possible combinations of the available global climate models and emission scenarios.

by averaging the ensemble of estimates for all combinations of percentage increase and emission scenarios. As indicated in Figure 1.15, global warming will likely reduce the return periods of maximum wind gusts for Delhi.

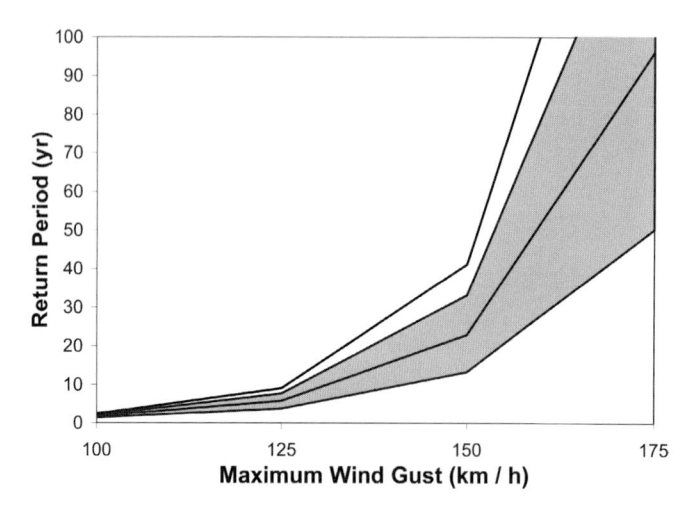

Fig. 1.15. Relationship between peak wind gust and return period for Delhi, India, for present day (black line) and 2050 (blue lines). The uncertainty envelope shows the maximum and minimum estimates of return periods for 2050, based on all possible combinations of the percentage increases and emission scenarios.

1.6
Risk Management

Based on Clarkson et al. (2006), there are six requirements that must be met if farmers are to manage risks related to climate extremes, variability and change. These include:
- awareness that weather and climate extremes, variability and change will impact on farm operations;
- understanding of weather and climate processes, including the causes of climate variability and change;
- historical knowledge of weather extremes and climate variability for the location of the farm operations;
- analytical tools to describe the weather extremes and climate variability;
- forecasting tools or access to early warning and forecast conditions, to give advance notice of likely extreme events and seasonal anomalies; and
- ability to apply the warnings and forecasts in decision making.

Farmers have many options for managing the risks they face, and most use a combination of strategies and tools. Some strategies deal with only one kind of risk, while others address multiple risks. Some of the more widely used strategies include (USDA 2006b):
- **Enterprise diversification:** this is based on the assumption that incomes from different crops and livestock activities are not perfectly correlated, meaning that when some activities produce low incomes other activities will likely offset this decreased earning by producing higher income;

- **Financial leverage:** this refers to the use of loans to help finance farm operations; higher levels of debt, relative to net worth, are generally considered riskier; the optimal amount of leveraging depends on several factors, including farm profitability, the cost of credit, tolerance for risk, and the degree of uncertainty in income;
- **Vertical integration:** this can decrease risk associated with the quantity and quality of inputs or outputs since a vertically integrated firm retains ownership or control of a commodity across two or more phases of production and/or marketing, thereby spreading risk;
- **Contracting:** this can reduce risk by way of guaranteed prices, market outlets, or other terms of exchange which are settled in advance; contracts that set price, quality, and amount of product to be delivered are called marketing contracts, or simply forward contracts; contracts that prescribe production processes to be used and/or specify who provides inputs are called production contracts;
- **Hedging:** this uses futures, or options, contracts to reduce the risk of adverse price changes prior to an anticipated cash sale or purchase of a commodity;
- **Liquidity:** this refers to the farmer's ability to generate cash quickly and efficiently in order to meet financial obligations; liquidity can be enhanced by holding cash, stored commodities, or other assets that can be converted to cash on short notice without incurring a major loss.;
- **Crop yield insurance:** this pays indemnities to producers when yields fall below the producer's insured yield level; coverage may be provided through such instruments as private insurance or government subsidized multiple peril crop insurance;
- **Crop revenue insurance:** this pays indemnities to farmers based on gross revenue shortfalls instead of just yield or price shortfalls; for example, in most areas of the United States several federally subsidized revenue insurance plans are available for major crops; and
- **Household off-farm employment or investment:** this can provide a more certain income stream to the farm household to supplement income from the farming operation.

Most producers use a mix of tools and strategies to manage risks. Since the willingness and ability to bear risks differ from farm to farm, there is usually variation in the risk management strategies used by producers.

Specific risk management measures that can be used include irrigation and water allocation strategies; shelter from wind or cold; shade from excessive heat; anti-frost and anti-erosion measures, soil cover and mulching; plant cover using glass or plastic materials; artificial climates of growth chambers or heated structures; animal housing and management; climate control in storage and transport; and efficient use of herbicides, insecticides, and fertilizers. Weather and climatic conditions often determine the type of pests and diseases that will have to be controlled in a given growing season, as well as the efficacy of any control procedures.

1.7
Conclusions

Variations in the weather, climate, yields, prices, government policies, global markets and other factors can cause wide swings in farm production and, in the case of commercial agriculture, in farm income. Risk management involves choosing among strategies that reduce the social and financial consequences of these variations in production and income.

Extreme weather events, and climatic anomalies have major impacts on agriculture. In both the developing and developed worlds risk characterization and management are important aspects of farming. High preparedness, prior knowledge of the timing and magnitude of weather events and climatic anomalies and effective recovery plans will do much to reduce their impact. When user-focused weather and climate information are readily available, and used wisely by farmers and others in the agriculture sector, losses resulting from adverse weather and climatic conditions can be minimized.

In recent decades major advances in short-term and seasonal weather forecasting, as well as in long-term climate modelling, have yielded major improvements in early warnings and advisories as well as in longer-term planning. This is resulting in increasing emphasis on proactive rather than reactive management of the risks to agriculture resulting from extreme weather events and anomalous climatic conditions on agriculture.

There is a well established approach to characterizing and managing risks. The risk-based methodology makes explicit the link between weather- and climate-related risks and the actions required to reduce them to acceptable levels.

Farmers have many options for managing the risks they face, and most use a combination of strategies and tools. Some strategies deal with only one kind of risk, while others address multiple risks. Most producers use a mix of tools and strategies to manage risks. Since the willingness and ability to bear risks differ from farm to farm, there is usually variation in the risk management strategies used by producers.

Acknowledgements

The assistance of Dr M.V.K. Sivakumar, Chief of the Agricultural Meteorology Division, and of his colleagues in the World Meteorological Organization, is gratefully acknowledged. Dr Madhavan Nair Rajeevan, Director of the National Climate Centre, India Meteorological Department, facilitated access to the meteorological data for Delhi and Pune.

References

Aggarwal PK (2002) Climate Change and Agriculture: Information Needs and Research Priorities. International Conference on Capacity Building for Climate Change 21 October 2002 www.ficci.com/media-room/speeches-presentations/2002/oct/oct-climate-aggarwal.ppt

Chopra P (2006) Drought Risk Assessment using Remote Sensing and GIS: A case study of Gujarat. M.Sc. Thesis, Indian Institute of Remote Sensing, National Remote Sensing Agency (NRSA), Department of Space, Dehradun, India & International Institute For Geo-Information Science and Earth Observation, Enschede, The Netherlands, 81 pp.

Clarkson NM, Abawi GY, Graham LB, Chiew FHS, James RA, Clewett JF, George DA and Berry. D (2000) Seasonal streamflow forecasts to improve management of water resources: Major issues and future directions in Australia, Proceedings of the 26th National and 3rd International Hydrology and Water Resources Symposium of The Institution of Engineers, Australia. Perth, Australia. pp 653–658.

Del Ponte EM and Yang XB (2006) Understanding the risk of soybean rust in Texas for soybean production in Iowa. Integrated Crop Management 496(5) http://www.ipm.iastate.edu/ipm/icm/2006/4–3/rust.html

George DA, Birch C, Buckley D, Partridge IJ, Clewett JF (2005) Assessing climate risk to improve farm business management. Extension Farming Systems Journal vol 1(1).

Gommes R (1998) Special: Agroclimatic concepts. Production variability and losses CAgM/MMG-2/Doc.5 http://www.fao.org/sd/EIdirect/agroclim/riskdef.htm

Government of India (2004) India's First National Communication to the United Nations Framework Convention on Climate Change. Ministry of Environment and Forests, Government of India, New Delhi, India.

Intergovernmental Panel on Climate Change (IPCC) (2001) IPCC Third Assessment. Climate Change 2001. A Report of the Intergovernmental Panel on Climate Change. Cambridge University Press.

Schreiner C (2004) What will a warmer world be like? Centre for International Climate and Environmental Research, Oslo, Norway. http://www.atmosphere.mpg.de/enid/1wb.html

Marvi HS and Tupper GJ (2004) Agrometeorology Principles and Applications of Climate. Studies in Agriculture, Food Products Press, New York.

Stigter CJ, Das HP and Murthy VRK (2003) Beyond climate forecasting of flood disasters. Invited Lecture on the Opening Day of the Fifth Regional Training Course on Flood Risk Management (FRM-5) of the Asian Disaster Preparedness Center (Bangkok) and the China Research Center on Flood and Drought Disaster Reduction (Beijing).

USDA (2006a) Farm Risk Management: Risk in Agriculture, United States Department of Agriculture. http://www.ers.usda.gov/Briefing/RiskManagement/RiskinAgriculture.htm

USDA (2006b) Farm Risk Management: Risk Management Strategies. United States Department of Agriculture. http://www.ers.usda.gov/Briefing/RiskManagement/Strategies.htm

CHAPTER 2

Preparedness and Coping Strategies for Agricultural Drought Risk Management: Recent Progress and Trends

Donald A. Wilhite

2.1 Introduction

In many countries, drought is responsible for the greatest loss of agricultural production. For example, in the United States, drought was the predominant source of indemnities paid because of crop losses between 1970 and 2003. These losses totaled more than $15 billion (USDA/RMA). In China during the period from 1949 to 2000, drought affected an average of 21 million hectares. More than 60 million tons of grain was lost in China as a result of the drought of 2000, the highest loss in 51 years (Zhang et al. 2005). Recent droughts in Europe, Brazil, Mexico, Australia, Southern Africa, and many other regions have also resulted in devastating impacts in the agricultural sector. With growing pressure on water and other natural resources because of population increases and other factors, there is an increasing need to reduce both the impacts of drought on agriculture and other sectors and the demand for government- or donor-sponsored drought assistance programs. These programs are costly and largely ineffective in reducing societal vulnerability to future drought episodes.

Agricultural producers in both developed and developing countries have many options available to them to lessen their drought risk. Irrigation has long been an important mitigation measure, but increasing pressure on limited water supplies is reducing opportunities for further expansion and placing more pressure on agriculture to use water more efficiently. Supplemental irrigation offers tremendous opportunities to increase yields if water is applied strategically during critical periods of crop growth and development during both drought and non-drought periods (Oweis 2005). Water harvesting has also been used successfully for centuries to sustain plant growth in arid or semi-arid regions and is being widely promoted today as a drought risk reducing strategy. Other agricultural practices such as changes in crop type, the timing of planting, and changes in cultivation practices can also lessen vulnerability to drought if applied in a timely manner. Given new computer tools and techniques such as Geographic Information Systems (GIS), satellite-based remote sensing products, and the ability of the Internet to deliver timely and reliable information to agricultural producers, natural resource managers, and other decision makers, there are now added opportunities to further mitigate agricultural drought impacts. However, the development of appropriate decision-support tools for agriculture requires improved communication between the developers of these tools and users throughout the development process. To be successful, this process must be based on a greater awareness and understanding of the drought hazard and how changes in management can alter vulnerability to reduce risk.

2.2
Agricultural Drought Risk Management: Understanding the Hazard and Societal Vulnerability

Many people consider drought to be largely a natural or physical event. Although all types of drought originate from a deficiency of precipitation, it has both a natural and social component. The risk associated with drought for any region is a product of both the region's exposure to the event (i.e., probability of occurrence at various severity levels) and the vulnerability of society to the event. The natural event (i.e., meteorological drought or the drought hazard) is a result of the occurrence of persistent large-scale disruptions in the global circulation pattern of the atmosphere. Exposure to drought varies spatially and there is little, if anything, that we can do to alter drought occurrence. Vulnerability, on the other hand, is determined by social factors such as population changes, population shifts (regional and rural to urban), demographic characteristics, land use, environmental degradation, environmental awareness, water use trends, technology, policy, and social behavior. These factors change over time and thus vulnerability is likely to increase or decrease in response to these changes. Preparedness planning, risk assessments, and improved early warning systems can greatly lessen societal vulnerability to drought.

2.3
Drought as Hazard

Drought differs from other natural hazards in several ways. First, drought is a slow-onset natural hazard often referred to as a creeping phenomenon (Gillette 1950). Because of the creeping nature of drought, its effects accumulate slowly over a substantial period of time. Therefore, the onset and end of drought is difficult to determine, and scientists and policy makers often disagree on the basis (i.e., criteria) for declaring an end to drought. Should drought's end be signaled by a return to normal precipitation and, if so, over what period of time does normal or above-normal precipitation need to be sustained for the drought to be declared officially over? Do precipitation deficits that emerged during the drought event need to be erased for the event to end? Do reservoir and ground water levels need to return to normal or average conditions? Impacts linger for a considerable period of time following the return of normal precipitation, so is the end of drought signaled by meteorological or climatological factors or diminishing impacts? The first two examples above are natural elements of drought – i.e., they are part of the concept of meteorological drought. However, the latter examples are complicated by how water is used and managed – i.e., how we manage reservoirs and ground water will not only determine how rapidly water levels will decline in response to reduced inflows or infiltration but also how rapidly they will recover from an extended period of drought.

Second, unlike most other natural hazards, drought does not have a precise and universally accepted definition. This fact adds to the confusion about whether or not a drought exists and, if it does, its degree of severity. Realistically, definitions of drought must be region and application (or impact) specific. This is one explanation for the hundreds of definitions that exist. For this reason, the search for a universal definition of drought is of little value. Policy makers are often frustrated by disagreements among scientists on whether or not a drought exists and its degree of severity. This problem is largely a function of the different drought types – i.e., meteorological, agricultural, hydrological, and socioeconomic, each of which reflects a different disciplinary perspective.

Third, drought impacts are nonstructural and spread over a larger geographical area than are damages that result from other natural hazards. Quantifying the impacts and providing disaster relief (what some would refer to as "political" drought) are far more difficult tasks for drought than for other natural hazards. Drought impacts are largely "invisible" and, therefore, difficult to quantify. Agricultural impacts are often unknown until the growing season is over and the harvest complete. Even during the latter stages of the growing season and in the post-harvest season, estimates of the economic impacts of drought may change to reflect changing prices, estimates of harvested hectares or land abandoned, or the percent of crop that may be cut for silage. Methodologies for estimating drought impacts are noticeably non-standardized from one estimator or estimate to another.

These characteristics of drought have hindered development of accurate, reliable, and timely estimates of severity and impacts – delaying the formulation of drought preparedness plans. They affect both the way we measure and perceive exposure to drought and our vulnerability to it.

2.4
Drought: Understanding Vulnerability

Recent droughts in developing and developed countries and the concomitant impacts and personal hardships that resulted have underscored the vulnerability of all societies to this "natural" hazard. This appears to be a clear sign of increasing societal vulnerability resulting from unsustainable resource use and growing pressures on natural resources. As noted previously, many factors are contributing to this trend. Adding to the concern regarding increasing societal vulnerability is distress over how the threat of global warming may increase the frequency, severity, and duration of extreme climate events in the future. As pressure on finite water supplies and other limited natural resources continues to build, more frequent and severe droughts are cause for concern in both water-short and water-surplus regions where conflict between water users has increased dramatically. Conflict has also increased within and between countries because transboundary water issues are exacerbated during water-short periods. Reducing the impacts of future drought events is paramount as part of a sustainable development strategy.

Drought impacts vary on both spatial and temporal scales. Each region or watershed is unique, and the societal characteristics for that area or basin are dynamic in response to many factors. A drought event today may be of similar intensity and duration as a historical drought event, but the impacts will likely differ markedly because of changes in societal characteristics. Thus, the impacts that occur from drought are the result of interplay between a natural event (precipitation deficiencies because of natural climatic variability) and the demand placed on water and other natural resources by human-use systems. For example, societies can aggravate the impacts of drought by placing demands on water and other natural resources that exceed the supply of those resources (i.e., overdevelopment or overappropriation) or through a degradation of the natural resource base. The literature is replete with examples of this situation in many countries. Societies often expect or plan for normal or above-normal water supplies, ignoring the natural variability of climate and the challenges of adapting to a significant reduction in supply, especially when this reduction extends over multiple seasons or years and drought effects are magnified by a rapidly increasing population, urbanization, land degradation, or other factors.

According to Randolph Kent (1987), a disaster occurs when a disaster agent, such as drought, exposes the vulnerability of a group or groups in such a way that their lives are directly threatened or sufficient harm has been done to economic and social structures, inevitably undermining their ability to cope and survive. The goal of drought risk management is to impose management and policy changes between hazard events such that the risk associated with the next event is reduced through the implementation of well-formulated policies, plans, and mitigation actions that have been embraced by stakeholders.

2.5
Drought Types and Vulnerability

All types of drought originate from a deficiency of precipitation (Wilhite and Glantz 1985). When this deficiency spans an extended period of time (i.e., meteorological drought), its existence is defined initially in terms of these natural characteristics. However, the other common drought types (i.e., agricultural, hydrological, and socioeconomic) place greater emphasis on human or social aspects of drought, highlighting the interaction or interplay between the natural characteristics of the event and human activities that depend on precipitation to provide adequate water supplies to meet societal and environmental demands (Fig. 2.1). For example, agricultural drought is defined more commonly by the availability of soil water to support crop and forage growth than by the departure of normal precipitation over some specified period of time. There is not a direct relationship between precipitation and infiltration of precipitation into the soil. Infiltration rates vary according to antecedent moisture conditions, slope, soil type, and the intensity of the precipitation event. Soils also vary in their characteristics, with some soils having a high soil water holding capacity while others have a low water holding capacity. Soils with a low water holding capacity are more drought-prone.

Chapter 2: Preparedness and Coping Strategies for Agricultural Drought Risk Management

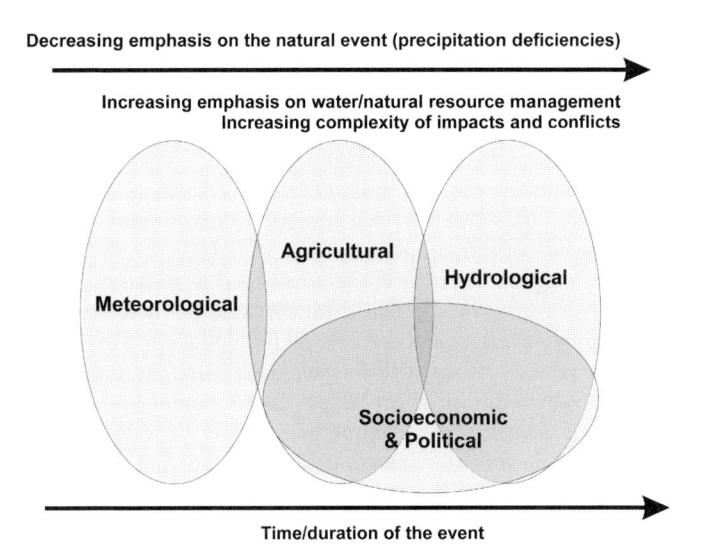

Fig. 2.1. Interrelationships between meteorological, agricultural, hydrological, and socio-economic drought. (Source: National Drought Mitigation Center, University of Nebraska-Lincoln, USA.)

Hydrological drought is even further removed from the deficiency of precipitation since it is normally defined in terms of the departure of surface and subsurface water supplies from some average condition at various points in time. Like agricultural drought, there is not a direct relationship between precipitation amounts and the status of surface and subsurface water supplies in lakes, reservoirs, aquifers, and streams because these components of the hydrological system are used for multiple and competing purposes (e.g., irrigation, recreation, tourism, flood control, hydroelectric power production, domestic water supply, protection of endangered species, and environmental and ecosystem preservation). There is also considerable time lag between departures of precipitation and the point at which these deficiencies become evident in the components of the hydrologic system. Recovery of these components is also slow because of long recharge periods for surface and subsurface water supplies. In areas where the primary source of water is from snow pack, such as in the western United States, the determination of drought severity is further complicated by infrastructures, institutional arrangements, and legal constraints.

Socioeconomic drought differs markedly from the other types because it associates the supply and demand of some economic good or service with elements of meteorological, agricultural, and hydrological drought. Socioeconomic drought is associated directly with the supply of some commodity or economic good (e.g., water, maize, hay, hydroelectric power) that is dependent on precipitation. Increases in population can alter substantially the demand for these economic goods over time. This concept of drought supports the strong symbiosis between drought and its impacts on human activities. Thus, the magnitude of drought impacts could increase because of a change in the frequency of meteorological drought, a change in societal vulnerability to water shortages, or both.

The interplay between drought and human activities raises serious questions about our attempts to characterize it and define it in a meaningful way. Drought results from a deficiency of precipitation from expected or "normal" that is extended over a season or longer period of time and is insufficient to meet the demands of human activities and the environment. Conceptually, this definition assumes that the demands of human activities are in balance or harmony with the availability of water supplies during periods of normal or mean precipitation. If development demands exceed the supply of water available, the result can be that demand exceeds supply even in years of normal precipitation. This can result in a situation of human-induced drought that is separate from the drought types previously discussed. When this situation exists, development can only be sustained through mining of ground water and/or the transfer of water into the region from other watersheds. Is this practice sustainable in the long term? Should this situation be defined as "drought" or unsustainable development? Some would define this as a "water shortage" drought because it is not necessarily the result of a deficiency of precipitation but rather the result of an overallocation of water supplies.

2.6
Decision-Support Tools for Drought Risk Management

The mission of the National Drought Mitigation Center (NDMC) at the University of Nebraska-Lincoln is to lessen societal vulnerability to drought through the application of appropriate risk management techniques, including development of preparedness plans and improved drought monitoring and early warning systems and the adoption of appropriate drought mitigation measures. The NDMC was formed in 1995, and it has been working with local, state, and national government in the United States; foreign governments; international organizations; and others to build awareness of drought and to implement strategies to lessen risk. The NDMC's mission is much broader than just the agricultural sector. Agriculture is certainly one of the most drought-affected sectors, and it is one in which substantial reductions in vulnerability can be achieved through changes in management. However, in many cases, this requires access to better and more timely information. For this reason, the NDMC is focusing considerable attention on development of web-based decision support tools for agricultural producers and other decision makers. Examples of these tools are discussed below.

2.6.1
U.S. Drought Monitor

Recent efforts to improve drought monitoring and early warning in the United States and other countries have provided new early warning and decision-support tools and methodologies in support of drought preparedness planning and policy development. The lessons learned can be helpful models for other countries to follow as they try to reduce the impacts of future droughts. An effective monitoring, early warning, and delivery system continuously tracks key drought and water supply indicators and climate-based indices and delivers this information to decision makers. This allows for the early detection of drought conditions and timely triggering of mitigation and emergency response measures, key ingredients of a drought preparedness plan.

Until recently, a comprehensive, integrated drought monitoring, early warning, and delivery system did not exist in the United States. Until the late 1990s, the country relied principally on the Palmer Drought Severity Index (Palmer 1965) to track moisture conditions and anomalies across the country. This index, while having some advantages, has a number of key disadvantages, including the fact that it is slow to detect emerging drought conditions. A series of severe drought years between 1987 and 1992 highlighted both the vulnerability of the agricultural and other sectors to drought and the inability of the PDSI to adequately monitor these conditions.

Although drought is a normal part of the climate of all regions of the United States and occurs somewhere in the country each year, drought conditions have been especially severe in the period from 1996 to 2006. However, the drought episodes from 1996 to 1999 brought considerable attention to the drought issue because of the magnitude of the impacts experienced, the largely ineffective response of state and federal government, and the deficiencies of the nation's drought monitoring system. A partnership emerged in 1999 between the NDMC at the University of Nebraska-Lincoln, the National Oceanic and Atmospheric Administration (NOAA), and the U.S. Department of Agriculture (USDA) with the goal of improving the coordination and development of new drought monitoring tools. The U.S. Drought Monitor (USDM) became an operational product on August 18, 1999 (Svoboda et al. 2002). The USDM is a weekly product that is posted on the website of the NDMC (http://drought.unl.edu/monitor/monitor.html). This website has become a web-based portal for drought and water supply monitoring.

The development of the USDM was timely because many regions of the country have been affected over several consecutive years since 1999 and on more than one occasion. Some regions of the country experienced as many as 5 to 7 consecutive drought years, and drought conditions are ongoing today in many parts of the country. The USDM successfully integrates information from multiple parameters (i.e., climate indices and indicators) and sources to assess the severity and spatial extent of drought in the United States on a weekly basis. It is a blend of objective analysis and subjective interpretation. This map product has been widely accepted and is used by a diverse set of users to track drought conditions across the country. It is also used for policy decisions on eligibility for drought assistance. The USDM represents a weekly snapshot of current drought conditions. It is not intended to be a forecast and is illustrated in Fig. 2.2.

This assessment includes the 50 U.S. states, Pacific possessions, and Puerto Rico. The product consists of a color map, showing which parts of the United States are suffering from various degrees of drought, and accompanying text. The text describes the drought's current impacts, future threats, and prospects for improvement. The USDM is by far the most user-friendly national drought monitoring product currently available in the United States. Currently, the Internet is the pri-

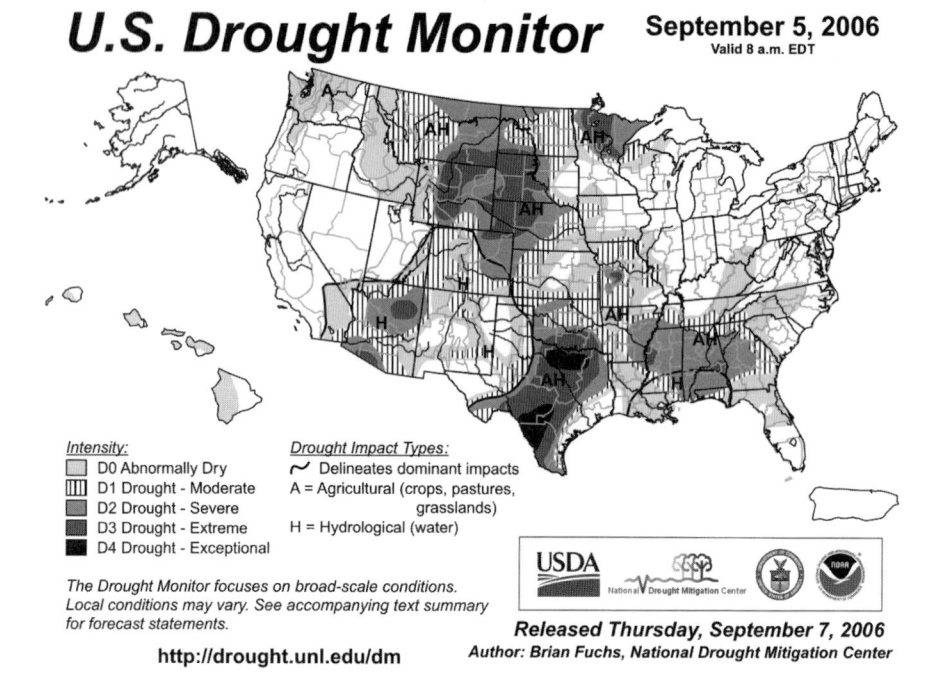

Fig. 2.2. U.S. Drought Monitor, September 5, 2006, showing the spatial extent and severity of drought in the United States.

mary distribution vehicle, although the map also appears in local and national newspapers and on television. A single weekly map illustrates the drought pattern in each year. All USDM maps since 1999 are archived on the website and available to users for comparison.

Because no single definition of drought is appropriate in all situations, agricultural and water planners and others must rely on a variety of data or indices that are expressed in map or graphic form. The authors of the USDM rely on several key indicators and indices, such as the Palmer Drought Severity Index, the Standardized Precipitation Index, stream flow, vegetation health, soil moisture, and impacts. Ancillary indicators (e.g., Keetch Byram Drought Index, reservoir levels, Surface Water Supply Index, river basin snow water equivalent, and pasture and range conditions) from different agencies are integrated to create the final map. Electronic distribution of early drafts of the map to field experts throughout the country provides excellent ground truth for the patterns and severity of drought illustrated on the map each week.

The USDM classifies droughts on a scale of one to four (D1-D4), with D4 reflecting an exceptional drought event (i.e., 1 in 50 year event). A fifth category, D0, indicates an abnormally dry area. The USDM map and narrative identify general drought areas, labeling droughts by intensity from least to most intense. D0 areas (abnormally dry) are either heading into drought or recovering from drought but still experiencing lingering impacts.

The USDM also shows which sectors are presently experiencing direct and indirect impacts, using the labels A (agriculture–crops, livestock, range, or pasture) and W (water supplies). For example, an area shaded and labeled as D2 (A) is in general experiencing severe drought conditions that are affecting the agricultural sector more significantly than the water supply sector. The map authors are careful to not bring an area into or out of drought too quickly, recognizing the slow-onset characteristics of drought, the long recovery process, and the potential for lingering impacts.

The methodology associated with the USDM has now been applied to the production of the North American Drought Monitor (NADM), a collaborative project between the United States, Mexico, and Canada. The partnership began in 2002 in an attempt to map drought severity and spatial pattern across the North American continent. Multiple indices and indicators are used to map drought conditions, similar to the procedure used to create the USDM. Responsibility for this product is shared by NOAA's National Climatic Data Center, the U.S. Department of Agriculture, and the National Drought Mitigation Center at the University of Nebraska in the United States; the National Water Commission in Mexico; and Environment Canada and Agriculture Canada. This product is prepared on a monthly basis and is an excellent example of international cooperation on drought monitoring.

2.6.2
Drought Monitor: Decision Support System

A project is currently underway to enhance the U.S. Drought Monitor by making a more robust drought portal to better address the needs of decision makers. This project is directed at providing users with a robust set of user-friendly web-based tools to visualize and assess drought conditions from national to local scales. Utilizing state-of-the-art technology and information delivery, the site will be a comprehensive one-stop portal for drought, with an approach centered on a more interactive U.S. Drought Monitor map. Advances in spatial and temporal resolution of various indicators now allow for better planning and assessment down to the local level. This suite of links and tools will enable the user to integrate and visualize various large databases, models, GIS techniques, and spatial analysis/visualization tools. The ultimate goal is to capitalize on this technology by developing a delivery mechanism for users of all skill levels, with the intent of improving their knowledge and decision-making abilities. Providing a simple interface with an emphasis on integration and application will allow users to get the answers they need in a timely manner and at the level of detail that is tailored to their needs.

Specific tasks that will be undertaken through this project include: (1) conducting a comprehensive review of the literature and web on other drought-specific decision support systems around the world; (2) expanding on and enhancing the current USDM model; (3) designing and building a user-friendly, interactive interface (web/CD-based delivery systems) that allows for drilling down to the local level to

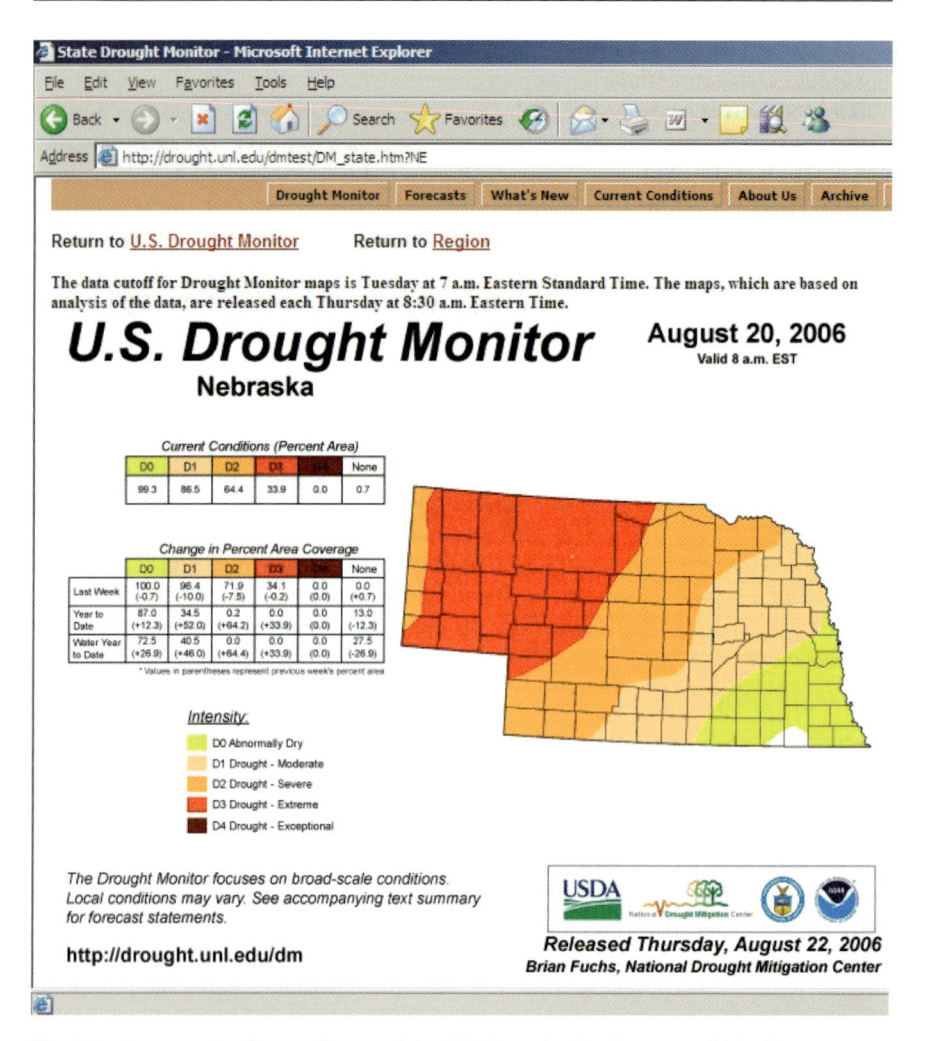

Fig. 2.3. An example of a new feature of the U.S. Drought Monitor map which allows users to drill down from a national map to a more precise local representation of drought conditions. (Source: National Drought Mitigation Center, University of Nebraska-Lincoln, USA.)

assess drought; and (4) fostering a continual process of user feedback, evaluation, assessment, and dissemination of the tools for agricultural producers and others.

One specific example of a more interactive product is to provide users with the capability to 'drill-down' from the national level to the state and county level to obtain a higher resolution depiction of the drought areas affected for their area (Fig. 2.3). These higher resolution maps will be accompanied by tables that provide a breakdown of the percent area in various drought severity classes. This type of information is extremely useful for policy decisions and the state and national level and also for the media.

2.6.3
The Drought Impact Reporter:
A Web-based Impact Assessment Tool and Database

The widespread drought episodes over the past two decades in the United States have emphasized the need to better assess the magnitude of drought impacts, sectors affected, and their spatial dimensions. According to the Palmer Drought Severity Index (PDSI), severe to extreme drought covered more than 25% of the United States in 2000, 2002, 2003, 2004, and 2006. At the end of July 2002, drought or dryness was affecting all 50 states at the same time, and parts of 26 states were classified under "severe", "extreme", or "exceptional" designations, according to the U.S. Drought Monitor (http://drought.unl.edu/dm). In spite of the widespread severity of recent drought years, there has been no comprehensive assessment of economic, environmental, or social impacts. There is also no national database of drought-related impacts. Without more timely and precise estimates of impacts across the multitude of sectors affected by drought, policy and other decision makers are reluctant to allocate money and resources to mitigation and preparedness, according to the Council of Governors' Policy Advisors (Brenner 1997). These state officials have a general understanding that "mitigation makes sense," but their desire was for quantitative proof. In fact, this report identified the "lack of information" as the major obstacle to adopting mitigation strategies. Wilhite and Buchanan-Smith (2005) also identified the lack of a comprehensive impact assessment methodology as an obstacle to activating effective drought mitigation and response programs. To overcome this obstacle, timely and quantitative assessments of the impacts and economic losses associated with drought must be compiled.

Because of the number of affected groups and sectors associated with drought, the geographic size of the area affected, and the difficulties in quantifying environmental damages and personal hardships, the precise determination of the financial costs of drought is a formidable challenge. These costs and losses are also quite variable from one drought year to another in the same place, depending on timing, intensity, and spatial extent of the droughts.

In July 2005, the NDMC launched a prototype web-based Drought Impact Reporter (DIR) to present real-time information on current drought impacts and serve as a national drought impacts database. The DIR has two main components: (1) a comprehensive database or archive of drought impacts and (2) an interactive map delivery system that provides quick access to the archive. The drought impacts archive is the backbone of the DIR. NDMC staff began entering drought impact information during summer 2005. When the DIR was launched in July, the web-based tool was still in its earliest development phase. The NDMC has now received additional funding to broaden and enhance the scope of the DIR and the interactive map delivery system so it is more efficient and user-oriented. The NDMC is also developing additional linkages with governmental agencies, non-governmental organizations, university research groups and extension programs, and others to provide impact reports to ensure a comprehensive collection of drought impacts across all potential sectors and scales. User evaluations and feedback are also important components of the DIR system. The NDMC will continue to foster linkages with a broad range of users as it enhances the DIR. The DIR has been constructed so its primary elements are consistent with an increased emphasis on drought impact assessment and mitigation and the need for an interactive web-based system to deliver information for all users, as called for in the report on the National Integrated Drought Information System (NIDIS), prepared by the Western Governors' Association (2004).

The sources of the drought impact data for the DIR are:

- An online clipping service that provides daily drought impact-related news articles and scientific publications. The NDMC began subscribing to this service in March 2005.
- Articles on drought impacts, collected routinely since 1997. The NDMC now has an internal archive of more than 11,000 articles. These articles will also be reviewed for drought impact information and entered into the database.
- Drought impact information from reports and other materials from historical drought periods, such as the 1930s, 1950s, 1970s, and late 1980s to early 1990s, and other shorter-duration drought events will be reviewed and entered.
- User-entered drought impact information directly through the website by government officials, water utilities, water and natural resource managers, agricultural producers, and others. This information is reviewed and verified by NDMC staff and is characterized as "submitted" reports.

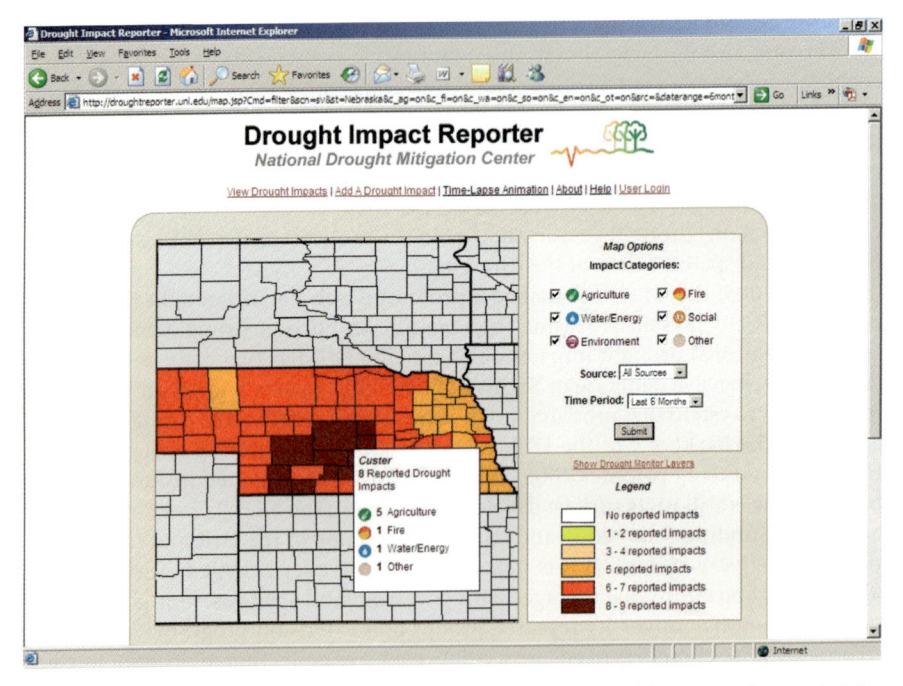

Fig. 2.4. The Drought Impact Reporter, a new feature on the website of the National Drought Mitigation Center that provides a spatial and sectoral representation of drought impacts in the country. (Source: National Drought Mitigation Center, University of Nebraska-Lincoln, USA.)

The DIR has been developed and is supported by an ArcGIS/IMS architecture. As this tool evolves, enhancements to the delivery system will be needed. In addition, since the Drought Impact Reporter is one tool in the larger National Drought Impacts Reporting Strategy, it is envisioned that there will be a suite of web-based products and interactive features that will also be supported as part of the same delivery system.

The DIR can be accessed through the NDMC's website (http://drought.unl.edu) or directly at http://droughtreporter.unl.edu. When the tool is accessed, the first default screen displays a map of the United States illustrating the number of drought impacts reported during the past month (Fig. 2.4). The legend appears in the lower right corner of the page. In the upper right corner is a list of impact categories. All categories will be displayed initially, but the user can select only those categories of interest. The user can also select the sources of information (e.g., media, public), but all sources are shown initially. The user can also select the time period for the impacts [Note: NDMC staff have, at this writing, entered impacts reported through news articles back to 1995]. The default for the map is the past month. After making the selections for sources and time period, the user can click the "select" button to generate a new map.

By positioning the cursor over a state, a box appears with a listing of the total number of impacts for the selected period and how these break down by sectors. Clicking on the state will produce a map of that state depicting counties (Fig. 2.4). By placing the cursor over a county, a box appears again depicting the number of impacts for that county with a breakdown by sector. Clicking on the county will reveal the sources for this information (queried from the database), allowing the user to learn more about the impacts reported. This "drill down" technique is a critically important feature of the DIR, allowing users to interrogate to the local or county scale to identify specific impacts.

The user also has the option to overlay the various categories of drought severity from the U.S. Drought Monitor map (http://drought.unl.edu/dm). Overlaying the drought categories on the Drought Impact Reporter map gives users the option of visually correlating impacts with drought severity levels. Currently, this option is only available for the most recent U.S. Drought Monitor, but plans are to expand this capability in the future. Overlaying the Drought Monitor categories on the DIR map will also help users understand and appreciate the lag characteristics of drought impacts, since dry conditions may persist for long periods. For example, the northern Great Plains and northern Rocky Mountain states have been in various drought severity levels for the past seven years.

Other features of the DIR include an option to animate the impacts over a time period and also for users to add drought impacts. To add an impact, the user clicks on this feature and then enters the requested information, including the selection of the impact categories and describing the impact. Information entered is quality-checked by NDMC staff before it is added to the database. To date, about 10% of the impacts entered have been from the public, but this number is expected to increase significantly as user groups become more aware of the DIR and the archive becomes more comprehensive.

Numerous sectoral impacts have been added to the database since it was first launched. The total number of impacts added for the period 1995 to present is nearly 4,000. Although this represents only a small fraction of the impacts that have occurred during this period, it does illustrate both the diversity of impacts and the relative importance of these impacts by sector. As one would expect, the largest number of impacts reported is in the agricultural sector, but significant impacts have also been reported in the water, energy, and fire sectors. Social impacts, which are usually underappreciated for drought, are significant over the period of 1995 to present.

Many benefits are expected for policy and other decision makers, the scientific community, and the general public as a result of improved drought impact assessments and the creation of an impact archive. First, this project is the first step toward development of national and regional assessments of drought conditions across the United States. For example, Canada was able to make a rough national assessment of the 2001–02 drought, estimating losses at approximately Canadian $5.7 billion (Saskatchewan Research Council 2003). Although not perfect, the Canadian drought assessment placed the losses in context for officials and provided a basis for making adjustments and improving on this assessment in future drought events. Second, the archived collection of drought impacts within the large database will be freely and easily accessible to researchers, as well as to decision makers requiring information for policy and management options. Initial reaction to the DIR has been extremely positive, and ongoing efforts to enhance this product will further heighten its use and popularity. For example, since the DIR was launched on July 27, 2005, it has received more than 25,000 users accounting for more than 142,000 page views and more than 1.25 million hits (as of December 31, 2005). The NDMC is actively publicizing this product and engaging a wide range of user groups in building the archive and obtaining user feedback. Important bridges will be built between research and user communities that will ultimately increase the capacity for better drought mitigation and response activities across the country.

Third, the project builds a foundation for development of standardized methodologies of identifying, collecting, and quantifying drought impacts on national, regional, state, and local levels, as well as the methods for estimating economic losses at these levels. The NDMC will continue to pursue development of these methodologies in collaboration with other research entities. Future enhancements to the DIR will include linking this tool to databases such as agricultural statistics at the state and local levels to compare reported impact information with specific production losses as well as to information on drought disaster declarations by federal agencies. Fourth, the DIR will provide a platform for identifying and reporting drought impacts in under-reported sectors, such as livestock, timber, recreation, tourism, and energy. It is likely that the recent drought years from 1996 to 2005 across the United States resulted in impacts in these sectors greater than or equal to crop production losses, which are the most frequently quantified economic impact of drought. Fifth, discussions have been held between the NDMC and NOAA/National Weather Service (NWS) and U.S. Department of Agriculture personnel about using the Drought Impact Reporter and its data entry format as the tool for entering and documenting "drought incident reports" similar to storm reports that are filed on severe weather events. This would provide NWS offices with a uniform format for reporting drought conditions and impacts, and would provide an additional dissemination method for these reports, and the drought impacts taking place, through the Drought Impact Reporter's map-based delivery system. Finally, this project supports both NIDIS and the National Drought Preparedness Act. The interactive map delivery system will easily connect with other drought-related decision-support tools now being developed by the NDMC, government agencies, and other organizations.

2.6.4
Drought Risk Atlas

The goal of the drought risk atlas is to provide users with a comprehensive assessment of the history, frequency, intensity, duration, and trend of droughts over the past century on a site-specific basis. The intent is to provide users with a tool to help them better understand and visualize their drought risk in order to make better long-term management decisions. An interactive web-based interface will allow producers, water managers, and decision makers to tap into a database containing precipitation, stream flow, and drought indices based on daily data that goes back to the late 1800s and early 1900s. Historical analyses will allow users to examine seasonal, monthly, weekly, and daily patterns, along with the spatial distribution and characteristics of precipitation and drought. Users will be able to display and print their results in table, graph, or map format. Tutorials and libraries of information explaining the results will accompany both the web-based tool and an envisioned CD-based product that will be distributed free to users.

Specific tasks that will be accomplished through this project include: (1) conducting a comprehensive review of the literature and web on other drought-specific atlas products; (2) expanding on the comprehensive collection of drought risk information in order to build a time series database containing the frequency, trends, and magnitude of precipitation/droughts on a station location basis; (3) building and enhancing an interactive interface (web/CD-based delivery systems) to the data and visuals so that it is as efficient and user-oriented as possible; and (4) fostering a continual process of user feedback, evaluation, assessment, and dissemination of the tool, especially among agricultural producers.

2.6.5
Vegetation Drought Response Index (VegDRI)

Drought monitoring is challenging because of the considerable variability in the duration, intensity, and spatial extent among specific drought events. Climate and weather data traditionally have been used to monitor drought, but the information has lacked the spatial detail required for local livestock and forage producers to use for drought planning. In response, the NDMC is developing two tools, the Vegetation Drought Response Index (VegDRI) and the Vegetation Outlook (VegOut), which utilize a combination of climate, satellite, oceanic, and biophysical (land cover and soils information) data to map and monitor the impact of drought on general vegetation conditions. Both tools produce maps at 1-km^2 spatial resolution that are updated every two weeks to reflect the changing vegetation conditions. The goal is to create tools that provide timely and more spatially precise drought-related information that can be used by livestock and forage producers in their decision-making process. The NDMC will provide these tools and information in readily accessible and usable formats that will be determined from producer feedback.

The NDMC is working collaboratively with other researchers at the University of Nebraska–Lincoln (UNL) and the United States Geological Survey (USGS) to develop the VegDRI tool, which maps the spatial patterns of drought impact on the current vegetation conditions. VegDRI classifies seven categories of relative drought severity, ranging from extreme drought to an extremely moist spell. The tool can be used to monitor drought conditions over large areas, but also has enough spatial precision to be used at county to sub-county levels.

The NDMC is also working with UNL researchers to create the VegOut tool, which provides an outlook of the general vegetation conditions several weeks in advance. VegOut will provide 2-, 4-, and 6-week outlook maps that reflect the projected level of vegetation stress based on current vegetation conditions and climatic trends. The three VegOut maps will be produced and updated in conjunction with the VegDRI maps.

2.6.6
Ranching Drought Plan: A Drought Planning Tool for Livestock and Forage Producers

According to the National Drought Policy Commission report, *Preparing for Drought in the Twenty-First Century* (2000), many agricultural producers do not have access to information to develop and implement a drought plan, and even fewer producers are receiving technical assistance to help them develop and implement such plans. In order to address this problem, the NDMC is developing a model drought planning process and web-based educational delivery system for forage and rangeland producers.

The NDMC is working collaboratively with the University of Nebraska–Lincoln's (UNL) Cooperative Extension Service and Department of Computer Science and Engineering, livestock and forage production consultants, and individual ranchers to identify essential planning components and develop a generic drought planning process that can serve as a template for producers to follow.

Tasks being completed for this project include:
- Conducting a review of existing literature on livestock production and drought planning;
- Developing new information on aspects of livestock production and drought planning such as climate and weather, grazing systems, alternative forage production, herd management, supplemental feeding, financial management strategies, and federal assistance programs;
- Utilizing video technology to record and display drought management strategies;
- Organizing relevant information into a drought planning process that meets the needs of livestock and forage producers; and
- Developing a web-based educational delivery system for mass distribution.

2.7
Summary

Drought is a pervasive natural hazard that is a normal part of the climate of virtually all countries. It should not be viewed as merely a physical phenomenon. Rather, drought is the result of interplay between a natural event and the demand placed on water and other natural resources by human-use systems. These systems can significantly exacerbate the impacts of drought through the unsustainable use of natural resources.

Numerous opportunities exist to mitigate the impacts of drought in the agricultural sector if appropriate management practices are adopted in a timely manner. Developing more comprehensive and integrated drought monitoring and early warning systems is an essential component of a more proactive, risk-based management system. However, equally important is an effective user-driven delivery system, the availability of helpful and efficient decision support tools, and the training of users on how to apply this information at critical decision points before and during the growing season. The NDMC is directing an increasing share of its efforts toward these tasks, with the goal of reducing the vulnerability of the agricultural sector to future episodes of severe drought.

References

Brenner E (1997) Reducing the impact of natural disasters: Governors' advisors talk about mitigation. Council of Governors' Policy Advisors, Washington DC

Gillette HP (1950) A creeping drought under way. Water and Sewage Works, March, pp 104–105

Kent RC (1987) Anatomy of disaster relief: the international network in action. Pinter Publishers, New York and London

National Drought Policy Commission (2000) Preparing for drought in the 21st century. Washington DC [http://www.fsa.usda.gov/drought/finalreport/accesstoreports.htm]

Oweis T (2005) The role of water harvesting and supplemental irrigation in coping with water scarcity and drought in the dry areas. In: Wilhite DA (ed) Drought and water crises: science, technology, and management issues. CRC Press, Boca Raton FL, pp 191–213

Palmer WC (1965) Meteorological drought. Research Paper No. 45. US Weather Bureau, Washington DC

Risk Management Agency (RMA) (2004) Crop insurance indemnities paid for drought. Personal Communication with James Callan, August 5, 2004. Risk Management Agency, US Department of Agriculture, Washington DC

Saskatchewan Research Council (2003) Canadian droughts of 2001 and 2002: climatology, impacts and adaptations. SRC Publication No. 11602–1E03, Saskatoon SK (unpublished)

Svoboda M, LeComte D, Hayes M, Heim R, Gleason K, Angel J, Rippey B, Tinker R, Palecki M, Stooksbury D, Miskus D, Stephens S (2002) The Drought Monitor. Bull Am Meteorol Soc 83:1181–1192

Western Governors' Association (WGA) (2004) Creating a drought early warning system for the 21st century: the National Integrated Drought Information System. Denver CO [http://www.westgov.org/wga/publicat/nidis.pdf]

Wilhite DA, Buchanan-Smith M (2005) Drought as a natural hazard: understanding the natural and social context. In: Wilhite DA (ed) Drought and water crises: science, technology, and management issues. CRC Press, Boca Raton FL, pp 3–29

Wilhite DA, Glantz MH (1985) Understanding the drought phenomenon: the role of definitions. Water Int 10:111–120

Zhang HL, Dan, KL, Zhang SF (2005) Drought and water management: can China meet future needs. In: Wilhite DA (ed) Drought and water crises: science, technology, and management issues. CRC Press, Boca Raton FL, pp 319–343

CHAPTER 3

Challenges to Coping Strategies with Agrometeorological Risks and Uncertainties in Africa

Elijah Mukhala, Adams Chavula

3.1 Introduction

In sub-Saharan Africa, 90% of agricultural production is rainfed and only 10% of the arable land is irrigated. At the same time, the continent is susceptible to interannual rainfall variability. These statistics strengthen the argument that weather and climate are one of the biggest production risk and uncertainty factors impacting on agriculture systems' performance. The Southern African region faces well-documented challenges in maintaining and improving food security in the face of multiple stresses. Climate stress in particular has compromised the ability of the region's agricultural sector to sustain production. Such a situation is particularly concerning in the light of the projected climate stress under future climate change due to, for example, the increasing frequency of extreme precipitation events (IPCC 2001). The paper discusses the use of integrated sustainable agriculture in Africa that takes into account preparedness, monitoring, assessments, mitigation and adaptation that address issues of extreme climatic events including severe droughts, floods and cyclonic systems. The paper also discusses efforts in the use of improved climate knowledge and technology, including monitoring and response mechanisms to current weather to reduce the uncertainties in agrometeorological risks.

In Africa, the challenges that farmers face are beyond agricultural related activities and go further to include marketing, access to loans, HIV/AIDS and lack of inputs exacerbated by poverty. Inadequate policies also subject farmers to a life of perpetual poverty (ECA 2006). Literature is replete with information indicating the fact that in Africa, extreme climatic events such as droughts, floods and cyclones have been known to strongly impede sustainable agricultural production and development. Even with this kind of information available, appropriate policies are hard to come by that try and put in place measures that will mitigate these extreme climatic events.

In Africa, decision making in agricultural production is a complex process which requires a lot of information in order to assist informed decisions. The paper tries to identify and assess farmers' coping strategies with risks in southern Africa, particularly extreme climatic events such as droughts, floods and cyclones. The paper also discusses the inadequate attention to agroclimatic characteristics of a location and lack of timely information on weather and climate risks and uncertainties. The also discusses major challenges of access to technological advances in Africa as well as preparedness and response strategies and training of intermediaries between National Meteorological and Hydrological Services (NMHSs) and farmers etc. Finally the paper focuses on the opportunities for farmers in Africa to mitigate risks and uncertainties using structural measures such as irrigation and water harvesting and non-structural measures such as seasonal climate forecasts as well as medium-range weather forecasts for strategic and tactical management of agriculture. The use of crop insurance strategies and schemes to reduce the vulnerability of the farming communities to agrometeorological risks is also addressed.

Sub-Saharan African economies are especially susceptible to climate variations due to their predominately agrarian structure. In Ethiopia, agriculture accounts for about 40% of GDP, 80% of export earnings and 85% of employment in diverse traditional subsistence systems for production mainly of cereals, oilseeds and livestock (Hertz 1996). Population and land tenure pressures have led to reduced productivity, increasing the vulnerability of the predominantly rain-fed agricultural systems to rainfall variations. Despite a trend towards urbanization, the majority of poverty remains in the rural areas, where households have limited assets to withstand climatic, disease or income shocks. The impact that climate variability can have on such agrarian economies is well reflected in the case of Ethiopia, where economic growth and food imports closely track variations in rainfall (Figure 3.1). (Grey and Saddoff 2005)

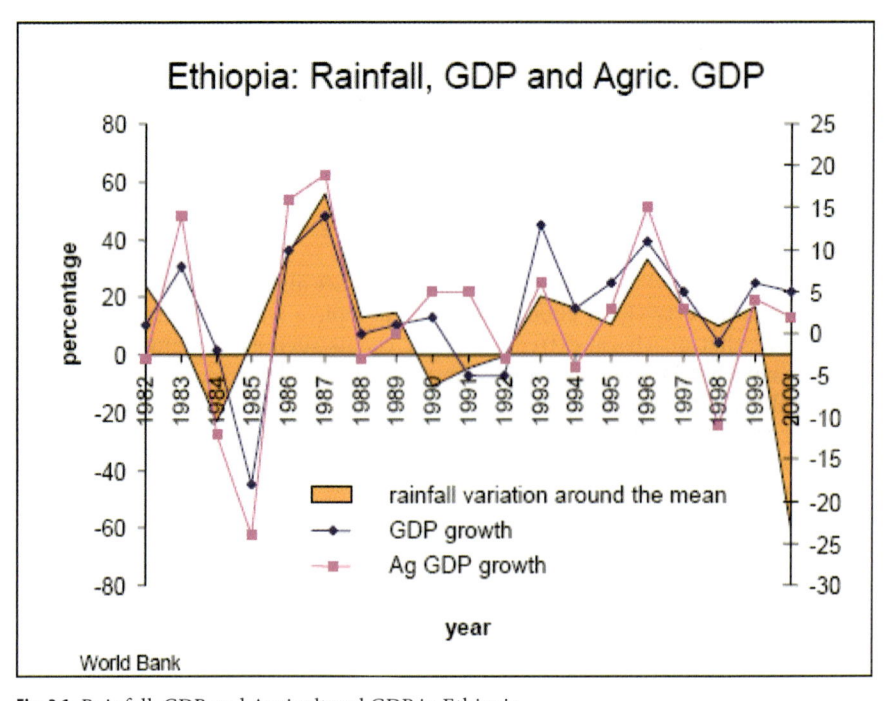

Fig. 3.1 Rainfall, GDP and Agricultural GDP in Ethiopia

3.2
Farmers' coping strategies

Farmers in Africa have been subjected to persistent unfavorable climatic conditions. The level of assets and wealth determine how much adjustments one can make to sustain agricultural production. In Zambia, a recent World Bank Survey reported an increase in poverty levels in the rural areas from 70% in 1994 to 90% in 1995 (Balat and Porto 2005). These statistics reinforce the fact that farmers have to find coping strategies to survive. At the institutional level, ineffective extension and research services and inappropriate agricultural policies, which have relied excessively on maize production, have been cited as contributory factors (Siacinji-Musiwa 1999).

Literature is replete with information regarding the declining performance of the agricultural sector in Africa and the reasons advanced are due to changes in climatic and economic circumstances (FAO GIEWS 2005a). The severe droughts experienced recently, particularly in the southern Africa region were characterized by both a decline in overall precipitation and increasingly erratic distribution patterns (FAO GIEWS 2005b). When drought occurs, future draft power is affected due to poor health of the livestock in most cases leading to death. The decimation of draft oxen, the reduction in active farm labour, and disruptions in input supply and marketing arrangements all have a negative impact on productivity, income and most importantly, food security (Siacinji-Musiwa 1999). On the positive side there is evidence that farmers themselves are attempting to adopt strategies to cope with these problems. Among the strategies that are being adopted include; crop diversification, the use of drought tolerant varieties, the adoption of reduced tillage methods and an increase in off farm income-generating activities.

Conservation farming is one method that is being used widely in southern Africa to sustain agricultural production and mitigate the impacts of intra-season rainfall variability. The benefits of conservation farming are well proven and offer smallholders the opportunity to increase their productivity, safeguard their land and reduce the risks of total crop failure in drought years (Siacinji-Musiwa 1999). Sustainable agriculture takes into account a series of farming operations that take care of the "whole" farming system in such a manner that farming can be sustained over a long period of time. One such method that the farmers use is minimum tillage (MT) which refers to reducing tillage operations to the minimum required for crop development. When using hand tools or animal draft, farmers plough out the row where the crops are to be established, leaving the rest of the land untouched before planting. MT is not a new concept and has always been a traditional way of planting. Farmers who wait for the rain while they have planted in holes are basically exercising MT. The main benefits of MT are that farmers are able to plant a larger area and can plant early.

Conservation Tillage (CT) includes all operations which:
a) Protect the soil from the damaging effects of rain splash;
b) Reduce runoff and keep more of the rain on the fields (rain harvesting);
c) Make the best use of costly fertiliser and seed and
d) Allow farmers to finish land preparation well before the rains.

Therefore, Conservation Farming (CF) incorporates MT and CT and is a term used to describe a range of husbandry and conservation practices which, when used in combination, bring about the benefits stated above. Conservation Farming also means crop diversification and rotations so that at least 30% of the land is occupied each year by a legume. Farmers who practice CT and also use rotations are doing conservation farming. Essentially, CF combines sound husbandry and management practices, which arrest soil exhaustion, reduce the impact of intra-season rainfall variability, increase productivity, and enable farmers to spread out labour demand and get their work done on time. The technology can be applied to a wide range of farming groups from resource poor to commercial with good results.

3.3
Provision of climatic information

Climate plays a significant role in agricultural production in Africa. Humanitarian organizations have highlighted this fact most vividly in the electronic media. In sub-Saharan Africa, research has shown clearly that critical gaps exist in the ability of climate information to be applied in the agricultural sector (IRI 2006). In southern Africa, agrometeorological activities are well organized through the sub-regional organization, Southern African Development Community (SADC). There is a continuous analysis of climate data. There is also a regular annual agrometeorological meeting prior to commencement of the rainy season to discuss the implications of the most recent seasonal climate forecast. The discussions cover the impact of the seasonal forecast on agricultural production, probabilities of the onset and cessation of rainfall, probabilities of dry spells and other challenges in the provision of agrometeorological information.

In Nov 2002, the SADC Regional Remote Sensing Unit (RRSU) convened an annual Agrometeorological Workshop in Harare, Zimbabwe entitled 'Application of climate information to sustain agricultural production and food security in the SADC region". The workshop was attended by representatives of Agronomy, Agrometeorological and National Meteorological and Hydrological Services (NMHSs) active in the National Early Warning Units (NEWU's) of SADC Member States. As part of the workshop, stakeholders present were interviewed and requested to prepare detailed responses regarding their assessment as to the extent to which the climate information system currently served the agricultural sector in their respective countries. In all, 12 countries responded and these include: Angola, Botswana, Lesotho, Malawi, Mauritius, Mozambique, Namibia, South Africa, Swaziland, Tanzania, Zambia and Zimbabwe. Specifically, NEWU participants were asked to answer four overarching questions:
- What are the specific forecast needs for agricultural decision-making, given the specific characteristics of your agricultural sector?
- To what extent are such forecast needs currently being accommodated in the country's forecast system?
- What are the specific gaps in the forecast system (as it serves the agricultural sector)?
- Identify three strategies to close these gaps?

The following is a list of challenges that were identified by the representatives of the twelve Member States during the workshop.

3.3.1
Intra-seasonal distribution

In agricultural production, intra-season rainfall distribution is more important than cumulative rainfall. Respondents indicated that measures of intra-seasonal rainfall distribution or 'seasonal quality' be predicted. This would help the farmers plan properly in terms of timing of planting to avoid the crops reaching critical stages at times when there are high probabilities of dry spells. Studies in other parts of Africa have indicated that intra-seasonal distribution is a climatic parameter necessary for decision-making in the agricultural sector (Usman et al. 2005). In Kenya, research reports that total seasonal rainfall may be enough to sustain crop production, but its distribution and occurrence of intra-season dry spells (ISDS) and off-season dry spells (ODS) affect crop production. Rainwater harvesting (RWH) and management, especially on-farm storage ponds for supplemental irrigation offers an opportunity to mitigate the recurrent dry spells (Ngigi et al. 2005).

3.3.2
Language/terminology and communication

Communication is deemed effective if there is a response to the message conveyed. This is often unsuccessful when the language used is unfamiliar to the targeted audience. Challenges of language and terminology were specifically highlighted by ten of the responding country teams. This is a substantial hurdle in forecast outreach – it may also marginalize particular communities and areas of a country. Mukhala (2000) reported that there is a need to research the effectiveness of other channels of communicating meteorological information. These include farm demonstrations, farm discussions, farmers' days, meetings and other farmers. Whatever media or channels are used, the time-tested adage 'know your audience' is the best starting point. It also has been reported that Governments and meteorological bodies have not yet developed effective methods of communicating probabilistic forecasts through conventional channels such as extension services (Blench, 1999).

3.3.3
Capacity of key institutions

The respondents indicated that they did not have sufficient capacity to conduct agrometeorological analysis that would meet the farmers' demands. This meant that the skills and equipment were not adequate. In many African meteorological services, funding has become a really big challenge for operational purposes or indeed for capacity building. The remuneration of staff is also not very competitive and hence the occurrence of high staff turn-over. The lack of capacity in institutions has been echoed by other researchers. The International Research Institute for Climate and Society (IRI) Gap analysis report indicates that at present, there is the lack of effective institutional arrangements to facilitate the generation, analysis and systematic integration of relevant climate information with other pertinent information in a form that planning and operational agencies can use (IRI 2006).

3.3.4
Stakeholder awareness/training

In some cases, the resource poor farmers are not aware of the availability of information that could be used in their decision-making. The respondents indicated that there is a need for user/stakeholder training and awareness activities (SADC-RRSU 2002). Two countries found that the SARCOF regional forecast consensus process itself had key weaknesses relating to their ability to develop human resources in the region that would apply climate information effectively to the agricultural sector. For example, Botswana representatives observed that each year a different staff member attended SARCOF – something that neither the Botswana nor the Namibia team (who reiterated this concern) considered an effective capacity building strategy. Further, both teams identified weak dissemination of function and capacity down to the national (NEWU) level from SARCOF as a key concern (SADC-RRSU 2002).

3.3.5
Tailored forecasts

The applicability of the climate forecasts depends so much to the extent to which the producers of the information had the users in mind while generating the information. In the current state, it is left to the general users to try and fit in climate forecasts into their activities. Hence the respondents highlighted the need for tailored forecasts that would make it easier for the users to make use of the information. These would imply that forecasts would aim at greater and more specific utility for particular users. Research reports that climate forecasts need to be integrated with other aspects of infrastructure and input supply; for instance, at present a prediction of drought is unlikely to be linked to increased availability of appropriate seeds (Blench 1999).

3.3.6
Key relationships

For agricultural purposes, the producers of climate information do not have access to farmers directly. Agricultural extension officers are mandated to interact with farmers, however, there is a weak link between NMHSs and the extension services or other agricultural expert intermediaries (SADC-RRSU 2002). Field studies on the impact of recent forecasts in southern Africa suggest that there is a considerable gap between the information needed by small-scale farmers and that provided by the meteorological services (Blench 1999).

3.3.7
Timely issuance

The timing of the release of the climate forecast is critical for effective use of the forecast. Several country teams (Lesotho, Mozambique and Swaziland) found that timely issuance remains a key weakness in climate information systems. The Mozambique team, for example, observed that at present the forecast is provided too late for planting decisions in parts of southern Mozambique (SADC-RRSU 2002). Other scientists have also stated that if the meteorologists are to respond to the needs of users they also need to meet the time requirements specified by the users. If the meteorologist takes time to understand the process of the users then they can ensure that the flow of weather information is also provided in a timely manner. This is a vital aspect of weather forecasts as the meteorological information is considered a "perishable" item i.e. today's weather forecast is of no use tomorrow (Walker 2001).

3.4
Key recommendations to address identified weaknesses

A focus on identified weaknesses (many of which are already well documented and well known) would be sufficient. The weaknesses are thus distilled into concrete recommendations intended to address one or more of the weaknesses. Figure 3.2 illustrates the weaknesses which are listed in order, with most frequently identified weaknesses at the bottom of the figure. Concrete actions around such recommendations, each shown by arrows to address one or more of the identified weaknesses, appear in the middle.

It is essential to note that in the case of many of these recommended actions, work is already under way. Gaps remain, however, and the stakeholder identified weaknesses shown in this paper ensure that closing these gaps is not merely a research task. As stated earlier, the tradeoff here is between what stakeholders require, and what can be robustly provided. The recommendations, we propose, show a middle ground – much of which, despite existing work, is yet to be attained in the SADC region.

3.4.1
Forecast improvement

Given that many weaknesses refer to different aspects of forecasts themselves, three key sets of actions around forecast improvement can be identified. Firstly, models

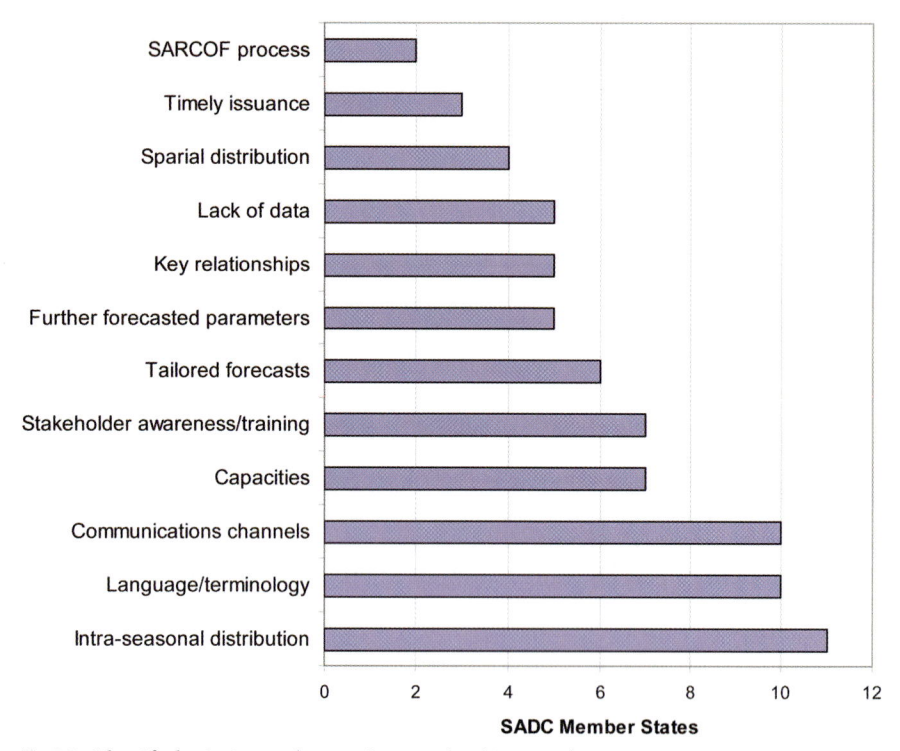

Fig. 3.2 Identified priority weaknesses/gaps in the climate information systems

used in forecasting can (and are being) improved in a number of ways. Ongoing model improvement to provide forecasts in a more timely manner while increasing the skill is required. Another essential task to be done is further analysis of how climate information could serve the agricultural sector better. However, the main challenge here is to balance what users in the agricultural sector require with what scientists can confidently provide, given the existing level of technology. Investigation undertaken during the 2002/3 rainy season under regional conditions of elevated disaster risk showed a number of weaknesses and gaps in the climate information systems in the Southern African region, and making it more challenging to benefit key sectors, particularly agriculture (SADC-RRSU 2002, Archer et al. 2007).

In an effort to improve the provision of climate information, a project 'Mitigating the Effects of Hydroclimatic Extremes in Southern Africa,' funded by the U.S. Agency for International Development (USAID) was established. The project focused on problems of climate information dissemination and interpretation in the region as a strategy to contribute to mitigating current (and future) climate risk. Essentially, the project attempted to answer the question "What would constitute an improved role for climate prediction in contributing to sustaining agricultural production and food security in Southern Africa?". Key gaps seen as a priority included the lack of information on intra-seasonal distribution, insufficient translation in terms of language and terminology and a range of challenges around stakeholder and institutional capacity. Recommendations are made in the areas of forecast improvement (e.g. greater focus on forecasting intra-seasonal distribution) and increased national and non-traditional investment in outreach and applications, amongst others (SADC-RRSU 2002, Archer et al. 2007).

3.4.2
Access to technological advances

There are also major challenges of access to technological advances in Africa as well as preparedness and response strategies and training of intermediaries between NMHSs and farmers etc. In most developing countries, women make up the majority of the population working in agriculture, but they are marginalized with respect to access to ICTs for economic and social empowerment (Odame et al. 2002). Due to this marginalization, the farmers find it very difficult to access climate information which could be very vital in planning their agricultural activities. At the same time it is acknowledged that telecommunications connectivity in developing countries is usually available only within the capital and in major secondary cities. Yet the majority of the population lives outside of these cities (Odame et al. 2002). But, for most poor farmers in Africa, production depends on the vagaries of the climate. They have no way to obtain modern farming technology, or credit to buy much needed inputs. They often cannot even reach vital markets because of poor infrastructure (Diouf 1997).

3.4.3
Structural and non-structural measures to mitigate risks and uncertainties

There are also opportunities for farmers in Africa to mitigate risks and uncertainties using structural measures such irrigation and water harvesting and non-structural measures such as seasonal climate forecasts as well as medium-range weather forecasts for strategic and tactical management of agriculture. Research indicates that the agrohydrological challenge in semi-arid and dry subhumid tropics is not necessarily related to inadequate cumulative rainfall - at present basically only 1/8 – 1/3 of the rain is used in crop production on average. Instead the challenge is to manage the unreliable distribution of rainfall over time, and minimize non-productive water flow in the water balance (SIWI 2001). The degree of acceptance of irrigation scheduling technology through extension depends directly on the literacy levels of the farming community. Unfortunately, most of the traditional irrigation systems are located in areas where educational standards are low. It is essential that reliable information is developed and disseminated in a simplified manner understandable to the trainers and end-users (Hasan et al. 1993). Government agencies, along with universities and the private sector, must provide the required training.

3.4.4
Crop insurance activities

Food security and weather risk management are inextricably linked: weather risk management, or the lack of it, determines the level of systemic risk in the food security system. The exposure to weather risk drives overall food insecurity. At the farm level, weather-based index insurance allows for more stable income streams and could thus be a way to protect peoples' livelihoods and improve their access to finance. Weather-based insurance instruments provide financial protection based on the performance of a specified index in relation to a specified trigger and they offer protection against the uncertainty in revenue accruement that results from volume volatility.

Weather based index insurance is slowly gaining recognition as one of the methodologies that can be used to sustain livelihoods and reduce poverty as part of the Millennium Development Goals (MDGs). A few countries in Africa are piloting the methodology and among the countries include: Malawi and Ethiopia. In Malawi the pilot drought insurance program has been introduced for local groundnut farmers. The main aim is to help mitigate the risks associated with periodic droughts as the country is prone to drought. Due to high levels of poverty, the farmers are not credit worthy and hence they cannot access loans to purchase inputs. Therefore, the insurance helps farmers obtain financing necessary to obtain certified seeds, which produce increased yields and revenues as well as greater resistance to disease. The program is currently being utilized through the pilot program by nearly 900 farmers in four areas. Once the project has achieved the intended goals, it will be scaled up to other crops and other areas of Malawi and Africa.

The stakeholders include National Smallholder Farmers' Association of Malawi (NASFAM), Insurance Association of Malawi and with technical assistance from the World Bank and Opportunity International Network financed by the Swiss State Secretariat for Economic Affairs. The stakeholders have designed the index-based weather insurance contract that would pay out if the rainfall needed for groundnut production was insufficient. If there is a drought that affects production, the calculated index then triggers a pay out from the insurance contract. In the event of a drought that affects the crops, the payout funds will be paid directly to the financial lending institution to pay off the farmers' loans. If there is no drought, the farmers will benefit from selling the higher value production in the marketplace, hence ensuring their food security and livelihoods (World Bank 2005).

The activity has been welcomed by the Malawi Government as they continue to explore innovative ways to manage weather and price risks to contribute to the food security needs of the country. Malawi is one of the pioneers in Africa to implement such index-based weather insurance policies that have been sold to smallholder farmers. A similar pilot in India in 2003 has now expanded from an initial 230 farmers to more than 250,000 farmers who have access to weather insurance (World Bank 2005).

Ukwe Farmers Association in Lilongwe are happy with the pilot project and said, "It is good to note that in case of severe drought I do not have to worry about paying back loans in addition to looking for food to feed my family. In future I hope to send my children to school with income from this project." The Insurance Association of Malawi also notes that, "Drought index insurance is a real breakthrough, as it does not only avail the potential for re-accessing the commercial farming community, but also accessing rural farming folk who need it most" (World Bank 2005).

The Crop Production Director for the NASFAM commented that, drought is one of the major risks in rainfed agricultural production. In the event of a drought the farmer may face low yields, or even total crop failure. If the farmer uses production loans, he/she may not be able to pay for the loan. The Drought Insurance Pilot Project has offered an option so that he/she will be covered by the insurance. A further advantage is that by covering the risk of drought, the micro finance institutions will be more amenable to providing loans on otherwise risky crops, hence more farmers will have access to micro financing (World Bank 2005).

A World Bank Country Manager for Malawi, was pleased with the role the Bank and other partners were playing with the Government of Malawi in piloting this program, and is optimistic that it can play an important role in supporting rural agriculture in the country. "Before the pilot, farmers had little cash and no access to finance, and thus could not afford to purchase certified seed. Banks were unwilling to lend to these farmers for a variety of reasons, but primarily because of the risk that farmers would not be able to repay their loans if there was drought. This program can mitigate this risk and bring needed resources to this crucial sector of the Malawi economy" (World Bank 2005).

3.5
Conclusions

Africa faces great challenges in terms of improvement of the use of climate information for agricultural use and food security. The challenges at hand are not really impossible to achieve but feasible only if the Governments of the day can put in place appropriate policies and allocate sufficient resources for relevant institutions to conduct their mandate effectively. Key gaps seen as a priority included the lack of information on intra-seasonal distribution; this requires resources to fund research to develop technologies or models for forecasting intra-seasonal rainfall distribution.

The other challenge of insufficient translation in terms of language and terminology requires capacity building in communication science by those mandated to communicate information to the users. Challenges around stakeholder and institutional capacity also require urgent attention. Recommendations are made in the areas of forecast improvement (e.g. greater focus on forecasting intra-seasonal distribution) and increased national and non-traditional investment in outreach and applications, amongst others.

Weather based insurance is an upcoming strategy that has proven its worth in places such as India and it is important that it given the attention it deserves as helps improve the food security of communities especially the resource poor.

References

Archer E, Mukhala E, Walker S, Dilley E, Masamvu K (2007) Sustaining agricultural production and food security in southern Africa: an improved role for climate prediction? Climate Change (In press).

Arthur AA (2005) Value of Climate Forecasts for Adjusting Farming Strategies in Sub-Saharan Africa Geojournal, 181–189. Volume 62, Numbers 1–2

Balat JF, Porto GG. (2005) Globalization and complementary policies: Poverty impacts in rural Zambia. The World Bank Report, July 2005.

Blench R (1999) Seasonal Climatic Forecasting: Who Can Use It And How Should It Be Disseminated? ODI, Natural Resource Perspectives. Number 47. http://www.odi.org.uk/NRP/47.html

ECA (2006) Mitigating the impact of HIV/AIDS on smallholder agriculture, food security and rural livelihoods in Southern Africa: Challenges and action plan. Economic Commission for Africa, Addis Ababa, Ethiopia.

Diouf J (1997) 'Agriculture should be the foundation' From Africa Recovery, Vol.11#2 (October 1997), page 12 (part of special feature on Agriculture in Africa)

FAO GIEWS (2005a) Special Report FAO Crop And Food Supply Assessment Mission To Burkina Faso. 10 January 2005. http://www.fao.org/giews/english/alert/index.htm

FAO GIEWS (2005b) Special Report FAO Crop And Food Supply Assessment Mission To Mozambique. 20 June 2005. http://www.fao.org/giews/english/alert/index.htm

Grey D, Saddoff C (2005) Water for growth and development: a theme document for the 4th World Water Forum. Washington, World Bank.

Hasan R, Tollefson L, El Gindi A, Moustafa A (1993) Country and Subregional action programme – Egypt. In: International Action Programme on Water and Sustainable Agricultural Development. FAO, Rome. pp. 1–37.

Hess U, Syroka J (2005) Weather-based insurance in southern Africa: The case for Malawi. Agriculture and Rural Development Discussion paper No 13, World Bank, Washington DC.

Hertz KO (1996) Funding agricultural research in selected countries of sub-Saharan Africa. From: "Report of the Expert Consultation on Funding of Agricultural Research in Sub-Saharan Africa", Kenya Agricultural Research Institute (KARI), 6–8 July 1993.

IFPRI (2006) A Gap Analysis for the implementation of the Global Climate Observing System Programme in Africa. IRI Technical Report Number IRI-TR/06/1, International Food Policy Research Institute, Washington DC, USA.

IPCC (2001) Climate Change 2001, Impacts, Adaptation, and Vulnerability, Contribution of Working Group II to the Third Assessment Report of the Intergovernmental Panel on Climate Change, Cambridge University press.

IRI (2006) A Gap Analysis For The Implementation Of The Global Climate Observing System Programme In Africa. International Research Institute for Climate and Society, Palisades, New York, USA.

Mukhala E (2000) Meteorological services and farmers in Africa: Is there shared meaning? SD: Knowledge – Communication for development http://www.fao.org/sd/CDdirect/CDre0051.htm

Ngigi SN, Savenije HHG, Thome JN, Rockström J, Penning De Vries FWT (2005) Agro-hydrological evaluation of on-farm rainwater storage systems for supplemental irrigation in Laikipia district, Kenya. Journal of Agricultural water management, Vol 73, No 1, pages 21–41

Odame HH, Hafkin N, Wesseler G, Boto I (2002) Gender And Agriculture In The Information Society. Briefing paper 55. ISNAR CTA

SADC-RRSU (2002) Proceedings of "Application of Climate Information to Sustain Agricultural Production and Food Security in the SADC Region", SADC-RRSU, Harare,11–15 Nov. 2002.

Siacinji-Musiwa JM (1999) Conservation tillage in Zambia: Some technologies, indigenous methods and environmental issues. In: Kaumbutho P.G. and Simalenga T.E. (editors), 1999.

Conservation Tillage with Animal Traction. A resource book of Animal Traction Network for Eastern and Southern Africa (ATNESA). Harare. Zimbabwe.

SIWI (2001) Water Harvesting for Upgrading of Rainfed Agriculture:Problem Analysis and Research Needs. Stockholm International Water Institute, Report 11, Sveavägen 59SE–113 59 Stockholm, Sweden

Usman M, Archer ERM, Johnston P, Tadross MA (2005) A Conceptual Framework For Improving Rainfall Forecasting For Agriculture in Semi-arid and dry sub-humid Environments. Natural Hazards 34:111–129.

Walker S (2001) Mechanisms To Promote Used Satisfaction To Achieve Recognition Of The Value Of The Meteorological Services. WMO CLIPS Workshop

World Bank (2005) Malawi Pilots Drought Insurance Coverage With Local Farmers. World Bank, Washington DC, USA.

CHAPTER 4

Challenges to Coping Strategies with Agrometeorological Risks and Uncertainties in Asian Regions

L.S. Rathore, C.J. Stigter

4.1 Introduction

In the four coping strategies that can be distinguished as disaster preparedness, mitigation practices, contingency planning and responses, and disaster risk mainstreaming (see Stigter et al., 2007), we have reduced every strategy to certain aspects of preparedness. In this view, challenges to coping strategies are challenges to preparedness strategies; with respect to the hazards and the vulnerabilities that together lead to the disasters producing risks and their consequences and with respect to the uncertainties producing possible damages (see also Medury in Sahni and Aryabandu 2003).

One may argue that
- preparedness itself as a coping strategy is preparedness for what cannot be prevented, for what must be seen as inevitable. Hazards have causes that cannot be mitigated and vulnerabilities have aspects that cannot be warded off. This preparedness is very much connected to the reception of contingency responses and ideas on the future after the disaster;
- mitigation practices as a coping strategy mean preparedness to prevent what can be prevented, in advance. The occurrence of hazards themselves may sometimes be reduced by measures, such as in the case of fires, but it is most often the vulnerabilities that can be seriously reduced by temporary or permanent measures of impact reduction;
- contingency planning and responses belong to preparedness for what should be done after the events that couldn't be prevented. The hazards and the vulnerabilities are important for the details involved but it is mainly about fast actions to shorten and reduce the consequences of the hazards;
- disaster risk mainstreaming is by definition preparedness for disasters in development planning exercises, bringing into practice permanent adaptation strategies that reduce the vulnerabilities to hazards. However, in everyday practice coping with disasters remains "events based" rather than "development based" (e.g. Harichandan in Sahni and Ariyabandu 2003). To change this situation is the largest challenge of all.

These four aspects need extensive examination and in-depth study even with the presently available technologies. Priorities for each of the strategies need to be determined. This involves administrative and organizational management (apart from scientific and technical items) and work on reviewing "implementation mechanisms for the developed strategies". Formulation of action plans at local level, dissemination channels, supervision and management, organization of post strategy monitoring, feed back information, updating and analysis as detailed above need to be given attention. These are all interdependent entities to be mould into a system through integration of the sub-systems.

One of the major problems in dealing with preparedness strategies to deal with extreme weather events in agriculture, rangelands and forestry is the lack of systematic and standardized data collection from disasters. There is no recognized and acceptable international system for disaster-data gathering, verification and storage (Sivakumar 2005). For some disasters like drought, lack of appropriate definition of natural disaster itself is a problem. Definitions of natural disasters are based on the need to respond to development and a humanitarian agenda. Different disasters can be classified as different types by different databases.

For example, a flood which was a consequence of severe wind storm, may be recorded as one or the other (Sivakumar 2005). Flood (as related to rains) and drought may be long-lasting (Gommes and Nègre 1992), but in Asia their shorter durations are more common. The disasters dealt with below are those relatively short-term extreme events chosen by Salinger et al. (2000): (i) high intensity rainfall and floods, (ii) tropical storms, tornadoes and strong winds, (iii) extreme temperatures including heat waves and cold waves, (iv) droughts and (v) wildfires and bushfires. Challenges to disaster risk mainstreaming and to mitigation practices need to be dealt with for each of these events while challenges to contingency planning and to basic preparedness have so much in common for these events that they can be dealt with in more general terms. The paper will end with challenges to some available methodologies in disaster science. Our cases come largely from Asia but we will learn from some useful results from Africa and Latin America.

4.2
Challenges to disaster risk mainstreaming

4.2.1
Adaptation strategies

Seven principles and five steps of/in an interesting Adaptation Policy Framework, including a "bottom-up" or vulnerability-driven risk assessment approach to adaptation were dealt with by Burton and Lim (2005). Agrometeorological adaptation strategies to consequences of increasing climate variability and climate change were discussed by Salinger et al. (2000). They argued that historically changing economic conditions, technologies, resource availabilities and population pressures were the most important factors that made adaptation necessary and that adaptation to climate trends and fluctuations have to be seen in these contexts.

Diamond (2005) expressed this differently in arguing from historical research that collapse of societies can historically be explained from lack of adaptation strategies to (i) damage caused to their production environments; (ii) climate change; (iii) enemies; (iv) changes in relations with trading partners; (v) political, economical and social consequences from the other four factors. Both approaches show the overwhelming importance of resource availabilities in determining long term vulnerabilities, so also in adaptation strategies. In a recent address, British Defense Secretary John Reid warned that global climate change and dwindling natural resources are combining to increase the likelihood of violent conflict over land, water and energy (Klare 2006), demanding local and international assistance for adaptation strategies before more resource wars break out.

4.2.2
High intensity rainfall and floods

Vulnerabilities to flood hazards can best be prevented by not using flood prone areas for agricultural production and related habitation. Flood hazard maps are essential tools for land use planning. They appear often unsuccessful but when followed up by actual management decisions on land use, these monitoring exercises are invaluable.

As in many other cases of natural disasters, there are flood prone areas that nevertheless have to be used for agricultural production. Flood control and management are the starting points of any flood preparedness initiatives in development planning (e.g. Lohani and Acharya in Sahni and Ariyabandu 2003). The most obvious improvements are large scale flood water detention or flood diversion attempts for agricultural purposes (Stigter et al. 2003a). It must also be clear that such calamities as for example happen in China around its Yangtze river with a not negligible frequency (Winchester 1996) can hardly be met with any production adaptation strategy although annual flooding can be agriculturally used (Stigter et al. 2003a). Flood resistant construction techniques are discussed by Dhameja in Sahni and Ariyabandu (2003) and Ariyabandu indicates in Sahni and Ariyabandu (2003) that a demand driven approach to address substantial issues in vulnerabilities of communities like that of livelihood security in response to floods is in the offing. These are challenges that go far beyond monitoring and early warnings schemes.

A serious effect of floods is landslides. The counter-measures as an aspect of large scale land use planning that have to be taken in advance demand for direct expenditures in civil engineering but may also compete with agricultural land. In such cases priority should be given to the engineering aspects although keeping enough good quality land for agricultural production has been recognized as a priority policy issue (Stigter et al. 2000).

4.2.3
Tropical storms, tornadoes and strong winds

In development planning in areas where wind catastrophes are limited, the wind climate can be most successfully used as a source of energy (e.g. Wisse and Stigter 2007). Agroforestry solutions will often do in wind protection under such conditions (Stigter et al. 2002). But south, southeast and east Asia are among the areas susceptible to hurricanes and typhoons, again with track areas that have to be used for agricultural production. Like floods they are among the highest relative intensities of natural disasters, and wind calamities are often occurring in combination with floods (Viet 2002). However, floods are getting much more attention than winds in planning for coping with damage due to cyclones, due to higher vulnerability to floods in most instances, with forests as an exception (Viet 2002). This is in line with little attention for damage to buildings in wind disasters in Africa (e.g. Wisse and Stigter 2007), but Dhameja gives several pages in Sahni and Ariyabandu (2003) of points that are of importance for constructions that may be exposed to cyclones.

Tropical cyclone warnings are issued with certain limitations. The ranges of intensity, size and path of the tropical cyclone are so large that each storm must for the forecasting be carefully treated as an individual event. However, there are common elements of tropical cyclone structure, mechanism and life cycle and there is some regularity in the seasonal variation of the track of the tropical cyclone. Therefore awareness among the public, government bodies and voluntary organizations of such behaviour of cyclones can allow them to take proper advantage of the warnings in the various forms of preparedness that we distinguished in the introduction. This development mainstreaming issue was dealt with by Mandal (2001).

4.2.4
Extreme temperatures including heat waves and cold waves

Extreme temperatures are no considerations yet in development planning for agriculture in developing countries, with the exceptions of vulnerabilities to large scale wildfires and bush fires (see section 4.2.6) in cases of long term heat combined with drought and of large scale losses from low temperatures and frost damage (Salinger et al. 2000). However, the same source warns for rising temperatures in all dryland regions in all seasons as a challenge.

4.2.5
Droughts

Stigter et al. (2007) have summarized contributions of the meteorological community to improved coping with drought in India supporting development efforts related to monsoon water use as a resource. Viet (2002) rates drought at a medium relative disaster rate intensity in the most vulnerable central highlands and the south of Vietnam. Changing cropping calendars and patterns appear the only solutions envisaged there in present planning exercises (Stigter et al. 2007). Sahni reports in Sahni and Ariyabandu (2003) that the High Powered Committee on Disaster Management of India has recommended that with the frequent changes in land use, irrigation development, cropping patterns and agricultural practices, it is necessary to frequently update mapping of drought prone areas for development planning.

This Committee also concluded that currently there is no operational procedure to forecast the impending drought conditions with respect to area of impact, extent and duration. Such difficulties were also among the reasons for the failure reported by Lemnos et al. (2002) to use seasonal forecasts for emergency drought relief in north-eastern Brazil. Stigter (2004) showed, however, that main bottlenecks here were insufficient considerations of the actual conditions of the livelihood of farmers and therefore the development of inappropriate support systems. This was in line with Walker's (1991) general conclusion that special government interventions in response to drought are fraught with inefficiencies and seldom do an adequate job of selecting for those in need. Sahni and Ariyabandu (2003) have examples in several chapters of local government and NGO development activities to prevent such errors in India by actually reducing local vulnerabilities.

This principle is independent of the hazard concerned but most literature examples are drought related. Mungai et al. (1996) illustrate such an approach in research education in semi-arid Africa. Abdalla et al. (2002) report on development related research on improved traditional storage of sorghum grain, quantifying and better understanding local innovations to overcome vulnerability to longer periods of drought in Sudan. Difficulties of development related farmer differentiation and upscaling related to such drought induced cases have been reported by Onyewotu et al. (2003) and Bakheit and Stigter (2004). These are all examples that bear lessons for Asia as well (Stigter 2001).

4.2.6
Wildfires and bushfires

Salinger et al. (2000) have reviewed the situation in that frequent large scale fires are related to inappropriate environmental management, wasteful logging practices, poor fire prevention and fire-fighting systems if not purposely induced. However, burning is a land planning technique widely used in the diminishing shifting cultivation and extensive livestock–keeping to clear land. When well timed and controlled, the natural vegetation will respond with new healthy growth. Ill-timed or poorly controlled burning can seriously reduce the amount of organic matter on the soil surface and leave it exposed to erosion by wind and water (Reijntjes et al. 1992).

Related to this disaster is haze pollution and the transboundary component has led to an international agreement in Asia, including assistance in combating forest fires. The related deforestation is diminished in Brazil by satellite surveillance of the Amazon area and other successful experiments to slow and even reverse environmental degradation, and this system has been proposed for Indonesia as well (Leitmann 2004), to reduce its contribution to the "Asian Brown Cloud". These are all preparedness measures for resource protection as parts of adaptation strategies that are challenges met in disaster risk mainstreaming.

4.3
Challenges to contingency planning and responses

Because the nature of different hazards is not involved as a primary issue and the organizational challenges have so much in common, we do not differentiate here between the hazards concerned. However, when drought becomes a long-term creeping phenomenon it needs a different treatment, but we deal in this paper with relatively short-term phenomena. There is a consensus in the newest literature that emergency relief measures are an important part of preparedness strategies to reduce suffering after the events, the same way insurances work, but that the other preparedness approaches should reduce emergency relief necessities as much as possible (e.g. Smolka in Sahni and Ariyabandu 2003; Stigter et al. 2003a). In India the armed forces always play a major role, are involved in contingency planning and coping with disaster response and are prepared to be provided to the civil government.

Swamy (in Sahni and Ariyabandu 2003) summarizes the primary objectives of the emergency relief response mechanism as to undertake immediate rescue and relief operations. The mechanism requires planners to identify disasters and their probability, evolve signal/warning mechanisms, identify the activities and sub-activities, define the level of response, specify authorities, determine the response kind, work out individual activity plans, have quicker response teams, undergo preparedness drills, provide appropriate delegations and have alternative plans. This has to be organized identically but must have different contents for each type of disaster.

There must exist a relation between these challenges and those to preparedness for the inevitable that will be our last subject below after the challenges to mitigation practices.

4.4
Challenges to mitigation practices as a coping strategy

4.4.1
Impact reductions

Vulnerabilities to hazards can be seriously reduced by temporary or permanent measures leading to impact reduction. Ariyabandu reports in Sahni and Ariyabandu (2003) that in India for the first time in the last fifty years several state governments are dealing with drought in different ways, moving away from relief to mitigation, while in two states where this started decades ago it brought visible results. He also indicates that mitigation planning and development planning in effect share common goals. Reporting on the Traditional Techniques of Microclimate Improvement (TTMI) project, that was funded by Netherlands Government bilateral development collaboration in Africa, Stigter and N'gang'a (2001) illustrate this with several examples.

Uncertainties in agrometeorology are part of everyday farmer conditions and Stigter et al. (2005a) have for example extensively dealt with traditional methods and indigenous technologies to cope with such consequences of climate variability. That such variability is increasing makes it more important to improve and extend the mitigating practices involved and pay attention to farmer innovations and to products from NMHSs, research institutes and universities that can be absorbed by farmers to better cope with increasing uncertainties and disasters.

4.4.2
High intensity rainfall and floods

Although implementation of afforestation programmes in the upper parts of rivers are generally advised and forest depletion discouraged, also in India (e.g. Lohani and Acharya in Sahni and Ariyabandu 2003), there are many good reasons to plant and protect trees. But do not expect trees to stop floods or landslides immediately. Choosing trees according to the farmer's preference will go a long way towards improving watershed quality while also providing them with a livelihood, like that is also the case in wind protection (e.g. Onyewotu et al. 2003).

The main causes of increased flooding are changes in riverbeds, destruction of wetlands, loss of groundcover, compaction of the soil around houses and on roads, and loss of temporary storage areas (Meine van Noordwijk, CIFOR, personal communication 2006). In contrast with this, people in the eastern floodplains of the Ganga and Brahmaputra in India and Bangladesh had a long tradition of living in harmony with floods. The living style, the habitations, the crops grown were all evolved taking into consideration the climate and the flood-proneness of the area. Ancient people inhabiting the floodplains took care not to block the natural drains, preserved the natural beds and depressions, and cultivated only those crops which could stand submergence (H.P. Das, personal communication, 2006). In order to evolve an alternative development paradigm, the "modern" must assimilate the merits of the "traditional" (e.g. Dhameja, 2001).

As already indicated by Stigter et al. (2005a), there are few agrometeorological components other than soil cover mitigation in high intensity rainfall impact and flood control. Their impact reduction applies in our fields to techniques of using inputs, soil conditions and planting densities, choices of cropping systems and varieties, applications of (improved) protection strategies in crop/tree space and applications of other multiple cropping microclimate management and manipulation techniques (Stigter et al. 2003a). A detailed example for agriculture on sloping land has been given by Kinama et al. (2007). They proved a combination of contour hedgerows and mulches most suitable to limit water run off and soil loss in heavy showers, but the largest challenge is to minimize competition between hedges and crops. Other challenges are in preventing the mentioned flood causes under abundant rainfall conditions to become serious.

4.4.3
Tropical storms, tornadoes and strong winds

Wind induced stem and root lodging of crops is too widely occurring but understanding of the complex interaction between husbandry as a mitigating factor, weather and soil has just begun (Sterling et al. 2003). Variety choice, sowing date, seed rate, drilling depth, soil fertility and the widely used application of plant growth regulating chemicals are the best known mitigation factors in large scale cereal growing (Berry et al. 2003; Sterling et al. 2003). Wind reduction in agriculture as far as very strong winds are concerned have mainly to do with microclimate manipulation using forestry and non forest trees (e.g. Stigter et al. 1989; Stigter et al. 2002; Ruck et al. 2003; Stigter et al. 2003b).

Even mildly strong winds can be disastrous, as the wind erosion problems in northern China illustrate, but again mitigation with ground cover and plantings has the solutions (e.g. Zhao Caixia et al. 2006). On the other hand there are positive effects from wind damage of forests as well (e.g. Ruck et al. 2003). This also applies to the humid and sub-humid tropics (e.g. Zhao Yanxia et al. 2005).

4.4.4
Extreme temperatures including heat waves and cold waves.

Based on IPCC's Third Assessment Report, there is high confidence that, in the tropics, where some crops are near their maximum temperature tolerance, yields will decrease generally with even minimal changes in temperature. Rathore et al. (unpubl.) found a decrease by 15–17% in wheat yield over the northwest Gangetic plains due to unusual warming (4–6 degree above normal) in the months of January and February 2006 coinciding with the booting and anthesis stages. The simulated results also indicated that wheat crop matures slightly earlier in the season under study as compared to normal weather. Higher minimum temperatures will be detrimental to crops such as rice in lower latitudes (Zhao Yanxia et al. 2005). The same authors indicate that crop diseases will have to be combated also more seriously because higher temperatures in winter are highly favorable to pathogen survival rates and warmer wetter periods to their development and spread.

Changes in crops, growing periods, planting dates, varieties, irrigation and fertilizers as well as crop diversification, intercropping, growing off-season crops and preference for the more resistant traditional varieties are some of the farm level mitigations possible (e.g. Gommes and Nègre 1992) together with microclimate management and manipulation (Stigter et al. 1992; Stigter 1994). The very likely higher maximum temperatures, more hot days and heat waves over nearly all land area will also give increased heat stress in livestock and wildlife (Zhao Yanxia et al. 2005) for which shade and other protection facilities will have to be expanded.

4.4.5
Droughts

The just given mitigation approaches for temperature extremes are also successfully used in relatively short term drought conditions, although Walker (1991) believes that such chances are larger in Asia than in Africa. As an example of microclimate manipulation, Onyewotu et al. (2004) made understandable how multiple shelterbelts protected crops from advected hot dry air under conditions of limited water supply on soils reclaimed from desertification in Nigeria. As an example of crop selection, Das et al. (2003) give the case of using a perennial xerophyte cotton variety surviving drought with lower yields and compensating with good production in normal years.

An even more recent illustration was provided by H.P. Das (personal communication) for a prolonged dry period resulting in water deficiency in all the districts of Assam, India, from the 1st week of July up to the second week of August 2006, although the seasonal rainfall total had been forecasted as "normal" in that region. In fact a drought situation was declared by the state government. Along with this dry period, unusual high temperature was prevailing almost throughout the state. In view of this, farmers were advised to stop sowing "Sali" rice, as the delayed sowing of this would cause severe moisture stress in the plants, and to start sowing short duration pulses with minimum irrigation. They implemented the advice. Subsequently after onset of new rains, the farmers were advised to sow more pulses. The state of the crops according to extension officers was satisfactory till the last date covered, by early September. It was noted that farmers are very conscious of their needs and appeared to be more interested in the onset of rains and dry spells. In this particular case, close monitoring of the rainfall situation rather than long range forecasts of the seasonal rainfall led to giving the right agrometeorological service. This illustrates the challenge of finding for each problem the simplest methods of support to mitigation possibilities instead of forcing methodologies to fit problem solving.

Although rainfall in Asia is generally expected by climate change scenarios to increase, the also increasing variability has already given yield diminishing moisture stresses due to prolonged dry spells in combination with heat stress (Zhao Yanxia et al. 2005). This makes the mitigation of, and for, dry spells the most challenging issue for the future. A culture of small scale water impoundment (Stigter et al. 2005a) may assist if the health dangers of standing water can be overcome.

4.4.6
Wildfires and bushfires

The necessary ingredients to maintain a fire are given in a fire triangle concept; this concept portrays a triangle with each side sequentially labelled fuel, oxygen and heat. The absence of fuel, oxygen or the heat produced causes the fire to burn out. Fire-fighting methods are based on breaking the triangle by cooling the heat component, smothering the oxygen or removing fuel. The fuel component of the fire triangle merits consideration because it lends itself to modification at all times of the year. Oxygen is always available but its supply is enhanced by certain stability and wind situations; the source of heat is generally an imposed factor, say by lightning, and does not lend itself to overall control (WMO 1993 and 2007).

The climate of a region determines the type, amount, distribution and state of fuel available for the outbreak of fires. Fuels are found in almost infinite combinations. Every fuel has an inherent flammability potential which can generally be realized, mainly depending upon the amount of water in the fuel. In the tropics, most fires are used by humans as an important tool in land management. In the sub-tropics, lightning is one of the main causes of fire outbreaks, whilst in the tropics fires started from lightning are rare (WMO 1993 and 2007). From the mitigation point of view, it are the amounts and distribution of the most flammable fuels around that could be influenced most. However, in practice this is not possible on a large scale and therefore the only mitigation applied is related to minimizing and where necessary managing the imposed sources of fires. Fire fighting itself as a form of mitigation needs a lot of knowledge of which the agrometeorological aspects have recently again been reviewed (WMO 2007).

4.5
Challenges to preparedness as a coping strategy

Stigter et al. (2003a) put forward the opinion that coping with flood disasters would gain from the same change of emphasis in the approach as took place in Japan with respect to earth quakes. They thought that in such an approach additional measures had to get focus and different questions would have to be asked with respect to preparing victims for the occurrence of floods. It was concluded from the Orissa super cyclone disaster in India that trying to build preparedness models may be counterproductive, because of the occurrences and effects being extremely location specific. Furthermore, lessons learned from very exemplarious villages fitted in three preparedness categories: (i) livelihood-focused support, (ii) participation perspectives, and (iii) community perspectives (Stigter et al. 2003a). The challenges of giving contents to such an approach would apply to the other disasters that we have considered as well.

As to the first category, beyond contingency and response planning there needs to be basic contemplation on whether there are alternatives for the present livelihood situation. These could include changing place, changing subsistence activities, changing income generating activities, for the individual/family, part of the village, and the whole village. Insurances may be a way out. Governments may offer such possibilities, villages/families/individuals have to respond; conditions may force people to leave or to accept worsening conditions. For the second category, all other preparedness aspects we discussed in this paper have participation issues. The differentiation discussed by Stigter et al. (2007) shows that such issues depend on education, income, occupation, and information channels. It appeared that the more the participation, the better are the chances of solving problems or living with problems. Finally, it was shown that where a community was able to organize itself, the third preparedness category, with an eye on a common (still differentiated) future, offers increased chances for a more successful future. Such factors are essential in the total preparedness picture that we wanted to give. The challenges involved are getting people organized by themselves, NGOs or the government in preparations for future disasters in these ways.

We will now end this paper with some challenges to methodologies in disaster science and development of an Agrometeorological Advisory Service (AAS). The complexity of the problems necessitates cooperation between research scientists in various disciplines. It is becoming important that we rapidly identify gaps in our knowledge and initiate research aimed at operationally increasing the adaptability of agriculture in the face of climatic change.

4.6
Challenges to methodologies in disaster science to support preparedness

4.6.1
Early warning systems for assessing agrometeorological risks

Use of improved climate and weather information and forecasts along with efficient early warning systems would contribute to the preparedness for extreme weather events. New technologies have brought about an accelerated increase in our knowledge of the climate system. Today the accuracy of forecasts of large-scale weather patterns for seven days in advance is the same as those for two days in advance only 25 years ago. The accuracy of tropical cyclone track forecasts and the timeliness of warnings have been steadily improving in the past few years. When properly communicated and absorbed, early warnings may empower farmers and communities threatened by natural hazards to prepare themselves in sufficient time and in an appropriate manner so as to minimize the risk of the impending hazard. Technologically oriented early warning, integrated with field data on crop and livestock conditions, price movements, human welfare etc. is for example crucial for tracking drought, its onset, its impact and farmers response to it. Primary policy decision makers, resource generators, and relief and mitigation workers need information about early warning of onset of drought events, estimation of area, intensity and duration, long term and short term plans for coping with droughts etc. It is a challenge to have this information operationally provided.

A definition of warning stages (e.g. normal, alert, alarm, emergency as is prevalent for cyclones in India) should be generated by the early warning system to trigger government and other responses. The effective warning system should have meteorological/agricultural information, production estimates, price trends of food and feed, availability of drinking water and household vulnerability, so that a variety of indices related to production, exchange and consumption could be addressed. The challenge is that information on the spatial extent and duration of risk events, time of occurrence with reference to crop calendar and severity of the events could operationally help in the preparations of coping strategies.

Several approaches are employed to estimate the impact of weather conditions on plant diseases. One can provide predictions based on previously established empirical relations between the population density of pathogens, vegetation status and climatic variables. But, faced with the complexity of the problem, it is much more efficient to use epidemiological models: epidemiological development is described in the form of a functional model where each biological process is linked to climatic parameters. These models when coupled to crop simulation models may provide a reasonable forecast on likely infestation. However, the establishment of these models necessitates the acquisition of various observed data and knowledge acquired by experimenting on the disease. At the present time, very few of these models are available and it is a great challenge that progress is made in this direction.

4.6.2
Remote sensing for spatial information

To study certain impacts of meteorological hazards on agriculture and forestry and improve our understanding of certain preparedness issues, use of remote sensing data is a precious tool in obtaining spatial information on areas of interest where ground measurements are difficult. Moreover, additional information on the land may be essential in establishing its sensitivity to water excess or deficit, water and wind erosion, and the risks of soil degradation. In recent years, many investigations have demonstrated the capability of satellite-borne sensors to provide information on various crop indicators, which help to monitor and identify crop stress more effectively. For example during drought conditions, physiological changes within vegetation may become apparent. The National Remote Sensing Agency (NRSA) in India is using a vegetation index to determine vigor of vegetation. Condition of the crop is affected by factors such as supply of water and nutrients, insect/pest attack, disease outbreak and weather conditions. These stresses cause physiological changes, which alter the spectral properties of leaf/canopy. The task of crop condition assessment requires (i) detection of stress, (ii) differentiation of stressed crop from normal crop at a given time (iii) quantification of extent and severity of stress, and (iv) assessment of the production loss. Crop stress conditions are often better characterized through the use of spectral Vegetation Indices (VI) in comparison to use of individual spectral bands.

Many factors affect crops, most of which are time and space-dependent so that they can be represented as maps e.g. physiography, soil type, soil fertility, depth-to-ground water table, slope of the area, date of sowing and application of irrigation. All these factors can be represented in the form of maps. These maps can be combined (integrated) and analyzed using GIS to find the potentiality of the area for a particular crop and expected crop yield. It is a challenge that suitability of various crops can be estimated by integrating various factors affecting them under GIS environment, simultaneously evaluating agrometeorological risks and suggesting alternative crops or cropping systems for an area.

4.6.3
Data analysis in research

To prepare agriculture and forestry for adapting to meteorological risks, another challenge is that efforts must be made in research, based on the knowledge of climate data currently available, and orienting it towards the development of the most useful techniques. These developments must be accompanied by efforts in agronomic research which take the hypothesis related to climatic extremes and variability into account in research in plant genetic improvement and development of sustainable cropping systems to attain the delivery of operational applications regarding adaptation strategies. It then becomes indispensable to single out two types of adaptation depending on the final user: those which can be implemented by the farmer himself (modification of sowing dates, varietal choice, use of seasonal forecasts, etc.) and those for decision-makers, land and natural resource managers which necessitate investment in development and construction infrastructures.

The concept of the Markov Chain probability model (Robertson 1976) on initial and transitional probabilities of dry and wet spells has been found a very useful tool for crop planning and drought monitoring. The Markov Chain model can be fitted to weekly rainfall totals to obtain sequences of dry and wet spells. These initial and transitional probabilities can be used for answering several questions concerning the expected frequencies of sequences of dry and wet weeks. The challenge here is that understanding sequences of wet and dry spells should help in preparing better agro-advisories, which should help farmers and agricultural scientists to take appropriate decisions for farm operations (Biswas 1994).

4.7
Agrometeorological Advisory Service (AAS)

The major challenge to coping strategies is the development of well differentiated and sufficiently scaled up operational services supporting preparedness strategies (e.g. Stigter et al. 2007). In India, the National Centre for Medium Range Weather Forecasting has for example developed an AAS in close collaboration with the India Meteorological Department, the Indian Council for Agricultural Research and the State Agricultural Universities. General Circulation Models (T-80 and T-170) constitute the basic tool for preparing location specific forecasts in the medium range. The model output is subjected to statistical (Perfect Prog. Model) and synoptic interpretation for improving the skill of weather forecasts. In relation to the forecasts currently available, progress is expected by users on enhancing the skill and range of meteorological variables. It would be necessary to obtain information not only on the average values, but also on the extreme values (for example, for rainfall or wind speed) and exceeded threshold values (the case of frost and heat waves). The center is providing agro-climatic zone specific day to day weather forecasts for next 4–5 days twice a week i.e. Tuesday and Friday along with cumulative weekly rainfall. The weather forecast is made in quantitative terms for rainfall, cloud cover, maximum temperature, minimum temperature, wind speed and direction. These zone-specific forecasts are tailored in the light of current weather observations received from the AAS unit in real time. The forecast is disseminated to the AAS units in real-time using fast communication facilities like Internet/Telefax etc. IMD is providing support to maintain an observational network of observatories at AAS units.

On receipt of the forecasts at the AAS units, they prepare the medium range weather forecast based agrometeorological advisories in vernacular language in consultation with a panel of experts in various subject matters of agriculture. These agro-advisories which are crop specific, weather event specific and farm operation specific are disseminated to farmers through all possible mass media like newspaper, radio, television and also through personal contacts by extension workers. The advisories are kept as simple as possible both in terms of the language and the terminology keeping in view the literacy level of the local farmers. In addition to the farmers, these bulletins are also provided to authorities of concerned departments like those of agriculture, horticulture, irrigation, soil conservation, animal husbandry etc. to enable them to take necessary measures for effective utilization of the advisories. These are examples of what in the literature now more generally is called agrometeorological services (e.g. Stigter et al. 2005b; WMO 2006).

4.8
Conclusion and recommendation

The four coping strategies with preparedness from a different perspective have been explained and illustrated. The key challenge is in the combination of these strategies and in facing a combination of challenges to each of them. The stronger governments and/or NGOs are at local levels, but supported by higher up levels, the more chances of coping with disasters there are. Much suffering will remain in coping with disasters, because fate has to be faced and more resource wars loom. But it has to be recommended that the preparedness strategies are taken serious because local, federal and international support can also be better absorbed and used when more challenges to coping strategies that we discussed are met within the local possibilities of communities, families and individuals.

References

Abdalla AT, Stigter CJ, Bakheit NI, Gough MC, Mohamed HA, Mohammed AE, Ahmed MA (2002) Traditional underground grain storage in clay soils in Sudan improved by recent innovations. Tropicultura 20:170–175

Bakheit NI, Stigter K (2004) Improved matmuras: effective but underutilized. LEISA Mag Low Ext Input Sust Agric 20 (3):14

Berry PM, Sterling M, Baker CJ, Spink J, Sparkes DL (2003) A calibrated model of wheat lodging compared with field measurements. Agric For Meteorol 119:167–180

Biswas BC (1994) Role of agrometeorology to develop sustainable agriculture. Fert Newsl 39(12): 215–220

Burton I, Lim B (2005) Achieving adequate adaptation in agriculture. Clim Change 70:191–200

Das HP, Adamenko TI, Anaman KA, Gommes RG, Johnson G (2003) Agrometeorology related to extreme events. Technical Note No 201, WMO No 943, Geneva

Dhameja A (2001) Droughts and floods: a case for "dying wisdom". In: Sahni P, Dhameja A, Medury U (eds) Disaster mitigation: experiences and reflections. Prentice-Hall of India, New Delhi, pp 76–91

Diamond J (2005) Collapse – How societies choose to fail or to survive. Allan Lane, London

Gommes R, Nègre T (1992) The role of agrometeorology in the alleviation of natural disasters. Agrometeorology Series Working paper No 2, FAO, Rome

Kinama JM, Stigter CJ, Ong CK, Ng'ang'a JK, Gichuki FN (2007) Contour hedgerows and grass strips for erosion and runoff control on sloping land in semi-arid Kenya. Arid Land Res Managem, accepted for public. 21:1–19

Klare MT (2006) The coming resource wars. www.energybulletin.net, also available at the INSAM web site (www.agrometeorology.org) under "News and Highlights"

Leitmann J (2004) Lessons from Brazil in tackling deforestation. Jakarta Post, 28 September, p. 6

Lemnos MC, Finan TJ, Fox RW, Nelson DR, Tucker J (2002) The use of seasonal climate forecasting in policymaking: lessons from northeast Brazil. Clim Change 55:479–507

Mandal GS (2001) Tropical cyclones and their forecasting and warning systems in India. In: Sahni P, Dhameja A, Medury U (eds) Disaster mitigation: experiences and reflections. Prentice-Hall of India, New Delhi, pp 171–199

Mungai DN, Stigter CJ, Ng'ang'a JK, Coulson CL (1996) New approach in research education to solve problems of dryland farming in Africa. Arid Soil Res Rehabil 10:169–177

Onyewotu L, Stigter K, Abdullahi Y, Ariyo J (2003) Shelterbelts and farmers' needs. LEISA Mag Low Ext Input Sust Agric 19(4):28–29

Onyewotu LOZ, Stigter CJ, Oladipo EO, Owonubi JJ (2004) Air movement and its consequences around a multiple shelterbelt system under advective conditions in semi-arid northern Nigeria. Theor Appl Climatol 79:255–262

Reijntjes C, Haverkort B, Waters-Bayer A (1992) Farming for the future. An introduction to low-external-input and sustainable agriculture. Macmillan, London

Robertson, GW (1976) Dry and wet spells. UNDP/FAO, Tun Razak Agric. Res. Center. Sungh: Tekam, Jerantut, Pahang, Malaysia, Project Field Report, Agrometeorology (A.6)

Ruck B, Kottmeier C, Mattheck C, Quine C, Wilhelm G (eds) (2003) Wind effects on trees. University of Karlsruhe, Germany

Sahni P, Ariyabandu MM (2003) Disaster risk reduction in south Asia. Prentice-Hall of India, New Delhi

Salinger MJ, Stigter CJ, Das HP (2000) Agrometeorological adaptation strategies to increasing climate variability and climate change. In: Sivakumar MVK, Stigter CJ, Rijks DA (eds) Agrometeorology in the 21st century - needs and perspectives. Agric For Meteorol 103:167–184

Sivakumar MVK (2005) Impacts of natural disasters in agriculture, rangeland and forestry: an overview. In: Sivakumar MVK, Motha RP and Das HP (ed) Natural disasters and extreme events in agriculture. Springer, Heidelberg Berlin pp 1–22.

Sterling M, Baker CJ, Berry PM, Wade A (2003) An experimental investigation of the lodging of wheat. Agric For Meteorol 119:149–165

Stigter CJ (1994) Management and manipulation of microclimate. In: Griffiths JF (ed) Handbook of Agricultural Meteorology, Oxford University Press, pp 273–284

Stigter K (2001) Operational agrometeorology in Africa and Asia: Lessons for sustainable agriculture. In: ICAST (ed) Promoting global innovation of agricultural science & technology and sustainable agriculture development. International Conference on Agricultural Science and Technology, Session 2: Sustainable Agriculture (1), Part III: Land, Water and Nutrient Management, Beijing, pp 276–280

Stigter CJ (2004) The establishment of needs for climate forecasts and other agromet information for agriculture by local, national and regional decision-makers and users communities. In: Applications of Climate Forecasts for Agriculture. Proceedings of the RA I (Africa) Expert

Group Meeting in Banjul, the Gambia (December 2002). AGM-7/WCAC-1, WMO/TD-No 1223, WMO, Geneva, pp 73–86

Stigter CJ, Ng'ang'a, JK (2001) Traditional Techniques of Microclimate Improvement African Network (TTMI/AN) Project for research training. Successful research training partnership case study for GFAR-2000, Global Forum on Agricultural Research, Rome. Available on CD-ROM from FAO

Stigter CJ, Darnhofer T, Herrera Soto H (1989) Crop protection from very strong winds: recommendations in a Costa Rican agroforestry case study. In: Reifsnyder WE, Darnhofer, T (eds) Meteorology and Agroforestry. Proceedings of an ICRAF/WMO/UNEP Workshop on Application of Meteorology in Agroforestry Systems Planning and Management, Nairobi, Kenya. ICRAF, Nairobi, pp 521–529

Stigter CJ (coord & ed), with contributions from Karing PH, Stigter CJ, Chen Wanlong, Wilken GC (1992) Application of microclimate management and manipulation techniques in low external input agriculture. CAgM Report No 43, WMO/TD-No 499, Geneva

Stigter CJ, Sivakumar MVK, Rijks DA (2000) Agrometeorology in the 21st century: workshop summary and recommendations on needs and perspectives. In: Sivakumar MVK, Stigter CJ, Rijks DA (eds), Agrometeorology in the 21st century – needs and perspectives. Agric For Meteorol 103:209–227

Stigter CJ, Mohammed AE, Al-Amin NKN, Onyewotu LOZ, Oteng'i SBB, Kainkwa RMR (2002) Agroforestry solutions to some African wind problems. J Wind Engn Industr Aerodyn 90:1101–1114

Stigter CJ, Das HP, Murthy VRK (2003a) Beyond climate forecasting of flood disasters. Invited Lecture on the Opening Day of the Fifth Regional Training Course on Flood Risk Management (FRM-5) of the Asian Disaster Preparedness Center (Bangkok) and the China Research Center on Flood and Drought Disaster Reduction, Beijing. Available from ADPC (Bangkok) on CD-ROM

Stigter CJ, Al-Amin NKN, Oteng'i SBB, Kainkwa, RMR, Onyewotu LOZ (2003b) Scattered trees and wind protection under African conditions. In: Ruck B, Kottmeier C, Mattheck C, Quine C, Wilhelm G (eds) Wind effects on trees. University of Karlsruhe, pp 73–80

Stigter CJ, Zheng Dawei, Onyewotu LOZ, Mei Xurong (2005a) Using traditional methods and indigenous technologies for coping with climate variability. Clim Change 70:255–271

Stigter CJ, Kinama J, Zhang Yingcui, Oluwasemire KO, Zheng Dawei, Al-Amin NKN, Abdalla AT (2005b) Agrometeorological services and information for decision-making: some examples from Africa and China. J Agric Meteorol (Japan) 60:327–330

Stigter CJ, Tan Ying, Das HP, Zheng Dawei, Rivero Vega RE, Nguyen van Viet, Bakheit NI, Abdullahi, YM (2007) Complying with farmers' conditions and needs using new weather and climate information approaches and technologies. In: Sivakumar MVK, Motha R (Eds.) Managing Weather and Climate Risks in Agriculture. Springer, Berlin Heidelberg, pp. 171–190.

Viet NV (2002) Some measures to cope with the impacts of climate disasters (extreme climate events) on agriculture in Vietnam. Agrometeorological Centre, Institute for Meteorology and Hydrology, Ministry of Natural Resources and Environment, translation of internal reports

Walker TS (1991) Improved technology in relation to other options for dealing with climatic risk to crop production in the semiarid tropics. In: Muchow RC, Bellamy JA (eds) Climatic risk in crop production: models and management for the semi-arid tropics and sub-tropics. CAB International, Wallingford, pp 511–525

Winchester S (1996) The river at the center of the world. A journey up the Yangtze, and back in Chinese time. Henry Holt and Company, New York

Wisse J, Stigter K (2007) Wind engineering in Africa. J Wind Engin Industr Aerodyn, accepted for public. 21 pp, in press

WMO (1993, additions to the second edition; and 2007, third edition) Applications of meteorology to forestry (and non-forest trees). Ch 8 in: Guide to Agricultural Meteorological Practices. WMO No 134, Geneva

WMO (with Baier W, Motha R, Stigter K) (2006) Commission for Agricultural Meteorology (CAgM). The first fifty years. WMO No 999, Geneva

Zhao Caixia, Zheng Dawei, Stigter CJ, He Wenqing, Tuo Debao, Zhao Peiyi (2006) An index guiding temporal planting policies for wind erosion reduction. Arid Land Res Managem 20:233–244

Zhao Yanxia, Wang Chunyi, Wang Shili, Tibig LV (2005) Impacts of present and future climate variability on agriculture and forestry in the humid and sub-humid tropics. Clim Change 70:73–116

CHAPTER 5

Challenges and Strategies to face Agrometeorological Risks and Uncertainties – Regional Perspective in South America

Constantino Alarcón Velazco

5.1 Introduction

South America is one of the regions in the world, most exposed to a wide range of hydro-meteorological hazards, according to the EM-DAT, an international database on disasters, that the Office of US Foreign Disaster Assistance (OFDA) is in charge of. The Center for Research on the Epidemiology of Disasters (CRED) estimates that between 1980 and 2005, almost 80% of the natural disasters, 30% of the loss of human life and 75% of economic loss that took place in the Region were caused by hydro-meteorological conditions and hazards.

In South America where agriculture, fishery and forestry contribute 12% of the Gross Domestic Product (GDP), the success of these activities depends basically on a favourable climate. A good rainy period, together with adequate temperature conditions and the absence of extreme weather, guarantee the success of agricultural production. Despite its importance, agriculture is the sector most vulnerable to natural disasters.

Natural disasters that pose the most dramatic threat to agriculture in the region are: floods, droughts, strong winds, landslides, volcanic eruptions, and climate change. While the first four have a catastrophic character, the last two represent a risk that could exert a pressure towards a change in the land-use and the implementation of some adaptation measurements. This situation threatens the economic development and food security in the countries.

One aspect important to the region is a better understanding of the links between climate variability, with an special emphasis on El Niño/Southern Oscillation (ENSO), and the extreme hydro-meteorological events, since this is fundamental to the design of appropriate preventive measurements for disaster risk reduction.

These climatological phenomena severely affect the social, economic and political life of the affected countries. Heavy rain, prolonged droughts, intense canicula, polar cold, shortage of agricultural products, abundance of some other unexpected products, floods and others, constitute a worrisome reality.

5.2
Climatology and Geography of South America

A fundamental aspect for being able to face risks and uncertainties, which South America confronts, is an understanding of its geographical location and climatology (IADB 2001).

The geographical location provides the main explanation for the extreme natural phenomena that cause great disasters in the region. This region is extremely prone to earthquakes and volcanic eruptions, because its territory is located on some tectonic plates, along the Pacific Ring of Fire.

Climate variability is extreme. Evidence of it are the occurrence of severe droughts, floods, cold waves and winds caused by the El Niño events, by the annual North-South displacement of the Inter-tropical Convergence Zone (ITCZ) and by the entrance of frontal systems (Fig. 5.1). Recent climate changes seem to have aggravated climate variability in the region.

The degree of vulnerability in the region, in relation to extreme natural phenomena is not only determined by its geographic location and climate patterns, but also by several socio-economic factors that increase the vulnerability to these natural hazards, i.e. settling of populations in some places inappropriate to agriculture, environmental degradation, etc.

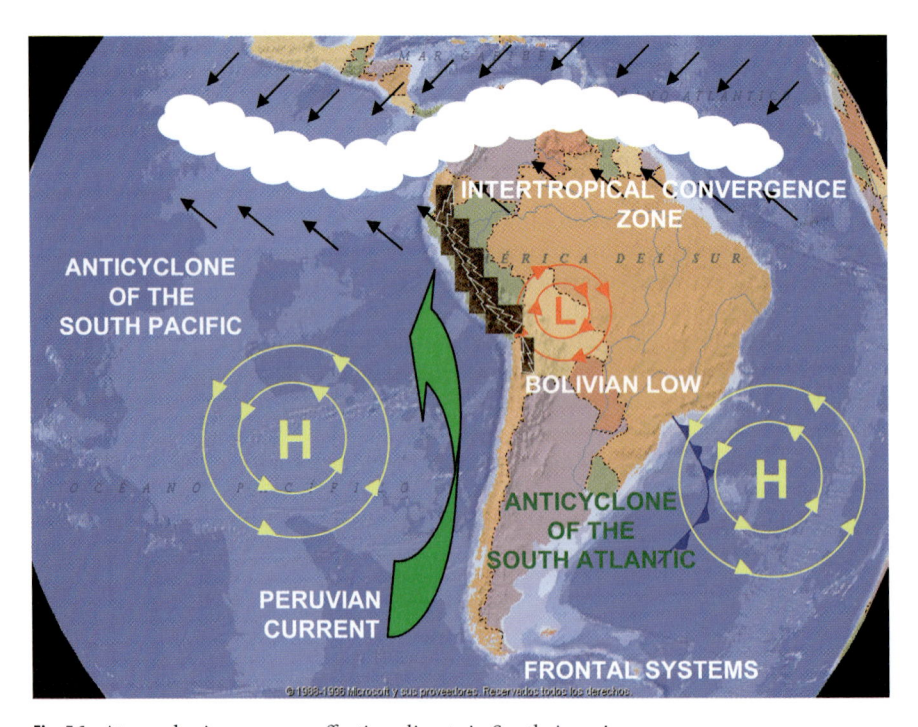

Fig. 5.1. Atmospheric processes affecting climate in South America

5.3
Natural Phenomena that affect Agriculture in South America

The main natural phenomena affecting agriculture in the region are the following (FAO 2000):

5.3.1
Landslides

Due to the topography of the region, most of the landslides generally occur on the hillsides, and in specific places, where deforestation and improper management of the soil are the main factors for the occurrence of these factors.

5.3.2
Volcanic Eruptions

At present several existing volcanoes in the area are active and they cause a serious impact on the national economies and the health of the communities. Volcanic eruptions block solar radiation causing cooling and diminishing of luminosity. These impacts together with the deposit of minerals that are in high concentrations are toxic to animals, can generate great harm to agriculture and livestock.

5.3.3
Floods

Floods occur in areas near the riverside where farmers, wanting to make good use of the land, start sowing in the lowlands. When a heavy rain occurs, it causes an overflow in the rivers and devastates the crops.

Some of the problems that arise whenever there is a flood, are the following:
- Heavy precipitation causes an "increase" in the river water level. This is more evident when there is an El Niño event.
- Flood-plain inundation and damage to the houses, transportation, farm production, and in some cases with loss of human lives.
- Slow drainage of inundated areas which results in pools of stagnant water. This situation generates health problems for the population.
- Flow attacks to the banks of the main riverbed, causing permanent changes in the river course and loss of productive areas.

5.3.4
Droughts

Drought is considered a natural disaster that originates from a deficiency of precipitation over an extended period of time, causing harm to the different activities of the population. The damages caused by effect of droughts are much stronger in areas of extreme poverty.

Drought causes negative impacts such as: reduced water availability for irrigation purposes; delay in the sowing dates and less crop yield; loss of productivity in natural prairies and dry-land crops; increased erosion on plains and high areas; environmental stress in hydromorphic areas; soil salinization due to the reduction in the volume of water for irrigation and drying up of wet areas; effect of frosts due to the delay of the sowing in drylands; intensification of freatic water tables and drying up of wells; and advance of desertification in the arid, semi-arid and sub-humid ecosystems.

The degree of severity of droughts can be intensified due to some other climate factors depending on each region and these include high temperatures, strong winds and low relative humidity.

5.3.5
Extreme temperatures including heat waves and cold waves

During periods of heat waves, the extreme high temperatures together with a high vapor pressure deficit can generate intense evapotranspiration. This situation is generally intensified with the occurrence of the El Niño events.

Frosts occur only in the Inter-Andean region where it affects the small-scale crops. Generally farmers carry out their activities in the frost-free periods in order to avoid the negative effects caused by the occurrence of this phenomenon.

5.3.6
The Climate Change

The effects of projected climate change on crop yields in the region are uncertain. The effects include both positive and negative impacts as follows:
- The changes in the number of humid months induced by the changes in precipitation and evapotranspiration could mean greater changes in the spatio-temporal distribution of agricultural systems.
- The increase in the area of climate types described as semi-arid and sub-humid dry, implies greater risks of land degradation when inadequate agricultural practices are performed.
- In relation to temperature, the impacts are positive when it increases in a colder climate, as well as the fertilization with atmospheric CO_2. In warmer climates the increase of temperature could be critical when the thermal stress increases and it accelerates the development cycles of crops. On the other hand, the in-

Chapter 5: Challenges and Strategies to face Agrometeorological Risks and Uncertainties

crease of night time temperatures implies an increase in respiratory losses and consequently diminished yields.
- With regard to precipitation, reduction in the arid, semi-arid, and sub-humid dry zones could reduce crop yields, while in the humid zones where excessive precipitation is the norm, a reduction in the amount of precipitation could be beneficial.

5.3.7
El Niño – La Niña Phenomenon

The main climatic characteristics in South America, during the El Niño event are frequent anomalies (Fig. 5.2). Together with an increase in the temperatures in the western coasts of the Pacific Ocean, it modifies the atmospheric circulation patterns, pressure, precipitations, rivers discharge and the water level of the lakes. El Niño causes above normal rains, and droughts in several places.

5.3.8
Strong winds

High speed winds not only favor excessive evapotranspiration but also can cause the breakage of some parts of the plants and harm structures of the plant that actively contribute to the crop yield.

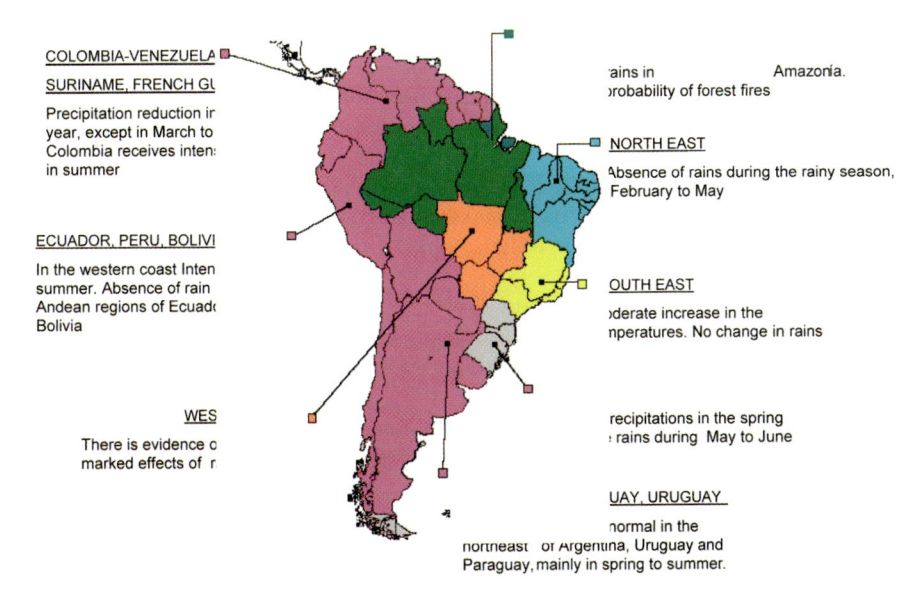

COLOMBIA-VENEZUELA, SURINAME, FRENCH GU...
Precipitation reduction ir
year, except in March to
Colombia receives inten:
in summer

ECUADOR, PERU, BOLIVI...
In the western coast Inten
summer. Absence of rain
Andean regions of Ecuad
Bolivia

WES...
There is evidence o
marked effects of r

...ains in ... Amazonia.
...robability of forest fires

NORTH EAST
Absence of rains during the rainy season,
February to May

...OUTH EAST
...oderate increase in the
...nperatures. No change in rains

...recipitations in the spring
...rains during May to June

...UAY, URUGUAY
...normal in the
northeast of Argentina, Uruguay and
Paraguay, mainly in spring to summer.

Fig. 5.2 Impacts of El Niño – Southern Oscillation (ENSO) phenomenon in South America

5.4
Vulnerability of the Region

Vulnerability in the Region is aggravated because of the location of human activities in some places of great risk, natural resources subject to excessive pressure of poverty, lack of environmental management policies, excessive centralization, little agricultural technology, and lack of education of the population to prevent and face risks (OAS 1993).

Main causes of vulnerability in the region are the fast and un-regulated urbanization, rural and urban poverty, deterioration of natural resources, inefficient public policies, and the delays and mistakes in infrastructure investments. In the region, there is little investment concerning mitigation of natural hazards and the response is mainly under emergency situations.

Vulnerability of natural forests, mountain ecosystems and agriculture is reflected in the following aspects (WMO 2007):
- Deforestation of rainy tropical forests is altering the hydrological cycle in the region causing a shortage in precipitation as a consequence of reduced evapotranspiration. These deforested areas have lost all protection to face climate variability, making them more vulnerable to floods and soil degradation.
- In the subtropical forests of the semi-arid regions, where precipitation is expected to be reduced, the loss of vegetative cover is generating desertification.
- The vegetation in the prairie ecosystems is highly dependent on precipitation. Overgrazing and inadequate management practices are generating high desertification areas.

The mountain ecosystems play an important role in South America because:
- They constitute places with considerable human settlements (Bolivia, Colombia, Ecuador Peru and the Highlands of Chile)
- They perform an hydrological regulating function and at the same time the altitudinal gradient formed by the Andes mountains is a source of rich biodiversity.
- At present the mountains are threatened by global warming which is altering the runoff regimens of the rivers, making the irrigated areas vulnerable to the occurrence of droughts (Peru, Bolivia, Chile y Argentina).

Most of the agricultural lands is going through a process of degradation due to inappropriate management and use. Non-irrigated agricultural land depends on precipitation which makes them vulnerable to seasonal to inter-annual variations of precipitation.

Inappropriate management of soils has led to a stage where agricultural activities are taking place in fragile areas that are not suitable for such practices. This has resulted in a situation in which almost all the countries have a generalized and irreversible erosion in terms of agricultural productivity. The irrigation systems, not efficient enough, are responsible for the salinization of the river waters, sedimentation of the river beds and reservoirs. All these practices are increasing the vulnerability of agriculture.

5.5
Capacities and Resources in the region to face Agrometeorological Risks and Uncertainties

Capacities and resources in the region vary from one country to another, however in facing natural disasters there are some aspects in common between the countries in general, which include the agro-meteorological phenomena.

5.5.1
Economic status of the countries

In the region most of the countries have reached a certain degree of macro-economic stability which is enabling governments to better respond to the impacts of disasters and they are investing more in the prevention and reduction of risks and uncertainties.

5.5.2
Government policies

As a consequence of the increasing incidence of disasters, which has made the region extremely vulnerable, several governments have included disaster prevention in their political agenda. This political commitment is demonstrated in the assumed compromises in relation to the International Strategy for Disaster Reduction (ISDR).

5.5.3
Creation and strengthening of specialized institutions

The new political orientation is leading some countries to create new inter-institutional and sector integrated systems for disaster prevention and response. Other countries are initiating modernization of the national agencies specializing in disaster mitigation.

Some governments have created regional institutions, such as the International Center for the Research of El Niño Phenomenon (CIIFEN), aimed at promoting, complementing and starting scientific and applied research projects, necessary to improve the understanding and early warning of the ENSO and the climate variability at a regional scale. Projects such as the Disaster Prevention in the Andean Community (PREDECAN) contribute to vulnerability reduction of the populations and of the material goods that are exposed to danger and natural risks in the countries of the Andean Community of Nations. The Network of Social Studies for Disaster Prevention in Latin America (The Network), which is constituted by non-governmental institutions and researchers in all the region, informs and advises the governments about their policies and gives advise to other regional and international organizations.

The United Nations agencies also provide support to the region to improve prevention and response to disasters, and they constitute a considerable resource. It is worth mentioning the work of the Economic Commission for Latin America (CEPAL) in the assessment of the economic impacts of natural disasters. These evaluations provide important information for planning reconstruction and prevention purposes.

Institutions such as UNDP, UNESCO, WMO, FAO, CAF, IDB and the World Bank are promoting the necessary scientific capacity building to reduce risks. With important bilateral assistance from Europe, Japan, Canada, the United States and some other countries, national efforts are being promoted by means of regional, national and local projects to evaluate risks, and for preparedness in case of an emergency situation and prevention.

The local non-governmental organizations and the civil society of the Americas are playing an important role in the prevention and response to natural disasters, and in many cases with support from international NGOs.

5.5.4
Risk identification and analysis

Countries in the region identify the main natural disasters to which they are most exposed and assess their frequency, intensity, duration and localization (risks maps). The identification of certain risks is being used to define the necessary measurements for prevention and mitigation and planning of the socio-economic activities in each country (WMO 1993).

5.5.5
Monitoring networks and early warning

Most countries have monitoring and early warning networks, however it is important to point out the necessity to modernize the hydro-meteorological information network; modernize Meteorological Services and the means by which forecasting is made in the region; improve early warning systems and improve communication between the scientists responsible for evaluating the atmospheric and hydrological conditions and also between those in charge of providing early warnings.

5.5.6
Information on the risks for decision making

Most of the Governments in the region have understand that it is necessary to have:
- The projections on the incidence and the estimation of impacts of natural disasters in order to prioritize preventive actions.
- The information on risks for the adoption of preventive policies and preparedness and for the establishment of specific objectives and priorities related to investment.

- The information on risks and vulnerabiltiy of natural disasters to develop private insurance and other market tools to combat risks.
- Research carried out in the region, with the purpose of analyzing the existing assessment methodologies, their use and to provide relevant and precise information to the persons in charge of formulating policies, action plans for the preventions and institutional development.

5.6
Defining Policies and Tools to face Agrometeorological Risks and Uncertainties

The efficient management and planning of agricultural activities requires policies and tools that allow communities to face agro-meteorological risks and uncertainties (WMO 2006). Among these we can consider the following:
- Mitigation of the impacts of agro-meteorological risks and uncertainties requires multi-sectorial and multi-disciplinary actions.
- Defining adequate strategies for land use based on forecasted risks, selection of varieties, and types of crops or change the sowing dates to reduce crop loss.
- Improve irrigation techniques and most efficient use of water, likewise, to manage water more efficiently and to prevent flooding and soil erosion.
- Adopt appropriate soil management systems to minimize the erosion during both rainy and dry events.
- Improve prairie management and adjust livestock numbers on the grazing lands according to the provisions of climate risks.
- Restoring degraded ecosystems.
- Promote the creation of early warning systems of adverse agro-meteorological phenomena
- Carry out evaluations and agro-climate zoning to determine the most favorable conditions for agriculture as well as extreme conditions that are anomalous and recommend appropriate measures to control them.

Several of these actions can be improved by means of an early warning system for adverse climate phenomena to reduce uncertainties through a better understanding of the dimensions of risk. On the other hand, all the information about the generation and evolution of climate phenomena generated daily by different institutions should be used. At present several institutions provide forecasting systems for El Niño events which include NCEP of the National Oceanic Atmospheric Administration, USA; MM5 from the University Corporation for Atmospheric Research, USA etc.

5.7
Strategies to cope with Agrometeorological Risks and Uncertainties

Given the high recurrence of extreme agro-meteorological events, the South American region must generate strategies to prevent and mitigate their impacts. In this sense, the agro-economic planning at a short and long term and at a local, regional or national scale, should be formulated more rationally including among its variables, the agro-meteorological and agro-climatic information (WMO 2004). For example, real time climate data can be used to predict production 2 or 3 months in advance, as this provides sufficient advance information for decision making in buying/sales processes and for distribution and commercialization of products. The agro-climatic analysis on a long term scale makes it possible to carry out crop zonation and identify the most suitable periods for their cultivation. They can also efficiently contribute to the planning and management of risks, by balancing the water requirements with water availability in the region.

When putting into operation the strategies one should recognize that the most important element in long term is prevention. Within this principle the following strategies are proposed.

- Implement and strengthen early warning systems for agriculture by establishing and strengthening national and regional monitoring and surveillance systems that allow the identification and dissemination, in advance and in a reliable manner, of the imminent occurrence of meteorological events that may cause harm to the agricultural sector.
- Strengthen international cooperation by implementing and operating these systems by means of surface networks, satellites, communication, exchange of expertise and experts.
- Develop and apply methods to evaluate the vulnerability of the countries by integrating biophysical, socio-economic and historical information and create charts for risks, vulnerability, potential impacts and risks reduction strategies. For this purpose, it is important to carry out some activities to identify the most critical zones concerning floods, droughts and wildfires, so that the necessary actions to reinforce or install early warning systems can be identified.
- Strengthen the capacity to analyze the agro-meteorological risks and uncertainty information.
- Develop scenarios and climate forecasts at the short, medium and long term at the regional level for the agricultural sector in order to define climatic anomalies scenarios related to climate variability or to climate change, in terms of intensity, distribution, seasonality, considering the most representative variables for the agricultural sector in the region.
- Establish and standardize regional methodologies to evaluate the influence of climate variability and climate change in the productions of crops.
- Promote efficient irrigation systems and soil management systems to maximize the use of precipitation and to reduce the risks of erosion.
- Evaluate socio-economic impacts of agro-meteorological risks and uncertainties through a knowledge of the phenology of the plants, an inventory of the most frequent impacts of the different climate variables and possible scenarios of the effects or positive impacts of climate anomalies on crops. It is necessary to have the information on the rise in production costs, loss of infrastructure and goods, loss of agricultural land, reduction of farm income, migration from the countryside to the city and impacts on employment and labor.
- Implement a Regional Information System for agriculture through the generation of data and information for decision making and agriculture planning

demands to have coordination mechanisms specially between regional institutions. The present technologies of communication such as the Internet provide a possible way of support to have access and to use important information for different users in the agricultural sector for prevention and mitigation of agrometeorological risks..
- Participate in the implementation of regional and national strategies to fight against drought and desertification, especially in the matters related to regional plans which up to the current moment have not been developed due to the lack of international cooperation capacity.
- Develop and implement regional training programmes in using, applying and operating geographic information systems (GIS) and remote sensing applications to monitor the crops, forecast extreme events, carry out agro-climatic modeling, interpret satellite images oriented to agriculture, handle agro-meteorological data and analytical tools, and transfer of methodologies among others.

5.8 Conclusions

In the last few years governments of the countries in South America are giving an increasing recognition to the social and economic benefits related to activities of prevention and reduction of climate risks. This recognition is reflected in issuing new policies, budgetary allocations to carrry out these activities and the development of the capacity of the organizations in charge of prevention and mitigation activities. Most countries in the region are showing a trend to leave behind mechanisms of just responding to disasters originating from climate and favor the multidisciplinary and multi-sectorial approaches for prevention in the long and medium term.

On the other hand, in this region there are some regional and sub-regional institutions that have proved to be crucial in promoting an inter-disciplinary approach and to support the country members to undertake the global practice of risk reduction and institutional development in this field. In this sense, there is an increasing interest in the education sector to face risks and uncertainties of climate origin, since there is a growing number of universities that offer post-graduate and masters courses on Risk Management and Reduction of Disasters and there are many countries, in which the reduction of disasters is being included in many school programs and at different levels.

References

FAO (2000) Efectos de los fenómenos climatológicos adversos en la producción y el comercio de los alimentos. 26a Conferencia regional de la fao para america latina y el caribe. mérida, méxico, 10 al 14 de abril del 2000.
IADB (2001) Geografía y desarrollo en América Latina http://www.iadb.org/res/publications/pubfiles/pubB-2001_3585.pdf

OAS (1993) Manual Sobre el Manejo de Peligros Naturales en la Planificación para el Desarrollo Regional Integrado. Departamento de Desarrollo Regional y Medio Ambiente Secretaría Ejecutiva para Asuntos Económicos y Sociales Organización de Estados Americanos. Washington, D.C.

WMO (1993) Practical use of agrometeorological data and information for planning and operational activities in agriculture. CAgM Report N° 60, World Meteorological Organization, Geneva, Switzerland.

WMO (2007) Actas del seminario tecnico de "prevencion y mitigacion de desastres naturales" para la AR-III. SENAMHI- OMM. Lima 6 de setiembre de 2006.

WMO (2004) Servicios de Información y Predicción del Clima (SIPC) y Aplicaciones Agrometeorológicas para los Países Andinos. AGM N°6/WCAC N° 2, World Meteorological Organization, Geneva, Switzerland.

WMO (2006) Strengthening operational agrometeorological services at the national level. Proceedings of the Inter-regional Workshop March 22–26, 2004, Manila, Philippines. AGM – 9 . WMO/TD No. 1277, World Meteorological Organization, Geneva, Switzerland.

CHAPTER 6

Agrometeorological Risk and Coping Strategies – Perspective from Indian Subcontinent

N. Chattopadhyay, B. Lal

6.1 Introduction

The Indian subcontinent has been exposed to disasters from time immemorial. The increase in the vulnerability in recent years has been a serious threat to the overall development of the country. Subsequently, the development process itself has been a contributing factor to this susceptibility. Coupled with lack of information and communication channels, this had been a serious impediment in the path of progress. India's vulnerability to various disasters has led to mounting losses year after year. Mammoth funds were drawn to provide post disaster relief to the growing number of victims of floods, cyclones, droughts and the less suspecting landslides and earthquakes.

Considering the vast area of the Indian landmass, around 57% of the land is vulnerable to earthquakes, 28% to droughts, 12% to floods and 8% to cyclones. Added to this is the susceptibility of various men-made hazards. Figuratively speaking, around one million houses are damaged annually.

Indian agriculture is passing through a critical phase as the rate of increase in crop production is barely keeping pace with the increase in population rates. The Prime Minister of India has rightly called for a doubling of crop outputs in 10 years' time. As more land cannot be diverted to agriculture, increase in unit area productivity of crops is called for. Our recent experience is that the strategy of erring on the safe side through over-irrigation, over-protection and over fertilization of crops has been counter productive, leading to a decrease in rates of crop production even under irrigation and degradation of soil and air environments and pollution of surface and groundwater reserves. The challenges facing agriculture in the country are ever increasing. In the first place agriculture is highly weather dependent and hence subject to its variabilities. Secondly, the possible impacts of climate change may pose major challenges. Finally, the very sustainability of intensive agriculture using present technologies is being questioned in the context of Global Change debate. The problem therefore has to be addressed collectively by scientists, administrators, planners and the society as a whole.

In the present paper the extreme weather events causing the agrometeorological risks in different parts of the country during last 100 years have been discussed. Besides the impacts of these weather events on crop and plausible coping strategies for management of these agrometeorological risks for better crop production have been mentioned.

6.2
Extreme weather events and its impacts on Indian agriculture

The year 1999 witnessed a super cyclone striking the eastern coast of India (Orissa State). It was a major natural disaster affecting the subcontinent in recent years. Droughts of 1972 and 1987, the heat waves in 1995 and 1998 and the cold wave in 2003 killing several hundred people are still fresh in public memory. The drought and failed monsoon of 2002, in particular, an unusually dry July, is matter of concern for scientists and planners (De et al. 2005).

Year to year deviations in the weather and occurrence of climatic anomalies/extremes in respect of these four seasons in the country are
i) Cold wave, Fog, Snow storms and Avalanches
ii) Hailstorms, Thunderstorms and Dust storms
iii) Heat wave
iv) Tropical cyclones and Tidal waves
v) Floods, Heavy rain and Landslides, and
vi) Droughts

6.2.1
Cold wave

Occurrences of extreme low temperature in association with incursion of dry cold winds from north into the sub continent are known as cold waves.

The cold waves mainly affect the areas to the north of 20 °N. Cold wave conditions are sometimes reported from States like Maharashtra and Karnataka as well.

The maximum number of cold waves occur in Jammu and Kashmir followed by Rajasthan and Uttar Pradesh (Table 6.1). It may be seen that number of cold waves in Gujarat and Maharashtra are almost one per year though these states are located in a more southern location. In the states of Uttar Pradesh and Bihar, the number of deaths from extreme events in the cold weather season during 1978 and 1999 was 957 and 2307 respectively. These two states rank the highest in terms of casualties from cold wave. West Madhya Pradesh experienced most frequent cold waves/severe cold waves and highest number of cold waves/severe cold waves days during the decade 1971–80 (Pai et al. 2004). In the first week of January 2000, the deadly cold spell resulted in the death of 363 persons of which 152 were from Uttar Pradesh and 154 were from Bihar (IDWR 2000). Severe cold wave conditions prevailed over most parts of Bihar and adjoining parts of Orissa during December 2001 and claimed around 300 human lives. During the 1st to 3rd weeks of January 2003, the northern states were under the grip of a severe cold wave and in all 900 people died of which 813 were from Uttar Pradesh alone.

Chapter 6: Agrometeorological Risk and Coping Strategies

Table 6.1 Number of Cold Waves

State	Epochs					
	1901–10	1911–67	1968–77	1978–99	2000–06*	1901–2006*
West Bengal	2	14	3	28	3	50
Bihar	7	27	8	67	14	123
Uttar Pradesh	21	51	8	47	28	155
Rajasthan	11	124	7	53	28	223
Gujarat, Saurashtra & Kutch	2	85	6	6	7	106
Punjab	3	34	4	19	48	108
Himachal Pradesh	–	–	4	18	31	53
Jammu & Kashmir	1	189	6	15	43	254
Maharashtra	–	60	4	18	–	82
Madhya Pradesh	9	88	7	12	24	140
Orissa	4	5	–	–	12	21
Andhra Pradesh	2	–	–	–	–	2
Assam	1	1	–	–	–	2
Haryana, Delhi & Chandigarh	–	–	4	15	22	41
Tamil Nadu	–	–	–	–	2	2
Karnataka	–	10	–	–	17	27
Telangana	–	5	1	–	8	14
Rayalaseema	–	3	–	–	3	6

6.2.1.1
Effect of extreme cold weather/frost

Long periods of extreme cold weather combined with other meteorological phenomena result in the loss of winter crops, fruit crops and vineyards due to frost injury. Low soil temperature at the depth of plant roots causes frost injury. Such reduction in soil temperature occur with strong frosts, in the absence of snow cover and with deep freezing of the soil. Most frost injury to winter crops takes place in the first half of winter before sufficient snow cover has formed. In the second half of winter, frost injury happens in regions with unstable snow cover. Under low temperatures basically a plant dries out and the protoplasm (the living part of cells) dies. Damage to the part of a plant does not always result in damage or destruction of the whole plant. A determining factor is the degree of frost injury to a tillering node, if it is heavy the whole plant will perish. The winter crops most frequently destroyed by frost are those grown on uplands, where snow cover is less and the depth of soil freezing is greater. The main agrometeorological factor influencing frost damage in winter crops is low soil temperature at the depth of the tillering node. Long (three days or more) and intensive cooling causes complete devastation of the crops.

6.2.1.2
Protection of crops from cold injury and frost

A key factor in protection of crops from cold injury is stable air temperature and snow covers throughout the winter. Thaws, resulting in packing or disappearing snow cover, worsen dormancy conditions and reduce or destroy the protective properties of snow cover. The prevention of crop damage by frost can be controlled by breaking up the inversion that accompanies intense night time radiation. This may be achieved by heating the air by the use of oil burners which are strategically located throughout the agricultural farm.

Other methods of frost protection include sprinkling the crops with water, brushing (putting a protective cover of craft paper over plant) and the use of shelterbelts (windbreaks).

6.2.2
Drought

Droughts have an immediate effect on the recharge of soil moisture resulting in reductions of stream flow, reservoir levels and irrigation potential and even the availability of drinking water from wells. The acreage planted to food crops is also affected by land quality. The cultivation of lands subject to a high degree of rainfall variability makes them extremely susceptible to wind erosion (and desertification) during prolonged drought episodes, as the bare soil lacks the dense vegetative cover necessary to minimize the effects of aeolian processes.

A study by Chowdhury et al. (1989) have ranked the year 1918 as the worst drought year of the last century- a year when about 68.7% of the total area of the country was affected by drought. It is of interest to note that the year 1917 had exceptionally high seasonal rainfall. Likewise the severe drought years of 1877 and 1987 were followed by flood years of 1878 and 1988. In the last century, the droughts of 1987 and 1972 are the next in order of severity (Table 6.2). Occurrences of droughts in consecutive years have been reported in 1904–05, 1951–52 and 1965–66. These pairs of years were associated with moderate droughts, where at least 25% of the country was affected. During 1999, 2000 and 2001 drought conditions prevailed over some parts of India, not affecting the country as a whole significantly.

Table 6.2 Years of Drought in India

Year	Area affected (x10^6 km^2)	% area of the country affected	Dt value	category
1915	2.16	68.7	3.64	Calamitous
1877	2.03	64.7	3.38	Calamitous
1899	1.99	63.4	3.31	Calamitous
1987	1.55	49.2	2.37	severe
1972	1.39	44.4	2.05	severe
1965	1.35	42.9	1.95	Moderate
1979	1.24	39.4	1.72	Moderate
1920	1.22	38.8	1.69	Moderate
1891	1.15	36.7	1.54	Moderate
1905	1.09	34.7	1.41	Moderate

6.2.2.1
Impact of drought on Farmers

Due to the uncertainty of rains during the drought, farmers sometimes make several attempts at sowing of seeds leading to a drastic reduction in seed reserves, which in due course are neither sufficient for planting nor for consumption. The farmer is then obliged to borrow by offering labour or perhaps portion of the future harvest as payment for the loan. In the event of a prolonged drought, farmers, either alone or with the entire family, may abandon their land in search of work and food in nearby cities. Fewer and weaker family members remain to till the land, affecting the area under cultivation. The acreage planted to food crops is also affected by land quality.

6.2.2.2
Drought management

In arid, semi-arid and marginal areas there is a probability of drought incidence once in ten years. It is important for those responsible for planning of land-use, including agricultural programmes, to seek expert climatological advice regarding rainfall expectations. Drought impact is the result of the interaction of human pattern of land use and the rainfall regimes. There is thus an urgent need for a detailed examination of rainfall records of these regions. Agricultural planning and practices need to be worked out with consideration to the overall water requirements within an individual agroclimatic zone. Crops which need a short duration to mature and require relatively little water, need to be encouraged in drought prone areas. Irrigation, through canals and groundwater resources, needs to be monitored with optimum utilization avoiding soil salinity and excessive evaporation loss. A food reserve is needed to meet the emergency requirements of up to two consecutive droughts. A variety of policy decisions on farming, human migration, population dynamics, livestock survival, ecology etc. must be formulated. Sustainable strategies must be developed to alleviate the impact of drought on crop productivity.

6.2.2.3
Sustainable strategies to alleviate the impact of drought

In areas of recurring drought, one of the best strategies for alleviating drought is varietal manipulation, through which drought can be avoided or its effects can be minimised by adopting varieties that are drought-resistant at different growth stages. If drought occurs during the middle of a growing season, corrective measures can be adopted; these vary from reducing plant population to fertilization or weed management. Rainfall can be harvested in either farm ponds or in village tanks and can be recycled for lifesaving irrigation during a prolonged dry spell. The remaining water can also be used to provide irrigation for a second crop with a lower water requirement, such as chickpea.

6.2.3
Fog

Immediately after the passage of a western disturbance (WD), a lot of moisture is available in the atmosphere and the regional and synoptic scale conditions provide the trigger for the formation of fog. Even though this phenomenon is not directly related to the extreme weather events it has an effect in all forms of transport and in particular aviation. This has an indirect effect on the economy of aircraft operations and air passenger inconvenience.

De and Dandekar (2001) studied the visibility trends during winter season for 25 aerodromes over a period of 21 to 31 years and concluded that most of the north Indian airports show a significant increasing trend in the poor visibility days (due to fog) amounting to 90% i.e. almost everyday. The airports in south India show only 20 to 30% days with poor visibility.

6.2.4
Thunderstorm, Hailstorm and Dust storm

The arid regions are characterised by frequent and strong winds which are partly due to considerable convection during the day. The usually sparse vegetation is not capable of slowing down air movement, so that dust and sand storms are frequent. Winds in dry climatic zone also affect growth of the plant mechanically and physiologically. The sand and dust particles carried out by wind damage plant tissues. Winds also cause considerable losses by inducing lodging, breaking the stalks and shedding of grains and ultimately decreasing the yield.

As winter season gives way to spring, the temperature rises initially in the southern parts of India, giving rise to thunderstorms and squally weather which are hazardous in nature. While the southernmost part of the country is free from dust storm and hailstorm, such hazardous weather affects the central, northeastern, north and northwestern parts of the country. Records indicate that the largest size hailstorm occurred in association with a thunderstorm in April, 1888 at Moradabad, a town near Delhi.The hailstorm frequencies are highest in the Assam valley, followed by hills of Uttar Pradesh now known as Uttarakhand.

6.2.4.1
Protection of crops from dust storm / sand storm

In most countries, afforestation of fields is the main measure to protect the soil from dust storms. Improving soil resistance to erosion can be achieved by careful selection of cultivation methods, applying mineral and organic fertilizers, sowing grass and spraying various substances which enhance soil structure. It is also important to reduce the areas where a dust can gather, especially in tracts characterized by erosion. One major protection strategy is to establish well developed plant cover before the dust storms period. This can encourage a reduction in the wind speed in the layer above the ground by forming an effective buffer.

When assessing the impacts of the dust storms on agricultural crops, it is necessary to take into account the degree of the development of the plants. On well-tilled crops, the deposition of soil moved by airflow is observed more often than soil carried by wind erosion over long distances.

When looking at the conditions under which dust storms develop and by examining the data on storm-induced damage, it is evident that measures to reduce the wind speed at the soil surface and to increase the cohesion of soil particles are both crucial. Such measures include the establishment of tree belts and wind breaks. Leaving stubble in fields, avoiding ploughing with mould boards, application of chemical substances promoting the cohesion of soil particles, soil-protective crop rotation using perennial grasses and seeding of annual crops are also important.

In regions with intensive wind erosion, especially on slopes or on light soils, strip cultivation may be used. On fallow lands, bare fallow strips of 50–100 m can be alternated with strips of grain crops or perennial grasses and spring crops can be alternated with winter crops.

6.2.5
Heat waves

Extreme positive departures from the normal maximum temperature result in heat waves during the summer season. The rising maximum temperature during the pre-monsoon months often continues till June, even in rare cases till July over the northwestern parts of the country.

Raghavan (1966) made an extensive study of the heat wave spells of the last century for the period from 1911 to 1961. His study indicated that the maximum number of heat waves occur over East Uttar Pradesh followed by Punjab, East Madhya Pradesh and Saurashtra & Kutch in Gujarat. Notably the period roughly coincided with last two decades of the twentieth century which witnessed unprecedented high temperatures globally as a result of the global warming. Number of heat waves from 1901–2006 are given in Table 6.3.

6.2.6
Tropical Cyclones

Though several studies by De and Joshi (1995, 1999); Srivastava et al. (2000) showed a decreasing trend in the frequency of Tropical Cyclones (TCs) and Monsoon Depressions (MDs) over the north Indian Ocean (The Bay of Bengal and the Arabian Sea) in recent years, their potential for damage and destruction still continues to be significant. A severe super cyclonic storm with winds of up to 250 km/h^{-1}, crossed the coast in Orissa on October 29, 1999. This proved to be the worst cyclone of the century in the Orissa region and was responsible for as many as 10,000 deaths (Table 6.4), for rendering millions homeless and for extensive damage (WMO 1999). Over the past decades, the frequency of tropical cyclones in the north Indian ocean has registered significant increasing trends (20% per hundred years) during November and May which account for maximum number of intense cyclones (Singh et al. 2000).

6.2.6.1
Impact of cyclones on agriculture

The losses affecting cash crops in the cyclone prone regions, which constitute a major source of export earnings, is rather high. Horticultural crops, particularly plants with soft stems, suffer a direct loss of fruits and mechanical damage. In coastal areas, in addition to the battering effects of winds on existing crops, there is the additional damage caused by airborne sea salt which occurs within a few hundred metres of the coast. Winds which blow from seas spray a lot of salt on coastal areas, making it impossible to grow crops sensitive to excessive salt.

6.2.6.2
Cyclone preparedness in agricultural systems

Disaster preparedness for impending cyclones, as is known, refers to the plan of action needed to minimize loss to human lives, damage to property and agriculture. Preparedness for cyclones in the agricultural systems can include early harvesting of crops (if mature), safe storage of the harvest etc. Irrigation canals and embank-

Table 6.3. Heat wave episodes in India

State	Epochs**					
	1901–10	1911–67	1968–77	1978–99	2000–2006*	1901–2006*
West Bengal	–	76	9	28	2	115
Bihar	–	105	6	23	12	146
Uttar Pradesh	–	27	3	42	32	104
Rajasthan	–	43	1	7	39	90
Gujarat, Saurashtra & Kutch	–	–	2	–	9	11
Punjab	–	–	1	–	20	20
Himachal Pradesh	–	–	–	–	3	3
Jammu & Kashmir	–	26	5	35	9	75
Maharashtra	–	82	4	13	14	113
Madhya Pradesh	–	32	4	15	38	89
Orissa	–	25	8	18	13	64
Andhra Pradesh	–	21	–	3	14	38
Assam	–	–	4	19	1	24
Haryana, Delhi & Chandigarh	–	–	1	2	28	31
Tamil Nadu	–	5	–	2	6	13
Karnataka	–	–	–	–	6	6
Telangana	–	–	–	–	9	9
Rayalaseema	–	31	2	28	5	66

ment of rivers in the risk zone should be repaired to avoid breaching. Beyond this, as the storm approaches the area, nothing can be done.

6.2.6.3
Protection of crops against wind

Crop damage by winds can be minimised or prevented by the use of windbreaks (shelterbelts). These are natural (e.g. trees, shrubs, or hedges) or artificial (e.g. walls, fences) barriers to wind flow to shelter animals or crops. Properly oriented

Table 6.4. Major cyclones in India and the neighbourhood

Year	Location	No. of deaths	Storm surge (height, in ft)
1737	Hoogli, West Bengal (India)	3,00,000	40'
1876	Bakerganj (Bangladesh)	2,50,000	10'40"
1885	False point, Orissa (India)	5,000	22'
1960	Bangladesh	5,490	19'
1961	Bangladesh	11,468	16'
1970	Bangladesh	2,00,000	13'17"
1971	Paradeep, Orissa (India)	10,000	7'20"
1977	Chirala, Andhra Pradesh (India)	10,000	16'18"
1990	Andhra Pradesh (India)	990	13'17"
1991	Bangladesh	1,38,000	7'20"
1998	Porbander cyclone	1173	–
1999	Paradeep, Orissa (India)	9,885	30'
2000	Meghalaya, Tamil Nadu (India)	12	–
2001	Andhra Pradesh (India)	108	–
2002	West Bengal, Orissa (India)	2	–
2003	Andhra Pradesh (India)	81	–
2004	Gujrat (India)	9	–

and designed shelterbelts are very effective in stabilizing agriculture in the regions where strong wind causes mechanical damage and imposes severe moisture stress on growing crops. Windbreaks prevent the loose soil being lifted by erosion and increase the supply of moisture to the soil in spring.

6.2.6.4
Tropical cyclones and storm surge

Coffee and bananas suffer the direct loss of fruits and mechanical damage due to tropical cyclones. Nonetheless, food crop losses were estimated to be higher (35%), while the livestock sector was less affected (8%, of which one fifth was poultry). The effects of strong winds in coastal areas are seen in stunted and often very sculpted trees providing the evidence of the direction of the strong winds. Fields inundated by the storm surge suffer a loss of fertility due to salt deposition, even after the sea water has receded. The affected land takes a few years to regain its original fertility.

6.2.7
Floods

Floods and droughts over India are the two aspects of weather associated with the abundance or deficit of monsoon rains. A large number of studies are available on various aspects of floods and droughts (Table 6.5).

6.2.7.1
Effects of flood on agriculture

Very intense (extreme) rainfall can result in catastrophic flood damage even though it occurs for a relatively short period of time (Table 6.6). In general, the greatest damage to agriculture results from high intensity rainstorms with sufficient duration as opposed to the low intensity, long duration storms. Direct damage to growing plants from floods is most often caused by depletion of oxygen available to the plant root zones. Flooding creates anaerobic soil conditions that can have significant impacts on vegetation. Root and shoot asphyxia, if prolonged, typically leads to plant death. Chemical reactions in anaerobic soils lead to a reduction in nitrate and the formation of nitrogen gas. The denitrification can be a significant cause of loss of plant vigour and growth following flooding.

6.2.7.2
Mitigation of damage on agricultural sector due to flood and heavy rainfall

Soils that are saturated prior to an extreme weather event are more likely to be affected severely by a damaging flood than soils that are relatively dry. Fields that have recently been tilled and are devoid of vegetation are much more susceptible to water erosion. Vegetation that is able to use much of the water and that can act as a barrier to moving water (horizontally and vertically) can reduce flood severity and impacts. Water storage systems (rivers, lakes, reservoirs, etc.) that are able to capture and hold most of the incoming water are usually effective in reducing flood damage.

6.2.7.3
Flooding and heavy rainfall

Soil erosion, disruption to critical agricultural activities, the logging of crops, increased moisture leading to increased problems with diseases and insects, soil moisture saturation and runoff, soil temperature reduction, grain and fruit spoil-

Table 6.5 Flood years and their category

Year	Area affected (x10^6 km^2)	% of the area affected	MFI value	Category
1961	1, 795	57, 166	3, 614	Exceptional
1971	1, 427	45, 446	2, 668	Exceptional
1878	1, 513	48, 185	2, 889	Exceptional
1975	1, 268	40, 382	2, 260	Exceptional
1884	1, 175	37, 420	2, 021	Exceptional
1892	1, 162	37, 006	1, 987	Exceptional
1933	1, 145	36, 465	1, 943	Exceptional
1959	1, 135	36, 146	1, 918	Exceptional
1983	1, 030	32, 803	1, 648	Exceptional
1916	1, 025	32, 604	1, 635	Exceptional

age and transportation interruption are among the more significant agricultural impacts from heavy rainfall (Table 6.7). Direct damage to growing plants from flood is most often caused by depletion of oxygen available to the plant root zones. Flooding creates anaerobic soil conditions that can have significant impact on vegetation. Chemical reactions in anaerobic soils lead to a reduction in nitrate and the formation of nitrogen gas. The denitrification can be a significant cause of loss of plant vigour and growth following flooding. There is often a balance needed between retaining enough water for agricultural production and environmental health and maintaining enough available storage volume to capture incoming water and prevent floods. Crops like rice that can function effectively in saturated and even submerged conditions are appropriate for locations that flood regularly and the system becomes dependent upon regular flooding. Many other crops (e.g. corn) would not be adaptable to such conditions and would not be appropriate alternatives to rice.

6.3
Crop Insurance

In India crop insurance was considered by the central government as early in 1947–48. The Indian government constituted an expert committee headed by Professor Dharam Narain, the Chairman, Agricultural Price Commission in 1970. A beginning in crop insurance was finally made in 1972 by implementing an experimental scheme for Hybrid-4 cotton in few districts of Gujarat State. The scheme

Table 6.6 Major Rainstorms in India

S.No	Date	Area	Casualty & Damage
1	01–03 July 1930	Maharashtra	Damage to agriculture and property was extensive.
2	01–03 Oct. 1961	Bihar	Damage to agriculture and property was extensive.
3	28–30 Aug. 1982	Orissa	Severe flooding to Mahanadi. Considerable damage to crops, property and loss of lives reported.
4	July to Aug.1st week1988	i) Andhra Pradesh	Paddy crop in 300,000 ha hectares completely damaged.
	Aug.3rd week to Sept.1988	ii) Assam	Standing Ahu and Sali paddy crops in 25,000 ha damaged.
5	June1^{st} & 2nd week 1994 and 14–16 July 1994	Kerala	Crop worth Rs. 1.445 billions damaged.
6	26–28 Aug. 2000	Andhra Pradesh, Hyderabad	Paddy, chilly crop worth hundreds of crores damaged.

was based on the individual approach and a uniform guaranteed yield was offered to selected farmers and the same was continued till 1979. Based on the recommendation of Prof. V.M Dandekar of GIC, another pilot scheme on area insurance was introduced in 1979. Participation in the scheme was voluntary but was open only to farmers who had received short term crop loans from financial institutions. The pilot crop insurance scheme was replaced by a Comprehensive Crop Insurance Scheme (CCIS) which was introduced from April, 1985 by the Government of India with the active participation of the State Government. A new crop insurance scheme entitled National Agricultural Insurance Scheme (NAIS) was introduced in the country starting from the Rabi (post rainy) season of 1999–2000. The NAIS modifies the CCIS in some crucial aspects. First, it provides for greater coverage for farmers as non-borrowing farmers are allowed to purchase insurance. Second, it provides greater coverage in terms of crops as well. The NAIS, at present, is implemented by 19 States and 2 union territories.

6.4
Strategies adopted in areas with high weather risk

The agrometeorological impacts on natural disasters have pervasive societal ramifications, particularly in the developing countries. Associated human misery underscores the vulnerability of the communities to these natural hazards. A disas-

Table 6.7 Weather factors which affect agriculture negatively

Weather factor	Negative effects on agriculture of extreme Values (both direct and indirect)
Rainfall	Direct damage to fragile plant organs, like flowers, soil erosion, water logging, drought and floods, land slides, impeded drying of produce conditions favourable to crop and livestock. Pest development, negative effect on pollination and pollinators.
Wind	Physical damage to plant organs or whole plants (e.g. defoliation, particularly of shrubs and trees), soil erosion, excessive evaporation. Wind is an aggravating factor in the event of bush or forest fires.
Air moisture	High values create conditions favourable for pest development, low values associated with high evaporation and often one of the most determinant factors in fire outbreaks.
High temperatures	Increased evapotranspiration, induced sterility in certain crops, poor vernalization, survival of pests during winter. High temperatures at night are associated with increased respiration loss. "Heat waves", lengthy spells of abnormally high temperatures are particularly harmful.
Low temperatures	Destruction of cell structure (frost), desiccation, slow growth, particularly during cold waves, cold dews.
High cloudiness	Increased incidence of diseases, poor growth.
Hail	Hail impact is usually rather localized, but the damage to crops particularly at critical phenological stages and infrastructure may be significant. Even light hail tends to be followed by pest and disease attacks.
Lightning	Lightning causes damage to buildings and the loss of farm animals. It is also one of the causes of wildfire.
Snow	Heavy snowfall damages woody plants. Unseasonable occurrence particularly affects reproductive organs of plants.
Volcanic eruptions, avalanches and earthquakes	The events listed may disrupt infrastructure and cause the loss of crops and farmland, sometimes permanently. A recent example of carbon dioxide and hydrogen sulphide emissions from a volcanic lake in Cameroon caused significant loss of human life and farm animals.
Air and water pollution	Air pollutants affect life in the immediate surrounding of point sources. Some pollutants, like ozone, are however known to have signifcant effects on crop yields over wide areas. In combination with fog, some pollutants have a more marked effect on plants and animals. Occurrences of irrigation water pollution have been reported.

trous event does not pose much of the threat and ceases to be a disaster, if suitable and adequate mitigation measures are adopted well in advance. Prevention of the formation of tropical cyclones is not in the realm of possibility, but much of their disastrous impacts can be reduced, limiting thereby the loss of human life and loss of property, by adopting appropriate strategies and taking timely precautions on receiving early warnings. Analysis of climatological data helps in the advance preparation of long-range policies and programmes for disaster prevention.

6.5 Conclusions

Government of India is concerned about improving the agricultural economy of the country irrespective of the existing status of infrastructure in a given area and to a certain extent, irrespective of the vagaries of weather too. It is essential that more inputs would be required for more vulnerable areas if development were to be carried out in a balanced manner across the country. All the existing services must be geared for that purpose. The best form of risk management is planning to ensure that any ensuing risk is manageable. Agroclimatic analyses can help in selection of crops and cropping practices such that while the crop weather requirements match the temporal march of the concerned weather element(s), endemic periods of pests, diseases and hazardous weather are avoided. Such agronomic planning on a micro scale to suit local climate is an essential step in crop-risk management.

Environmental planning would be necessary to avoid or mitigate losses from disasters, by using instruments such as land-use planning and disaster management. Natural disaster reduction measures are in place in a significant number of the nations surveyed and ongoing research and development to improve and expand these measures are also a feature of many national strategies to minimize adverse effects of extreme events on agriculture. Steps are being taken to significantly reduce the vulnerability of people and their communities to natural disasters; this can only be done through mitigation.

References

Chowdhury A, Dandekar MM, Raut PS (1989) Variability of drought incidence over India: A Statistical Approach. Mausam 40:207–214.
De US, Dandekar MM (2001) Natural Disasters in Urban Areas, The Deccan Geographer 39:1 –12.
De US, Dube RK, Prakasa Rao GS (2005) Extreme Weather Events over India in the last 100 years, I. Ind. Geophys. Union 9:173–187.
De US, Joshi KS (1995) Genesis of cyclonic disturbances over the North Indian Ocean–1891–1990, PPSR, 1995/3 issued by India Meteorological Department, Pune, India.
De US, Joshi KS (1999) Interannual and interdecadal variability of tropical cyclones over the Indian seas., The Deccan Geographer 37:5–21.
De US, Sinha Ray KC (2000) Weather & climate related impacts on health in Megacities, WMO Bulletin 44:340–348.
IDWR (2000) Indian Daily Weather Report, 2000/January, issued by India Meteorological Department, Pune, India.

India Meteorological Department (1888) Report of the Meteorology of India, 69–70.
Indian Institute of Tropical Meteorology (1994) Severe rainstorms of India.
Pai DS, Thapliyal V, Kokate PB (2004) Decadal variation in the heat & cold waves over India during 1971–2000. Mausam 55:281–292.
Philip NM, Daniel CEF (1976) Hailstorms in India, IMD Meteorological Monograms Climatology No. 10.
Raghavan K (1966) A climatological study of severe heat waves in India, Indian J. Met. Geophys. 4:581–586.
Rao GSP, Jaswal AK, Kumar MS (2004) Effects of urbanization on meteorological parameters, Mausam 55:429–440.
Singh OP, Alikhan TM, Rahaman MS (2000) Changes in the frequency of tropical cyclones over the north Indian ocean. Meteorol. Atmos. Phys. 75:11–20.
Srivastava AK, Sinha Ray KC, De US (2000) Trends in frequency of cyclonic disturbances and their intensification over Indian seas. Mausam 51: 113–118.
WMO (2004) WMO statement on the states of the global climate in 2003, WMO No. 966

CHAPTER 7
Challenges to coping strategies in Agrometeorology: The Southwest Pacific

James Salinger

7.1 Introduction

The climate system in the southwest Pacific provides a large source of interannual to multidecadal fluctuations beneath a theme of regional climate warming. These provide challenges especially to coping strategies for agrometeorology in the region. The El Niño-Southern Oscillation (ENSO) provides a large source of seasonal to interannual variability across the region promoting seasons of floods and droughts, and warmer and cooler seasons at higher latitudes (Trenberth and Caron 2000).

The Interdecadal Pacific Oscillation (IPO) (Trenberth and Hurrell 1994; Deser et al. 2004) is an important source of multidecadal climate fluctuations. These cause shifts in climate across the region. With this better understanding of the climate system of the region, these modes place a range of natural variability on the anthropogenic factors that will promote warming in the region during the 21st century. The latest IPCC projections (IPCC 2001) from the entire range of 35 IPCC scenarios place temperature increases in the range of 1.4 to 5.8°C by the end of the 21st century, with likely increases in heavy rainfall events and drought. It is the impact of the sources of variability and change on extreme events, such as floods, droughts, tropical cyclones and heatwaves that are significant. These will pose challenges for agriculture and forests to cope with future variability and change in the southwest Pacific.

7.2 El Niño-Southern Oscillation (ENSO)

El Niño-Southern Oscillation events are a coupled ocean-atmosphere phenomenon. It is a natural feature of the climate system. El Niño involves warming of surface waters of the tropical Pacific in the region from the International Date Line to the west coast of South America, with associated changes in oceanic circulation. It is accompanied by large changes in the tropical atmosphere, lowering pressures in the east and raising them in the west, in what is known as the "Southern Oscillation". The total phenomenon is generally referred to as ENSO. El Niño is the warm phase of ENSO and La Niña is the cold phase. Historically, El Niño (EN) events occur about every 3–7 years and alternate with the opposite phases of below average temperatures in the equatorial Pacific (La Niña). A convenient way of measuring ENSO is in terms of the east-west pressure difference, the Southern Oscillation Index, or SOI, which is a scaled form of the difference in mean sea-level pressure between Tahiti and Darwin. A graph of the SOI over the past 30 years is shown in Figure 7.1.

ENSO may be thought in terms of a slopping back and forth of warm surface water across the equatorial Pacific Ocean. The trade winds, blowing from the east towards the west, normally help to draw up cool water in the east and to keep the warmest water in the western Pacific. This encourages low air pressures in the west and high pressures in the east. An El Niño event is when the warm water "spills out" eastwards across the Pacific, the trade winds weaken, pressures rise in the west and fall in the east. Eventually, the warm water retreats to the west again and "normality" is restored. The movements of water can also swing too far the other way and waters become unusually cool near South America, resulting in what is termed a "La Niña", where the trade winds are unusually strong while pressures are lower than normal over northern Australia. As an El Niño event develops, the compensating shifts of the globe's weather zones and rainfall patterns result in widespread droughts in some regions, heavy flooding in others, and associated regions of warming and cooling. The regions most affected are the tropical and subtropical regions of Indonesia, Australia, and the Pacific Islands.

Figures 7.2 and 7.3 show relationships with temperature and precipitation throughout the region. Essentially ENSO events cause increased temperature in the tropical South Pacific from just west of the Date Line to the South American coast, whereas temperatures are decreased in ENSO events over Papua-New Guinea south east into the subtropical South Pacific, and New Zealand. The reverse temperature anomalies occur during La Niña episodes.

Precipitation anomalies can be much more dramatic (Figure 7.3). ENSO events can bring drought and decreased precipitation anomalies over the Phillipines, In-

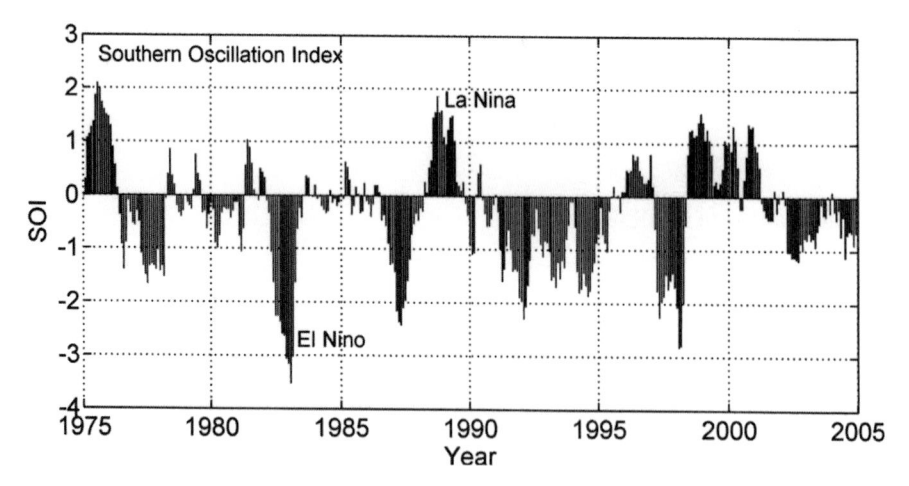

Fig. 7.1. The Southern Oscillation Index (SOI) for the last 30 years. Negative excursions indicate El Niño events, and positive excursions indicate La Niña events. The irregular nature of ENSO events is evident in the time sequence

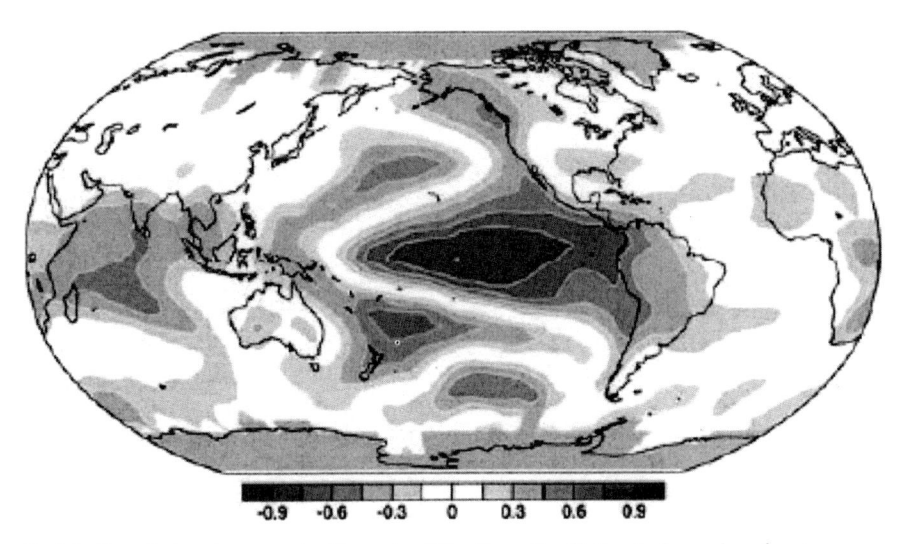

Fig. 7.2. Correlations between the May – April Southern Oscillation Index and surface temperature 1958-2004, after Trenberth and Caron (2000)

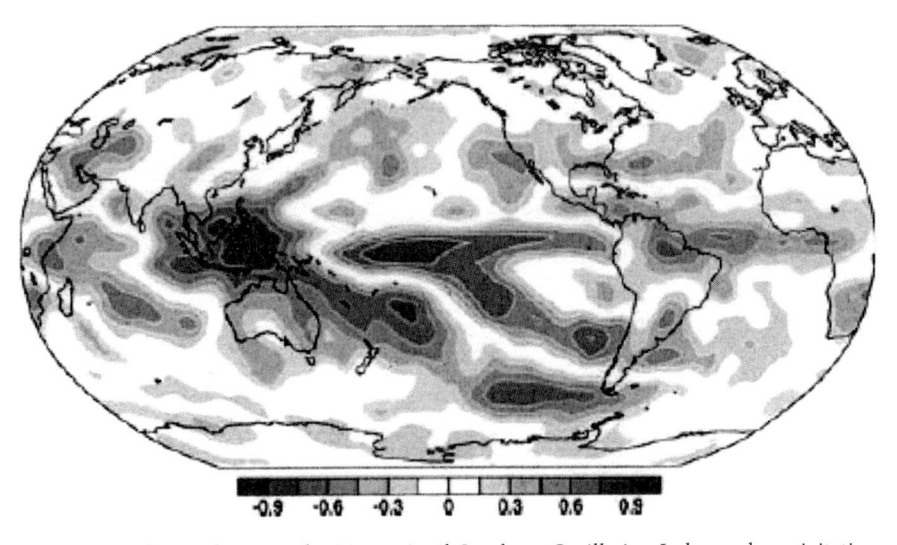

Fig. 7.3. Correlations between the May – April Southern Oscillation Index and precipitation 1958-2003, after Trenberth and Caron (2000).

donesia, northern and eastern Australia, the subtropical Southwest Pacific and the north east of New Zealand. Increased precipitation occurs in the equatorial Pacific from Kiribati (west of the Date Line) through to the Galapagos Islands. The reverse anomalies occur in La Niña episodes.

Tropical cyclones develop in the South Pacific over the wet season, usually from November through to April. Peak cyclone occurrence is usually during January, February and March based on historical tropical cyclone data analysis. Those countries with the highest risk include Vanuatu, New Caledonia, Fiji, Tonga and Niue. Taken over the whole of the South Pacific, on average nine tropical cyclones can occur during the November to April season, but this can range from as few as three in 1994/95, to as many as 17 in 1997/98, during the last very strong El Niño. The mean frequency of tropical cyclones for the 1970 – 2000 period for El Niño episodes is 11.5, and La Niña events 8.6 per season. The tropical cyclone track densities vary depending on the ENSO state (Figure 7.4). During El Niño episodes a higher frequency of tropical cyclone tracks occur near Vanuatu and Fiji, and their occurrence spreads further east to 160°W to affect the Cook Islands and most of French Polynesia. In contrast, during La Niña events the maximum occurrence is largely confined to the Coral Sea area of the Southwest Pacific centering on 160°E, 20°S and affecting New Caledonia in particular.

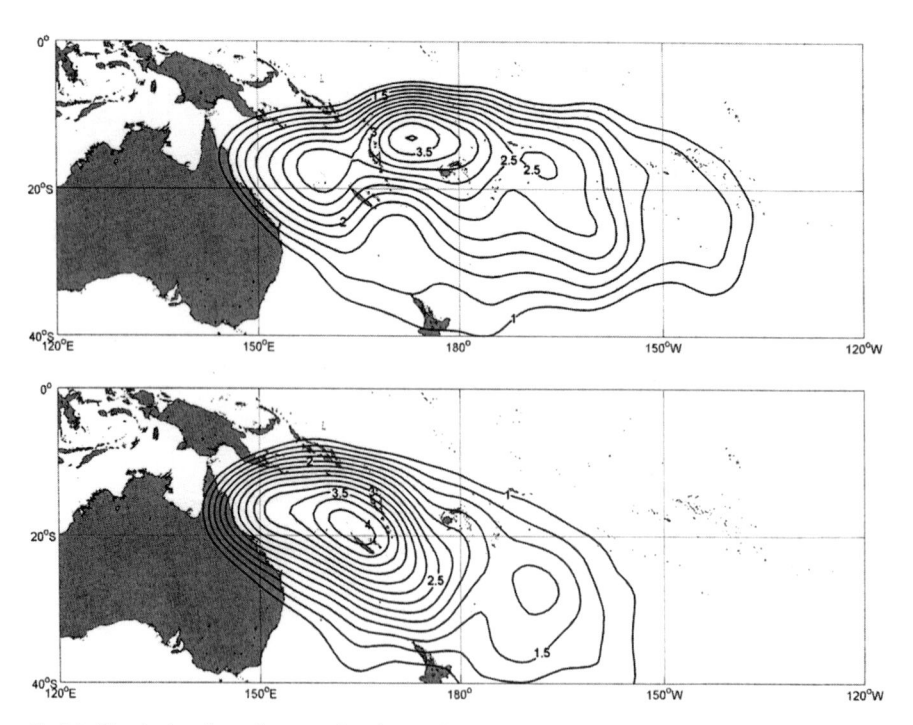

Fig. 7.4. Tropical cyclone densities for El Niño (upper) and La Niña (lower) seasons in the Southwest Pacific. Contour interval is 0.25, starting at 1.0

7.3
Decadal Variability

Decadal-to-interdecadal variability of the atmospheric circulation is most prominent in the North Pacific, where fluctuations in the strength of the wintertime Aleutian Low (AL) pressure system co-vary with North Pacific sea surface temperatures (SST) in the Pacific Decadal Oscillation (PDO). These are linked to decadal variations in atmospheric circulation, SST and ocean circulation throughout the whole Pacific Basin in the IPO. This is an 'ENSO-like' feature of the climate system that operates on time scales of several decades. The main centre of action in SST is in the north Pacific centred near the Date Line at 40°N. Three phases of the IPO have been identified during the 20th century: a positive phase (1922–1946), a negative phase (1947–1976) and the most recent positive phase (1977–1998). The IPO has been shown to be a significant source of decadal climate variation throughout the South Pacific (Salinger et al. 2001), New Zealand and Australia (Power et al. 1999)

Folland et al. (2002) showed that the IPO significantly affects the movement of the South Pacific Convergence Zone (SPCZ) in a way independent of ENSO. The South Pacific Convergence Zone (SPCZ) is one of the most significant features of the subtropical Southern Hemisphere climate (Kiladis et al. 1989; Vincent and Dias 1999; Vincent 1994). It is characterized by a quasi-permanent band of low-level convergence, enhanced cloudiness and precipitation originating from the west Pacific warm pool in the Indonesian region and trending south-east towards French Polynesia. The SPCZ when it is active during the November – April South Pacific wet season produces a good proportion of the precipitation. In the negative phase of the IPO the SPCZ is displaced southwest producing a drying in climates to the north east, and increases in precipitation to the southwest. The east and north of New Zealand becomes wetter, whilst the west and south drier. In Australia rainfall tends to increase in eastern Australia. The positive phase reverses the climate anomalies: the SPCZ is displaced north east increasing precipitation to the northeast and making the subtropical southwest Pacific drier. The north and east of New Zealand becomes drier, whilst the south and west wetter. Similarly eastern Australia tends to be drier.

7.4
Regional Warming

Climate model results show that globally average surface temperature is projected to increase by 1.4 to 5.8°C over the period 1990 – 2100 (IPCC 2001). Since the Third Assessment Report (TAR) (IPCC 2001), future climate change projections have been updated regionally (Ruosteenoja et al. 2003). Lal (2004) have applied these to the South Pacific. Models all project increases in temperature: for the 2080s increases of between 1.0 and 3.1°C are indicated with the 2100 surface air temperature to be at least 2.5°C more than in 1990. The models only simulate a marginal increase or decrease in annual precipitation (10 percent), with a drying in the subtropical South Pacific whilst the equatorial areas become wetter. During summer more precipitation is projected while an increase in daily rainfall intensity causing more heavier rainfall events is likely.

For Australia and New Zealand, scenarios based on more-detailed projections have been developed by CSIRO (2001) and NIWA (Wratt et al. 2004), respectively. Within 800 km of the Australian coast, a mean warming of 0.4 to 6.7°C is likely by the year 2080, relative to 1990. A tendency for decreased rainfall is likely over most of Australia, except Tasmania and New South Wales. A tendency for less run-off is also likely. In New Zealand, a warming of 0.5–3.5°C is likely by 2080s. The mid-range projection for the 2080s is a 60% increase in annual mean westerly winds (Wratt et al. 2004). Consequently, precipitation is likely to be biased towards increases in the south and west, and decreases in the north and east.

Global warming from anthropogenic forcing is likely to increase extreme events. Extreme temperatures above 30 and 35°C are likely throughout the region. In Australia the number of days over 35°C increases sihnificantly by 2020 with a 10 to 80% decrease in days below 0°C (Suppiah et al. 2007). In New Zealand there is likely to be a 50–100% decrease in frosts in the lower North Island, and a 50% decrease in the South Island, and a 10–100% increase in the number of days over 30°C (Mullan et al. 2001).

Under $3 \times CO_2$ conditions, there is a 56% increase in the number of simulated tropical cyclones over north-eastern Australia with peak winds greater than 30 ms^{-1} (Walsh et al. 2004). Maximum tropical cyclone wind intensities could increase 5 to 10 percent by around 2050 (Walsh 2004) with peak precipitation rates likely to increase by 25 percent as a result of increases in wind intensities.

For Australia, increases in extreme daily rainfall are likely where average rainfall increases, or decreases slightly. For example, the intensity of the 1-in-20 year daily-rainfall event increases by up to 10% in parts of South Australia by the year 2030 (McInnes et al. 2002), 5 to 70% by the year 2050 in Victoria (Whetton et al. 2002), up to 25% in northern Queensland by 2050 (Walsh et al. 2001) and up to 30% by the year 2040 in south-east Queensland (Abbs 2004). In New Zealand the frequency of high intensity rainfall is likely to increase, especially in western areas.

In the South Pacific, projected impacts include extended periods of drought. Projected changes in rainfall and evaporation have been applied to water balance models, indicating that reduced soil moisture and runoff is very likely over most of Australia and eastern New Zealand. Two climate models simulate up to 20% more droughts (defined as soil moisture in lowest 10% from 1974–2003) over most of Australia by 2030 and up to 80% more droughts by 2070 in south-western Australia (Mpelasoka et al. 2005). By the 2080s in New Zealand, severe droughts (the current one-in-twenty year soil moisture deficit) are likely to occur at least twice to four times as often in the east of both islands, and parts of Bay of Plenty and Northland (Mullan et al. 2005). The drying of pastures in eastern New Zealand in spring is very likely to be advanced by a month, with an expansion of droughts into spring and autumn.

Finally, an increase in fire danger in Australia is likely to be associated with a reduced interval between fires, increased fire-line intensity, a decrease in fire extinguishments and faster fire spread (Cary 2002;Tapper 2000;Williams et al. 2001). In south-east Australia, the frequency of very high and extreme fire danger days is likely to rise 4–25% by 2020 and 15–70% by 2050 (Hennessy et al. 2006). By the 2080s, 10–50% (6–18) more days with very high and extreme fire danger are likely in eastern areas of New Zealand, the Bay of Plenty, Wellington and Nelson regions (Pearce et al. 2005), with increases of 1–5 days in some western areas. Fire season length is likely to be extended, starting earlier in August and finishing in May in many parts of New Zealand, compared with the current October to April season.

7.5
Challenges to Agriculture and Forestry

7.5.1
ENSO

Because ENSO modulates the climate throughout the Pacific, this leads to much wetter and drier seasons respectively and often floods and drought. The consequences of these place challenges to agriculture and forestry, demonstrated by examples in this section.

The impact of ENSO climate variability on Australia has been highlighted by events during the early 1990s. In 1990–91 the wet season produced abundant rains, yet it failed almost completely the following year as drought set in across Queensland and New South Wales. While drought continued in some areas through 1992 and 1993 in southeast Australia there were the floods of spring 1992 and spring 1993, and the cool summers which followed. These were all connected with the Southern Oscillation. Rural productivity, especially in Queensland and New South Wales, is linked to the behaviour of the Southern Oscillation. Australian wheat yield (trend over time removed) have fluctuated with variations in the Southern Oscillation. Negative phases in the oscillation (drier periods) tend to have been linked with reduced wheat crops, and vice versa (Figure 7.5).

The relationships between wheat yields and the SOI are directly a consequence of ENSO effects on the Australian climate. ENSO tends to be accompanied by less rainfall and often droughts in Australia, especially during the winter in the interior of eastern Australia, and during the northern Australian monsoon. La Niña years are often wetter, with floods. For instance the 1998–99 La Niña brought record level rains to parts of central Western Australia. Lake Eyre in the northeast of South Australia is dry except for occasional major flooding, usually during La Niña years, due to extra rainfall in southwestern Queensland.

El Niño events produce widespread impacts on communities across the Southwest Pacific, as instanced by the 1997–98 event (Shea et al. 2001). Drought severely affected Fiji, Papua-New Guinea, the Solomon Islands, Tonga and the Marquesas Islands of French Polynesia. In Fiji, by October 1998, 54,000 people were receiving food supplies and 400,000, half the population, were receiving water deliveries (Hamnett et al. 2002, Lightfoot 1999). In contrast Kiribati was wetter than normal. Fisheries were impacted with a general shift of the catch to the east, followed by a shift westwards (Lehodey et al. 1997, Lefale et al. 2003). Catches declined for Papua-New Guinea and the Solomon Islands, and increased for Kiribati. Extended drought conditions were recorded in Niue in the 1982/83 ENSO event, which lasted

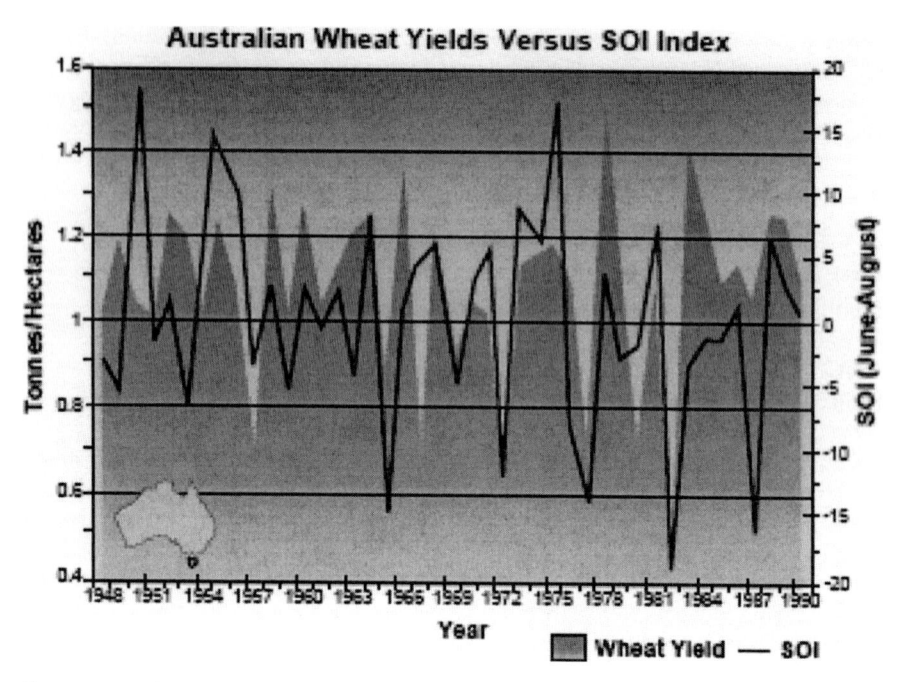

Fig. 7.5. Australian wheat yields and the Southern Oscillation Index (Rimmington and Nicholls, 1993).

18 months from July 1982 to December 1983. During this event exports went down, and local cows were slaughtered for food and imports of farm produce increased. The drought also created conditions of leafhopper and aphid invasions on important exports such as taro (Government of Niue 2000).

In El Niño years, New Zealand tends to experience stronger or more frequent winds from the west in summer, leading to drought in east coast areas and more rain in the west. In winter, the winds tend to be more from the south, bringing colder conditions to both the land and the surrounding ocean. In spring and autumn southwesterlies tend to be stronger or more frequent, providing a mix of the summer and winter effects. The La Niña events which occur at the opposite extreme of the Southern Oscillation Index cycle have weaker impacts on New Zealand's climate. New Zealand tends to experience more northeasterly winds, which bring more moist, rainy conditions to the northeast parts of the North Island. The impacts are instanced by the 1997–98 event (Basher 1998). It was much drier than normal in the east from July 1997 onwards – with drought very extensive throughout eastern areas up to the end of March 1998. In late April 1998 the New Zealand Ministry of Agriculture and Forestry estimated the likely cost of the drought on farm gate returns would be NZ$256 million for the year ending June 30 1998, and NZ$169 million for the following year, giving a total of NZ$425 million. Given the impact on downstream value-added agricultural production, the likely total cost to the country was estimated to be in excess of NZ$1 billion.

7.5.2
Tropical Cyclones

Of the extreme events in the Southwest Pacific, tropical cyclones are particularly significant (Krishna et al. 2001). Tropical cyclone Ofa caused an estimated US$120 million damages in Samoa in 1990 (or about 25% of its GDP) while tropical cyclone Val caused damages of about US$200 millions or 45% of its GDP in 1991. These two cyclones alone set back the development of Samoa by at least twenty years. In Fiji, tropical cyclone Kina in 1993 caused an estimated US$120 million damages, about 2.4% of Fiji's GDP. The damage caused by tropical cyclone Heta in January 2004, a category 5 tropical cyclone, is only now being assessed. Property, crops, roads and bridges were destroyed or damaged on Samoa with losses equivalent to US$226 million on American Samoa with roads washed away, and other damage to property and infrastructure. In the Cook Islands, 6 metre swells affected the west coast of Rarotonga. However, Heta devastated the tiny island nation of Niue (population 1200). One person died, much property was damaged or destroyed, roads were destroyed, infrastructure cut and crops destroyed because of intense rainfall and high winds (last recorded speed of 150 km/h and gusts to 200 km/h before equipment failure).

In Australia, tropical cyclones can also have severe impacts on cropping. In March 2006 Tropical Cyclone Larry tracked into Queensland at Innisfail, with winds of up to 290 km/h^{-1} (180 mph), the category five storm destroyed sugar and banana crops. Sugar cane and banana crops were severely impacts with an estimated 80% of the banana crops destroyed. This area produces 25% of the sugar crop.

7.5.3
Regional Warming

Regional warming in the Southwest Pacific will pose significant challenges to agriculture and forestry. Although warming in the lower range will be beneficial, challenges will be posed if water is limiting. Warming in the upper range will pose significant challenges and it is these that are highlighted here.

Warming and associated rainfall changes in Australia give varied impacts on wheat production regionally. Studies show that Western Australian regions are likely to have significant yield reductions by 2070 (Reyenga et al., 2001). In southern Australia cropping is likely to become unviable at the dry margins where rainfall is reduced substantially. Warming is likely to make a major pest, the Queensland fruit fly, a significant threat in southern Australia (Sutherst et al. 2000). Warming suggests a large expansion in its normal range across many non-arid areas of Australia. Pastoral and rangeland farming is very significant in Australia. A 20% reduction in rainfall is likely to reduce pasture productivity by 15% (Crimp et al. 2004). Climate change is also likely to increase major land degradation problems of erosion and salinization with the potential spread of weeds (Kriticos et al. 2003). Heat stress on livestock is also likely to increase (Howden et al. 1999b) with pests such as cattle tick likely to increase and spread southwards (White and Sutherst 2003). Reductions in production from this source are likely to be very high in Queensland.

For forestry in Australia warming increases fire risk and pest damage in southern Australia. Those areas where water resources, such as run-off, decline combined with an increased fire risk will face reduced production, and increases in rainfall intensity are likely to exacerbate soil erosion during forestry operations (Howden et al. 1999a).

For New Zealand crops, the principal effect will be the drying in eastern regions: this will be dependent on the availability of irrigation. The increased frequency of drought has already decreased pasture growth for dryland farms, and warming has meant that lower feed quality subtropical pastures continue to spread south. Warming is likely to increase the range and incidence of many pests and diseases. Plantation forestry in east coast areas is likely to experience growth reductions with rainfall decreases.

In the South Pacific islands there is little forestry, and traditionally agriculture has been based on subsistence and cash crops for survival and economic development. Subsistence agriculture has existed for several hundreds of years. The projected impacts of climate change include both extended periods of droughts, and the loss of soil fertility as a result of increased precipitation, both which will impact on agriculture. The consequences of the loss in agricultural productivity for both subsistence and cash cropping, including other impacts has been estimated at 2–3 percent of Fiji's 2002 GDP, and 17–18 percent of Kiribati's 2002 GDP (World Bank 2002). Fisheries are important for many island states in the South Pacific. Tuna fishing is particularly important, and warming is likely to produce a decline in total stock and a migration westwards, both of which will lead to changes in catches in different countries.

7.6
Discussion and Conclusions

In the south west Pacific, climate and extreme climatic events dominate in providing challenges for coping with agrometeorological risks and uncertainties. ENSO imposes large seasonal to interannual climate variability throughout the south west Pacific. It generally modulates the rainfall climate and either causes floods or droughts and displaces tropical cyclone tracks. Droughts made up one of the largest components and the experiences during the 1997–1998 El Niño event highlight the significant consequences that such climate-related extreme events can have for the southwest Pacific. These provide risks and uncertainties for coping strategies including crop yields and quality, water management, soil salinity, water erosion and runoff.

Tropical cyclones are one of the most devastating risks for agrometeorology on the small island developing states in the region. These generally cause large-scale destruction to crops and infrastructure through high intensity rainfall and severe winds. Although warning systems can predict the tropical cyclone tracks, presently little can be done to protect crops and agriculture from the full impacts of the extreme rainfall and hurricane force winds.

Chapter 7: Challenges to coping strategies in Agrometeorology: The Southwest Pacific

On the medium to longer term, although the IPO can cause shifts in the mean climate state (temperature and rainfall), this is less of a threat to coping strategies. Of more immediate concern are the impacts that global warming will have on regional climates. Firstly for the developed countries (e.g Australia, New Zealand) coping strategies are more sophisticated and involve both structural and non-structural measures to reduce the impacts of change on crop and livestock production. In these countries it will be the rate of change that will pose the risks. During the course of the 21st century, scientific evidence points to global-average surface temperatures are likely increasing by 2 to 4.5°C as greenhouse gas concentrations in the atmosphere increase. Any warming above about 2°C (Figure 7.6) is likely to be outside the immediate coping range for agriculture and forestry. At the same time there will be changes in precipitation, and climate extremes such as hot days, heavy rainfall, drought and fire risk are expected to increase throughout the

Fig. 7.6. The risk of adverse impacts increase with the magnitude of climate change. Global mean annual temperature is used as a proxy for the magnitude of climate change (IPCC WG2, as modified by Mastrandea and Schnedier, 2004).

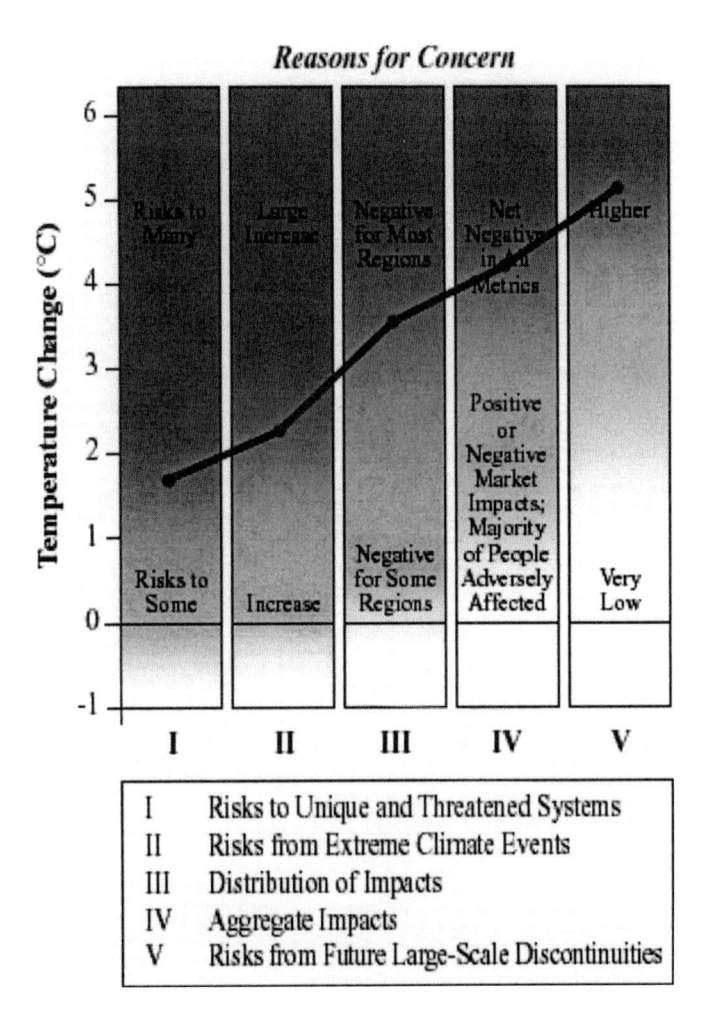

I Risks to Unique and Threatened Systems
II Risks from Extreme Climate Events
III Distribution of Impacts
IV Aggregate Impacts
V Risks from Future Large-Scale Discontinuities

region. The greatest challenge associated with any warming will be those areas where rainfall decreases especially in arid or semi-arid areas with will pose additional stresses on an already finely managed agriculture and forestry activities. The potential increase and spread of pests and diseases will also be important.

For developing countries, especially small island developing states in the south west Pacific, the largest threats to agriculture will be posed by impacts of warming on tropical cyclones and high intensity rainfall. Warming is likely to strengthen tropical cyclones, and increase maximum wind speeds and high intensity rainfall produced by these systems.

References

Abbs DJ (2004) A high resolution modelling study of the effect of climate change on the intensity of extreme rainfall events. Staying afloat: Floodplain Management Authorities of NSW 44th Annual Conference, Coffs Harbour, NSW. Floodplain Management Authorities of NSW, 17–24.

Basher RE (1998) The 1997/98 El Niño event: impacts, responses and outlook for New Zealand : a review prepared for the Ministry of Research, Science and Technology. Wellington, 28 pp.

Cary GJ (2002) Importance of a changing climate for fire regimes in Australia. In: Bradstock RA, Williams JE, Gill AM (eds) Flammable Australia: Fire Regimes and Biodiversity of a Continent, Cambridge University Press, pp 26–49.

CSIRO (2001) Climate Change Projections for Australia.CSIRO Atmospheric Research, 8pp pp.

Deser C, Phillips AS, Hurrell JW (2004) Pacific interdecadal climate variability: Linkages between the tropics and the north Pacific during boreal winter since 1900. J. Climate 17: 3109–3124.

Folland CK, Renwick JA, Salinger MJ, Mullan AB (2002) Relative influences of the Interdecadal Pacific Oscillation and ENSO on the South Pacific Convergence Zone. Geophys Res Lett 29(13): 10.1029/2001GL014201, 21-1 – 21-4.

Government of Niue (2000) Niue Initial Communication to the UNFCCC

Hamnett M, Anderson CL, Gosai, A (2002) Pacific Islands Summary. Pages 38–42 in Preparing for El Niño: Advancing regional plans and interregional communication. International Research Institute for Climate Prediction, New York, USA.

Hennessy K, Lucas C, Nicholls N, Bathols J, Suppiah R, Ricketts J (2006) Climate change impacts on fire-weather in south-east Australia.Consultancy report for the New South Wales Greenhouse Office, Victorian Department of Sustainability and Environment, Tasmanian Department of Primary Industries, Water and Environment, and the Australian Greenhouse Office. CSIRO Atmospheric Research and Australian Government Bureau of Meteorology, 78pp.

Howden SM, Reyenga PJ, Gorman TJ (1999a) Current evidence of global change and its impacts: Australian forests and other ecosystems.CSIRO Wildlife and Ecology Working Paper 99/01, 23 pp.

Howden SM, Hall WB, Bruget D (1999b) Heat stress and beef cattle in Australian rangelands: recent trends and climate change. People and Rangelands: Building the Future. Proceedings of the VI International Rangeland Congress, Townsville, Australia pp 43–45.

IPCC (2001) Climate Change 2001: The Scientific Basis. Contribution of Working Group 1 to the Third IPCC Scientific Assessment. Houghton JT, et al. (eds.) Cambridge University Press , Cambridge, United Kingdom, 881 pp.

Kiladis GN, von Storch H, van Loon H (1989) Origin of the South Pacific convergence zone. J Climate 2:185–1195.

Kriticos DJ, Sutherst RW, Brown R, Adkins SW, Maywald GF (2003) Climate change and the potential distribution of an invasive alien plant: Acacia nilotica spp. indica in Australia. J Appl Ecol 40:11–124.

Krishna R, Lefale PF, Sullivan M, Young E, Pilon JC, Schulz C, Clarke G, Hassett M, Power S, Prasad R, Veitch T, Turner K, Shea E, Taiki H, Brook R (2001) A needs analysis for the strengthening of Pacific Island Meteorological Services, Australian Agency for International Development (AusAID), Published by the South Pacific Regional Environment Programme (SPREP).

Lal M (2004) Climate change and small island developing countries of the South Pacific. Fijian Studies, Special issue on Sustainable Development 2(1) 15–31.

Lefale PF, Lehodey P, Salinger MJ (2003) Impacts or ENSO on Tuna stocks and coral reefs of the Pacific Islands. NIWA Report AKL 2003–16.

Lehodey P, Bertignac M, Hampton J, Lewis J (1997) El Niño/Southern Oscillation and tuna in the western Pacific, Nature 389: 715–718.

Lightfoot C. (1999) Regional El Niño social and economic drought impact assessment and mitigation. Disaster Management Unit, South Pacific Geoscience Commission, Suva, Fiji, 39 pp.

McInnes KL, Suppiah R, Whetton PH, Hennessy KJ, Jones RN, (2002) Climate change in South Australia: report on assessment of climate change, impacts and possible adaptation strategies relevant to South Australia.CSIRO Atmospheric Research, Aspendale, 61pp.

Mpelasoka FK, Hennessy KJ, Jones R, Bathols J (2005) Projected changes in Australian droughts due to greenhouse warming.CSIRO Technical Report.

Mullan AB, Salinger MJ, Thompson CS, Porteous AS (2001) The New Zealand climate – Present and future. In: Warrick RA, Kenny GJ, Harman JJ (eds) The effects of climate change and variation in New Zealand: An assessment using CLIMPACTS, University of Waikato, pp11–31.

Mullan AB, Porteous A, Wratt D, Hollis M (2005) Changes in drought risk with climate change. NIWA Report WLG2005–23, 58pp.

Pearce G, Mullan AB, Salinger MJ, Opperman TW, Woods D, Moore JR (2005) Impact of climate variability and change on long-term fire danger. Report to the New Zealand Fire Service Commission, 75 pp.

Power S, Casey T, Folland C, Mehta V (1999) Interdecadal modulation of the impact of ENSO on Australia. Climate Dynam 15: 319–324.

Reyenga PJ, Howden SM, Meinke H, Hall WB, (2001) Global change impacts on wheat production along an environmental gradient in South Australia. Environ Int 27: 195–200.

Rimmington GM, Nicholls N (1993) Forecasting wheat yields in Australia with the Southern Oscillation Index Aust J Agr Res 44(4): 625–632

Ruosteenoja K, Carter TR, Jylha J, Tuomenvirta H (2003) Future climate in world regions: an intercomparison of model-based projections for the new IPCC emissions scenarios. The Finnish Environment 644, Finnish Environment Institute, Finland, 83 pp.

Salinger MJ, Renwick JA, Mullan AB (2001) Interdecadal Pacific Oscillation and South Pacific climate. Int J Climatol 21: 1705–1721.

Shea EL, Dolcemascolo G, Anderson CL, Barnston A, Guard CP, Hamnett MP, Kubota ST, Lewis N, Loschnigg J, Meehl G (2001) Preparing for a changing climate: The potential consequences of climate variability and change. A report of the Pacific Islands Regional Assessment Team for the U.S Global Change Research Program. East-West Center, 102 pp.

Suppiah R, Hennessy KR, Whetton PH, McInnes K, Macadam I, Bathols J, Ricketts J Correct text: (2007) Australian climate change projections derived from simulations performed for the IPCC AR4. Australian Meteorological Magazine (in press).

Sutherst RW, Collyer BS, Yonow T (2000) The vulnerability of Australian horticulture to the Queensland fruit fly, *Bactrocera (Dacus) tryoni*, under climate change. Aust. J Agr Res 51: 467–480.

Tapper N (2000) Atmospheric issues for fire management in Eastern Indonesia and northern Australia. In. Proceedings of an international workshop: Fire and sustainable agricultural and forestry development in Eastern Indonesia and northern Australia. Northern Territory University, Darwin, Australia, 13–19 April 1999, pp 20–21.

Trenberth KE, Caron JM, (2000) The Southern Oscillation revisited: Sea level pressures, surface temperatures and precipitation. J Climate 13: 4358–4365.

Trenberth KE, Hurrell JW (1994) Decadal atmosphere–ocean variations in the Pacific. Clim Dynam 9: 303–319.

Vincent DG (1994). The South Pacific Convergence Zone (SPCZ): a review. Mon Weather Rev 122: 1949–1970.

Vincent DG, Silva Dias PL (1999) In Karoly DJ, Vincent DG (eds) Meteorology of the Southern Hemisphere. American Meteorological Society, Boston MA pp 101–117

Walsh K, Hennessy K, Jones R, McInnes KL, Page CM, Pittock AB, Suppiah R, Whetton P (2001) Climate change in Queensland under enhanced greenhouse conditions – third annual report 1999–2000.CSIRO consultancy report for the Queensland Government, 108 pp.

Walsh KJE, Nguyen KC, McGregor JL (2004) Finer-resolution regional climate model simulations of the impact of climate change on tropical cyclones near Australia. Clim Dynam 22: 47–56.

Whetton PH, Suppiah R, McInnes KL, Hennessy KJ, Jones RN (2002) Climate change in Victoria: high resolution regional assessment of climate change impacts.CSIRO consultancy report for Department of Natural Resources and Environment, 44 pp.

White N, Sutherst RW (2003) The Vulnerability of the Australian Beef Industry to Impacts of the Cattle Tick (*Boophilus microplus*) under Climate Change. Climatic Change 61: 157–90.

Williams AA, Karoly DJ, Tapper N (2001) The sensitivity of Australian fire danger to climate change. Climatic Change 49: 171–191.

World Bank (2002) Cities, Seas and Storms: Managing change in Pacific Island economies. World Bank Washington D.C.

Wratt DS, Mullan AB, Salinger MJ, Allan S, Morgan T (2004) Climate change effects and impacts assessment: A guidance manual for local government in New Zealand. New Zealand Climate Change Office Ministry for the Environment, 140 pp.

CHAPTER 8

Challenges to agrometeorological risk management – regional perspectives: Europe

Lučka Kajfež Bogataj, Andreja Sušnik

8.1
Introduction

Agriculture is one of Europe's largest land users and as such highly dependent on environmental conditions. Inter-annual climate variability is one of the main sources for uncertainty in crop yields. In the EU-25, 162 million hectares are under agricultural use, which amounts to roughly half the Union's land. Farming plays a key role for the health of economies in rural areas, and continues to be a determinant of the quality of the countryside and the environment, although it has become less important for the national economies. The contribution of the agricultural sector to the GDP of EU-25 was around 1.6% in 2004. Regionally and nationally, however, the contribution may be substantial, particularly in southern and central European countries where agriculture represents a more significant sector for employment and GDP.

Food production is still the major concern of agriculture and Europe is one of the world's largest and most productive suppliers of food and fibre, with 21% of global meat production and 20% of global cereal production in 2004. About 80% of this production occurred in EU25. The productivity of agriculture is generally high, in particular in western Europe; average cereal yields in the EU are over 60% higher than the world average. The total value of EU agricultural production is around 200 billion € and in particular the EU production of wheat is around 100 million tons, making EU the second largest producer in the world after Asia (FAOSTAT, 2005). The area of forests in Europe is increasing and annual fellings are considerably lower. Policies today promote multiple forest services at the expense of timber production. The role of farmers today include not only food production, but also countryside management, nature conservation, tourism and biomass production, mainly renewable energy sources. There are concerns that the expansion of biomass production may lead to a further intensification of European agriculture (European Commission, 2005b).

As European agriculture is highly intensive, weather remains the main source of uncertainty for crop yield assessment and crop management (Metzger et al. 2006). Since 1998, Europe has suffered more than 100 major floods, causing extensive damage. According to the estimation of IPCC (2001), it is likely that there has been up to 4% increase in the frequency of heavy precipitation in the mid and high latitudes of the Northern Hemisphere over the second half of the 20th century. Ironically, each year since 1990 the average land area and population affected by droughts has doubled. It can be seen that scarcely a year goes by without at least one country or some part of the Europe being affected by drought. Water scarcity is a problem that affects at least 14 Member States and around 100 million inhabitants in 26 river basin districts throughout Europe. Based on the results of trend analysis for the last 40 years, long dry periods in summer showed an increase in most stations in central Europe, the UK and southern Scandinavia, and long dry periods in winter increased in southern Europe.

There is an increasing demand for climate predictions at different time scales in Europe, because they carry valuable benefits for decision-making in the management of European Union agricultural production. But the key challenge for Europe today is climate change. The yields per hectare of all cash crops have continuously increased in Europe in the past decades due to technological progress, while climate change has had a minor influence. Today climate change already has considerable impacts on agriculture which are expected to become more severe in future. The European heat wave in 2003 had major impacts on agricultural systems and society by decreasing the quantity and quality of the harvests, particularly in Central and Southern European agricultural areas. In general it is believed that climate change and increased CO_2 concentration could have a beneficial impact on agriculture and livestock systems in northern Europe through longer growing seasons and increasing plant productivity. However, in the south and parts of eastern Europe the impact is likely to be negative. It is becoming increasingly clear that the EU target of limiting global temperature increase to no more than 2.0°C above pre-industrial levels is likely to be exceeded before 2050 (EEA, 2004). The assessment of climate change and its agricultural impacts in Europe is still subject to uncertainties and information gaps. Especially for the countries outside EU-25 only a small range of the potential consequences of climate change was studied. Better knowledge and understanding is still needed about the exposure and sensitivity of agriculture with respect to climate change.

This paper will focus on the latest developments in the field of seasonal weather forecasts for crop yield modelling in Europe and on climate change research and its connections with agrometeorological risk management.

8.2
Seasonal weather forecasts for crop yield modeling in Europe

Seasonal weather forecast for crop yield modeling in Europe is one of the applications of weather risk management. The variability of weather at different time scales, such as daily, monthly, seasonal and beyond, is one of the factors that determine the growth of field-grown crops. The key weather parameters for crop prediction are rainfall, temperature and solar radiation, secondary parameters being humidity and wind speed. Crop predictions require forecasts of these variables several weeks or even months ahead to enable informed management decisions. There is an increasing demand for climate predictions at different time scales in Europe, because they have valuable benefit for decision-making in the management of European Union agricultural production. Examples of this are monthly forecasts for emergency plans to reduce the impact of warm or cold spells, seasonal forecasts to cope with the remote effects caused by El Niño or La Niña events, inter-annual forecasts as a contribution for the management of food crisis and multi-decadal predictions to design plans to mitigate and adapt to climate change.

In the framework of the EU-funded DEMETER (Development of a European Multimodel Ensemble System for Seasonal to Inter-annual Prediction) project, led by the European Centre for Medium-Range Weather Forecasts (ECMWF, UK), ensembles of global coupled climate models have shown some skill for seasonal climate prediction (Palmer et al., 2004). Meteorological outputs of the seasonal prediction system were used in a crop yield model to assess the performance and usefulness of such a system for crop yield forecasting (Cantelaube and Terres, 2005). An innovative method for supplying seasonal forecast information to crop simulation models was developed. It consisted in running a crop model from each individual downscaled member output of climate models. An ensemble of crop yields was obtained and a probability distribution function (PDF) was derived. Preliminary results of wheat yield simulations in Europe using downscaled DEMETER seasonal weather forecasts suggest that reliable crop yield predictions can be obtained using an ensemble multi-model approach. When compared to the operational system, for the same level of accuracy, earlier crop forecasts are obtained with the DEMETER system. Furthermore, PDFs of wheat yield provide information on both the yield anomaly and the uncertainty of the forecast. Based on the spread of the PDF, the user can directly quantify the benefits and risk of taking weather-sensitive decisions.

This project officially ended in September 2003, but the data and results generated are being used by a steadily increasing section of the scientific community (Doblas-Reyes et. al. 2006). The use of ensembles of seasonal weather forecasts brings additional information for the crop yield forecasts and therefore has valuable benefit for decision-making in the management of European Union agricultural production. Skilful seasonal weather forecasts and related seasonal crop yield forecasts could generate an economical benefit for the CAP.

Seasonal forecasting in Europe is still more at the research level than at practical level. Obstacles for operational application, like mismatch between farmers' needs and the scale and relevance of available forecasts or insufficient understanding by farmers and their advisors will probably remain in near future. Agrometeorological community in Europe is challenged to help to close this gap in future.

8.3
Climate change as a challenge to agrometeorological risk management in Europe

8.3.1
The use of different high resolution climate models in Europe

One of the most urgent challenges to agrometeorological risk management in Europe today is timely adaptation to climate change effects. The prerequisite for intelligent, effective and efficient adaptation of agriculture to climate change is a good understanding of regional impacts. Regional modelling of climate development provides the essential basis in this regard. The use of different high resolution climate models linked to impact/crop models enables us to quantify the uncertainties of predictions and analyse how these uncertainties are transferred from the climate models into the crop models. Major scientific progress has been made in this field recently.

PRUDENCE was a recent EU project (PRUDENCE, 2005) using four Atmosphere General Circulation Models (AGCM) namely NCAR CCM3 (Italy), HadAM3H AGCM (United Kingdom), ECHAM AGCM (Germany) and eight Regional Climate Models (RCM), namely HIRHAM (Danmark), PROMES (Spain), ICTP RegCM (Italy), ARPEGE (France), CHRM (Switzerland), LM and CRCM-2 (Germany) and RCA (Sweden) to quantify the uncertainties associated with climate predictions and impacts of future climate changes on Europe. PRUDENCE is providing improved model representation of climate change scenarios by utilising high-resolution models (at spatial scales of ~50 km) for current (1961-1990) and future (2071-2100) climate, characterising the level of confidence in these scenarios, and assessing the uncertainty resulting from model formulation. Future scenarios correspond to the IPCC A2 and B2 CO_2 emissions (IPCC, 2001). For temperature, GCMs and RCMs behave similarly, except that GCMs exhibit a larger spread. The differences between GCM and RCM precipitation responses for some regions are significant. The spread of precipitation during summer period is larger for RCMs than for GCMs. For both, however, in terms of precipitation, the bias is twice as large as the response to climate change, when observed climate is used as a cross validation.

The scenarios indicate that European regions undergo substantial warming in all seasons in a range of 1 - 4 °C (B2 scenario) and 2.5 - 5.5 °C (B2 scenario) by 2071-2100. Over Northern and Eastern Europe, the warming is stronger in winter, and the reverse happens over Western and Southern Europe with stronger increases in summer (IPCC, 2007). Within Europe, the warming is estimated to be greatest over western Russia and southern countries (Spain, Italy, Greece), and less pronounced along the Atlantic coastline.

Across all scenario simulations, the results agree on a general increase in winter precipitation in Northern and Central Europe and on a general decrease in summer precipitation in Central and Southern Europe, a bit smaller in central Scandinavia (Raisanen et al. 2004). Over all, there is an annual increase in Northern Europe and an annual decrease in Southern Europe. Increased Atlantic cyclonic activity could lead to stronger precipitation (up to 15–30 %) in winter over Western, Central and Northern Europe, and in response to anticyclonic circulation to reduced precipitation in winter over Southern Mediterranean regions (Giorgi et al. 2004). In summer, a blocking situation caused by enhanced anticyclonic circulation over the Northeastern Atlantic could lead to decreases in precipitation (up to 30–45 %) over Western and Central Europe and the Mediterranean. Precipitation changes for spring and autumn are less pronounced than for winter and summer.

Notable changes are also projected for temperature and precipitation extremes in Europe. According to IPCC (2007), yearly maximum temperature is expected to increase much more in Southern and Central than in Northern Europe. According to EEA (2004), cold winters, which occurred on average once every 10 years in the period from 1961 to 1990, are likely to become rare in Europe and will almost entirely disappear by 2080. In contrast, by 2080 nearly every summer in many parts of Europe is projected to be hotter than the 10 % hottest summers in the current climate (EEA, 2004). Under high emission scenarios every second summer in Europe will be as hot or even hotter than 2003 by the end of the 21st century (Goodess 2005). In southern Europe, these changes are projected to occur even earlier. Extreme daily precipitation will even increase in most of those areas where the mean annual precipitation decreases (Raisanen et al. 2004). Risk of drought is likely to increase in central and southern Europe.

Uncertainty in projections of future precipitation is larger in comparison with temperature. This applies particularly to regional precipitation patterns and seasonal distribution of precipitation. But it should be stated that scientific confidence in the ability of climate models to estimate future precipitation is steadily increasing (IPCC 2007).

Climate change poses many challenges to agrometeorological risk management in Europe today, since extreme weather events, such as hot spells, heavy storms, intense rainfall and droughts, can severely disrupt crop production all over Europe. Especially the precipitation reduction in central, eastern and southern Europe is expected to have severe effects, e.g. more frequent droughts, with considerable impacts on crop production and availability of water resources.

8.3.2
Expected impacts of climate change in Europe during this century

Wide ranging impacts of changes in current climate have been documented in Europe in the last decades. The observed changes are consistent with projections of impacts due to anthropogenic climate change. The warming trend and spatially variable changes in rainfall have already affected managed ecosystems (EEA 2005b).

For instance the European heat wave in 2003 had major impacts on agricultural systems and society by decreasing the quantity and quality of the harvests, particularly in Central and Southern Europe. The winter crops already suffered from the effects of a harsh winter and late spring frost. The heat wave that began in early June accelerated crop development by 10 to 20 days, thus advancing ripening and maturity. Winter-spring cereals formed grain with insufficient soil moisture. The very high air temperature and solar radiation resulted in a notable increase in the crops' water consumption. This, together with the summer dry spell, resulted in an acute depletion of soil water and lowered crop yields. Even in Switzerland river withdrawals for agricultural use were banned, thus affecting producers of potatoes and tobacco. Over all of Europe, the main sectors hit by the extreme climate conditions were the green fodder supply, the arable sector, the livestock sector and forestry. Potato and wine production were also seriously affected. The fodder deficit varied from 30% (Germany, Austria and Spain) to 40% (Italy) and 60% in France and the livestock farmers suffered the most. In Switzerland, fodder had to be imported from as far away as Ukraine. The fall in cereal production in EU reached more than 23 million tonnes (MT) as compared to 2002. More than 26,000 fires were recorded in Portugal, Spain, Italy, France, Finland, Austria, Denmark and Ireland. The estimate is that some 70,000 hectares of forest area (not including agricultural areas) were burned. The global financial impact of the drought and the forest fires in Europe was estimated to amount to 14 billion €. Summer 2003 showed also the additional side effects, which were felt in the next year such as problems of soil erosion and flooding, effects on winter sowing, and the budding of trees (COPA COGECA 2003).

Under a changing climate, drier conditions and rising temperatures in the Mediterranean region and parts of eastern Europe may lead to lower yields. Bindi and Moriondo (2005) showed a general reduction in yield of agricultural crops in the Mediterranean region, under the IPCC SRES A2 and B2 scenarios by 2050 even when the fertilising effect of increased CO_2 is taken into account (Table 8.1). Similar yield reductions have also been estimated for eastern Europe, with increased variability in yield, especially in the steppe regions (Maracchi et al. 2004).

Climate-related increases in crop yields are mainly expected in Northern Europe. For example wheat yield increase is projected to be +2 to +9% by year 2020, +8 to +25% by year 2050 and +10 to +30% by year 2080 (Olesen et al. 2006, Auds-

Table 8.1 Changes of crop yields (%) for some Mediterranean regions by 2050 (modified from Bindi and Moriondo, 2005)

	A2	B2	A2	B2
	Without CO_2		With CO_2	
C4 summer				
N-W	0.2	5.8	4.2	8.8
N-E	−4.4	−2.5	−0.6	0.2
Legumes				
N-W	−24.9	−13.4	−14.4	−4.9
N-E	−18.6	−8.1	−7.2	1.0
C3 summer				
N-W	−21.8	−10.4	−12.4	−2.9
N-E	−15.6	−6.9	−5.4	1.0
Tubers				
N-W	−10.4	−4.2	4.9	7.5
N-E	−22.5	−6.8	−9.3	4.4
Cereals				
N-W	−11.0	−3.5	−0.3	4.7
N-E	−6.8	3.7	4.4	12.5

N-W = Portugal, Spain, France and Italy; N-E = Serbia, Greece and Turkey

ley et al. 2006, Alexandrov et al. 2002, Ewert et al. 2005). Another example is sugar beet yield increase of 14–20% until the 2050s in England and Wales (Richter and Semenov 2005).

Uncertainties in the projection of future precipitation complicate the estimates of future yield gains or losses. This is particularly true for southern and southeastern Europe, where water will be a critical factor for agriculture in the future. In these areas, model results diverge to a great extent, depending on the scenarios in use and the model itself. For central and northern Europe, where water supply is less critical, projections are relatively robust.

8.3.3
Increasing drought Risk with Global Warming in Europe

Increased drought risk associated with global warming and impacts on water resources are among the main concerns among agrometeorologists in Europe (Fig. 8.1). Several recent studies highlight the challenges that result from changes in water availability and water quality (EEA 2004, IPCC 2001, IPCC 2007, Schröter et al. 2005, EEA 2005a).

In Europe, large amounts of water are extracted from both surface and groundwater stocks for agriculture every year. For Europe as a whole (including New Member States and Accession Countries) some 38% of the extracted water is used for agricultural purposes. In Malta, Cyprus and Turkey almost 80% of the extract-

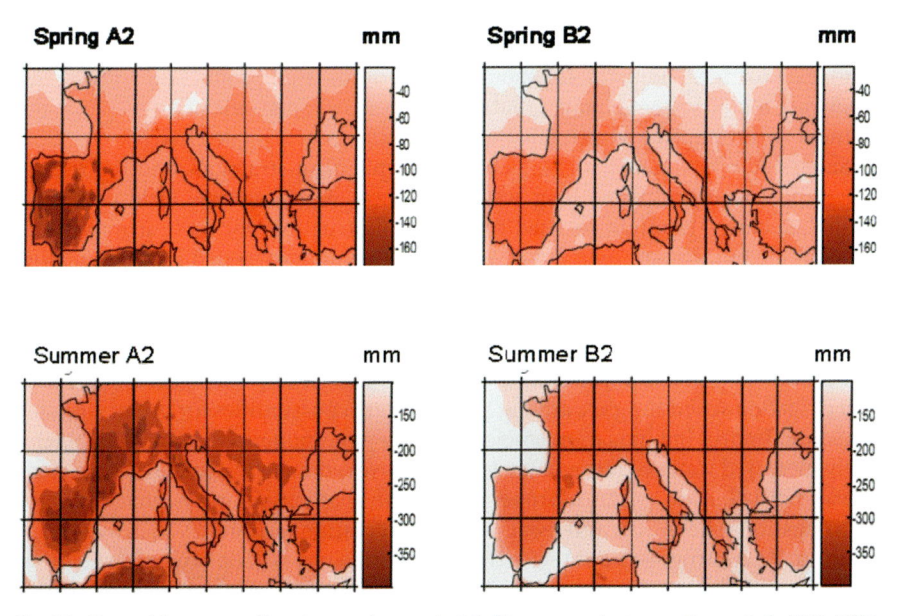

Fig. 8.1. Ensemble mean soil moisture changes in Mediterranean between the periods 1961-1990 and 2070- 2099 in spring and summer under the IPCC SRES A2 and B2 scenarios (PRUDENCE, 2005)

ed water is used for agriculture, in the southwestern countries (Portugal, Spain, France, Italy, Greece) about 46% and in the central and northern countries (Austria, Belgium, Denmark, Germany, Ireland, Luxembourg, Netherlands, UK, and Scandinavia), to the contrary, agricultural use of the extracted water is limited to less than 5% (Eisenreich, 2005). By far the largest part of the water used in agriculture is used for irrigation. This percentage approaches 100% in the southern European countries, which at the same time have the largest share of irrigated land in Europe. While the expansion of the irrigated areas, mainly in the Mediterranean, has raised concern about the overuse and depletion of water resources in the past, possible changes in climate and weather patterns as a consequence of global warming are at the focus of the discussion now.

Under climate change conditions, it is expected that irrigation water demand will further increase, aggravating the competition with other sectors whose demand is also projected to increase. In addition, an expected lowering of the groundwater table will make irrigation more expensive, which, in turn might have to be limited to cash crops. Extreme weather events such as heat waves will impact on peak irrigation requirements. As the evaporative demand will increase due to higher temperatures, it is expected that capillary rise will increase the salinisation of soils, having a major impact on irrigation management.

8.3.4
Options for future adaptation strategies

Strategies to adapt to climate change should not be seen as individual remedies since agriculture will compete for water allocation with other sectors affected by climate change. Short term adjustments should aim at optimising production without introducing major system changes, but for long-term adaptations heavier structural changes will take place to alleviate the adverse effects of climate change.

Suggested adjustments include changes in planting strategies and the use of more appropriate cultivars: long season cultivars might increase yield potential, while late cultivars might be used to prevent destruction due to heat waves and drought during the summer. However, the use of more extended growing season crops might increase seasonal irrigation requirements. In addition, with faster crop growth, farmers might tend to go for multiple cropping, also increasing water requirements.

Management practices, such as conservation tillage, drip and trickle irrigation, and irrigation scheduling are among the short-term possibilities for preserving soil moisture. Improving irrigation efficiency is a key component of combating potentially increased water requirements. It will involve reducing water losses from storage and distribution systems, proper maintenance of irrigation systems, optimising irrigation scheduling, and using water conservation techniques such as drip irrigation. Promoting such strategies by agrometeorology will be crucial since these practices, besides preserving soil moisture, will allow farmers to reduce the cost of production (Olesen and Bindi 2004).

Long-term changes include the change of land use to adapt to the new climate in order to stabilise production and to avoid strong inter-annual variability in yields.

This could be achieved through the substitution of existing crops with crops with a lower productivity but more stable yields (e.g., wheat replaced by pasture). For areas with increased water stress, it has further been recommended to use less water consuming and more heat resistant crops. Other measures include the change in farming systems since many farms are specialised in arable farming and, therefore, are tightly linked to local soil and climate conditions.

Changing or improving harvest insurance mechanisms to protect farmers from the economic impacts of flood or drought damage is also a necessary measure in future (Bindi and Howden 2004).

A need for further research exists with respect to spatial resolution in vulnerability mapping, technological and management-based adaptation measures, and the breeding of more drought-or heat-resistant crops.

8.3.5
European agrometeorological research needs

Agrometeorology has an important role in assisting national adaptation efforts in agriculture. Research needs can be identified in the following areas: agroclimate modelling, modelling of changes in soil- plant- water status and water resources (regional and local data to be merged with soil-plant-hydrological models and to improve the accuracy of the models). With respect to the observation of climate change trends there is a necessity of maintaining observation networks and use of improved methodologies, including remote sensing techniques in agrometeorological monitoring.

The need for research on the vulnerability of European agriculture to climate change impacts is felt by many of the European States. Research is needed also about several specific issues, for instance water related climate change impacts on individual agriculture sectors, the quantification of impacts, the relationship between climate change impacts and land use, research into the long term use of recycled water in agriculture, and desertification.

European agrometeorology can also help to develop climate change mitigation and adaptation measures in agriculture and assess their effectiveness and efficiency (European Commission, 2000). For instance, research should help design tools that demonstrate the economic benefit of adaptation at the local scale and develop indicators for successful adaptation measures. Appropriate communication strategies are also necessary to ensure that farmers and farm advisory services are sufficiently informed about impacts and adaptation strategies to take the necessary actions.

In spite of intensive research in the field of climate change impacts on agriculture there are still many knowledge gaps and research needs with regard to impacts, vulnerability and adaptation options. There is a need for regional agrometeorological studies because impacts and adaptations vary strongly within Europe. The following key questions remain:
- How will increasing CO_2 concentration affect the overall water use and water use efficiency of various crops under European growing conditions?

- How will a change in climatic mean and variability affect the water use, crop yields and applicability of various adaptation options?
- What are the possibilities for optimising water use by shifting cropping seasons, changing crops and adopting water conserving practices?

8.4. Conclusions

European farmers at the moment are living in times of change. For farmers change has always been part of life, but in the next couple of years European farmers will face at least two major changes of a very fundamental character. First of all there is the reform of the Common Agricultural Policy and the second is the need for climate change mitigation and adaptation practices. Unfortunately many EU policies, such as the common agricultural policy (CAP), do not yet include strategies or policies to explicitly address the current and future impacts of climate change.

What seem to be other drivers of European agriculture in the next decades? The first to mention is the growing demand for safe and quality food together with growing awareness for nutrition issues and healthy food. Due to climate change mitigation demand for renewable energy is increasing. We can expect also increased demand for biodegradable paper, fibres, polymers, lubricants, surfactants and solvents. Increased demand for bio-pharmaceuticals can be expected, as well. There is also challenge in what type of farming is needed in Europe in future. To produce domestically some of the food and raw materials that Europe needs, there is a much greater control over how these are produced. It is also needed that farming does its part in caring for European landscapes and even for the global environment. Europe should neither aim blindly for productivity at all costs on the one hand, nor treat farmers only as glorified park-keepers on the other hand. Certainly practical support should be given to both aspects of farming: the private goods for the market, and the public goods which Europeans also want.

Regarding climate change mitigation EU forestry strategy remains the maintenance and development of existing carbon stores and carbon sinks, the expansion of forest area where appropriate, the replacement of fossil fuels with fuel-wood from sustainably managed forests and the replacement of high-energy products (e.g. steel, aluminium and concrete) with industrial wood products (low-energy renewable raw material). Agriculture has also an important role, on the one hand renewable raw materials should be produced on farms (i.e. biomass, bio-fuel) but also animal manure can be used as a substitute for high energy fertilisers. European agriculture is already able to produce biofuels at a large scale and the use of biofuels must be considered as a strategy issue in the future policy concerning both climate change and energy.

Regarding climate change adaptation it is mostly needed in particularly vulnerable regions in Europe. Those where there is a large reliance on traditional farming systems and production of quality foods. Where such farming and production systems depend on favourable climatic conditions, climate change may cause large disruptions in rural society (Parry, 2000). To prevent or limit severe damage to the agriculture, society and economies, and to ensure sustainable development even under changing climate conditions, agriculture adaptation strategies are required at European, national, regional and local level. Adaptation needs the participation of all stakeholders who are involved in agricultural policy, business or service that is or will be affected by climate change. Agrometeorological community should help in this process to deal with the misconception that adaptation strategies and subsequent actions are always expensive to implement and that non-action is a cheaper alternative (Stern, 2006).

Climate change is likely to exaggerate the water resource differences between northern and southern Europe. As a consequence, the already existing pressures on water resources and their management in Europe are likely to increase over the next decades. This situation also calls for involvement of agrometeoroloy in long-term planning and pro-active management in order to ensure a sustainable use of water resources across Europe.

References

Audsley E, Pearn KR, Simota C, Cojocaru G, Koutsidou E, Rounsevell MDA, Trnka M, Alexandrov V (2006) What can scenario modelling tell us about future European scale agricultural land use, and what not? Environ Sci Policy 9:148–162.

Bindi M, Moriondo M (2005) Impact of a 2°C global temperature rise on the Mediterranean region: Agriculture analysis assessment. In: Giannakopoulos C, Bindi M, Moriondo M, Tin T. (Eds). Climate change impacts in the Mediterranean resulting from a 2°C global temperature rise, WWF, pp 54–66.

Bindi M, Howden M (2004) Challenges and opportunities for cropping systems in a changing climate. In: "New directions for a diverse planet". Proceedings of the 4th International Crop Science Congress, 26 Sept – 1 October 2004, Brisbane, Australia.

Cantelaube P, Terres J (2005) Seasonal weather forecasts for crop yield modelling in Europe. Tellus A, 57:476–487

COPA COGECA (2003) Factsheets: Assessment of the impact of the heat wave and drought of the summer 2003 on agriculture and forestry. http://www.copa-cogeca.be/en/dossiers.asp

DEMETER Final Report, November 2003. http://www.ecmwf.int/research/demeter/news/info/report_final.pdf

Doblas-Reyes FJ, Hagedorn R, Palmer TN (2006) Developments in dynamical seasonal forecasting relevant to agricultural management. Climate Res 33:19–26

European Commission (2000) Mitigation potential of Greenhouse Gases in the Agricultural Sector, European Climate Change Programme (COM(2000)88), Working Group 7 – Agriculture, Final Report.

European Commission (2004) Impact assessment of rural development programmes in view of post 2006 rural development policy, Final Report, November 2004.

European Commission (2005a) Communication from the Commission to the Council on risk and crisis management in agriculture. COM (2005) 74 final, Brussels, 9.3.2005.

European Commission (2005b) Communication from the Commission: Biomass action plan. COM(2005)628 final, Brussels, 7.12.2005.

EEA (2004a) Impacts of Europe's changing climate: an indicator-based assessment, European Environmental Agency Report No 2/2004, 107 pp.

EEA (2004b) EEA Signals 2004. A European Environmental Agency update on selected issues. 31 pp.

EEA (2005a) European Environmental Outlook. European Environmental Agency Report No. 4/2005, 85 pp.

EEA (2005b) Vulnerability and adaptation to climate change in Europe. European Environmental Agency Technical report No 7/2005, 79 pp.

Eisenreich SJ (2005) Climate Change and the European Water Dimension, European Commission- Joint Research Centre, Ispra, Italy. 253 pp.

FAOSTAT (2005) http://faostat.fao.org/faostat. Last accessed 01.01.2006.

Giorgi F, Bi XQ, Pal J (2004) Mean, interannual variability and trends in a regional climate change experiment over Europe. II: climate change scenarios (2071–2100). Clim Dynam 23(7–8): 839–858.

Goodess C (2005) STARDEX – Downscaling climate extremes, UEA, Norwich.

IPCC (2001) Climate Change 2001, Impacts, Adaptation and Vulnerability, Section, Hydrology and Water Resources, Report of Working Group II of the Intergovernmental Panel on Climate Change.

IPCC (2007) Climate Change 2007. The Physical Science Basis – Summary for Policymakers. Contribution of Working Group I to the Fourth Assessment Report of the Intergovernmental Panel on Climate Change. IPCC Secretariat.

Luterbacher J, Dietrich D, Xoplaki E, Grosjean M, Wanner H (2004) European seasonal and annual temperature variability. Trends and extremes since 1500. Science 303: 1499–1503.

Maracchi G, Sirotenko O, Bindi M (2004) Impacts of present and future climate variability on agriculture and forestry in the temperate regions: Europe. Climatic Change 70:117–135.

Metzger MJ, Leemans R, Schröter D, Cramer W, ATEAM consortium (2004) The ATEAM Vulnerability Mapping Tool. Quantitative Approaches in System Analysis No. 27. Wageningen, C.T. de Witt Graduate School for Production Ecology and Resource Conservation, Wageningen, CD ROM.

Olesen JE, Bindi M (2004) Agricultural impacts and adaptations to climate change in Europe. Farm Policy J 1: 36–46.

Olesen JE, Carter TR, Díaz-Ambrona CH, Fronzek S, Heidmann T, Hickler T, Holt T, Mínguez MI, Morales P, Palutikov J, Quemada M, Ruiz-Ramos M, Rubæk G, Sau F, Smith B, Sykes M (2006) Uncertainties in projected impacts of climate change on European agriculture and terrestrial ecosystems based on scenarios from regional climate models. Climatic Change (in press).

Palmer TN, Alessandri A, Andersen U, Cantelaube P, Davey M, Delecluse P, Deque M, Diez E, Doblas-Reyes FJ, Feddersen H, Graham R, Gualdi S, Gueremy J-F, Hagedorn R, Hoshen M, Keenlyside N, Latif M, Lazar A, Maisonnave E, Marletto V, Morse AP, Orfila B, Rogel P, Terres J-M, Thomson MC (2004) Development of a European Multi-Model Ensemble System for Seasonal to Inter-Annual Prediction (DEMETER)' B Am Meteorol Soc 85: 853–872.

Parry M (Ed.) (2000) Assessment of Potential Effects and Adaptations for Climate Change in Europe: The Europe ACACIA Project. Jackon Enviornment Institute, University of East Anglia, Norwich, United Kingdom, 320 pp.

PRUDENCE (2005) Prediction of Regional scenarios and Uncertainties for Defining EuropeaN Climate change risks and Effects. Final Report (http://prudence.dmi.dk)

Räisänen J, Hansson U, Ullerstig A, Doscher R, Graham LP, Jones C, Meier HEM, Samuelsson P, Willen U (2004) European climate in the late twenty-first century: regional simulations with two driving global models and two forcing scenarios. Clim Dynam 22: 13–31.

Richter G, Semenov M (2005) Re-Assessing Drought Risks for UK Crops using UKCIP02 Climate Change Scenarios. Final report of DEFRA Project CC0368.

Schröter D, Cramer W, Leemans R, Prentice IC, Araujo MB, Arnell NW, Bondeau A, Bugmann H, Carter TR, Gracia CA, De La Vega-Leinert AC, Erhard M, Ewert F, Glendining M, House JI, Kankaanpää S, Klein RJT, Lavorel S, Lindner M, Metzger MJ, Meyer J, Mitchell TD, Reginster I, Rounsevell M, Sabate S, Sitch S, Smith B, Smith J, Smith P, Sykes MT, Thonicke K, Thuiller W, Tuck G, Zaehle S, Zierl B (2005) Ecosystem service supply and vulnerability to global change in Europe. Science 310: 1333–1337.

Stern N (2006) The Economics of Climate Change. The Stern Review. Cambridge University Press. Available for download at http://www.hm-treasury.gov.uk/index.cfm.

CHAPTER 9

Methods of Evaluating Agrometeorological Risks and Uncertainties for Estimating Global Agricultural Supply and Demand

Keith Menzie

9.1 Introduction

The global food and fiber system – from the producer to the final consumer – is subject to a wide range of risks and uncertainties. Extensive research has been published in the fields of Economics and Agricultural Economics on every aspect of risk and uncertainty as it relates to agriculture, ranging from theoretical to highly-applied. Within the realm of applied research, much of the effort has been to provide producers with a framework for managing risk

9.2 Sources of Risk

Probably the most obvious effect of risk is the impact of seasonal weather variation directly on producers. However, the collective impact of weather-related production risk and uncertainty on individual producers and their management decisions affects every link in the marketing chain. Risk in agriculture can be broadly defined into several categories (USDA 2006). These include:

Yield risk. This type of risk is probably the most commonly thought of risk in agriculture, and it reflects directly the impact of weather on farm operations. Temperature and moisture variation are the typical causes of yield risk, with irrigation being one of the only significant approaches to minimizing the impact of hot or dry conditions.

Production risk. Production risk entails all of the factors that affect yield risk, plus the additional impact that adverse weather may have on producers' ability to plant. Two crop specific examples of planting risk include corn/soybeans in the United States and soybeans in India.

For the U.S., cool and wet spring weather may not allow for timely planting of corn. After a certain date, corn planting becomes infeasible due to the number of days to maturity and, as the spring days pass, the increased likelihood of the crop being affected by freezes in the fall. Initially, producers can consider shorter-season varieties of corn, but eventually, acreage that had been intended for corn may be shifted to soybeans. Conversely, good weather throughout the corn planting season often leads to increased corn plantings at the expense of soybean plantings. Potential corn production is reduced or increased and potential soybean production is increased or reduced depending on the weather.

Soybean planting in India can be significantly affected by the timing and consistency of the monsoon. If the monsoon begins early and rainfall remains relatively consistent, soybean planting often continues beyond initial expectations, with a resulting increase in total area planted to soybeans and a likely increase in expected production. Conversely, a late start to the monsoon with a more erratic rainfall pattern both in terms of timing and distribution may lead to reduced soybean plantings compared with initial expectations, and reduced expected production.

Price risk. Commodity prices are critically important in agriculture, not only to producers, but also to buyers up and down the supply chain. Agricultural commodity prices are highly volatile, subject to sharp increases and decreases over relatively short periods of time and over a wide geographical range depending on both local and global supply and demand conditions. Agricultural prices are determined both locally and in global markets, so agrometeorological conditions, adverse or favorable, in one part of the world can lead to uncertainty and price risk in distant markets.

Agricultural commodities often can be substituted for each other, so conditions favorable or adverse to one crop in one part of the world can affect prices in other commodity markets around the world, which in turn can affect production and consumption decisions in those markets. For example, corn, sorghum, barley, and oats can be substituted for each other in animal feed, and more than half a dozen oilseeds can all be used to produce vegetable oils. Price risk for one commodity in one local market can be increased or moderated depending on local and global market conditions for a variety of commodities over a geographical range of markets.

Income risk. Income is defined as the product of price and production minus the cost of producing. Income risk is caused by the three types of risk previously described, plus additional factors including variation in the price and availability of the inputs required for production. For example, if a particular variety of wheat planting seed is grown in Argentina, and producers in southern Brazil prefer that type of seed, then availability and price of the planting seed will have an impact on the incomes of Brazil's wheat producers. If adverse climate conditions in Argentina reduce production of the planting seed and raise its price, Brazil's wheat producers will have to pay more for the seed, thereby reducing income, or switch to another cheaper variety that may adversely reflect on yield, thereby reducing income. The global interconnectedness of agricultural markets increases the influence of risk beyond local markets.

While adverse conditions in one area or for one crop may adversely impact prices and incomes in another area, substitution by product and geography can also serve to limit the degree of income risk. For example, 25 years ago, South America was not a global source of soy protein. Livestock producers in Asian markets, for example, were dependent on U.S. production of soybeans to meet protein needs. With soybean production now more geographically dispersed, Asian livestock producers' risk is more limited since a short crop and higher prices in the U.S. may partly be offset by better supplies in South America.

Financial risk. Financial risk for agriculture includes the cost and availability of credit, which are impacted by changes in interest rates. Market forces beyond agricultural markets often influence the level of financial risk.

Institutional risk. Institutional risk reflects the legal and regulatory environment created by Federal, State, or local governments. Environmental laws and regulations, income and business tax laws, and Federal agricultural policy affecting production, trade, and consumption are sources of institutional risk. For example, in the U.S., farm policy in the 1930s defined a set of rules affecting the production of peanuts. The law specified the quantity of peanut production on a particular farm that would be eligible for a government-supported price that exceeded the world market price. Then in 2002, the law was changed to permit anyone to produce peanuts with the support price significantly reduced for all producers. The risk was that earlier benefits provided by Federal law were reduced for those who originally held quota rights, which had an effect on expected income for those producers.

For estimating global supply and demand for agricultural commodities, the key risk types are the first four types defined above: yield, production, price, and income risk.

9.3
Risk, Uncertainty, and the Agricultural Marketing System

While yield risk and its impact on the producer is the first and most recognized source of uncertainty, myriad related impacts affect economic conditions throughout the marketing chain and must be considered when estimating global supply and demand. Directly or indirectly, each economic agent at every level of the global agricultural marketing system is affected by one or several of the risk types. Price and income effects stemming from a particular set of meteorological conditions in a given crop cycle even influence cropping patterns in subsequent crop cycles, and these influences must be evaluated as global supply and demand estimates are made.

In order to highlight the scope of the potential impacts of meteorological events on global supply and demand, the following section summarizes a few of the implications of a hypothetical drought affecting the soybean crop in the western U.S. Corn Belt.

U.S. producers. The most direct impact of a drought is on the producers in the affected area. Lower yield reduces production for each producer, potentially putting each producer's income at risk. If the drought is widespread, national production levels may be reduced, resulting both in higher prices and in an atypical seasonal pattern of prices. Higher prices and atypical seasonal price patterns affect producers' decisions about quantity and timing of sales to maximize income.

Producers beyond the region immediately affected by drought are also affected. Decisions these producers make ultimately affect soybean supply and demand, first in local markets, but ultimately in global markets. Because drought affects the level and seasonal pattern of prices, but not production levels for producers beyond the geographical reach of the drought, these producers have an opportunity to maximize income by taking advantage of higher prices caused by the drought. However, because the drought affects the seasonal price patterns, causing a sharp departure from normal, these producers must also consider the timing of sales to maximize income.

Elevators/storage enterprises. The grain storage business can be affected by drought in different ways depending on the location of the storage. For storage facilities in the drought region, the main issue is whether there will be enough crop volume to efficiently utilize available storage. With too many facilities available for the limited regional supply of grain, decisions will have to be made regarding the best price to offer soybean producers as facilities compete for limited supplies. For storage facilities located in parts of the country not directly affected by the drought, the main issue becomes one of timing. While there is enough volume to utilize storage capacity efficiently, the impact of the drought is the level and pattern of cash and futures prices and how these impacts will affect the optimal timing of sales. A forecast of the price pattern over the remaining months of the marketing season is essential to make the best possible decisions about how long to store grain.

Transportation. The grain transportation system, like the storage system, depends on volumes of grain to maximize income. For the rail system, one of the main issues is positioning of rail cars during the harvest season and in the months that follow. When drought strikes a region, demand for rail cars may diminish, and may increase in parts of the country not directly impacted. Increased demand in the non-affected producing areas may come from domestic soybean crushers and exporters as these enterprises attempt to meet business needs. Water transportation services are affected in the same manner as rail transportation. Barges and ocean-going vessels may be less in demand due to reduced soybean supplies and increased prices that may reduce buyer demand. Profit and income is at increased risk for businesses that provide water transportation for grain.

Crushers. Soybean crushers account for almost 60 percent of the demand for soybeans in the U.S. market. Rising prices and diminishing volume available to meet crushing demand will affect profitability and income of crushers. As with other economic enterprises, crushers face increased competition for soybeans. Higher prices for soybean meal affect both soybean meal export demand and demand for soybean meal protein by livestock feeders. As soybean crush levels are reduced, soybean oil supply diminishes, vegetable oil prices increase, and demand for vegetable oil is affected both in the U.S. and abroad.

Livestock feeders. Higher soybean prices and reduced crush results in reduced soybean meal supplies and higher soybean meal prices. Since soybean meal is the main source of protein in livestock diets, higher soybean meal prices directly affect income levels of livestock production enterprises. Depending on the degree of impact on soybean supply and demand and on the price and supply of soybean meal, livestock producers may alter the mix of soybean meal and grain in rations, which may affect livestock production rates. At the extreme, livestock producers may liquidate some of their productive capacity (animal numbers), resulting in at first lower, then higher meat prices.

Foreign soybean producers. Global soybean supplies are distributed almost equally between the Northern and Southern Hemispheres, so a drought in the Northern Hemisphere likely leads to global supply adjustments 6 months later. South America accounts for more than 50 percent of exportable supplies of soybeans, soybean meal, and soybean oil. When global supplies are disrupted due to a drought in the U.S. Midwest, higher prices occur in global markets. Higher prices signal producers in South America to expand the area devoted to soybeans in the fall planting season subsequent to the drought. Higher production in South America leads to increased exports of soybeans and products from the Southern Hemisphere, and a shift in the demand for storage and transportation services from the region.

Other oilseed producers. In addition to affecting the global supply of soybeans, higher soybean prices affect producers' decisions regarding planting of substitute oilseed crops. Markets for rapeseed, sunflowerseed, and peanuts are all linked through protein and vegetable oil prices. All of these crops' primary products (protein meal and vegetable oil) substitute for one another, so higher prices for soybeans will lead to expanded area planted to other oilseeds, resulting in increased supplies and consumption. These economic reactions also help to offset the initial soybean price reaction stemming from the drought.

U.S. soybean producers. Completing the annual cycle, the impact of drought in one year in a part of the U.S. likely will result in increased area planted to soybeans and increased supplies in the subsequent marketing year.

9.4
Information - the Key to Efficient Market Function

Agrometeorological risk and uncertainty permeate the entire marketing system with far-reaching consequences for market participants. In order to optimize business decisions relative to these risks and uncertainties for every economic agent within the global agricultural production and distribution system, accurate, timely, consistent, and widely available information is essential. This information requirement can be met in part through periodic review and estimation of global supply and demand for agricultural commodities. The quality and usefulness of such estimation is contingent upon many factors, the single most important of which is accurate and timely assessment of crop production.

It was the recognition of the need for such information about crop production that led to the creation of the World Agricultural Outlook Board (WAOB) within the United States Department of Agriculture. The WAOB was established to provide a reliable source of timely, consistently reported, and widely available market information. The WAOB, with a staff of about 30 economists and meteorologists serves as USDA's focal point for agricultural market intelligence. Under WAOB direction, interagency committees of experts develop official forecasts of supply, demand, and prices for major agricultural commodities. Parallel to its commodity forecasting roll, WAOB's Joint Agricultural Weather Facility (JAWF) coordinates weather, climate, and remote sensing work among USDA agencies. In addition, and in support of its commodity supply and demand forecasting function, JAWF has operational responsibility for monitoring and analyzing the impact of global weather on agriculture. This activity is conducted jointly by WAOB/JAWF and the National Weather Service of the Department of Commerce.

A primary focus of public interest is the monthly publication of a report entitled, *World Agricultural Supply and Demand Estimates (WASDE)*. Its forecasts cover major commodities for the United States and the world and are considered authoritative because they are backed by USDA's unparalleled access to information.

A unique feature of the USDA's system for estimating monthly global supply and demand is the information provided by JAWF. On a daily basis, meteorologists track global weather developments and interpret the impact on crops in the world's major farming regions. This information is presented to USDA analysts and frequently forms the basis for monthly adjustments to global supply and demand estimates. By combining sophisticated technology and a scientific understanding of crop phenology, JAWF is able to provide early warning of emerging weather problems and potential crop shortfalls even in remote regions of the world. This type of analysis, in conjunction with economic analysis of agricultural markets, helps permit USDA to meet the critical market need for timely and consistent information.

In the next section, a brief overview is presented of several of the meteorological tools and methods of crop assessment typically applied in developing the monthly world agricultural supply and demand estimates at USDA. Examples are taken from analyses presented by JAWF to the Interagency Committee economists as part of the monthly process of estimating global commodity supply and demand.

9.5
Global Crop Assessment Methods and Risk Reduction – Tools and Analysis

Summary and analysis of temperature and precipitation data during crop cycles constitute the primary and most significant meteorological input into crop estimation each month. These data are presented in several ways to evaluate their impact on crops as they progress from planting to harvest. Data typically are presented for the year in question and compared against long-term norms and analog years. Selection of analog years can be very helpful to bridge the often qualitative link between observed data and predicted crop yields.

Figures 9.1 and 9.2 illustrate a review of temperatures during the corn growing season in the Po Valley of northern Italy. Figure 9.1 summarizes the entire season-to-date, while Figure 9.2 focuses in on the temperature extremes during July with an overlay of the crop development stage to provide insight into the timing of the temperature extreme. From these two figures, and the focus on temperature extreme during silking, analysts can draw conclusions about the potential impact of high temperatures on yield.

Several approaches are used to evaluate the timing and quantity of rainfall. The amount of rainfall is typically analyzed either as cumulative precipitation or as monthly total rainfall. Figure 9.3 illustrates simple cumulative rainfall for the season for Rio Grande do Sul in southern Brazil:

Chapter 9: Methods of Evaluating Agrometeorological Risks   131

Northern Italy - Po Valley
Extreme Max & Min Temperatures

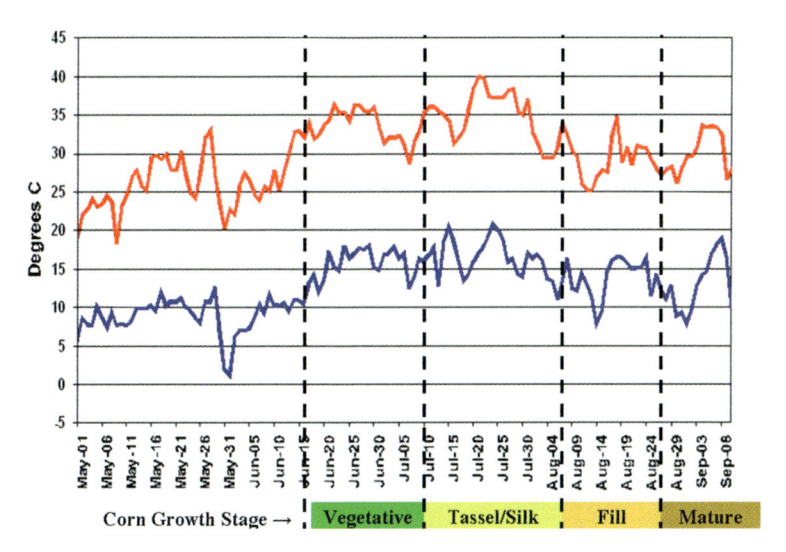

Fig. 9.1 Temperatures during the corn growing season in the Po Valley, Italy.

Northern Italy - Po Valley
Extreme Max Temperatures (30°C Threshold)

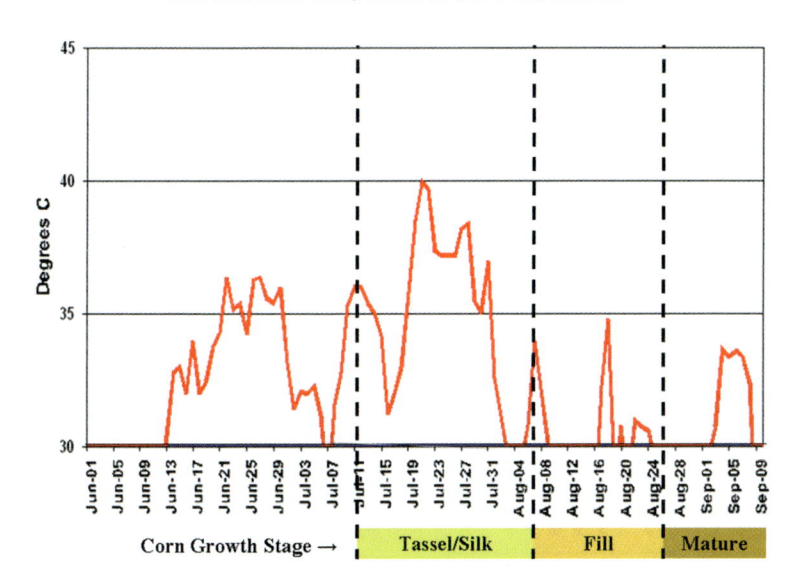

Fig. 9.2 Temperature extremes in the Po Valley, Italy.

Although Figure 9.3 indicated that by March 15 rainfall was slightly above normal, a closer look tells a much different story. Over 200 mm of rain fell in the later half of December, shifting the cumulative curve up. Because soils in Rio Grande do Sul are sandy and unable to hold moisture well, hence the timing of rainfall is critical. Two extended dry periods in early 2004 proved disastrous to yields. Additional meteorological analysis hinted at this outcome. Figures 9.4, 9.5, and 9.6 show monthly rainfall totals (low in 2004), days between rainfall (above normal in 2004), and a soil moisture analysis showing a precipitous decline beginning in January. All three indicators provided clear warning of a poor crop. (Clearly, the 2005 crop was also affected by drought).

Another recently employed analytical method, the CPC MORPHing technique (CMORPH), makes use of satellite data to provide an estimate of rainfall to enhance station data. CMORPH can be used to provide indications when station data are lacking or, in the case shown in the Figure 9.7, missing due to technical difficulties. It provides an additional source of information that can support or even replace other sources of information.

The importance of using multiple sources of weather information to assist meteorologists and economists at the WAOB was exemplified by a unique weather event in South Asia. During March 2006, an unseasonably strong mid-latitude storm system moved northeastward from the Arabian Sea into central and northern India, impacting India's maturing winter grains. Not only was the timing and location of the storm rare, but the storm intensity was also noteworthy. Satellite imagery showed a cloud pattern indicative of very heavy precipitation along with severe

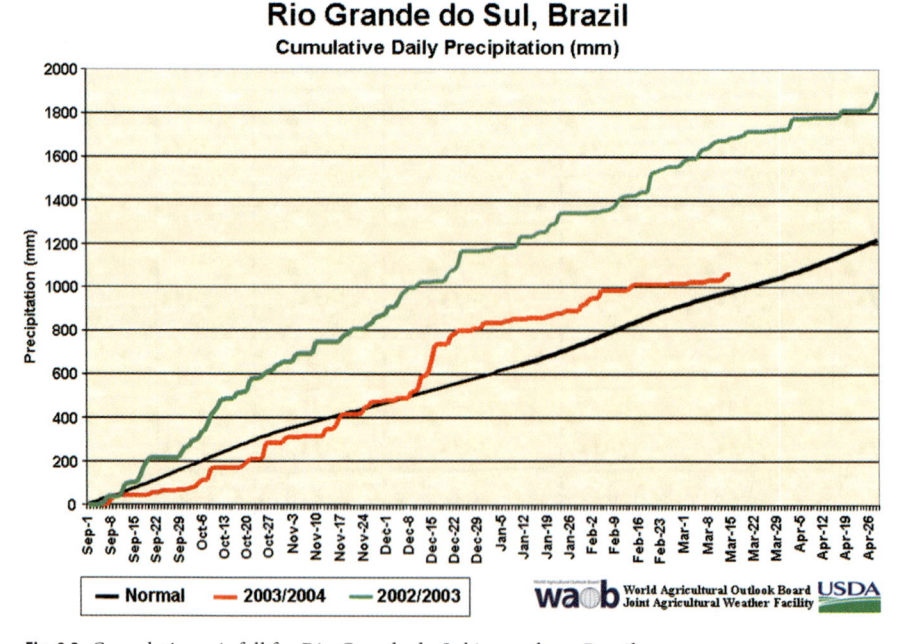

Fig. 9.3 Cumulative rainfall for Rio Grande do Sul in southern Brazil.

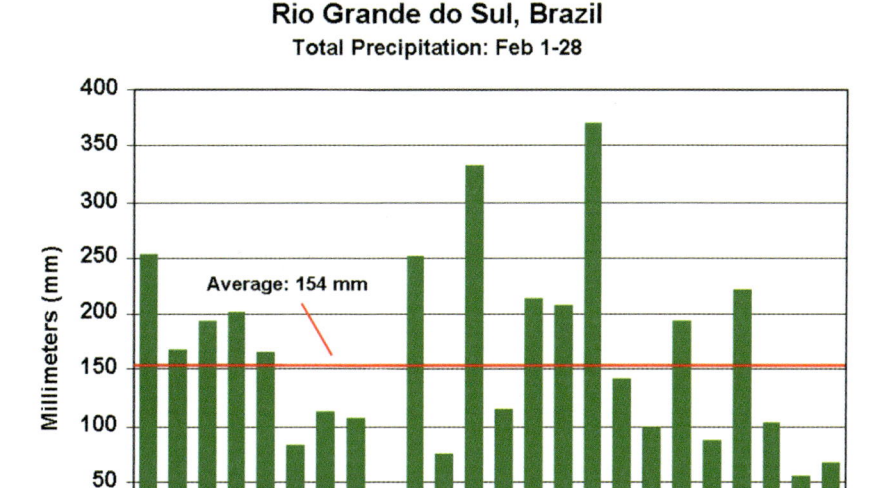

Chapter 9: Methods of Evaluating Agrometeorological Risks 133

Rio Grande do Sul, Brazil
Total Precipitation: Feb 1-28

Fig. 9.4 Total precipitation from February 1-28 in Rio Grande do Sul.

Rio Grande do Sul, Brazil
Average Days Between Rain: Feb 1-28

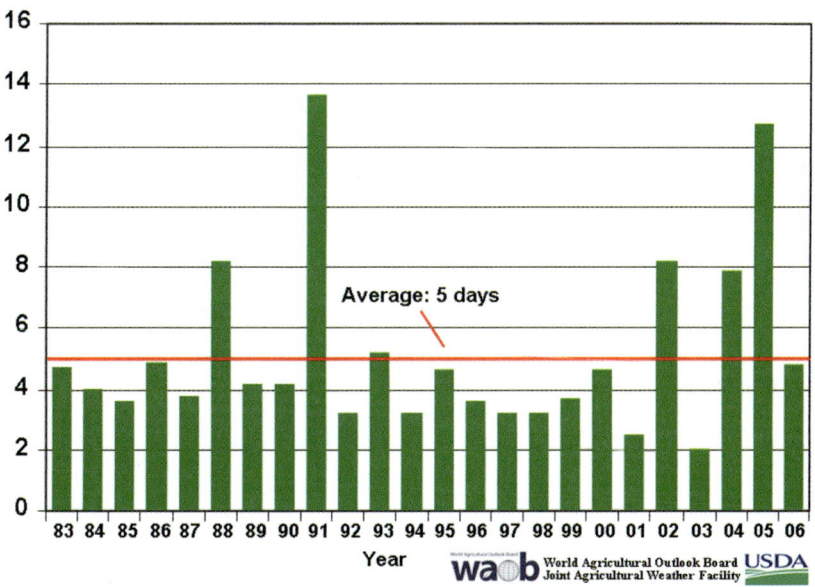

Fig. 9.5 Average days between rain from February 1-28 in Rio Grande do Sul.

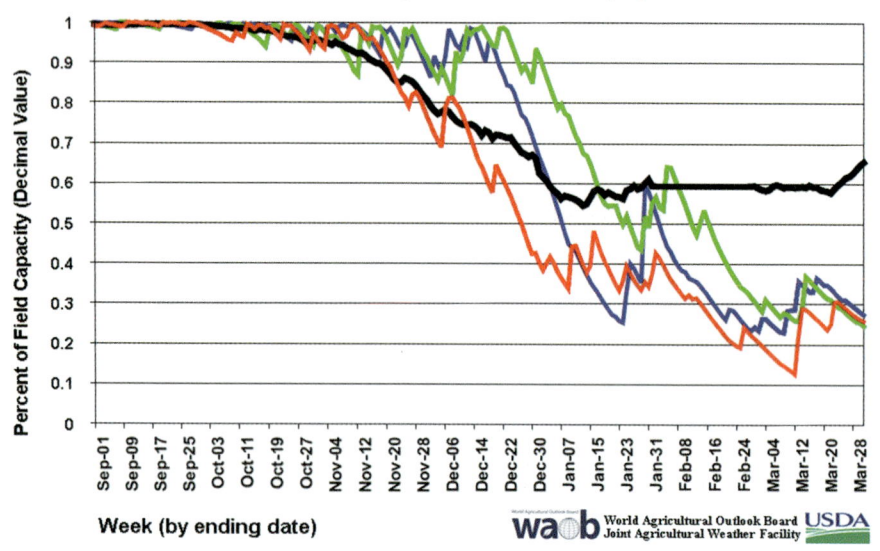

Fig. 9.6 Soil moisture in Rio Grande do Sul.

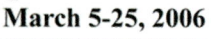

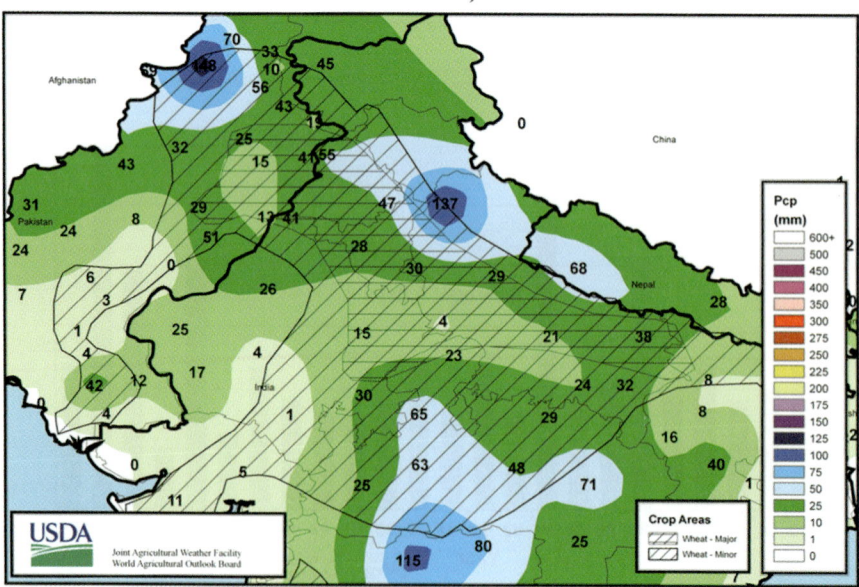

Fig. 9.7 Station data for March 5-25, 2006.

weather. However, precipitation totals for the event in primary wheat areas were very low, especially considering the storm's intensity and duration. A detailed assessment of WMO station data from portions of central and northern India indicated missing values, likely due to power outages or data transfer problems. However, as seen in Figure 9.8, satellite precipitation estimates (CMORPH) for the region indicated a large swath of heavy rainfall across much of central and northern India.

Despite its limitations, JAWF meteorologists felt compelled to use the CMORPH data to develop crop impact assessments; the untimely widespread heavy rain and severe weather was expected to cut into crop yields. This assessment was ultimately supported by field reports over the ensuing months. If the crop assessment had been based solely on station data, the negative impacts would not have been noted until much later.

These and other analytical tools are used each month of the crop cycle depending on the particular crop and country of interest. The next section demonstrates how a combination of these tools and analytical techniques are used to make crop estimates.

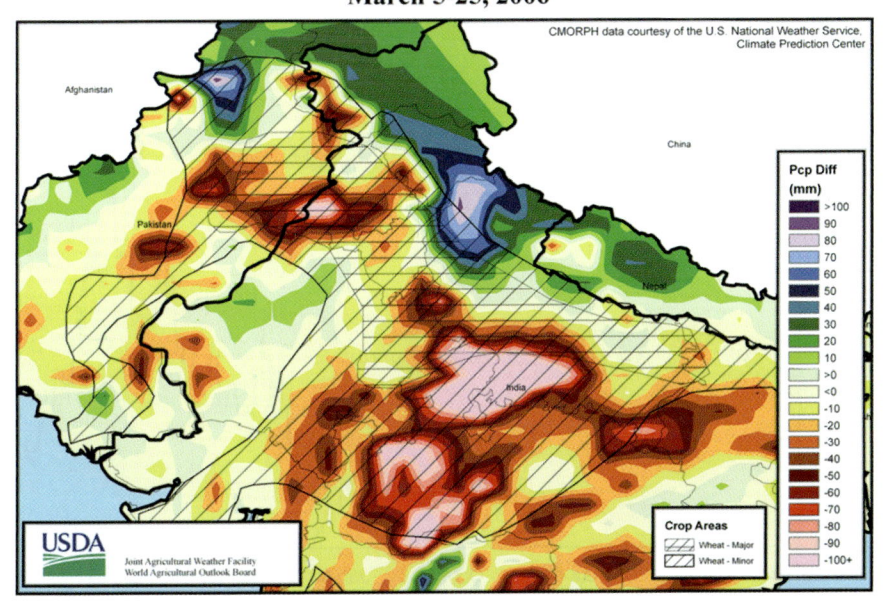

Fig. 9.8 Station data vs. CMORPH.

9.6
Global Crop Assessment Methods and Risk Reduction – the Case of Brazilian Soybeans

Global soybean production is valued at about $45 billion annually. About one-third of global soybean consumption is derived from imported soybeans. Brazil, the world's second largest soybean producer, accounts for 37 percent of global soybean exports, up from just 8 percent in 1990. Accurate and timely estimation of Brazil's soybean crop each year is critical for decision-making and planning throughout oilseeds, grains, and livestock markets around the world.

To highlight this point, Figure 9.9 shows the Chicago Board of Trade November futures price of soybeans tracked from April 2004 through September 2005. Prices are measured in U.S. dollars per bushel. The figure illustrates the potential for significant market impact due to weather events, and the critical need for timely information for buyers, sellers, and all other business ventures involved in oilseed markets around the world.

April 2004 through September 2004 is the period during which the Brazilian soybean producer makes planting decisions, with actual planting occurring mainly in October and November. This was generally a period of falling prices, indicated by a 15-percent decline in the November futures price. Despite declining prices, area expanded that year and there was market expectation of a record large Brazilian soybean crop. However, drought struck parts of Brazil during the main 2004/05 growing season. From February through mid-March 2005, CBOT prices rose almost 25 percent. Business enterprises throughout the oilseed sector stood to gain or lose significant income depending on the timing of decisions to buy or sell during this short period of rapidly escalating prices. Accurate global supply and demand estimates were needed by the market during this period, especially so with

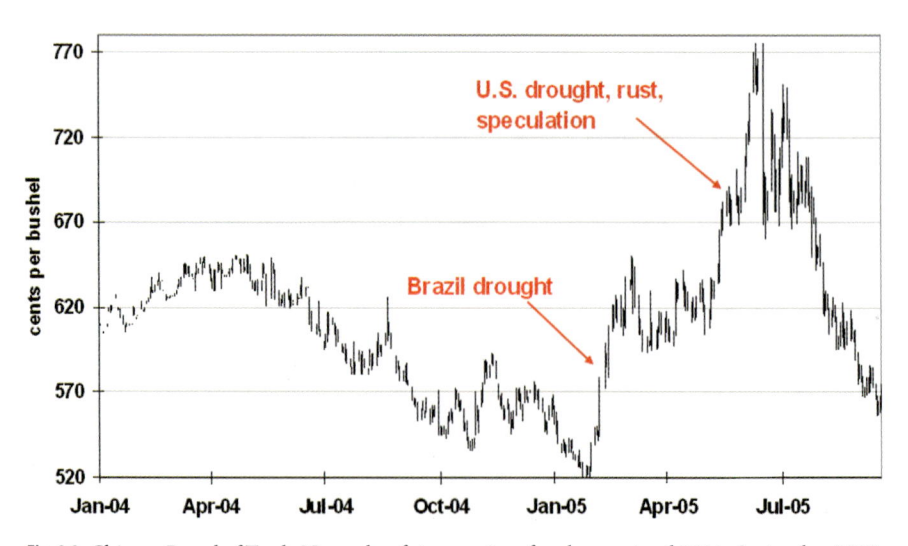

Fig. 9.9 Chicago Board of Trade November futures price of soybeans, April 2004 -September 2005.

hindsight knowledge of the fact that parts of the U.S. crop suffered dry conditions early in the season, pushing prices even higher.

Assessing the size of the Brazilian soybean crop is made especially difficult by the fact that there are many estimates at a point in time, including estimates from two different Brazilian government sources. During the critical month of March 2005, when prices were rapidly escalating, estimates ranged from 56 to 61 million tons, with Brazilian Government estimates of the crop ranging from 57 to 61 million tons. Timely clarification of the specific size of the crop was critical in the weeks approaching the U.S. planting season. Meteorological data and analysis provided critical insight into the unfolding situation in Brazil and enabled timely dissemination of information to the global marketplace.

In March 2006, yet another year in which drought was affecting the Brazilian soybean crop, the Brazilian Government released a yield estimate of 2.4 t ha^{-1} for the state of Mato Grosso do Sul, down from its previous estimate of 2.7 t ha^{-1}. USDA was at the time estimating the yield at 2.65 tons per hectare. Our issue was whether we should lower our production estimate for Brazil in light of the new Brazilian yield estimate.

A significant problem in evaluating this new estimate was the lack of meteorological data for the state. WMO stations and general soybean producing areas for Mato Grosso do Sul are depicted in Figure 9.10.

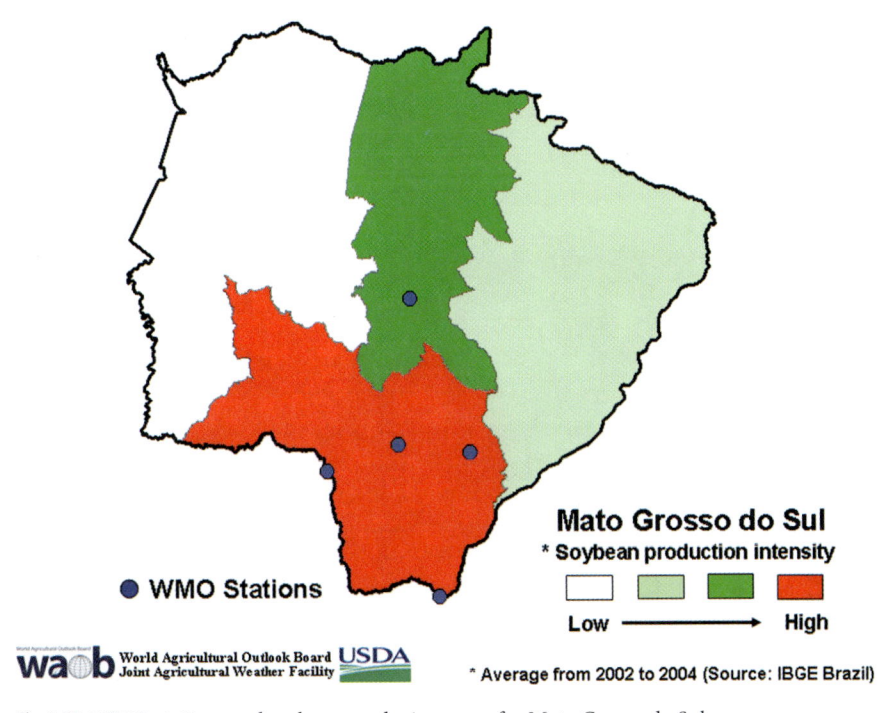

Fig. 9.10 WMO stations and soybean producing areas for Mato Grosso do Sul.

An accurate analysis would require a combination of all of the available tools.

Temperature and cumulative precipitation charts with comparisons to normal values and to 2002 (selected as an analog year) revealed nothing to suggest a significant yield reduction was appropriate at that stage of the crop cycle. The soil moisture profile depicted in Figure 9.11 shows a drying period from mid-January through early February, but recharge returned values to historic norms by early March. Again, no clear indication was presented to reduce yield. However, with the limited availability of WMO station data, more analysis was necessary to confirm the estimate. Weekly CMORPH data were analyzed, typified by Figure 9.11 for February 19-25:

With CMORPH data and analysis supporting conclusions from other weather indications, a case was developing to hold off on reducing the Brazilian crop due to yield deterioration in Mato Grosso do Sul.

Further analysis of the situation was conducted through application of satellite imagery crop masking techniques to more clearly identify soybean producing areas and comparative NDVI analysis of the region. These techniques are depicted in Figure 9.12.

With major soybean areas outlined in the south-central and northern regions, it is apparent that 2006 was faring better than 2002 in some cases, and in the areas with lower NDVI scores, soybean production was not highly concentrated.

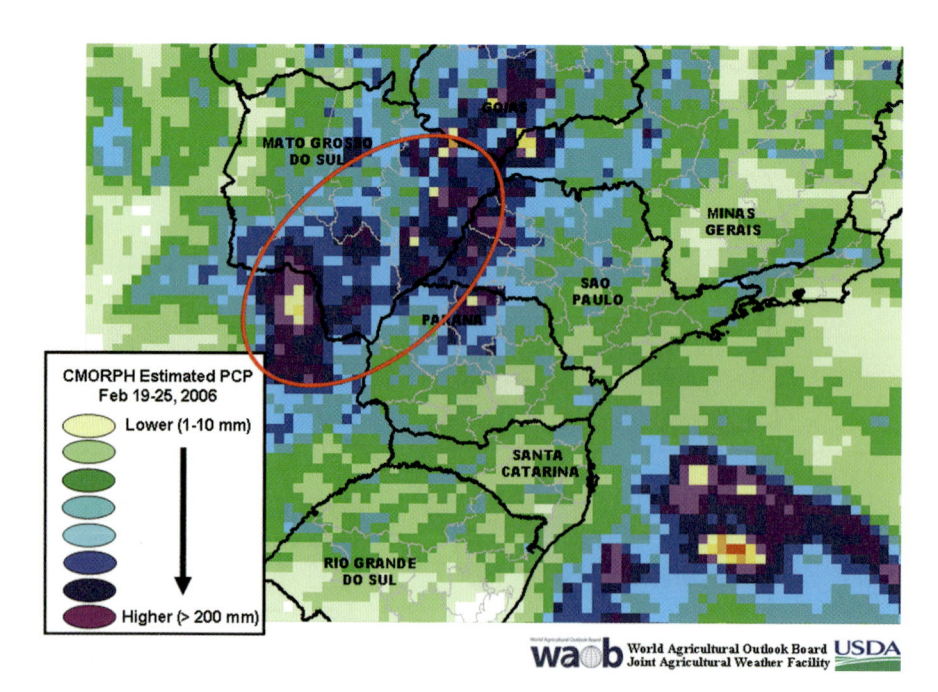

Fig. 9.11 CMORPH estimated PCP, February 19-25, 2006.

From the combination of analysis of data for Mato Grosso do Sul for early 2006, it was concluded to leave the yield unchanged at 2.65 t ha^{-1} for that month. However, the analysis continued throughout the crop cycle, leaving open the possibility that the crop estimate could be adjusted in later months.

9.7
Conclusions

Risk and uncertainty affects every aspect of the agricultural commodity marketing system – from producer to final consumer. Weather-related yield and price risk translate into income risk in agricultural markets around the world. Accurate, timely, consistent, objective, and widely available information including analysis of the impacts of weather on crop production is critical for economic enterprises to make optimal business decisions. This paper reviews methods used at USDA to assess impacts of weather-related risk and uncertainty on global crop production as a first step toward estimating global supply and demand for commodities. More timely and accurate estimates result when multiple analytical techniques are employed to evaluate the impact of seasonal weather conditions on crops.

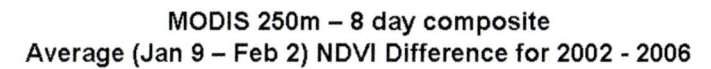

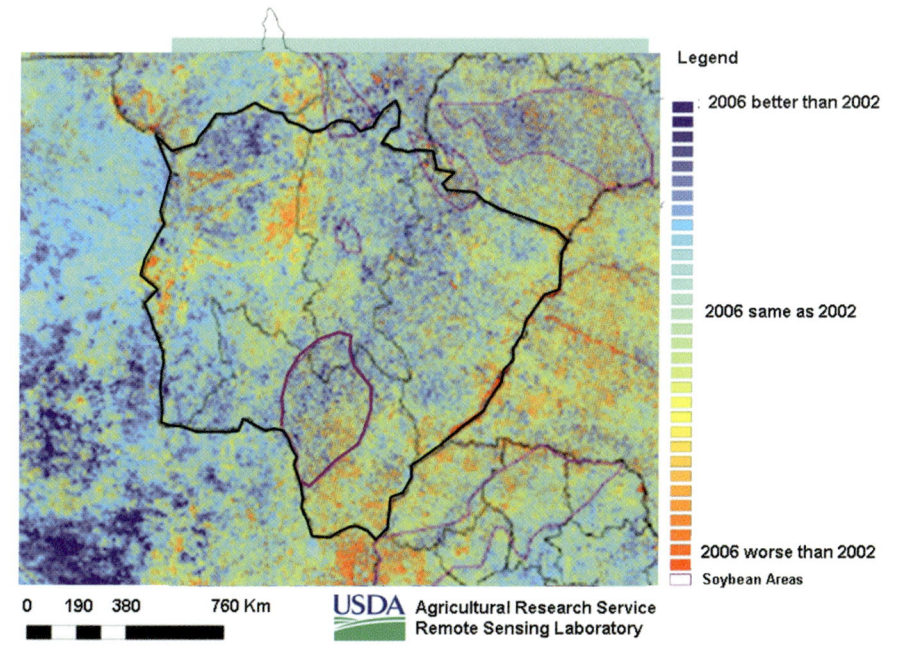

Fig. 9.12 NDVI difference for 2002 vs. 2006.

References

USDA (2006) Risk Management, USDA 2007 Farm Bill Theme Papers, May 2006. www.usda.gov/documents/Farmbill07riskmgmtrev.doc

CHAPTER 10

Weather and climate and optimization of farm technologies at different input levels

Josef Eitzinger, Angel Utset, Miroslav Trnka, Zdenek Zalud, Mikhail Nikolaev, Igor Uskov

10.1 Introduction

Weather and climatic conditions are the most important production factors for agriculture. Farmers within any agroecosystem therefore try to adapt to these conditions as much as possible (Adger et al. 2005; Smit and Yunlong 1996). Farm technologies play a major role in this adaptation process in both the short and the long term. Farm technologies are optimized for different purposes such as maximizing food production or profit. There is an urgent need, however, for such aims to be directed to permit sustainability of food production at the local level, which can be based only on stable agroecosystems (Fig.10.1). This has to be the basic strategy

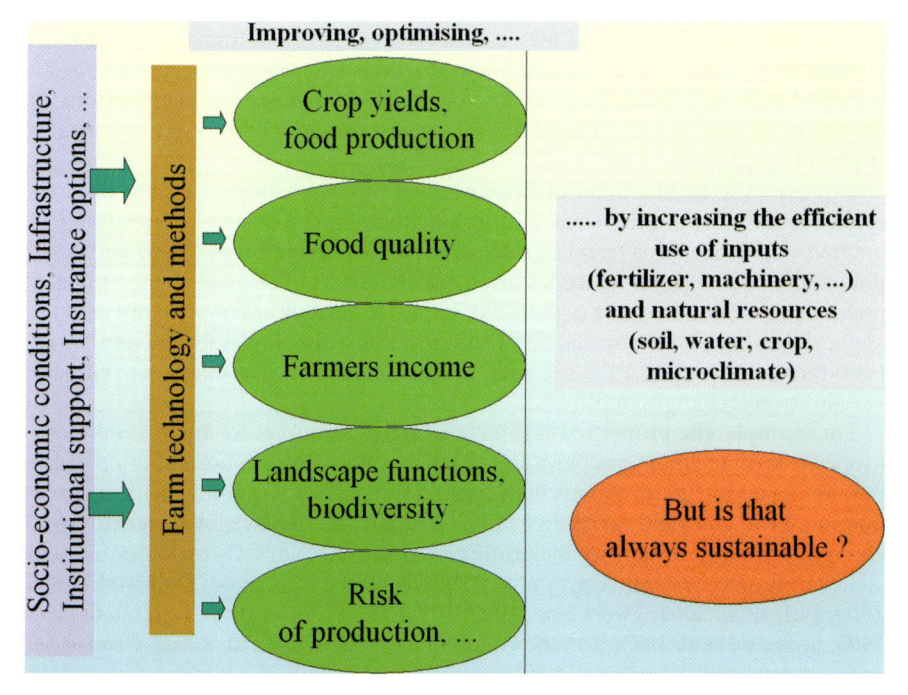

Fig. 10.1. The short and long term impact factors on farm management and its relation to resource management and sustainability of agricultural production.

for the long term as important resources for agricultural production such as water, land and soil resources are highly limited in our world. Moreover, these resources are also endangered in many regions by desertification and climate change.

New farm technologies and those that have been established for many generations – indigenous technologies – offer many opportunities to react or adapt to the given climatic and weather conditions. Because of climate variability and change, the optimization of farm technologies becomes even more important for the productivity of various agricultural production systems at different input levels (Sivakumar et al. 2005). Available farm technologies are often closely linked to specific management options, which will therefore be considered as well in the following analysis. These options for the various agricultural systems are always embedded within the given socioeconomic, policy and trading framework within and between countries and regions and these can vary widely. This framework is an important consideration when identifying measures to adapt to weather and climate conditions and has a strong influence on the adequacy of measures for adapting farm technologies (Chiotti and Johnston 1995). This background and impact are not considered in detail in our analysis but should be kept in mind when applying the general findings and examples to a region with specific agricultural systems and conditions.

Using available farm technologies to ensure sustainable production within given climatic and weather conditions often calls for the proper management of resources or conditions for a specific agricultural crop production such as water, soil (including nutrients), crops (including crop management) and microclimate (Iglesias et al., 1996; Karing et al., 1999; Rounsevell et al., 1999; Salinger et al., 2005). In all agroecosystems since farming began farmers have developed specific strategies, mainly the use of different farm technologies and related management options, to survive in the given environment, but for various reasons not always with sustainability in mind.

However, the development or improvement in farm technologies has been responsible for most of the increases in productivity and yields in agricultural production worldwide. This trend should continue (Rounsevell et al. 2005) and could potentially outrange, for example, any negative effects of climate change impacts on food production in many regions. For a specific agricultural system not only the applicability but also the availability of appropriate technologies for the local farmers is therefore crucial for the potential to optimize production or adapt to climatic variability and change conditions.

For example, the proper management of water resources by application of appropriate farm technologies plays and will play a major role in both developed and developing countries in regions with limited water resources for agricultural production. Yield and yield variability can be strongly affected by global warming and changing climatic variability including the direct effect of CO_2 on water use efficiency in agroecosystems (Curry et al. 1990; Downing et al. 2000; Dubrovsky at al. 2000; Erda et al. 2005; Ewert et al. 2002; Isik and Devadoss 2006; Kartschall et al. 1995; Semenov et al. 1993; Semenov and Porter 1995; Wolf et al. 2002). Crop water use and deficit in different climate scenarios and potential adaptation measures, however, depend on crops, soil and climatic conditions and have a mixed impact on crop yields (Easterling and Apps 2005; Izaurralde et al. 2003; Rosenzweig et al. 2004; Tao et al. 2003).

Studies of European agricultural systems conclude that there is strong evidence in climate change scenarios, especially for soils with low soil water storage capacity or no groundwater impact to the rooting zone, that irrigation or water-saving production techniques (e.g. by introducing mulching systems, adapting crop rotation), will remain important requirements in future climate conditions in Central European agricultural regions for crops to attain their yield potential (Eitzinger et al. 2003). Further they conclude that if the droughts frequency and duration increase further (Seneviratne et al. 2006; Pal et al. 2004) or soil and groundwater reserves decrease (e.g. by decreasing summer river flow from Alpine region) drought damage will become more common. Summer crops will be more vulnerable and dependent on soil water reserves, as the soil water or higher groundwater tables during the winter period cannot be utilized as much as by winter crops. Evapotranspiration losses during summer due to higher temperatures would increase significantly.

Negative yield effects for several crops and significant additional water use for irrigation (up to 60-90%) might be expected in the Mediterranean region (Marracchi et al., 2005; Tubiello et al., 2000) or regions with low soil water availability due to climate change. According to Olesen and Bindi (2002), reduced water availability in Mediterranean countries as a consequence of climate change and variability might be the most important climate risk for crop yields in Europe, especially if extreme weather events increase. A European study (EEA 2005) draws a similar conclusion, remarking on the need for future studies on the effectiveness of irrigated agriculture in Southern Europe.

The results of climate change impact and adaptation studies in agriculture give us a good insight into the effects on agricultural production of the optimization of farm technologies and management. They suggest several potential measures for adaptation of farm technology and management to changing climatic conditions. In many studies focusing on climate change impacts on crop production in temperate agricultural regions, only simple measures such as possible changes in sowing dates (earlier sowing dates) and cultivar selection (e.g. selecting slower maturing varieties) were investigated (Abraha and Savage 2006; Alexandrov et al. 2002; Reilly and Schimmelpfenning 1999; Parry 2000; Sivakumar et al. 2005), showing that these measures often have the potential to significantly reduce negative impacts on crop yields (Alexandrov et al. 2002; Baethgen and Magrin 1995; Gbetibouo and Hassan 2005; Luo et al. 2003). Adaptation of planting density and fertilizing can have similar effects (Holden and Brereton 2006; Cuculeanu et al. 1999).

Studies that focus more on adaptation confirm that simple and low-cost technologies can effectively reduce the negative effects of climate warming scenarios and extreme weather on crop yields (Easterling et al., 1993, 1996; Salinger et al., 2005).

However, many adaptation measures to current or changing climates in crop and animal production depend on the availability and costs of different farm technologies, related to the established agricultural system and socioeconomic and policy conditions (Giupponi et al. 2006). Technological research and development are among the most frequently advocated strategies for adapting agriculture to climate variability and change (Ewert et al. 2005; Perarnaud et al. 2005; Smithers and Blay-Palmer 2001).

In developed countries with high-input agriculture many farmers may be able to deal better with climate variability and change thanks to their available extensive "technological" tool-kit, but the long-term vulnerability and risk may increase as well (Bryant et al. 2000; Burton and Lim 2005). In low-input agricultural systems, on the other hand, the individual farmers depend to a large extent on low-cost technologies or on external input such as institutional support for more costly technologies. The concept of low external inputs sustainable agriculture (LEISA), which is well described by Stigter et al. (2005), is probably the only realistic option for many developing countries if they are to secure sustainable food production and welfare. Moreover, studies on climate change impacts on crops showed that there is enormous variability between areas (e.g. shown by Jones and Thorton (2003) for African maize production), which makes locally adapted technologies even more important.

In this paper we will try to give an overview of this complex picture by using examples from selected countries with different climatic conditions and agricultural systems. It discusses also the optimization of farm technologies as a means of ensuring sustainable agricultural production. This optimization may include stabilizing agroecosystems and providing an acceptable income for farmers.

10.2
Strategies for optimizing farm technologies in various agricultural systems

In order to analyse optimisation strategies in various agricultural systems considering sustainability (Fig. 10.2.) we propose to make a distinction between the most important and climate-sensitive agricultural resources to be managed, such as water, soil (including nutrients), crop (including management) and microclimatic conditions in relation to low, medium and high agricultural input systems. Of course, many farm technology optimization strategies can affect more than one of these resources at the same time. Low-input systems may be characterized as small farm structures and with low income in a less developed socioeconomic environment as is found in developing countries (almost no financial reserves for investment in farm technologies available). Medium input systems might be characterized as small farm structures with acceptable farm income in a good socioeconomic environment, as in small farms in Western Europe (limited financial reserves for investment in farm technologies available). High input systems might be characterized as farms with high income levels in any socioeconomic environment, where there is theoretically no limitation to investment in new farm technologies.

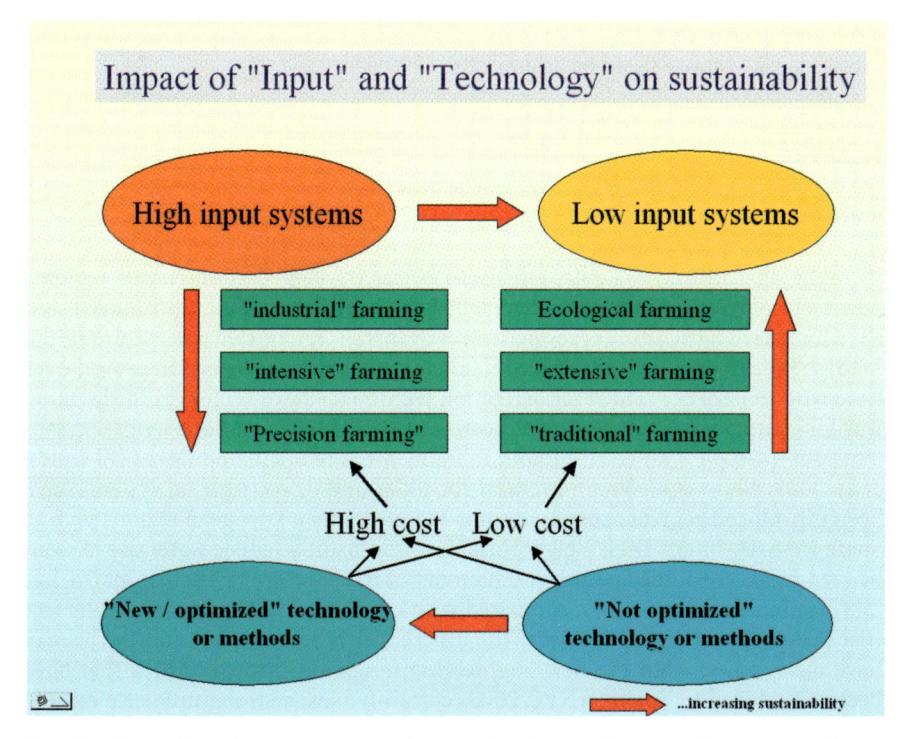

Fig. 10.2. Types of farming systems in relation to technology used and trend in sustainability.

10.2.1
Optimization of farm technologies and water resources

It is well known that on a global scale water is probably the most limited resource for agricultural production and directly sensitive to climate variability. Water resources can therefore vary strongly from year to year and within a single year. Extreme precipitation events and floods can be as devastating as droughts (Rosenzweig et al. 2002; Chang 2002), and these extremes could increase under climate change, depending on the region. Extreme precipitation can further lead to nitrogen leaching on sandy soils, which might be accelerated under increasing climate variability in more humid regions (Wessolek and Asseng 2006) and have implications for agricultural land use and management for groundwater recharge harvesting, for example in northern Germany.

However, water shortage and droughts are the most important devastating factors for agriculture and food production because of their large spatial extension, especially in many subtropical regions and developing countries. Over the centuries, mankind in semi-arid and arid regions have therefore developed technologies or systems for water harvesting or irrigation. Nowadays known as traditional methods or indigenous techniques, they are still in use in many parts of the world, not least because they are well adapted to local conditions and are often the only option because of their low costs or inputs. Examples are given in numerous publications such as those listed by Stigter et al. (2005).

10.2.1.1
The role of farm technologies in water management in developed regions or countries

In many agricultural systems, mostly in better developed countries or regions, new technologies for water management have been successfully introduced and have increased agricultural productivity. For instance, irrigated agriculture in the Mediterranean area was introduced in ancient times and has been improved over time with experience. However, irrigation techniques have been kept in the same way for centuries in most Mediterranean countries. Inefficient flood irrigation systems, for example, can be still found in many areas of Spain and Egypt (El Gindy et al. 2001; Neira et al. 2005). Modern sprinkler and drip-irrigation systems have been introduced at great expense in some Mediterranean European regions such as Spain (MAPA 2005). These new techniques significantly reduce water use. As can be seen in Fig. 10.3, the Spanish productivity of irrigated crops, such as maize, has increased in the last 15 years, compared with countries like Egypt, despite the fact that the total production is lower. The differences between Spain and Egypt may have many causes, but the new engineering irrigation infrastructures that have been introduced in Spain (ANPC 2003) certainly have a strong influence on this yearly yield increment.

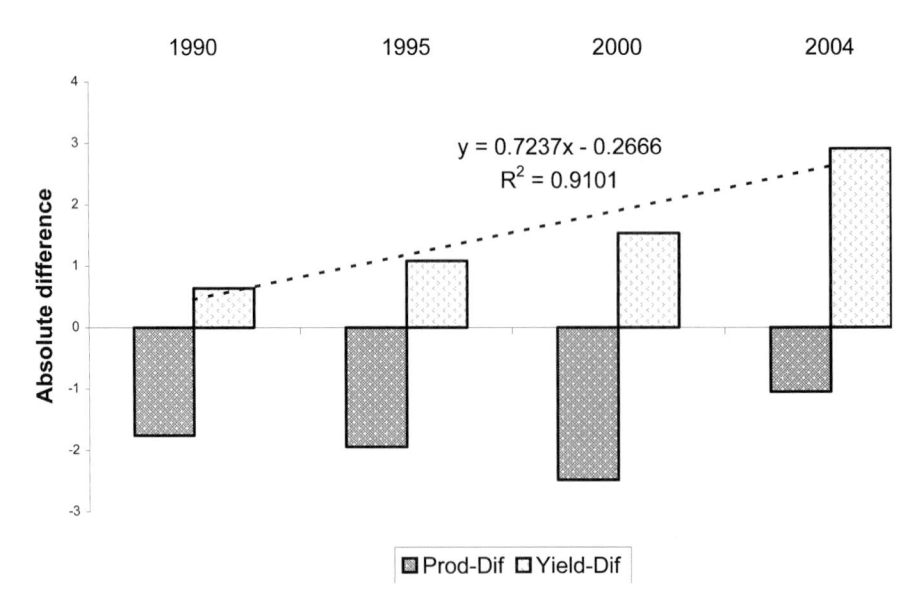

Fig. 10.3. Absolute differences between Egyptian and Spanish maize production (in BT) and yield (in t/ha).

Water availability could well be the most important agricultural constraint in Mediterranean agriculture in the future (Olesen and Bindi 2002; EEA 2005) and in many other agricultural regions worldwide. Adaptation tests by Rosenzweig et al. (2004) for several major agricultural regions worldwide have shown that few regions can readily accommodate an expansion of irrigated land in a changed climate, while others would suffer decreases in system reliability if irrigation areas had to be expanded. Timely improvements in crop cultivars, irrigation and drainage technology and water management are therefore required. Farmers in southern Europe, for example, must realize that techniques such as the "deficit Irrigation" should be considered as an option in the next decades, or irrigated agriculture will become unaffordable (Fereres 2005). Nevertheless, the success of deficit irrigation in a given year depends on weather behavior during that year (Farre 1998), which makes it difficult to introduce it into farming practice. The only practical solution for the extensive introduction of deficit irrigation and similar techniques to improve irrigation efficiency is through very local assessments, taking into account weather variability (Bastiaansen et al. 2004; Eitzinger et al. 2004; Utset et al. 2004; Utset 2005; Fereres 2005).

As an example of medium- and high-input farming systems, irrigation is being modernised in Spain on a fairly large scale with governmental support (Beceiro, 2003; MAPA, 2005) to replace flooding by sprinkler and by other more efficient technologies. It usually implies large investments and farmers cannot afford them on their own. However, irrigation must not only be kept but also enlarged if the Lisbon Strategy goals are to be met and rural living conditions improved (MAPA 2005).

Moreover, complete sprinkler coverage is very important in terms of personnel savings. Southern European and Spanish agriculture is mainly based on family businesses. The rural population has dramatically decreased in Europe. Complete coverage combined with automatic control devices therefore allows the manpower effort involved in irrigation to be reduced to a minimum. Furthermore, irrigation advisers in the form of local specialists should be accessible to farmers to accompany modernization. These local services must be able to provide help to farmers in dealing with the new available technology. The irrigation advisers in Spain and other European countries usually have modern laboratories for soil property analysis as well as a relatively dense network of agrometeorological stations and other high-input technologies. Besides, the specialists involved in such advice services could be trained in modern techniques such as simulation modeling tools and remote sensing interpretation.

Irrigation investments include channel designs, water distribution systems and pumping devices. The engineering effort involved is usually significant, costing several million euros. Complete sprinkler coverage usually involves underground PVC or metal tubes all over the agricultural field. Automatic control devices also need solar cells, modems, computer systems and other related technology. Furthermore, irrigation advice services call for government investment in trained personnel, as well as laboratory infrastructures and technological facilities. Despite the large investment involved in these three potential measures, they can be amortized in few years, particularly in view of the anticipated increase in water prices in Europe as a result of political measures. The irrigation advice service could also become independent and self-funding in few years. A government-directed effort, providing loans and supporting funds is absolutely essential in the first stage. The total amount involved is very high, which makes these potential measures affordable only by developed countries.

There are many examples in Spain and other southern European countries, but we will concentrate here on the results obtained in the irrigation community of Valladolid, Spain. The community comprises 610 ha, 209 farmers and 383 irrigated fields in Simancas, Geria and Villamarciel, in Valladolid province, central Castille, Spain (41:31N, 4:53W). Maize, sugar beet and alfalfa are the main crops. On-demand pivot and sprinkler irrigation (complete coverage) was recently provided to the farmers, according to the Spanish Plan Nacional de Regadíos. The investment included an underground pressure-based distribution network, a computer-based automatic open-close control system for the hydrants and on-field control of the used water, which is now being taken into account in the relevant invoices. Flood irrigation is no longer allowed following the modernization investment. Weekly crop water requirements are provided locally by a web service (www.inforiego.org) and by SMS. The Penman-Monteith daily evapotranspiration calculations are based on a network of agrometeorological stations installed in the last five years all over Castille using European funds.

The Castilla y León irrigation advice service made a survey of 26 per cent of the local farmers in Simancas-Geria-Villamarciel (Utset et al. 2006a,b). According to the survey, 77 per cent of the farmers take into account the particular season and the crop to manage irrigation. However, 86 per cent do not conduct a water balance to calculate the irrigation supply requirements. They recognize they could save more water and they might pay for irrigation advice, particularly considering soil differences and water-table fluctuation levels. TDR-based periodic soil-moisture monitoring is being carried out by the irrigation-advice service in Simancas-Geria-Villamarciel on several fields, providing the farmers with updated information on water use. In particular, the recurrence of soil water content higher than field capacity at deeper depths is an indication of water loss and potential pollution problems. Irrigation productivity calculations are also provided for farmers. This service is being supplemented by modeling assessments (Utset et al. 2005), which might help farmers to decide how to manage crop and water in a sustainable way in their respective farms.

10.2.1.2
The role of farm technologies in water management in developing countries

For many farmers in developing countries, however, these expensive new technologies are not affordable without external support and are therefore not applicable in low-input agricultural systems with weak infrastructure and poor socioeconomic conditions. The adaptation and use of traditional methods are recommended in these cases (Stigter et al. 2005). Indigenous techniques for agricultural water use in semi-arid regions are known for example from the Incas. *Chacras hundidas* are sunken pits or basins that allow crops to reach the groundwater table (Golte 1978).

Galerias filtrantes, artificial and complex surface and underground canals to collect groundwater, are also still used in places. Similar systems of ancient underground pipes for the transfer of irrigation water through arid areas can be found in Iran. Ancient surface canal systems and surface tanks over large areas for the transfer, distribution, collection and storage of water from the monsoon periods can be found in India (Das 2001 in Stigter et al., 2005) and Sri Lanka, which also still work effectively and are used for crop production in low-input farming systems. Stigter et al. (2005) reports several examples of traditional and newly adapted effective water use methods in Africa using planting pits with improved soil water storage through the addition of manure, for example.

Beside traditional or indigenous methods, new and low-cost technologies may be still a promising option for low-input farming systems, especially for countries in transition such as India or China. Eitzinger et al. (2005) showed that even simple low-cost technologies could significantly improve irrigation scheduling and crop water use compared with flood irrigation. These technologies, based on simple measurements and algorithms to estimate actual evapotranspiration for irrigation scheduling, have still to become more user-friendly, however. Moreover, a basic and stable infrastructure for local companies and technical support should exist or be built up, and this is not the case in many regions of developing countries, especially in Africa (Stigter, personal communication, 2004). This could also act as an incentive for technological change to be driven more by environmental objectives and farmer innovations operated through the market as recommended by Norse and Tschirley (2000), among others.

An important management option…" by

"An important management option for low (and all level) input farming systems regarding water resources is the change to crops with better water efficiency. This is especially important in regions where pressure on water reserves is increasing owing to human activities, climate change and variability. For example the change from wetland rice to dry land rice or other crops can have enormous effects on agricultural water reserves, as demonstrated in northern China (You 2001).

10.2.2
Optimization of farm technologies and soil resources

Soils and prevailing agricultural production systems strongly interact with climate and climate variations, so farm technologies and management options have to be adapted to maintain soil functions for crop production to secure sustainable agricultural production as a basis for the welfare of many countries. Soil types in their current form developed over many centuries are determined among other factors by the climatic conditions. Agricultural practices can strongly impact on soil functions in the short term, and farming technologies and management can play an important role in these processes. In many regions with extreme weather conditions, for example, soil functions can react very quickly to agricultural practices. Unfortunately this can lead to rapid and irreversible degradation of soil functions and further to desertification, which has become a significant problem in many agroecosystems in the world.

For example, improper irrigation schemes and use of salinated irrigation water can lead to increasing salinity of soils, making them unusable for agricultural production. This process is well known from badly adapted production systems as a result, for example, of long-term inappropriate policy (as was the case with Soviet cotton production concentrated in Central Asia). Other examples are overgrazing in the Sahel zone and other semi-arid regions for various reasons, leading to wind erosion and desertification. Crop productionof not suitable crops in warm semi-arid zones with frequently strong winds can easily lead to wind erosion triggered by soil degradation (for example, short-term profit-driven agriculture as with the wheat production in Western Australia in earlier times). In tropical regions the high soil temperatures combined with high precipitation cause high decomposition and leaching rates and an inappropriate change in soil use for agricultural crop production can lead to fast soil degradation (for example, with too few organic matter residues or manure), (Sivakumar et al. 2005). In climates with frequent extreme precipitation events, such as the Asian monsoon regions, soil water erosion, especially in hilly terrains, has already caused enormous soil degradation. This is the case especially with production systems where the soil surface is not always fully covered or there are no terrace systems, as in the tea plantations of Sri Lanka.

There are many other examples showing how agricultural practices that are not adapted to local climatic conditions have led to irreversible damage to agricultural soils. Under climate change and changing climate variability, these problems will become an even more significant threat for the soils in many agroecosystems through increasing evapotranspiration rates for irrigated regions (Yeo 1999), more frequent droughts or extreme precipitation, for example.

Since farming is carried out by humans, soil cultivation plays an important role in crop production. The first important aim was to control weeds and to optimize root growth conditions. This is still an important argument for ploughing in many agricultural areas and in ecological farming. However, because soil cultivation is an important cost factor, many options have been developed to decrease it, such as reduced soil cultivation or minimum to no soil cultivation and tillage systems. Furthermore, these systems can reduce soil water and wind erosion significantly and also increase soil water-holding capacity and infiltrability. It has been shown experimentally that increasing soil water-holding capacity by reducing soil cultivation in combination with mulch has had a significant positive yield effect (on cereals in the semi-humid region of eastern Austria in years with drought episodes, for example).

Similarly, simulation studies have shown that increased initial soil water content at the beginning of the growing season has a significant positive long-term yield effect (Trnka et. al., 2004); (Fig. 10.4). However, for larger field sizes these systems are mainly used with medium- and high-input systems where complex machinery is required. Often, soil fertility and functions could be improved and erosion reduced significantly by using these systems, which would make an important contribution to sustainable crop production. In ecological farming, however, ploughing still remains an important measure because of its weed control function. Nevertheless, minimum tillage and reduced soil cultivation can also be used under certain conditions in these systems.

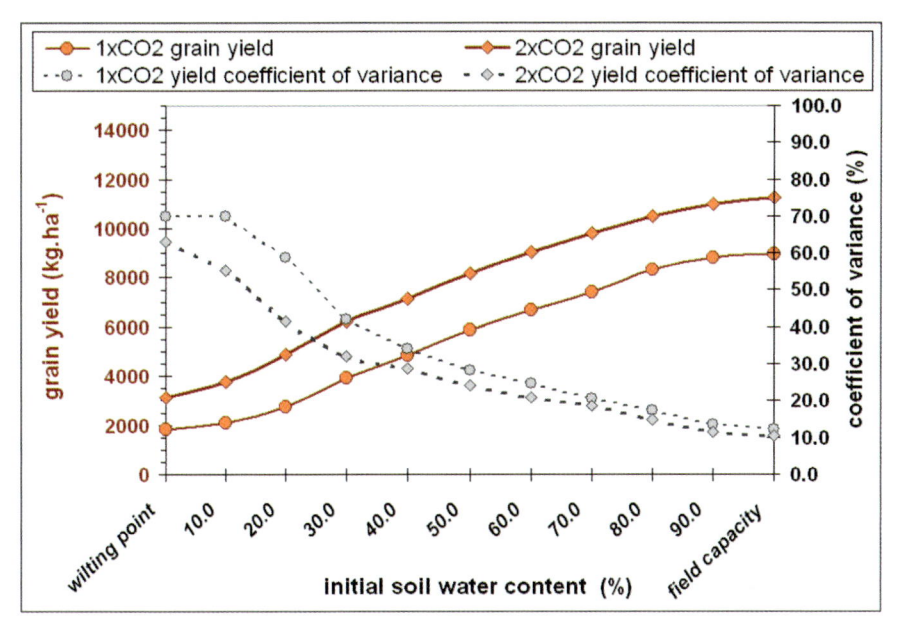

Fig. 10.4. Sensitivity of water-limited spring barley (cultivar Akcent) grain yield to different levels of initial available soil water (ISAW) in a highly fertile region under present and 2×CO_2 climatic conditions (both ambient CO_2 increase and changed climate were considered). Each point represents a 99-year simulation described by the mean and the value of the coefficient of variance.

The most widespread decrease in appropriate soil functions for crop production such as water-holding capacity and soil fertility is caused by wind or water erosion. Soil erosion can have multiple causes, one of which is climate and climate variability in the form of extreme precipitation or strong winds in dry conditions. Farming practices and technologies have a strong impact on the climate-based potential for soil erosion. There are many examples of soil erosion caused, for example, by overgrazing in semi-arid regions with sandy soils or by growing slowly developing crops causing reduced soil cover for long periods. This is the case with maize, soybean or sugar beet also in temperate regions with significant extreme precipitation events.

Changes in crops or crop rotation to adapt to changing climate and variability may therefore impact indirectly on soil erosion in vulnerable regions (Rounsevell et al., 1999). Climate change and climate variability may also indirectly affect soil erosion. For example, O'Neal et al. (2005) report that increasing precipitation and decreasing cover from temperature-stressed maize are important factors for increasing soil erosion in the Midwest of the United States of America. In almost all agroecosystems soil erosion, caused be various factors, leads to a decrease in soil fertility and hence to a reduction in crop productivity because of loss of organic matter, nutrients and lower water-holding capacity. In temperate regions with high-input systems heavy machinery, often in combination with slowly developing crops and soil cover, contributes to soil compaction, decreasing water infiltration, increasing runoff and therefore water erosion. In Europe these problems are apparent with sugar beet and maize, where soils are not covered for a long time in spring and heavy machinery has a devastating and often irreversible affect on soil structure during the frequently wet harvest periods in the autumn. This problem accelerates with increasing slopes of fields, as are frequently found in Europe.

Perennial crops in various climatic regions such as vineyards, orchards, tea or coffee, which are often grown in hilly regions, are also subject to water erosion, especially during extreme precipitation events. Mulching technologies such as grass or straw mulch or other crop residues are therefore often applied and are sometimes mandatory. In some cases, even the more costly or manpower-intense terrace systems have been re-established in order to stop long-term soil erosion.

10.2.3
Optimization of farm technologies and crop resources

Crop yield and crop production within a certain territory can be seen as an interaction of many factors. However, crops adapted to certain conditions are an important local resource for crop productivity with a significant influence on yield risk. Crops can respond nonlinearly to changes in their growing conditions, exhibit threshold responses and be subject to combinations of stress factors that affect their growth, development and yield. Thus, climate variability and changes in the frequency of extreme events are important for yield and the stability and quality from year to year. Higher temperature and precipitation variability increase the risk of lower yield, as many experimental and simulation studies have shown (Porter and Semenov 2005).

Over the generations farmers have selected the best cultivars for their use, creating locally well adapted crops, some of which are still in use in agricultural systems and are an important genetic resource for modern crop breeding. Farmers can not only change crops and cultivars but also modify crop management, by changing the sowing date according to the expected seasonal weather, for example. The seasonal precipitation pattern (onset of rain, duration of rainy season, distribution during crop growing period) is one of the most important pieces of information for farmers in semi-arid regions using rain-fed cropping, especially for low-input systems (Stigter et al. 2005; Ingram et al. 2002; Mati 2000) in developing countries, which enables them to adapt their sowing dates and crop selection. Seasonal forecasts, provided they are reliable enough, are already being successfully used in developed countries at the farm level to adapt seasonal crop planning (Meinke and Stone 2005), but there is still a deficit when it comes to making such information useable for farmers in low-input systems (Salinger et al. 2005). However, seasonal forecasts are already being used in developing countries for yield forecasting to support policy decision making (Hansen and Indeje 2004) or the MARS project of the European Union, which has been extended to the African regions (Rojas et al. 2005).

10.2.3.1
The role of crop modeling in farm technologies

Crop and whole farm system modeling can help farmers significantly in decision-making for crop management options and related farm technologies, provided it is used properly and infrastructural support of the standard in developed countries is available. An example is presented by Keating et al., (2003) applying the APSIM model for farming system simulation in Australia. Examples have also been presented for tropical regions such as Asia, where related user-friendly software has been developed (Aggarwal et al. 2006a,b).

However, for medium- and low-input systems in developing countries crop or agroecosystem modeling is currently used mainly to guide general decision-making on a higher institutional or farm-advising level. Matthews et al. (1997), for example, reported that for rice production in Asia the modification of sowing dates at high latitudes, where higher temperatures allowed a longer potential crop-growing season, permitted a transition from single cropping to double cropping in some locations, which could had a significant effect on regional production. Two shorter ripening varieties might be a better strategy than a longer maturing variety because the grain formation and ripening periods are pushed to less favorable conditions later in the season. Planting dates could also be adjusted to avoid high temperatures at the time of flowering (spikelet sterility). Spikelet sterility resistance of cultivars to temperature is another option for reducing the yield risk for rice under high or increasing climate variability. Further examples are changes to more heat resistance or earlier ripening cultivars, as it has been shown that heat stress can significantly reduce crop yield (Southworth et al. 2000; Soja et al. 2005). For crops in tropical regions, e.g. soybean in India, a delay in the sowing date has been recommended for similar reasons (Mall et al. 2004).

Climatic variability influences not only the production of individual crops but also the agriculture systems, which are composed of several interdependent segments. For example, with grassland or cereal production considerable variations owing to climatic factors might be of major importance to dairy farmers (or dairy unit of the farm) (Holden and Brereton 2002). Therefore the whole system must allow for the risk of unfavorable weather conditions. This requires stocking up necessary reserves and possibly purchasing forage if this cannot be produced locally. Prior knowledge of such a need (e.g. deficit or surplus in crop production) might enable the subject to obtain a better price or the state agency to prepare to intervene. The case of Austrian grassland production in 2001 and 2003 could serve us as an example, showing how such an "early warning system" can work (Fig. 10.5). As most of the yield variability is caused by climatic factors and their interaction with soil conditions, sward composition and management a relatively simple model of the whole system could be created (Trnka et al., 2006). If such system is then combined with appropriate GIS information it might be updated (and regionalized) in order to identify regions where forage growth has became critically low (Fig. 10.5). When such a model is coupled with a weather generator a probabilistic forecast can be issued early in the season, allowing farmer to better estimate the chance of a good/poor harvest (Fig. 10.5) or prepare to set up irrigation technologies.

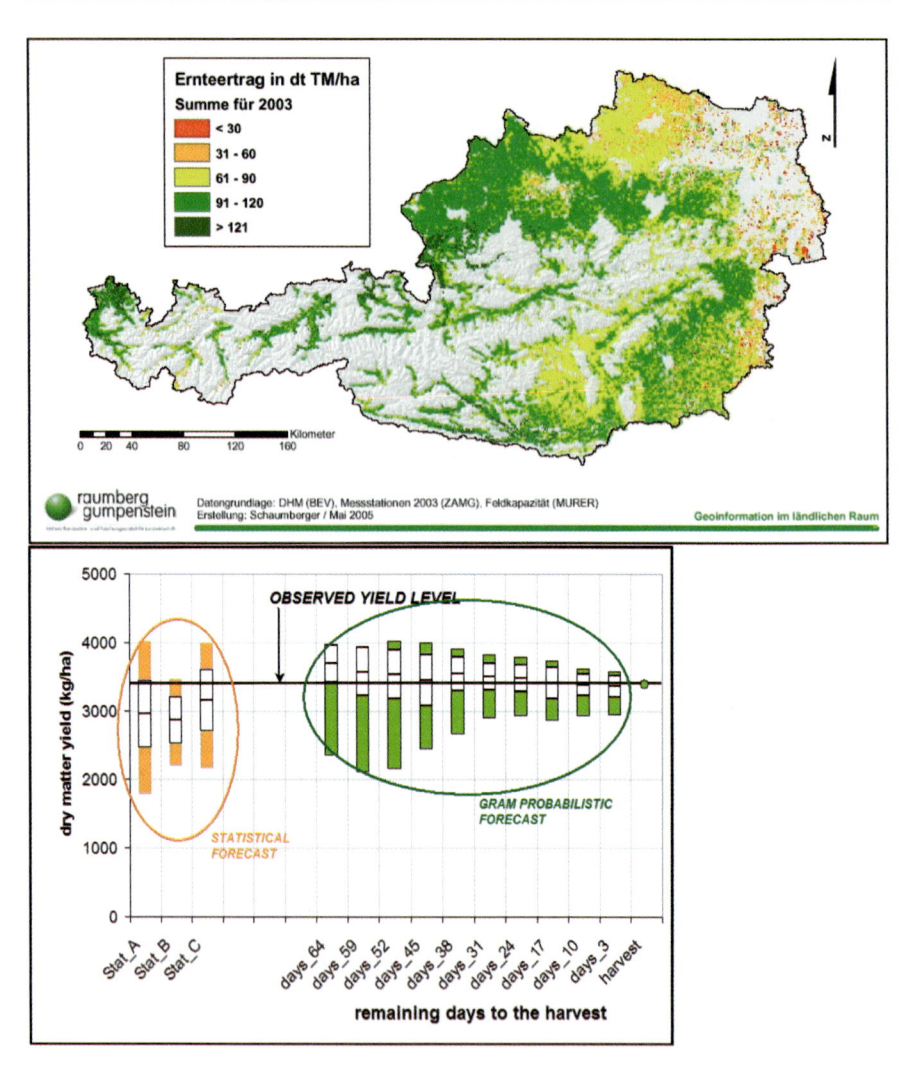

Fig. 10.5 The high resolution GIS map of Austrian grassland yields during the unusually dry year 2003 documents extremely low yield areas b) Example of the yield forecast for cut 2 (from 22 May to 26 July 2002) at Gumpenstein experimental station. The horizontal line represents the level of observed dry matter yield at the site. The orange vertical bars represent yield predictions based on three statistical forecasting methods. The green bars represent probabilistic forecasts, each based on the 99 GRAM model runs issued on the given day preceding harvest. The x-axis description depicts the number of days to the harvest. The lowest and highest parts of each bar represent minimum and maximum predicted yields. The white part of each bar indicates mean value ± standard deviation (SD).

In fact this method performs better than the standard "statistical" yield prediction and can be performed with reasonable accuracy relatively early in the season. Forecast precision could be improved by issuing a probabilistic forecast that incorporates a long-term weather forecast for the rest of the season.

10.2.3.2
Changing agricultural systems and the role of farm technologies

Crop response to environmental conditions is a complex problem. Beside the seasonal weather, crop characteristics and management, crop yield is influenced by soil and terrain properties, fertilization (especially nitrogen, phosphorus and potassium), pests and diseases pressure as well as soil cultivation. All these factors can alter with time and changing production systems and interact with farm technologies. For example, in the emerging farming systems with lower than optimum doses of mineral fertilizers (as in ecological farming systems) yield has been found to be more directly related to crop rotation schemes (for example to the percentage of perennial legumes used in case of cereals). One per cent use of forage crops (mostly legumes) caused growth in grain yield of 23 kg/ha^{-1} (Sroller et al. 2002). Ecological farming systems, for example, use more complex crop rotation, different soil cultivation and crop protection measures and finally different technologies than in conventional farming.

A comparative study of the past decades in Central Europe (including 10 Western and former Soviet countries) presented by Chloupek et al. (2004) claims that the influence of cultivars was relatively low in comparison with the influences of location, year, nitrogen application, use of growth regulators and fungicides (Síp et al. 2000) in relation to yield increase. However, their impact increased between 1962 and 1992 from 25–30 to 50 percent (Bares et al. 1995). In the 1950s, Fischbeck (1999) reported that yield growth for wheat in Germany was due to increased nitrogen fertilization, and later due to chemicals used to shorten straw and to fungicides. Factors influencing crop yields in the neighboring Czech Republic were summarized by Vrkoc (1992) for 1948–1990. He reported that the most important factors were the decreasing influence of inherent soil fertility (40–0%); the decreasing influence of weather (20–0%); the relatively stable influence of cultural practices (10–25%); and the increasing influence of varieties, fertilizers and plant protection practices (during the period: 5–30, 10–25 and 5–20%, respectively). This is a view also expressed by Chloupek et al. (2004) who claims increasing yield stability. Even though some additional findings support these claims, e.g. increases in minimum regional yields level or decrease in interregional differences during individual seasons (Fig. 10.6) the claims of decreasing weather influence even in high- and medium-input agriculture depend to a large extent on the responses of crop cultivars to the prevailing climate and weather conditions or their degree of adaptation to these conditions. For example, it has been shown that those crops exhibiting the highest increase in yield were also the most adaptable to inter-annual weather variability, cultivars grown and cultivation techniques used.

In a simulation study we found that a relatively large proportion of yield variability could be explained simply at least in some regions by the monthly drought

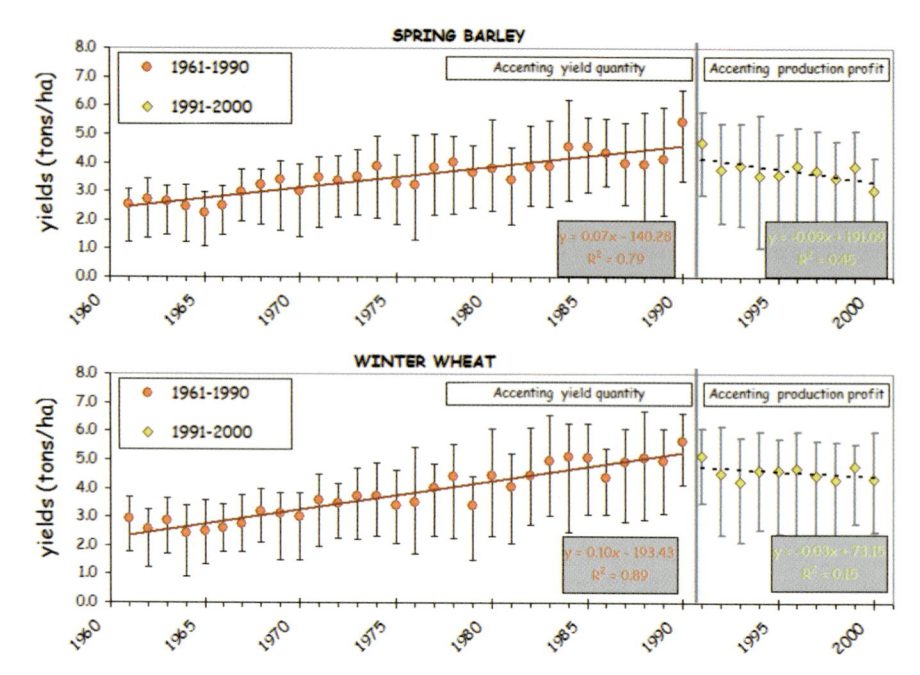

Fig. 10.6. Development of the mean national yield (dot) of two major cereal crops (spring barley and winter wheat) during the period 1961-2000. The bars represent the minimum and maximum yields obtained at the regional level (country composed of 78 individual regions). Linear trends are provided separately for 1961-1990 and 1991-2000.

index. It showed that too dry/wet conditions during the growing season significantly reduced yields (Fig. 10.7). In another study Trnka (2006) demonstrated that interseasonal permanent grassland yield variability was mostly a function of varying global radiation, temperature and soil moisture regimes even with relatively intense production under the temperate Central European conditions.

As mentioned above, weather and/or climate and optimisation of farming technologies are not the only drivers influencing agriculture systems, their productivity and even their sustainability. Optimisation strategies in agriculture might, for example, include changes in land use, shift of production areas because of climate shifts (Seguin, 2003), changing the size of farms and fields in combination with technology, soil amelioration and sustained improvements of inputs or methods (e.g. better adapted cultivars, machines, chemicals, fertilisers) additionally triggered by policy and governmental incentive measures. Combining these manyfold impacts might result in different sensitivities of the agriculture systems to weather/climate factors, with positive and negative effects .

The effect of these factors over extended periods, for example the past five decades, can be demonstrated by looking at farming systems in the Czech Republic (during the change from the Communist regime to a democratic system). The large farms and high investment (although in many cases ineffective) together with the

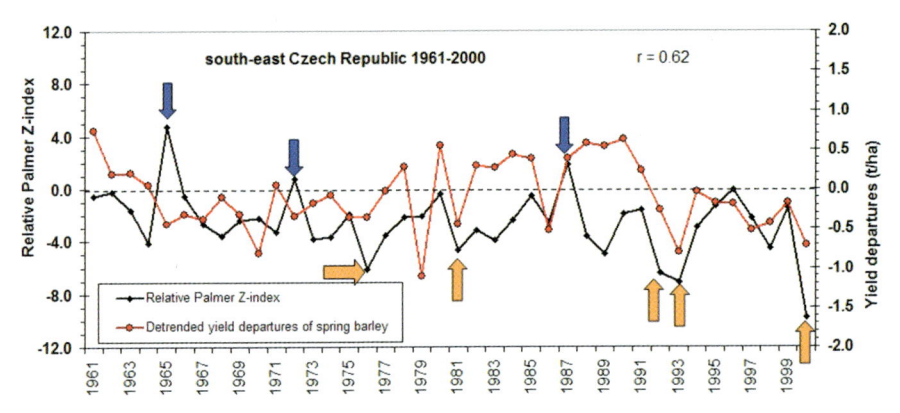

Fig. 10.7. Time series of the relative Palmer Z index averaged over the period April-June of each year and detrended spring barley yield for the south-eastern part of the Czech Republic. The blue arrows mark unusually wet seasons and red arrows for seasons with unusually dry growing periods.

intensive use of fertilizers and pesticides triggered by the policy-oriented yield-maximizing strategy produced a sustained increase in yields (as shown in Fig. 10.6 for cereals) in the Czech Republic, which from 1961 to 1990 was comparable with or higher than the mean of the EU-15, despite the comparatively worse climate and soil conditions (Chloupek et al. 2004). Even in the late 1980s when the intensity of the Czech Republic's agricultural production was at its peak, the effect of environmental factors on production remained high (as has also been experienced in other European countries with different agricultural structures), as can be seen from the yield variability at the district level compared with the national mean yield (Fig 10.6a-b). The change of the political system in 1989-1990 led to the introduction of market economy principles, which forced producers to put the emphasis back on productivity and sustainable production rather than maximizing yield. Some of the least fertile areas were thus turned back to grasslands or forests and the amount of fertilizers and pesticides decreased dramatically. This change led to a decrease in national mean yields, especially in the case of spring barley, where in Western countries like Austria a continuously increasing yield trend has been observed, driven mainly by productivity. Only during the past two decades has sustainability and environmental protection been forced either by policy measures or by changes in farmers' strategy in response to market demand, such as in ecological farming systems. The latter cases have led to a diversification of yield levels, depending on the intensity of production, and productivity (driven by market prices and government support).

As Bares et al. (1995) and Sroller et al. (2002) noted, the influence of new cultivars has been increasing over past 40 years, especially in high- and medium-input farming systems. The continuous effort of crop breeders in close cooperation with state authorities in charge of approval of newly bred cultivars can be regarded as one of the most important drivers of increasing productivity in EU and Central European agriculture in the past 40 years. As Chloupek et al. (2004) showed for the

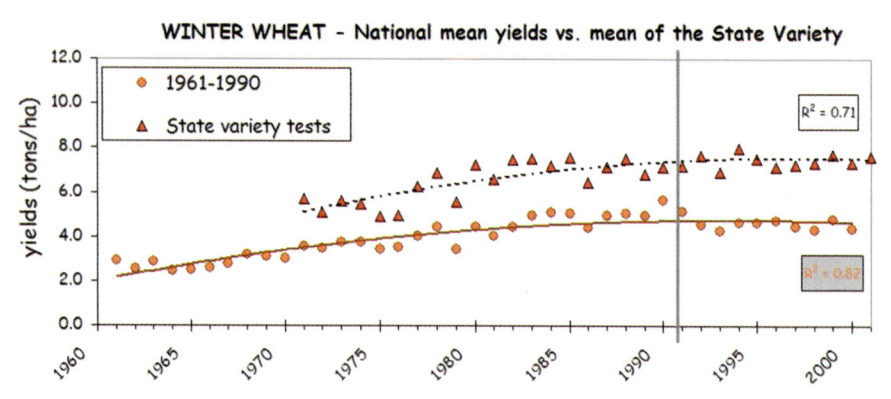

Fig. 10.8. Development of the mean national yield (circles) of winter wheat in the Czech Republic from 1961 to 2000 and the mean attainable yield (based on 8113 yield experiments from over 40 State Institute for Agriculture Supervision and Testing sites). Trends lines are provided to document development of both yield series over time.

Czech Republic, the yield of all major crops increased steadily during the period 1961–2000 (although there has been a depression during the past decade because of a system change as mentioned above).

This is partly explained by the other farming techniques such as appropriate application of nitrogen, pesticides and better technology, but all of these intensification factors can be utilized only when proper cultivars are used. For some crops (e.g. winter wheat) we noted that the positive trend in maximum attainable yields (i.e. level of yields from the mix of the newly introduced hybrids grown in near-optimum soil conditions with high standard of farming practices) was almost the same as the national mean yield trend (Fig. 10.8). Interestingly, from 1971 until the late 1980s there was quite a large increase in the maximum attainable yields whereas this breeding progress was much less pronounced in 1990s. The present national mean yield is only about 60 per cent of attainable experimental yields (Fig. 10.8), and thus 40 per cent of the cultivar production potential is theoretically not being utilized (in the case of winter wheat). This could be a result of the less ideal soil conditions compared with the experimental sites but is probably attributable for the most part to the reduced application of optimum conditions for reaching maximum yields (nutrient and pest/diseases constraints) in order to optimize productivity.

10.2.3.3
The role of crop management in farm technologies

The adaptation of the crop calendar also involves potential changes in the sowing dates. For centuries the proper setting of the sowing date within a particular season has remained as extremely difficult task in practice. In some seasons it is also impossible to keep the recommended sowing dates because of constraints imposed either by weather conditions (e.g. high soil wetness / low soil workability) or other factors (e.g. machinery not adapted to wet soil conditions). The experimental data show that even small shifts in sowing dates can result in extreme differences in the final yields because vulnerable stages are exposed to different environmental stresses.

In many cases crop models can be used either to find the most appropriate sowing "window" or to calculate the penalty for premature /late sowing (Trnka et al., 2004; Žalud and Dubrovský, 2002). The sowing date optimization procedure for a central European site is demonstrated in Figure 10.9a-b where two crop models were used to evaluate the effect of changing sowing dates. In the case of spring barley (Fig. 10.9a) it is clear that the overall mean yield is relatively insensitive to small changes in sowing date. Specifically, the median of the yields remains nearly constant if sowing varies within 20 days. Therefore there would not be any severe penalty in most seasons if the sowing were moved. A simple rule based on the crop model results suggests that an earlier sowing date results in higher/more stable yields, which farmers in this region are already well aware of. The model we used also shows that the possibility of even earlier sowing is restricted by the soil workability in most of the seasons (grey box). On the other hand, if the planting

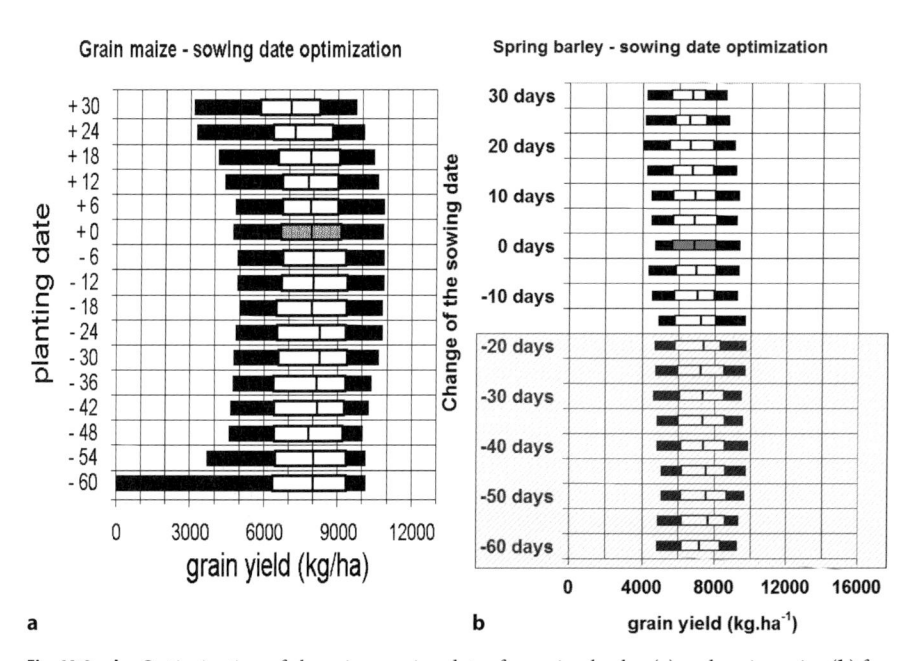

Fig. 10.9 a, b. Optimization of changing sowing dates for spring barley (**a**) and grain maize (**b**) for present climatic conditions as modelled by CERES-Barley and CERES-Maize. The shift is shown in terms of the deviation (in days) from the representative year's planting date (26 March-spring barley and 6 May). The bars represent quantiles (5th, 25th, median, 75th, 95th) of the model yields obtained in the 99-year crop model simulations for present and changed climate. The changed climate is represented by the AVG scenario. The shaded bars relate to the actual planting date. The shaded area in the case of spring barley marks unworkable soil conditions at the site.

date is delayed, the grain yields tend to decrease because of the shift of the vegetation period to months with higher temperatures and lower precipitation causing higher water stress during the grain-filling phase and a shortening of this phase. An examination of grain maize yields at the same site (Fig. 10.9b) shows that they are also fairly insensitive to small changes in sowing date, but that in the case of the earlier sowing date, the probability that the yield will be damaged by a spring frost increases. On the other hand, if the sowing date is delayed, the grain yields tend to decrease because of the occurrence of autumn low temperatures, which terminate the grain filling phase.

Under changed climatic conditions the appropriate sowing times will have to be assessed again, but this adaptation measure is quite straightforward in the case of annual crops. For perennials (e.g. vineyards or orchards) the onset of phenological stages and the higher chance of frost damage are very threatening (Chemielevsky et al. 2006) especially under changed climatic conditions, and for this reason frost protection measures should be introduced (see below).

10.2.3.4
Crop monitoring techniques – pests and diseases

Some of the gains in agricultural production under local current conditions or expected with higher production efficiency or change in climatic conditions could be cancelled out by the losses caused by pests or diseases (McCarthy et al. 2001; Cannon 1998). Even though appropriate agricultural practices and technology might help to control pest populations or diseases, these measures increase the overall costs of production and place further stress on the environment where pesticides or fungicides are used (Chen and McCarl 2001). In addition, some of the pest and disease control techniques (e.g. deep ploughing) could conflict with efforts to conserve soil water through minimum tillage systems. Similarly, the introduction of genetically modified crops remains problematic in EU countries, and it raises a lot of questions in its own right (Gutierrez and Ponsard 2006).

In the case of pests the ontogeny of poikilothermic insects is controlled mostly by temperature (with other weather factors reducing or enhancing survival rates), and this fact has been utilized by agrometeorologists and phytopathologists for several generations to increase crop protection efficiency. A wide variety of modeling techniques (Guisan and Zimmermann 2000) are used in ecology studies (Logan et al. 2006; Beaumont et al. 2005) and by farmers or consultant companies as part of expert or operational warning systems (Grünwald et al. 2000; Hijmans et al. 2000; Aggarwal et al. 2006a). Models of different complexities have also been used to determine the probability of a particular pest's establishment in a given locale or in the case of the unintentional introduction of alien species (Morrison et al. 2005; Gray 2004; Rafoss and Saethre 2003; Jarvis et al 2001).

Models capable of estimating the spatial extent of a climatically suitable area for a particular pest allow us not only to identify the species' current potential distribution but also to assess which regions will be climatically suitable under future climate scenarios. Theses models could be successfully used to monitor the development of the particular pest stage at a given locale or to provide farmers with timely information on the present status of the pest development. The latter could be done on the local scale (http://www.srs.cz/pas/mury/zavijec/index.php, 2006) or in the form of spatialized maps available to farmers in the given area (http://www.pestwatch.psu.edu/sweetcorn/tool/tool.html, 2006). Even on a larger scale monitoring can be useful. This is the case with locusts, a well known and highly destructive pest in many developing countries, which is sensitive to rain distribution during the early season. Monitoring on an international scale is necessary to establish an effective warning system so that appropriate measures can be taken. The fact that this aspect is underdeveloped at present can be seen only too clearly from the past events in the Sahel zone.

Similar monitoring techniques have been or are being developed for various diseases as well, but in many cases the interaction with climatic factors is much more complex and less understood than in the case of pests.

Recent studies (Trnka et al., 2007) have shown that the pest-crop-climate relationship is dynamic and that species that have not been considered important in specific regions in the past might become a major problem on account of climate shifts and could expand to new regions. One example is the observed occurrence (with apparent damage to the crop) of the European corn borer (ECB) in the Czech Republic between 1961 and 1990 and between 1991 and 2000 (Fig. 10.10). Whereas the climate mapping results suggested two potential niches for the pest between 1961 and1990 (Fig. 10.10a), the bulk of the ECB population was concentrated in the south-east of the country, not least because of the very low grain maize acreage in the other region. The decade from 1991 to 2000 saw a significant expansion of ECB (Fig. 10.10). The invasion of the pest has been blamed on the general increase in grain maize acreage, the overall decrease in the use of insecticides, the widespread use of minimum tillage technologies, and a general decline in the quality of farm-

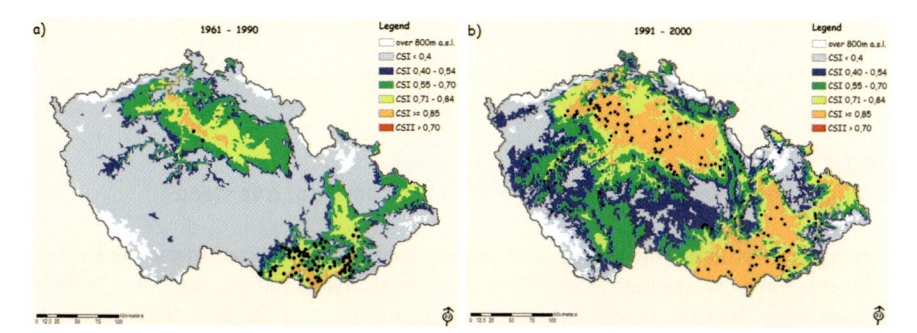

Fig. 10.10. Climatically suitable areas for the European corn borer (ECB) in the Czech Republic in terms of the climate suitability index for the monovoltine populations (CS_I). The CSI values > 0.71 mark regions with suitable conditions and those with CS_I > 0.85 those with excellent conditions. The ECB might also be found in unusually warm years in the area marked in green if maize is present. The CS_I value is shown for 1961-1990 (**a**) and 1991-2000 (**b**). The dots represent sites where the ECB occurrence in maize was observed under field conditions.
Notes: For better visualization Fig. 10.6 includes the whole territory of the country (excluding areas above 800 m above sea level) rather than the arable land only.

ing practices. These changes correlate with the overall social and economic reforms started in 1989 that led to a decade-long crisis in the agriculture sector.

The maps suggest, however, that the underlying cause of the ECB expansion was a major increase in the size and quality of the prime niche area (Fig. 10.10) and that eradicating the ECB populations (once they have become established) is virtually impossible. This claim is based on the past 70 years of experience with the pest in the south-east corn-growing region or the US Corn Belt, where despite all efforts the pest has never been eradicated. Farmers therefore have to adopt new strategies to keep the pest population below critical levels (e.g. appropriate crop rotation and tillage practices, timely use of insecticides or introduction/selection of resistant varieties) rather than invest in eradication programmes. In all cases the existence of real time monitoring programs allows more effective treatment of the exposed crops.

New technologies permit the monitoring of crop conditions on a much smaller scale. A related new emerging technology is "precision farming". This technology is still under development, currently applied only rarely and related to new technologies such as remote sensing, GPS and GIS. Because of high costs it is still available only for high-input farming (Pedersen et al. 2004; Godwin et al. 2003). It is based on observing spatial variabilities of several factors in crop fields, such as nitrogen content of leaves, drought status, disease occurrence or in-field yield variation. Using the observed information the farmer can apply measures based on the actual site-related status, considering field-level variations. This can significantly decrease costs for fertilizers and chemicals and enhance crop yield and productivity. Applications are also known for sprinkler irrigation of annual crops, applying water according to spatially changing soil conditions. The related equipment is still costly and not appropriate for low-input farming and small farms, but on a larger scale and on an institutional basis such technologies might be available at lower cost in the future following further development. Locally adapted crop management of low-input systems may use other options to precisely adapt management to spatially changing soil and crop conditions. In small, not technologically driven farms such options tend in any case to be based on the experience of the farmer.

10.2.4
Optimization of farm technologies and the microclimate of crop stands

Changes in climate variability and climate can affect microclimatic conditions is many ways (Sivakumar et al. 2005). Modifications to the microclimate of crop stands were used in ancient cultures such as the Incas in the Peruvian highlands (Vogl 1990). These ancient *camellones* and *qochas* are a combination of water-filled canals and plots designed to improve the microclimate (especially to decrease nocturnal cooling) and water availability of crops and are still in occasional use. Another examples is *kanchas* (stone fences around small fields) and terraces on slopes, which can increase both air temperature and water regime and reduce the wind speed of crop stands in these semi-arid and cold environment. Similar systems can also be found in other parts of the world, developed by ancient farmers on their own experience.

In current semi-arid low-input systems there are known examples not only for improving water resources but also for optimizing the temperature and radiation regimes of crop stands (Stigter 1988, 1994). A classic example is oasis agroecosystems with complex crop mixing and patterns to permit efficient use of radiation in a small area, to increase air humidity for the shaded crops and to avoid extreme diurnal temperature variations.

Agroforestry systems including shelterbelts are another farm management option to improve microclimatic conditions and not just to reduce wind and evapotranspiration. As crops respond especially to climatic extremes, any measure to reduce these extremes in most cases has had an accumulating positive effect on the yield level. Easterling et al. (1997), for example, show in a simulation study for the Great Plains that shelterbelts may provide nighttime cooling that could partially compensate the tendency of warming to shorten the growing season. This effect is even more significant under extreme climates or severe warming trends. On the other hand, heat stress on crops can be reduced by shading, which has been documented as a significant yield factor (Southworth et al. 2000, 2002).

Many examples of agroforestry systems in different climates and regions are known, all adapted to the specific characteristics of the relevant agroecosystem (climate, soils, crop production, farm input level, socioeconomic conditions). Such systems are already well established in many agricultural regions, especially in subtropical and tropical climates with extreme temperatures and/or weather variability. As Salinger et al. (2005) reported, soil surface heat extremes may surpass critical limits in many regions, especially with changing climate variability and extended drought periods. Under extreme climatic conditions, in semi-arid and arid tropics, for example, physiological critical temperature thresholds for crops were attained more frequently. The establishment of agroforestry systems as a long-term measure is probably the most effective option and solution to this problem.

Tree shading, on the other hand, can also prevent frost damage to crops and reduce nocturnal radiation cooling on the crop surfaces. This method is used not only in temperate regions but also in tropical highlands, for example in tea plantations in Sri Lanka. Other frost production methods, such as covering plants with sheets or foil, are also used in small plots for low-input systems. For orchards or large fields methods such as frost irrigation, foil covering, or applying aerosols are costly and are therefore found mainly in medium- and high-input farming and for cash crops. Long-term measures that are very important for avoiding damage to crops from radiation frost include planning of plantations in relation to topography in order to avoid impacts from cold air lakes. These measures are often ignored, especially when frost occurs seldom, but the effect on perennial crops can be more devastating than hail damage, as the whole crop can be damaged.

Hail is another danger, which can occur in almost all climatic regions. Although it is normally limited to a small region, its frequency can cause devastating damage. Protection against hail is not possible for annual crops on large fields (except by cloud injection, the effectiveness of which is uncertain). Only on perennial cash crops such as orchards is the investment on a hail net, for example, profitable. Hail insurance is probably the most effective protection against financial losses as a result of hail, but it is mainly used in developed countries and high-income farming systems. For low-input farmers in developing countries institutional support might be the only solution for hail protection.

10.3
Conclusions

Apart from the farm size or prevailing production systems, the optimization of farm technology and management plays an important role in reducing the negative impacts of climate variability and extreme weather events on crop and animal production. The relevant measures strongly influence the availability of the most important resources for agricultural production, namely water, soil (and nutrients), crops and microclimatic conditions. In low-, medium- and high-input levels various technologies are available ranging from traditional or indigenous methods to high-tech methods such as precision farming. Many authors have reported that high-input farming, especially in temperate regions, has the best prospect for adaptation to current or changing climate variability or for protection against extreme weather events. In these farming systems, usually located in developed countries with good infrastructures, technological developments and the availability of technologies in crop and animal production provide a rich toolkit enabling decision-makers to select measures from several options. Moreover, many agrometeorological forecasting, warning and monitoring services for farmers already exist.

Because of climate change and variability, however, beside low-cost options (e.g. change in planting date), long-term changes in agricultural production strategies (e.g. changes in perennial crops such as orchards or in irrigation technologies) could necessitate high investments and significantly increase the risk of production during transition periods.

Many low-input farming systems farmers depend on traditional methods or – in many cases not yet available – external inputs such as institutional forecasting and warning methods or investment in irrigation infrastructure. Both the re-establishment of locally adapted traditional (indigenous) farming technologies and warning/forecasting methods together with institutional support may help farmers in low-input agricultural systems to sustain or improve their productivity, food production and income. New low-cost technologies as used for irrigation scheduling may also be introduced in low-input farming systems once the basic infrastructure has been established and farmers have been trained to improve the often recommended demand-driven approach.

The optimization of farm technologies is often prompted by different aims such as maximizing food production or short-term profit. There is an urgent need, however, for such aims to be directed to permit sustainable food production at the local level, which depends on stable agroecosystems. Both the re-establishment of locally adapted traditional (indigenous) farming technologies and warning/forecasting methods together with institutional support may help farmers in low-input agricultural systems to sustain or improve their productivity, food production and income.

Acknowledgement

Research of Dr. Trnka and Dr. Zalud that contributed to this chapter was supported by the research plan No. MSM6215648905 "Biological and technological aspects of sustainability of controlled ecosystems and their adaptability to climate change", which is financed by the Ministry of Education, Youth and Sports of the Czech Republic

References

Abraha MG, Savage MJ (2006) Potential impacts of climate change on the grain yield of maize for the midlands of KwaZulu-Natal, South Africa. Agric Ecosyst Environ 115:150–160

Adger WN, Arnell NW, Tompkins EL (2005) Successful adaptation to climate change across scales. Glob Environ Chang Part A 15:77–86

Aggarwal PK, Kalra N, Chander S, Pathak H (2006a) InfoCrop: A dynamic simulation model for the assessment of crop yields, losses due to pests, and environmental impacts of agro-ecosystems in tropical environments. I. Model description, Agric Syst 86: 1–25.

Aggarwal PK, Banerjee B, Daryaei MG, Bhatia A, Bala A, Rani S, Chander S, Pathak H, Kalra N (2006b) InfoCrop: A dynamic simulation model for the assessment of crop yields, losses due to pests, and environmental impact of agro-ecosystems in tropical environments. II. Performance of the model. Agric Syst 89:47–67

Alexandrov V, Eitzinger J, Cajic V, Oberforster M (2002) Potential impact of climate change on selected agricultural crops in north-eastern Austria. Glob Chang Biol 8 (4):372–389

ANPC (2003) Gestión integrada del agua en el territorio desde una perspectiva económica. Aragon Nature-Protection Council Research Series 17, 29 pp.

Baethgen WE and Magrin GO (1995) Assessing the impacts of climate change on winter crop production in Uruguay and Argentina using crop simulation models. Climate change and agriculture: analysis of potential international impacts 207–228

Bares I, Dotlacil L, Stehno Z, Faberovà I, Vlasák M (1995) Original and registered cultivars of wheat in Czechoslovakia in the years 1918–1992. Czech Research Institute of Plant Production, Prague, 305 pp.

Bastiaansen WGM, Allen RG, Droogers P, D'Urso G, Steduto P (2004) Inserting man's irrigation and drainage wisdom into soil water flow models and bringing it back out: How far we progressed?. In Feddes RA, de Rooij GH, Van Dam JC (eds). Unsaturated-zone modelling: Progress, challenges and applications. Kluwer Academic Publishers, Wageningen.

Beaumont LJ, Hughes L, Poulsen M (2005) Predicting species distributions: use of climatic parameters in BIOCLIM and its impacts on prediction of species' current and future distributions. Ecol Model 186: 250–269.

Beceiro MS (2003) Legal considerations of the 2001 National Hydrological Plan. Water Int. 28 (3), 303–312.

Bryant CR, Smit B, Brklacich M, Johnston TR, Smithers J, Chiotti Q, Singh B (2000) Adaptation in Canadian agriculture to climatic variability and change. Clim Chang 45:181–201

Burton I, Bo Lim (2005) Achieving adequate adaptation in agriculture. Clim Chang 70: 191–200.

Cannon RJC (1998) The implications of predicted climate change for insect pests in the UK, with emphasis on non-indigenous species. Glob Chang Biol 4: 785–796.

Chang CC (2002) The potential impact of climate change on Taiwan's agriculture. Agric Econ 27:51–64

Chen C, McCarl BA (2001) An investigation of the relationship between pesticide usage and climate change. Clim Chang 50: 475–487.

Chmielewski F-M, Henniges Y (2006) Climate change and fruit growing in Germany (KliO) In: Proceedings of the 6th European Conference on Applied Climatology Ljubljana, Slovenia, 4 – 8 September 2006; Abstract no. EMS2006-A-00091

Chiotti QP and Johnston T (1995) Extending the Boundaries of Climate Change Research: A Discussion on Agriculture. J Rural Stud 11 (3):335–350

Chloupek O, Hrstkova P, Schweigert P (2004) Yield and its stability, crop diversity, adaptability and response to climate change, weather and fertilization over 75 years in the Czech Republic in comparison to some European countries. Field Crop Res 85 (2–3):167–190

Cuculeanu V, Marica A, Simota C (1999) Climate change impact on agricultural crops and adaptation options in Romania. Clim Res 12:153–160

Curry RB, Peart RM, Jones JW, Boote KJ, Allen Jr LH (1990) Response of crop yield to predicted changes in climate and atmospheric CO2 using simulation. Trans Amer Soc Agric Eng 33 (4):1383–1390

Dessai S, Lu X, Risbey JS (2005) On the role of climate scenarios for adaptation planning. Glob Environ Chang 15 (2):87–97

Downing TE, Harrison PA, Butterfield RE, Lonsdale KG (ed) (2000) Climate Change, Climatic Variability and Agriculture in Europe. An Integrated Assessment, Research Report No. 21, Brussels, Belgium: Commission of the European Union, Contract ENV4-CT95-0154, 445 pp

Dubrovsky M, Zalud Z, Stastna M (2000) Sensitivity of CERES-Maize yields to statistical structure of daily weather series. Clim Chang 46:447– 472

Easterling WE, Crosson PR, Rosenberg NJ, McKenney MS, Katz LA, Lemon KM (1993) Agricultural impacts of and responses to climate change in the Missouri-Iowa-Nebraska-Kansas (MINK) region. Clim Chang 24:23–61

Easterling WE (1996) Adapting North American agriculture to climate change in review. Agric For Meteorol 80:1–53

Easterling WE, Easterling MM, Brandle JR (1997) Modelling the effect of shelterbelts on maize productivity under climate change: An application of the EPIC model. Agric Ecosys Environ 61:163–176

Easterling., WE, Apps, M. (2005) Assessing the consequences of climate change for food and forest resources: a view from the IPCC. Clim Chang 70:165–189.

EEA (2005) Vulnerability and adaptation to climate change in Europe. European Environment Agency Technical report No 7/2005. EEA, Copenhagen, 84 pp.

Eitzinger J, Štastná M, Žalud Z, Dubrovský M (2003) A simulation study of the effect of soil water balance and water stress on winter wheat production under different climate change scenarios. Agric Water Manage 61:163–234

Eitzinger J, Trnka M, Hösch J, Žalud Z, Dubrovský M, (2004) Comparison of CERES, WOFOST and SWAP models in simulating soil water content during growing season under different soil conditions. Ecol Model 171:223–246

Eitzinger, J., Formayer, H., Gruszczynski, G., Schaumberger, A., Trnka, M. (2005) Evaluation of a decision support system for irrigation scheduling and drought management. In: European Meteorological Society: EMS Annual Meeting / ECAM 2005, September 12–16 2005, Utrecht, The Netherlands; CD-ROM, EMS Annual Meeting Abstracts; ISSN 1812-7053

El-Gindy AM, Abdel Maged HN, El-Edi MA, Mohamed ME (2001) Management of Pressurized irrigated Faba Bean in Sandy soils. Misr J Ag Eng 18: 29–44.

Erda L, Wei X, Hui J, Yinlong X, Yue L, Liping B, Liyong X (2005) Climate change impacts on crop yield and quality with CO2 fertilization in China. Philos Trans R Soc Lond B Biol Sci 360 (1463):2149–54

Ewert F, Rodriguez D, Jamieson P, Semenov MA, Mitchell RAC, Goudriaan J, Porter JR, Kimball BA, Pinter PJ, Manderscheid R, Weigel HJ, Fangmeier A, Fereres E, Villalobos F (2002) Effects of elevated CO2 and drought on wheat: testing crop simulation models for different experimental and climatic conditions. Agric Ecosys Environ 93:249–266

Ewert F, Rounsevell MDA, Reginster I, Metzger MJ, Leemans R (2005) Future scenarios of European agricultural land use I. Estimating changes in crop productivity. Agric Ecosyst Environ 107:101–116

Farre I (1998) Maize (Zea mays L.) and sorghum (Sorghum bicolor L. Moench) response to deficit irrigation. Agronomy and modelling. PhD. diss. University of Lleida, Spain. 150 pp.

Fereres E (2005) Deficit (supplemental) irrigation. In Proceedings of InterDrought-II Congress, Roma.

Fischbeck G (1999) Bedeutung der Resistenzzüchtung in der integrierten Pflanzenproduktion. Vortr. Pflanzenzüchtg 46, 7–29.

Gbetibouo GA, Hassan RM (2005) Measuring the economic impact of climate change on major South African field crops: a Ricardian approach. Glob Planet Chang 47:143–152

Giupponi C, Ramanzin M, Sturaro E, Fuser S (2006) Climate and land use changes, biodiversity and agri-environmental measures in the Belluno province, Italy. Environ Sci Pol 9:163–173

Godwin RJ, Richards TE, Wood GA, Welsh JP, Knight SM (2003) An economic analysis of the potential for precision farming in UK cereal production. Biosyst Eng 84 (4):533–545.

Golte W (1978) Grundwassernutzung bei den Küstenbewohnern des alten peru. Amerikanische Studien, Ed. Hartmann und Oberem, Band 1, p.182–193.

Gray DR (2004) The gypsy moth life stage model: landscape-wide estimates of gypsy moth establishment using a multi-generational phenology model. Ecol Model 176: 155–171.

Gutierrez AP, Ponsard S (2006) Physiologically based demographics of Bt cotton – pest interactions I. Pink bollworm resistance, refuge and risk. Ecol Model 191: 346–359.

Guisan A, Zimmermann NE (2000) Predictive habitat distribution models in ecology. Ecol Model 135: 147–186.

Grünwald NJ, Rubio-Covarrubias OA, Fry WE (2000) Potato late-blight management in the Toluca Valley: forecasts and resistant cultivars. Plant Disease 84: 410–6.

Hansen JW and Indeje M (2004) Linking dynamic seasonal climate forecasts with crop simulation for maize yield prediction in semi-arid Kenya. Agric For Meteorol 125:143–157

Hijmans RJ, Forbes GA, Walker TS (2000) Estimating the global severity of potato late blight with GIS-linked disease forecast model. Plant Pathol 49: 697–705.

Holden NM and Brereton AJ (2002) An assessment of the potential impact of climate change on grass yield in Ireland over the next 100 years. Irish J Agric Food Res 41 (2):213–226

Holden NM and Brereton AJ (2006) Adaptation of water and nitrogen management of spring barley and potato as a response to possible climate change in Ireland. Agric Water Manage 82:297–317

Iglesias A, Erda L, Rosenzweig C (1996) Climate change in Asia: A review of the vulnerability and adaptation of crop production. Water Air Soil Poll 92 (1–2):13–27

Ingram KT, Roncoli MC, Kirshen PH (2002) Opportunities and constraints for farmers of west Africa to use seasonal precipitation forecasts with Burkina Faso as a case study. Agric Syst 74:331–349

Isik M and Devadoss S (2006) An analysis of the impact of climate change on crop yields and yield variability. Appl Econ 38:835–844

Izaurralde RC, Rosenberg NJ, Brown RA, Thomson AM (2003) Integrated assessment of Hadley Centre (HadCM2) climate-change impacts on agricultural productivity and irrigation water supply in the conterminous United State. Part II. Regional agricultural production in 2030 and 2095. Agric For Meteorol 117:97–122

Jarvis CH, Baker RHA (2001) Risk assessment of nonindigenous pests: I. Mapping the outputs of phenology models to assess the likelihood of establishment. Divers Dist 7: 223–235.

Jones PG, Thornton PK (2003) The potential impacts of climate change in maize production in Africa and Latin America in 2055. Glob Environ Chang 13:51–59

Karing P, Kallis A, Tooming H (1999) Adaptation principles of agriculture to climate change. Clim Res 12:175–183

Kartschall T, Grossman S, Pinter PJJ, Garcia RL, Kimball BA, Wall GW, Hunsaker DJ, LaMorte RL (1995) A Simulation of Phenology, Growth, Carbon Dioxide Exchange and Yields under

Ambient Atmosphere and Free-Air Carbon Dioxide Enrichment (FACE) Maricopa, AZ, for Wheat, J Biogeo 22:611–622

Keating BA, Carberry PS, Hammer GL, Probert ME, Robertson MJ, Holzworth D, Huth NI, Hargreaves JNG, Meinke H, Hochman Z, McLean G, Verburg K, Snow V, Dimes JP, Silburn M, Wang E, Brown S, Bristow KL, Asseng S, Chapman S, McCown RL, Freebairn DM, Smith CJ (2003) An overview of APSIM, a model designed for farming systems simulation. Europ J Agron 18:267–288

Logan, D.J., Wolesensky W., Joern, A. (2006) Temperature-dependent phenology and predation in arthropod systems. Ecol Model 196: 471–482.

Luo Q, Williams MAJ, Bellotti W, Bryan B (2003) Quantitative and visual assessments of climate change impacts on South Australian wheat production. Agric Syst 77 (3):173–186

Mall RK, Lal M, Bhatia VS, Rathore LS, Singh R (2004) Mitigating climate change impact on soybean productivity in India: a simulation study. Agric For Meteorol 121:113–125

Maracchi G, Sirotenko O, Bindi M (2005) Impacts of present and future climate variability on agriculture and forestry in the temperate regions: Europe. Clim Chang 70 (1-2):117–135

Matthews RB, Kropff MJ, Horie T, Bachelet D (1997) Simulating the impact of climate change on rice production in Asia and evaluating options for adaptation. Agric Syst 54 (3):399–425

Mati BM (2000) The influence of climate change on maize production in the semi-humid-semi-arid areas of Kenya. J Arid Environ 46:333–344

McCarthy JJ, Canziani OF, Leary NA, Dokken DJ, White KS (Ed) (2001) Climate change 2001: impacts, adaptation and vulnerability. Contribution of Working Group II to the Third Assessment Report of the Intergovernmental Panel on Climate Change, Cambridge University Press, Cambridge/New York, 1031 pp.

Meinke H, Stone, R. (2005) Seasonal and inter-annual climate forecasting: The new tool for increasing preparedness to climate variability and change in agricultural planning and operations. Clim Chang 70: 221–253.

MAPA (2005) Plan Nacional de Regadíos. Ministerio de Agricultura y Pesca. http://www.mapa.es/es/desarrollo/pags/pnr/principal.htm.

Morrison LW, Korzukhin MD, Porter SD (2005) Predicted range expansion of the invasive fire ant, Solenopsis invicta, in eastern United States based on the VEMAP global scenario. Divers Dist 11: 199–204.

Neira XX, Alvarez CJ, Cuesta TS, Cancela JJ (2005) Evaluation of water-use in traditional irrigation: An application to the Lemos Valley irrigation district, northwest of Spain. Agric Water Manage 75:137–151.

Norse D, Tschirley JB (2000) Links between science and policy making. Agric Ecosyst Environ 82:15–26.

Olesen JE and Bindi M (2002) Consequences of climate change for European agricultural productivity, land use and policy. Eur J Agron 16:239–262

O'Neal MR, Nearing MA, Vining RC, Southworth J, Pfeifer RA (2005) Climate change impacts on soil erosion in Midwest United States with changes in crop management. Catena 61:165–184

Pal JS, Giorgi F, Bi XQ (2004) Consistency of recent European summer precipitation trends and extremes with future regional climate projections – art. no. L13202. Geophy Res Lett 31:13202–13202

Parry M (ed) (2000) Assessment of Potential Effects and Adaptations for Climate Change in Europe. The Europe Acacia Project. Jackson Environment Institute, University of East Anglia, Norwich, UK, 320 pp

Pedersen SM, Fountas S, Blackmore BS, Gylling M, Pedersen JL (2004) Adoption and perspectives of precision farming in Denmark. Acta Agric. Scand. Section B-Soil and Plant Science 54:2–8

Perarnaud V, Seguin B, Malezieux E, Deque M, Loustau D (2005) Agrometeorological research and applications needed to prepare agriculture and forestry to 21st century climate change. Clim Chang 70: 319–340.

Porter JR and Semenov MA (2005) Crop responses to climatic variation. Phil Trans R Soc Lond B Biol Sci 360 (1463):2021–35

Rafoss T, Sæthre M (2003) Spatial and temporal distribution of bioclimatic potential for the Codling moth and the Colorado potato beetle in Norway: model prediction versus climate and field data from 1990s. Agric For Ent 5: 75–85.

Reilly JM and Schimmelpfennig D (1999) Agricultural Impact Assessment, Vulnerability, and the Scope for Adaptation. Clim Chang 43 (4):745–788

Rojas, O., Rembold, F., Royer, A., Negre, T. (2005) Real-time agrometeorological crop yield monitoring in Eastern Africa. Agron Sust Dev 25 (1):63–77.

Rounsevell MDA, Evans SP, Bullock P (1999) Climate change and agricultural soils: Impacts and adaptation. Clim Chang 43 (4):683–709

Rounsevell MDA, Ewert F, Reginster I, Leemans R, Carter TR (2005) Future scenarios of European agricultural land use: II. Projecting changes in cropland and grassland. Agric Ecosys Environ 107 (2–3):177–135

Rosenzweig C, Tubiello FN, Goldberg R, Mills E, Bloomfield J (2002) Increased crop damage in the US from excess precipitation under climate change. Glob Env Chang 12:197–202

Rosenzweig C, Strzepek KM, Major DC, Iglesias A, Yates DN, McDluskey A, Hillel D (2004) Water resources for agriculture in a changing climate: International case studies. Glob Env Chang 14 :345–360

Salinger MJ, Sivakumar MVK, Motha R (2005) Reducing vulnerability of agriculture and forestry to climate variability and change: Workshop summary and recommendations. Clim Chang 70:341–362.

Seguin B (2003) Adaptation of agricultural production systems to climatic change. Comptes Rendus – Geosci 335 (6–7):569–575

Semenov MA, Porter JR, Delecolle R (1993) Simulation of the effects of climate change on growth and development of wheat in the U. K. and France. In: EPOCH Project: The Effects of Climate Change on Agricultural and Horticultural Potential in the EC. Final Project Report, University of Oxford-European Communities

Semenov MA and Porter JR (1995) Climatic variability and the modelling of crop yields. Agric For Meteorol 73:265-283

Seneviratne SI, Lüthi D, Litschi M, Schär Ch (2006) Land-atmosphere coupling and climate change in Europe. Nature 443/14:205–209.

Sip V, Skorpık M, Chrpova J, Sottnikova V, Bartova S (2000) Effect of cultivar and cultural practices on grain yield and bread-making quality of winter wheat. Rostl. Vy´r. 46, 159–167.

Sivakumar MVK, Brunini O, Das HP (2005) Impacts of present and future climate variability on agriculture and forestry in the arid and semi-arid tropics. Clim Chang 70:31–72.

Smit B and Yunlong C (1996) Climate change and agriculture in China. Glob Env Chang 6:205–214

Smithers J, Blay-Palmer A (2001) Technology innovation as a strategy for climate adaptation in agriculture. Appl Geogr 21:175–197

Soja G, Soja A, Eitzinger J, Gruszcynski G, Trnka M, Kubu G, Formayer H, Schneider W, Suppan F, Koukal T (2005) Analyse der Auswirkungen der Trockenheit 2003 in der Landwirtschaft Österreichs – Vergleich verschiedener Methoden. Endbericht von StartClim2004.C; in StartClim2004: Analysen von Hitze und Trockenheit und deren Auswirkungen in Österreich .Endbericht, Auftraggeber: BMLFUW, BMBWK, BMWA, Österreichische Hagelversicherung, Österreichische Nationalbank, Umweltbundesamt, Verbund AHP

Southworth J, Randolph JC, Habeck M, Doering OC, Pfeifer RA, Rao DG, Johnston JJ (2000) Consequences of future climate change and changing climate variability on maize yields in the Midwestern United States. Agric Ecosyst Env 82:139–158

Southworth J, Pfeifer RA, Habeck M, Randoflph JC, Doering OC, Johnston JJ, Rao DG (2002) Changes in soybean yields in the Midwestern United States as a result of future changes in climate, climate variability, and CO_2 fertilization. Clim Chang 53:447–475

Sroller J, Pulkrabek J, Novak D, Famera O (2002) The effect of perennial forage crop on grain yields in submontane regions. Rostl. Vy´r. 48, 154–158.

Stigter CJ (1988) Microclimate Management and Manipulation in Traditional Farming. CagM Report No. 25, WMO/TD-No. 228, World Meteorological Organization, Geneva, p.20, VI Appendices.
Stigter CJ (1994) Management and manipulation of microclimate. In Griffiths, J.F. (ed.), Handbook of Agricultural Meteorology, Oxford University Press, Chapter 27, pp273–284
Stigter CJ, Zheng Dawei, Onyewotu LOZ, Mei Xurong (2005) Using traditional methods and indigenous technologies for coping with climate variability. Clim Chang 70:255–271.
Tao F, Yokozawa M, Hayashi Y, Lin E (2003) Future climate change, the agricultural water cycle, and agricultural production in China. Agric Ecosyst Environ 95 (1):203–215
Trnka M, Dubrovsky M, Žalud Z (2004) Climate change impacts and adaptation strategies in spring barley production in the Czech Republic. Clim Chang 64:227–255
Trnka M, Eitzinger J, Gruszczynksi G, Burchgraber K, Resch R, Schaumberger A (2006) A simple statistical model for predicting herbage production from permanent grassland. Grass For Sci 61, 253–271.
Trnka M., Muška F., Semerádová D., Dubrovský M., Kocmánková E., Žalud Z. (2007): European Corn Borer Life Stage Model: Regional Estimates of Pest Development and Spatial Distribution under Present and Expected Climate. Ecological modeling, doi:10.1016/j.ecolmodel.2007.04.014
Tubiello FN, Donatelli M, Rosenzweig C, Stockle CO (2000) Effects of climate change and elevated CO_2 on cropping systems: model predictions at two Italian locations. Europ J Agron 13:179–189
Utset A, Farre I, Martínez-Cob A, Cavero J (2004) Comparing Penman–Monteith and Priestley–Taylor approaches as reference-evapotranspiration inputs for modelling maize water-use under Mediterranean conditions. Agric Water Manag 66:205–219.
Utset A (2005) Introducing tools for agricultural decision-making under climate change conditions by connecting users and tool-providers (AGRIDEMA). Commission of the European Communities. Research Directorate General. EC contract No 003944 (GOCE).
Utset A, Martínez-Cob A, Farré I, Cavero J (2006a) Simulating the effects of extreme dry and wet years on the water use of flooding-irrigated maize in a Mediterranean landplane. Agric. Water Manag. 85:77–84.
Utset A, Del Río B, Martínez JC, Martínez D, Provedo R, Martín JC (2006b) El plan de experimentación agraria desarrollado por ITACyL y los regantes de Castilla y León. Tierras del Norte de Castilla. (in press).
Vogl Ch R(1990) Traditionelle Andine Agrartechnologie. Diplomarbeit, Universität für Bodenkultur, Wien.
Vrkoc F (1992) Contribution of some factors to the development of crop production in the CSFR. Sci. Agric. Bohemoslovaca 24 (2), 125–131.
Wessolek G and Asseng S (2006) Trade-off between wheat yield and drainage under current and climate change conditions in northeast Germany. Europ J Agron 24:333–342
Wolf J, van Oijen M, Kempenaar C (2002) Analysis of the experimental variability in wheat responses to elevated CO2 and temperature. Agric Ecosyst Environ 93:227–247
Yeo A (1999) Predicting the interaction between the effects of salinity and climate change on crop plants. Sci Hort 78:159–174
You SC (2001) Agricultural adaptation of climate change in China. J Environ Sci (China) 13:192–197
Zalud Z, Dubrovsky M (2002) Modelling climate change impacts on maize growth and development in the Czech Republic. Theor Appl Climatol 72:85–102

CHAPTER 11

Complying with farmers' conditions and needs using new weather and climate information approaches and technologies

C.J. Stigter, Tan Ying, H.P. Das, Zheng Dawei, R.E. Rivero Vega,
Nguyen Van Viet, N.I. Bakheit, Y.M. Abdullahi

11.1
Introduction

The preparatory note of this Workshop (WMO/CAgM 2005) gives six specific objectives of the Workshop. Others will deal with identification and assessment of the components of farmers' agrometeorological coping strategies with risks and unceraintities, discuss the major challenges to these coping strategies (e.g. Rathore and Stigter 2007), review the opportunities that farmers have, to cope with agrometeorological risks and uncertainties, and provide examples. In this paper we particularly discuss and recommend suitable policy and policy support options to comply with farmers' conditions and needs that determine their vulnerabilities as well as their windows of opportunity. This includes the role of weather and climate information approaches and information technologies and whether new approaches and technologies have roles to play. If this is the case we should find out what determines the scope of the application of such developments. We belong to the schools that want to make a plea for achieving a "culture of disaster preparedness" (e.g. Sikka 2001; Rathore and Stigter 2007) and we feel that the term "risk management" should be abandoned for all but the richest farmers (Sahni and Ariyabandu 2003). The key-word in preparedness is not "management" but "resilience" (e.g. Reijntjes 2001; Björnsen and Gurung 2001). An analysis of farmers' agroecological resistance to drought in Africa and to hurricanes/cyclones in Central America and India indicated independently that resilience has a social as well as a technical dimension (ILEIA 2000; Holt-Gimenez 2001; Stigter et al. 2003).

Recently Lassa (2006) emphasized that disaster should be considered a forced marriage between a hazard and vulnerability. To cope with impact problems of frequently occurring disasters, the vulnerability of people should be reduced and the hazards should be mitigated, which therefore means fighting on at least two different fronts. What is often badly understood by those that have to carry out policies of disaster impact reduction is that there is a long process involved in for example a drought or flood hazard to produce a disaster (e.g. Brandt et al. 2001; Connelly and Wilson 2001; Stigter et al. 2003). In the context of complying with farmers' conditions and needs we will handle the four policy issues that Lassa (2006) distinguished: (i) mitigation practices; (ii) disaster preparedness; (iii) contingency planning and responses and (iv) disaster risk mainstreaming. These issues can also be recognized in Stigter et al. (2003). It should be realized that as agricultural scientists we have come closer than ever to farmers, but we are farther away than ever from policy makers (Stigter 2005a). However, the same appeared to apply in health services in developing countries where the analogies with agrometeorological services were striking when replacing the most terrible diseases affecting people with the most terrible climate and weather disasters (Stigter 2005a).

Another important introductional issue is the parallels found between problems with policy options for structural preparedness for and rehabilitation from disasters, such as in well selected agrometeorological services for such purposes, and some basic difficulties generally encountered in establishing services (e.g. Van Noordwijk et al. 2005, in the aftermath of the tsunami disaster in Aceh, Sumatra, Indonesia, after one year). There is little principal difference between what Stigter saw in April 2005 in Central Sudan (Stigter 2005b), revisiting an earlier research area (Bakheit et al. 2001, 2005) in the middle of its third year of drought, and the tsunami areas of Sumatra after the first relief had reached the victims, with the exception of the attention that at least the people in Aceh initially got (Stigter 2006a).

11.2
Complying with conditions and needs

According to Van Noordwijk et al. (2005), at a meeting of local governments, national and international agencies and NGOs in Meulaboh, Sumatra, there were five main reasons that became apparent for the problems encountered in assisting poor people in building or rebuilding a sustainable livelihood. A first basic problem in Aceh/Sumatra appeared to be one of appropriate need assessments. There was a call for more critical consideration of local needs. For more than 20 years now, we have been arguing and practising in agrometeorology the local bottom up determination of "which problems with agrometeorological components that farmers bring up need to be solved first" (e.g. Stigter et al. 2005c). This should replace the offers of agrometeorologists of what they are able to solve. In Aceh, as well as in Sudan, "we need to anticipate the broad range of people's needs in the recovery of infrastructure and help communities prepare for the future" (Van Noordwijk et al. 2005). Doing that, we should realize that "livelihood strategies emerge in response to opportunities, not from preconceived master plans or blueprints" (Van Noordwijk et al. 2005; see also Röling et al. 2004 and Hounkonnou et al. 2006; but also some failures wrongly talked into useful results by Bouma et al. 2006).

The second important issue is that the biggest challenge facing all organizations working in the affected regions in Sumatra is collaboration and coordination, between agencies and between actors at different levels (Van Noordwijk et al. 2005). For agrometeorology we recall Jacob Lomas' long time call for collaboration between relevant government organizations in agriculture and meteorology. He noted particularly the lack of cooperation between the institutions providing information and relevant advisories and those responsible for their transfer to the farming communities (e.g. Lomas et al. 2000), which one of us more recently again echoed (e.g. Stigter 2004).

Related to the above is the observed urgent necessity in Sumatra of attention to a "missing middle layer" in this co-ordination (Van Noordwijk et al. 2005). And this is again exactly the need for "intermediaries" between NMHSs, Research Institutes, Universities and agrometeorological extension (as another layer of such intermediaries) close to the farmers, which are advocated over already many years in agrometeorological services (Stigter et al. 2005a; WMO 2006a). The observed needs for capacity building in these directions, including better involvement of the lowest level local government agencies, that is presently absent in Sumatra/Aceh (Van Noordwijk et al. 2005), has again its parallel in the need for capacity building for agrometeorological services. This is in the observed insufficient involvement through education and training of the user community. The latter includes the farm advisory services that can provide relevant assistance, to be derived and adapted from more general weather information products (Lomas et al. 2000; Stigter 2004; WMO 2006a).

The three remaining issues in Aceh/Sumatra are all related to policy matters. Van Noordwijk et al. (2005) distinguish (a) environmental issues, (b) infrastructural and market issues and (c) issues related to the lack of base line data and the support to collect and collate these. Under (a) they argue for example that (a physical system for) preventing another tsunami to cause the same amount of damage has probably been too high on the public list. This is comparable to our pleas for having environmental monitoring, early warning, and other predictions of disasters in (agro)meteorology, always directly related to relatively easy and economically sound preparedness and mitigation possibilities within the livelihood of people (e.g. Stigter et al. 2003).

Under (b) it was indicated that many aid organizations forgot some important aspects of "macro" market chains and infrastructural necessities in Aceh/Sumatra. Well known parallels in agrometeorology are in examples where farmers do not exploit microclimate or other yield enhancing improvements because of a lack of roads to markets, lack of appropriate storage facilities or discouraging price ratios between added inputs and higher yields. In improvement of traditional underground sorghum grain storage in Central Sudan for example, mobility and economic aspects had to be taken into account as well (Bakheit et al. 2001, 2005).

The point (c) observed in Aceh/Sumatra finally is again only too well known to agrometeorologists. The slow process of trial and error in (changing) natural resource use, caused by the lack of base-line data and the support to collect and collate these, obviously applies to routine meteorological and agricultural data in rural and other remote areas. And also to data from fields stricken by pest or disease. However, it almost even more so applies to basic socio-economic data, causing completely wrong approaches in agrometeorological designs due to completely wrong assumptions (e.g. Onyewotu et al. 2003; Stigter et al. 2005b).

This confusion is also well illustrated in a recent paper by Verdin et al. (2005). They believe that creative coping strategies to manage climate change and increased climate variability requires fundamental technical capacities first, illustrated with Ethiopia as a case. They state that adaptation strategies cannot be developed and implemented until trends and shifts in climate have been identified using access to modern methods of data capture, data management, telecommunications, modelling and analysis. It is true that "it is in the hands of those with local knowledge that creative adaptation strategies will be forthcoming". But the wishful thinking of large scale transfer of advanced climate science and technology to African counterparts as condition sine qua non (Verdin et al. 2005), however sympathetically proposed, starts to come close to the naivety and lack of understanding of policy factors that was shown by Huntingford and Gash (2005) and criticized by Stigter (2005d). Local progress can be made differently through policies of improved response farming and other improved preparedness strategies under the present conditions (Stigter et al. 2005b). Further progress will come long thereafter based on further agrometeorological services established with the three components that were earlier distinguished (already in the title of Stigter 2005c).

With this last point we are back at what definitely is the largest and most important parallel between the experiences in Sumatra/Aceh and those in the introduction of agrometeorological services in poor rural areas: the lack of appropriate need assessments. For agrometeorology, we need to train a "middle level", "intermediaries", working as two-way guidance. They should simultaneously support highly needed actions of farmers at the production level as well as the generation of more relevant and better absorbable (more client friendly) products by NMHSs, Research Institutes and Universities. Without this, many weather and climate products, and related efforts, remain lost on those farmers that need our support most. This is well illustrated by the surprising results from Ahmed et al. (2006) in Pakistan for irrigated farming in Rechna Doab. They statistically found that farming experience was only a negative factor in farm economic land productivity, because more experienced farmers were too rigid and less experienced farmers too unskilled. Both need services for change. The pertinent differentiation and upscaling needed in these exercises are illustrated by the Chinese research we discuss in the following paragraphs.

Effective and accountable local authorities are the single most important institution for reducing the toll of natural and human induced disasters (Sahni and Ariyabandu 2003). In India, the country's day to day administration centres around the District Collector who is also in charge of all the relief measures at that level. There are sub-divisions and tehsils. The lowest unit of administration is the village. All these tiers of administration function as a team to provide succour to the people in the event of disaster (Sahni and Ariyabandu 2003). It would be helpful if establishment of agrometeorological services could be guided at the lowest administrative level. However, collected examples in China show how far agrometeorological information still was from farmers' conditions and actual services only ten years ago (Tai Huajie et al. 1997). Pilot projects presently underway have to show that this gap can actually be narrowed (Stigter 2006d).

11.3
Differentiated information needs and channels for various farmers

11.3.1
Information demands of different income levels in poor areas of China

Development and strength of a nation are very much determined by capacities of collecting, organizing and applying information in the right quantities and qualities (e.g. LEISA 2002). With appropriate macro-economic policies provided by the government, the existence of suitable market forces and sufficient attempts to accommodate the inescapable urbanization trends, the development of rural economies depends mainly on capacity building and services in rural communities (Tan Ying et al. in prep.). After improving, adapting and focusing rural information and education systems, information and communication technologies (ICTs) could play very important roles in such capacity building and services.

Essentially the process of information communication can be described as follows. Primary information is generated, on request or by supply, by various knowledge sources, supported by data, research, education/training/extension and policy systems (Stigter 2005c). This leads to informative products (in our fields from the Chinese Meteorological Administration (CMA), Universities and Research Institutes) that cannot yet be absorbed by most users in poor areas (Stigter 2006e). Suppliers or communicators of derived information encode it in a more client friendly form and disseminate the encoded information to receivers through some services channels and media. After decoding and using this information, ideally the receivers give feed back to the suppliers through some of the same or other channels. This is a continuously running loop. Sometimes suppliers and receivers interchange their roles with each other (Severn and Tankard 2000). Examples are farmer innovations. For information suppliers and communicators, some information receivers may also form a primary market for information. Therefore, study of the actual demands of information receivers is vital to the improvement of information services, as we already concluded from another approach above.

Tan Ying et al. (in prep.) found from experiments detailed below that 95% of farmers in parts of China where most poor farmers live think that information could bring them great profits. All except few elders admit that they need information and believe in its importance. These people think that it is very necessary to obtain information for their life and work, and that through news they could understand the world. They are concerned with social development. Many believe that by obtaining technological information, including weather and climate related information, and market information they could carve out or enlarge their production scale and raise profits. Most investigated farmers believe that China is in a period of economic structural adjustment and that the agricultural structure needs great alterations. Facing such an uncertain market and so many choices, they seem insatiable and panic. Therefore, they are eager to understand new things in order to make highly efficient low-risk decisions.

For technological and market information in agriculture, farmers and extension services are the basic receivers or target groups. Two bodies of the Ministry of Agriculture of China, the Department of Market Information and the Center for Agricultural and Rural Research, launched a sample survey by using their established long-term rural observation stations (Ministry of Agriculture of China 2000). The survey focused on the current situation of farmers' markets and the technological information they receive and use. It involved 31 provinces (including Beijing, Shanghai, Tianjin and Chongqing, the four cities under the direct jurisdiction of the central governance) and autonomous regions. The results from this survey showed that farmers are mostly concerned with two kinds of information, i.e. (i) information on practical technology with low investment and instant profits and (ii) information on market demands for agricultural products.

In 2001, the Institute for Technological Information Research in the Shandong Provincial Academy of Agricultural Science through synthesis analyzed the results of their research on agricultural information requirements from data provided by 12 local bureaus of 8 regions/cities (Gao Chunxin and Zheng Yan 2001). The analysis showed that among the various information, the technological information ranked first, averaging 67 percent, according to multiple selections by agricultural users. Market information came second, averaging 36 percent, confirming the above results.

Research on information demands for different farmer-users has been rare, especially for the poorer central and western regions of China. The central and western regions of China are mostly countryside including pastures, border areas and regions where minority groups are living. There are mainly mountains, tableland, remote deserts and abundant areas with poor natural resources as well as abominable working and living conditions. Information flow and communication are seriously impeded by severe natural conditions, complicated geographical environments and inconvenient traffic, which also directly affect the economic development of backward areas. It appears that 50% of the very poor are situated in western areas, and close to 60% of them are living in long-term poverty counties (China Agricultural University 2002). In 2002, the Institute of Science and Education, Northwest Normal University, brought forward that attention should be paid to farmers' actual demands for information, especially for the western regions (Zhu Fengli 2002). They suggested that although farmers are heterogeneous, both in their occupations and their information demands, this diversity had not been genuinely identified, and their detailed priority information demands had not been properly revealed.

11.3.2
Differentiation between income levels in poor areas of China

Ye (2002) defines a farmer initiative as the impetus that sufficiently and necessarily drives a farmer (or group of farmers) to formulate a realistic strategic plan, and to implement it in an attempt to create space for maneuver and to pursue change through changing social conditions. He lists the critical factors contributing to the process of various farmer initiatives in China as including trust, social networks, information derived from networks, past experiences, media and publications, calculations of cost-effectiveness, enlightenment from interaction with and influence of family members and the network of outsiders, information from the market, visits to successful cases, self-help and cooperation, reputation (respect, credibility), interests, beliefs, curiosity vis-à-vis the outside world, technology innovation, knowledge from publications and training, study visits, skills and technical capability, enlightenment from observation and favorable policies. Many of these are interrelated and some in fact can be grouped in broader categories, the broadest one being "social capital" as the mobilizer (Ye 2002).

More recently Tan Ying et al. (in prep.) posed the question whether and how the actual information needs of farmers are met given that new channels of information give lots of farmers in China new chances to choose the most suitable information for their use. They used information from about 400 farmer families, distributed over 30 villages in the provinces Yunnan, Shaanxi, Anhui, Hebei and Shanxi, Central and Western China. According to different situations of different areas, the methods varied. The research combined qualitative and quantitative analysis, including Participatory Rural Appraisal (PRA) surveys, questionnaire surveys (random and stratified sampling) and interviews.

A first experiment showed that four different income-levels of farmers treated the technological and related information differently and their levels of satisfaction were different too. Also, they appeared to receive the information largely through different channels. However, farms at the same income levels in different areas appear to have similar information needs. In addition, through the participative research it was understood that most farmers were not satisfied with the information that is provided by the mass media. From another experiment it followed that when the farmers had similar occupations (as planter, cultivator, businessman, village technician, village leader) their information requirements were close to each other. But different income type farmers used again different media channels to receive the information. Results of again another experiment implied that farmers with the same occupation often select similar information sources, while farmers from different jobs obviously make different choices, which were again income related.

This paper tries to make a sensible differentiation among rural people. Weather and climate information and their dissemination and use are here to be seen as part of technological information. No studies on these specific meteorological information needs and channels have been made. From the surveys qualitatively the following applies to the income differentiation in central and western China (Tan Ying et al. in prep.).

Very poor farmers. They have limited technological information demands and mainly obtain information from leaders, neighbors and relatives. Most are over 50 years old, illiterate, and only a few of them have studied in primary school. As to their family situation, many are solitary elderly people, widows/widowers or people in bad mental or bodily health. Their main problem is lack of labor. They generally make only use of local resources and expect help from the government. They are largely indifferent to multifarious information from various media, don't listen to/watch/read functional broadcasts/TV programmes/newspaper items. When watching TV, they are mostly interested in entertainment programs such as teleplays and films.

Low-income farmers. Most of them are planters and cultivators and have only had primary school. The main information channels for this type of farmers are mass media, leaders, able friends and relatives. They usually accept information passively and seldom seek technological and enriching information actively. They can only understand a little popularized science and few new technologies on TV, and a bit older farmers can hardly understand them at all. They have very unremarkable technological information demands, but are interested in many other information services, such as regarding rural policies and regulations, applied scientific and cultural information with good knowledge contents, sales and supply information of agricultural products. However, without special assistance they cannot easily articulate what the specific information is that they need most. All programs on TV are wonderful to them.

Middle-income farmers. Most of these farmers are somewhat larger planters and cultivators/growers whose information demands are comparatively strong. Besides TV & radio and personal communication, they begin to pay more attention to newspapers, brochures and books related to agricultural production. What they need most, usually are practical operational and market information services such as on the utilization of new technologies, weather forecasts, rural policies and on sales as well as enriching information from able villagers and so on. However they cannot use them with sufficient efficiency. They can find technological information services they need with the assistance of technicians, village leaders and able villagers. Most cannot select useful information themselves or, getting unsatisfied results after applying some new knowledge, hesitate to accept and use such information services. However, the informatization process in some areas is rather faster and there computers are becoming available in some villages. Several most educated farmers begin to get interested in this "new thing" and want to obtain some technological information from the internet. This picture is confirmed from elsewhere (LEISA 2002).

Richer farmers. Most of these farmers are 35-45 years old and influential planters, growers and traders (self-employed workers, entrepreneurs). The information channels for these farmers mainly are TV, the press, broadcasts and Internet, personal communication such as the marketplace, telephone, etc. Having studied in high school or taken adult education and new technological training after having been in agriculture already for some time, they are very sensitive to agricultural policies, market and farm product information with which they can create benefits quickly. Usually they can actively search various media, and spend more time on obtaining the latest information. They prefer to communicate with well-informed people (services, private information agencies). During the survey, it was found that most of these big planters, cultivators and businessmen benefited from the technological information. Several rich farmers even begin to use computers to sell their products.

11.3.3
Information channels for different income levels in poor areas of China

From the results of Tan Ying et al. (in prep.) it follows that in central and western regions of China, traditional modes of information flow and communication still occupy the main position. Qualitatively it followed from the survey contacts that from the point of view of utilization of technological information and acknowledgement/acceptance of its effects, 90 percent of farmers thought that the flow paths combining personal communication with mass media played an important role in farmers' information selection. They would rather transmit demands for new technology through personal communications such as with able villagers, model households, experts, technicians, etc., confirming that personal relations play a very important role in China's rural technological information initiatives.

This is confirmed by the results of Ye (2000). Therefore, although information service systems have been shaped and established, the scientific and technological requirements of investigated farmer households have not been met yet. For farmers in relatively rich regions, channels are more diverse, but farmers in comparatively poor regions can mainly get information through personal communication. This is again confirmed from sources outside China (LEISA 2002).

11.3.4
Demand and supply of information for different income levels in poor areas of China

The very poor farmer (family) pays little attention to technological information (Tan Ying et al. in prep.). According to its importance, the highest information demand for the category of low-income farmer families can be generally ranked as follows (acknowledging areal differences): new varieties, rural policies, utilization of new technologies and applied scientific knowledge. Information demand for anyone item of information is only 30% on average and 40% in some areas. Similarly, for the mid-income farmer families, the general order was as follows: utilization of new technologies, new varieties, the sale of products, applied scientific knowledge, market information and rural policies. The average information demand rate of four regions is just over 30%, not really higher than for the low-income group, but this category began to pay more attention to market information and the sales information of products.

As to the relatively-rich farmer families, the general order is as follows: market information, the sale of products, utilization of new technologies, applied scientific knowledge and rural policies. The average information demand of the four regions is 40 percent, higher compared with the mid- and low-income groups, but also still in need of much improvement (Tan Ying et al. in prep.). The low demand figures show that there is eventually still much work to do to make known the services that may be offered.

11.3.5
General implications of the findings for different income levels in poor areas of China

With the above in mind, four main conclusions on technological information services, including agrometeorological services, can be drawn. These are followed by some suggestions springing from the results obtained on the four income groups that we have differentiated (Tan Ying et al. in prep.).
1) **Very poor farmers can't use the existing technological information services and therefore have limited demands for such services.**
   Because this category of farmers is the last destination target group, government at all levels should give them sufficient support. Local government could

set up poverty-alleviation groups in the village committees and villager groups in order to strengthen their relations that are important in organizing communication and information for such farmers. In most rural areas this group is bound to disappear in the next development phase, so a structural solution is most often not needed but ad hoc assistance should be organized to relieve their plights.

2) **The awareness of low-income farmers of technological information services is too small.**
This category of farmers accounts for a relatively large target group in underdeveloped areas. Facing such farmer groups the chief tasks are to popularize knowledge, to derive suitable services based on knowledge and to increase an awareness of available technological information services. For example, an information station in Sanyuanzhen, Wuhu, Anhui province plays fully the government's services function. It sponsors information dissemination meetings regularly, in order to urge farmers listening to/watching technological programs and paying attention to technological information services. Meanwhile, they make use of lottery attached contests on the agriculture channel, as well as other programs, to stimulate and improve farmers' initiatives and awareness to information.

3) **Middle-income farmers can't utilize information services very efficiently.**
The leading role of able villagers and village leaders should be fully played. Generally speaking, able villagers are pioneers among progressive farmers and have relatively strong ability to obtain and use information services. Let them assist the others. Still, government's information services departments should adopt various training types to bring them services derived from applied scientific and technological knowledge. They should enhance their education so that they themselves can accumulate, analyze and utilize information services more efficiently.

4) **Rich farmers have greatest ability of utilizing effective information services.**
Relatively rich farmer families contain often able villagers and villager leaders. According to classical work by Rogers (1983), able villagers are often the first level destination, namely information pioneers; mid-income farmer families are the second level destination, namely early information receivers; while low-income farmer families are the third level destination, namely late information receivers. Case studies demonstrate that farmer initiatives are processes of "enlightenment", not only inspired by ideas, but more importantly, by engaging and learning from social interaction and everyday experience (Ye 2002).

11.4
Implications for information approaches and technologies

11.4.1
Poor farmers

For the poorest farmers, transfer and adaptation of simple innovative technology developed by others and of simple operational knowledge of all kinds to improve their conditions and income are possible information services for this group. Because of its temporary character, without investment of any kind being involved, the simplest information approaches and technologies will have to do. Because of the importance of social capital, being least developed within these groups, assistance to bring down the simplest successful agrometeorological services from one level higher up social classes with comparable farming systems or occupations appears to be the best approach.

Following the four policy issues of Lassa (2006) as earlier enumerated, as to (i) mitigation practices and (ii) disaster preparedness, such examples should this way be transferred downwards. In (iii) contingency planning and responses as well as (iv) disaster risk mainstreaming, the public domain should be involved with appropriate supportive policies.

11.4.2
Low-income farmers

In the long run, specific training of extension intermediaries, such as already existing village technicians and in-service trained members of Agricultural Extension Services and the NMHSs, in more to the point fields of agrometeorological services to such low-income farmers, will be a lasting solution. The actual needs of these farmers for such services have therefore to be studied and mapped out much more appropriately. This can partly be done in the training of such intermediaries. Field classes to train these farmers have been shown to be effective means that such intermediaries can use (Stigter et al. 2005a).

With this information approach and knowing the characteristics of the group and the importance of social capital, again the next higher social group should receive incentives to be involved in the field classes with examples of successful application of agrometeorological services. Rural radio and where possible TV appear to be the best instruments to structurally use at this level in addition to and as part of the field classes mentioned. This is the lowest level where we estimate that for agrometeorology improved Agrometeorological Bulletins (Sivakumar 2002) could already have impacts, but only with organized assistance for their interpretation and use.

Community based mitigation practices and disaster preparedness should be transferred using these media and training approaches (also Stigter et al. 2003). Contingency planning and responses as well as disaster risk mainstreaming may have the usual public as well as less common private aspects based on how the social capital can be locally organized. In other countries than China, such as India for example, NGOs may play a pivotal role (Morrow 2002a).

11.4.3
Middle-income farmers

Also here, specific training of the same extension intermediaries in to the point fields of applied services to such middle-income farmers will be a lasting solution.

The actual needs of these farmers for such services change much more dynamically. These needs have therefore to be permanently followed and to be dynamically met. This can again partly be done in the training of intermediaries but is even more demanding as to their level and flexibility. The same applies to the use of improved Agrometeorological Bulletins (Sivakumar 2002). According to the five periods of the diffusion of innovation adoption model of Rogers (1983) (knowledge diffusion period, persuasion period, decision period, using period and affirmation period), it may be argued that during the knowledge diffusion and persuasion periods extension intermediaries, field classes and mass media play a very important role. During the other periods in addition also able villagers' capacities (i.e. social capital) are very important. This has to be taken into account in the strategies chosen.

In addition to field classes, mass media and bulletins, other information technology can be introduced or stimulated, with mobile telephones appearing to become an obvious choice. This, however, means that establishment and use of agrometeorological services should be adapted to this information medium. It is also at this level that stakeholder-driven funding mechanisms for agricultural innovation may be part of the solution. This is about funding for technology development and dissemination interactively controlled and managed by stakeholders. Heemskerk and Wennink (2006) have indicated that for such innovative funding mechanisms to work, far-reaching institutional changes need to take place. This means enhancing client control over priorities and resources, expanding the range and skills of service providers, and making organizational changes in all stakeholder organizations, whether public sector, private sector or farmers' organizations (formal and informal).

A pre-condition is proper understanding of their problems in coping with risks, that is hazards and vulnerability, in the context of the four policy issues of Lassa (2006) dealt with earlier, and clarity/honesty in such policy issues (Wason 2002; Kaimowitz 2005). As to mitigation practices and disaster preparedness, training, media, new communication technologies and social capital are the critical resources for bringing change to rural communities in China at this level. However, unlike other kinds of capital, social capital cannot be inherited or passed on to others. It exists only when mobilized by specific social actors (Ye 2002). This includes scientists and technicians (Zheng Dawei et al. 2005; Zhao Caixia et al. 2005). Contingency planning and responses as well as disaster risk mainstreaming should be a mix of local government initiatives and private initiatives by these social actors.

11.4.4
Richer farmers

Generally, richer farmers have had (much) more formal education and are therefore also able to use newer communication technologies such as mobile telephones, computers and internet facilities. The commercialization of services, also agrometeorological services (Stigter 2006b), may start here, assisted by government support where richer farmers are able to play a role at lower social levels. For mitigation practices and disaster preparedness, the same applies. Contingency planning and responses as well as disaster risk mainstreaming should be a mix of private initiatives and local government guidance and services with financial implications for those making use of them successfully.

11.4.5
Other developing countries

General. For other developing countries a similar differentiation will definitely be valid, but the stories that belong to each of their income groups and rural occupations will differ and the implications also. This has been explicitly confirmed by the authors for the different social environments from Cuba, India, Nigeria and Sudan. Das et al. (2003) report from Brazil the development of a social differentiation approach in understanding drought effects that specifies target groups "instead of relying on government intervention through regional channels, which always strengthens the structures responsible for generating rural poverty and its consequent vulnerability to droughts". It should be noted that farming systems have a great influence on differentiation aspects and that other factors than income may become important as well or even more important uniquely within such farming systems as illustrated in Pakistan (Ahmed et al. 2006). Studies like those made in China and Pakistan and more success stories like those reported in LEISA (2001; 2002) for various groups of farmers and farming systems would be very helpful in understanding the needed services differentiation and what is necessary for scaling up services.

In India, getting the feedback from farmers has only very recently been better organized to reach IMD. From recent field visits, discussions and responses to questionnaires the necessity of regional and farmer differentiations became very clear. But in Nigeria neither the meteorological service nor the public extension system have the capacity to provide farmers with these services now nor have farmers the awareness to demand for it and to utilize it. They therefore continue to rely only on indigenous knowledge sources. In Sudan such services do exist, for example to advise tenants on irrigation amounts in the Gezira scheme (Ibrahim et al. 2002; Hussein Adam, private communication, 2005), but on a very limited scale.

Now that large countries like China, India, Brazil and several other Latin American countries have announced to be tackling the services problems in rural areas, while the same was done by donors with respect to Africa, agrometeorological services should be seen as part of these new approaches to rural services (including information approaches and technologies). Without such a general overhaul of the services climate also agrometeorological services will remain slow in contributing to poverty alleviation (Stigter 2006c).

India. Worries have been expressed in India about the loss of traditional adaptation strategies such as in the eastern floodplains of the Ganga and Brahmapoutra and in the use of disaster resistant indigenous wild crops as spare staple foods. Another serious gap is the absence of crop condition data on a sufficient spatial and temporal scale. This could be repaired by an improved collection system from the bottom up, using the District Meteorological Information Centres of IMD and SMS communication, supported by remote sensing data.

Of course where clear-cut cropping patterns exist, results may be easier to obtain. For example, the method developed by Chattopadhyay et al. (2003) for predicting the incidence of leaf spot disease on groundnut in India using simple meteorological parameters, as early as two weeks in advance, was successfully applied as an agrometeorological service in several groundnut producing states. In India, there are indeed growing demands for timely and effective agricultural weather information for a wide variety of agricultural management decisions, ranging from crop's response to daily weather to the crop's adaptation to changing climate. However, despite the recognition of the perceived benefits of weather data and climate forecasts, there is little firm economic analysis in India to support these notions. Nevertheless, recent studies show that when forecast information is made available in a timely fashion, farmers do indeed react by making strategic decisions about what crop to plant and how much area to sow.

In coping with drought in India the approchoaches were taken: (i) provision of improved long range forecast of all India seasonal rainfall before the beginning of the season. This provided time to the planners to adopt different strategies; (ii) close monitoring of the rainfall over different parts of the country on daily, weekly and monthly scales within the rainy season; (iii) delineation of different agro-climatological zones which helps in specific measures for agricultural planning on climatological basis; (iv) continued research efforts to enhance capabilities of forecasting monsoon rain on a local, regional and all India basis on different temporal scales. There was for example absence of rain for 20 days during June 2006 in most areas of Chhattisgarh state, India. As soon as monsoon rains returned, farmers were advised to select their crop(s) among the short duration varieties of rice, red gram, green gram, black gram, soybean and groundnut for sowing. The extension officers of the State Department of Agriculture were in constant touch with the progressive farmers to implement the advisories. The farmers in Raipur district of the state decided to sow rice for larger areas and soybean for the remaining areas. Recently it was confirmed that those short duration crops sown are in good condition due to subsequent monsoon rainfall. It may be noted here that the information had been delivered to the farmers well in advance, precise in space, coherent with available options and in a local language understandable to the farmers.

Another simple example comes from the northern dry zone of Karnataka state of India. In line with earlier reports on response farming from Africa (Stewart 1991), Venkatesh and Guled (2004) proposed a method for qualitative and quantitative assessment of seasonal rainfall variability of the region. The rainfall patterns of the growing season were such that high and low rainfall for June to August varied sequentially. Based on this, projection was made in the year 2000 that during the next few years, July rainfall would be higher than the rainfall of June, which proved true for the next four years. Farmers in the area were advised to choose crop sequences in tune with such rainfall patterns. Such agrometeorological services helped the farming community of the region in coping with the drought situation in that region.

Cuba, Nigeria, Sudan, Vietnam. In Nigeria experience teaches that social and other potentials of rural youths can serve as a good entry point in upscaling and diffusion of weather and climate information services if greater access would be established by organizing farmers and intermediaries (Auta et al. 2003). Experience from Sri Lanka shows that modern ICTs are appealing particularly to this group (UNESCO 2002). In Nigeria an additional observation is overlapping/duplication of organizations mandated with these services. Likewise there is no clear policy of collaboration between similar agencies which generate agrometeorological information products on varying scales, like the research institutes. This is confirmed by the experience of the Dutch funded TTMI-Project in Nigeria (e.g. Onyewotu et al. 2003).

In Sudan, a study of the Gezira Scheme field management systems revealed that since nationalization in 1950 and even after abolishment in 1981 of a cotton-sharing system with the government that had emphasized centralized decision making, hardly any attention was paid to development and education of tenants, virtually no extension existed and short-term production achievements got priority. In such conditions almost any central services to improve decision making are missing. This is confirmed by the experience of the Dutch funded TTMI-Project in Sudan (e.g. Bakheit and Stigter 2005).

In Cuba organizing the needs of different farmer groups should have to be planned scientifically by very well trained levels of intermediaries because of farmers' misunderstandings of limitations of modern technologies. Client friendliness was a determining factor in Cuba in the capacity building involved and differentiation within the groups of farmers helped much in getting information absorbed broadly. Successful examples from Cuba again contain a clear-cut farming system, that of sugar production and the highly needed guidance of large scale planting operations by a successful agrometeorological service in forecasting of suitable sowing conditions. Higher sugar production and better cost/benefit ratios resulted and intermediaries now bring such sowing information to various farmers also as agrometeorological services for other crops. Another example from Cuba is the calculation and use of comfort indexes in the poultry industry and related agrometeorological information used to manage water intake, ventilation and other protective measures in this industry. A radar-based agrometeorological service since May of this year provides targeted forecasts necessary to ensure the secure transport of live animals, preventing such transport to take place under bad road and transport conditions. These are all very specific and direct services with high economical significance.

Viet (2002) reports that from an analytical point of view, floods, droughts, typhoons, frosts have been regionally recognized and economically studied as disasters to agriculture in Vietnam. Solutions have been sought in changing cropping calendars and patterns which are provincially and regionally proposed, and in proposals on water and tree management. This has resulted in government planning and designs and a systematic approach to improve this in the short run and as a long term strategy. Especially sowing times for the ongoing season in the Central highlands and the Mekong delta as well as some permanent changes in cropping patterns with two to three rice crops annually, that replace rice one time for a rotation with maize, sweet potatoes, cassava have been successful (Viet 2002). Farmers forced to migrate due to dam building successfully used designs developed as agrometeorological services for the agroclimatologically most suitable production systems (Viet and Liem 2005). Also for example the design of water erosion prevention on sloping land by forage grasses and other permanent vegetation and of the stabilization of terrace banks and edges by grass strips have been successfully developed services for the hill farmers in Bavi District, Hatay Province (Viet 2005).

11.5
What WMO/CAgM should realize as implications of the above

We end this paper with what WMO/CAgM should realize as implications of the above for the future of weather and climate information approaches and technologies in agricultural production. WMO (2006a) has very recently indicated what it sees for the role of weather and climate information approaches and technologies as key to future activities of the Commission. It starts with the warning that in developing countries there remain risks that very few high-level agrometeorological personnel and limited resources are geared towards modern specializations. This situation is accentuated by low quality data and the limited absorption capacity of agricultural decision makers for such agrometeorological products (Gadgil et al. 2000; WMO 2006a). This can be confirmed from the above illustrations of the situation in China. The above results question the idea that "the key to future activities of CAgM will be how to take advantage of the rapid innovations in technology" (WMO 2006a).

This definitely is the case in richer countries with low and decreasing farming populations with a high level of education (Stigter 2006b). The Chinese results show that this also may be the case for a group of richer farmers distinguished there. But for all other farmers in developing countries this is for the time being only true in a very limited way. Technical difficulties and some solutions were explained by Morrow (2002b). There are other key factors here, depending on education, income level and occupation. The reports by Boulahya et al. (2005) on using new technologies in rural Africa for communicating drought information are giving hope but also show the limitations. Understanding the actual needs and scope for agrometeorological services, the bottlenecks in the establishment of agrometeorological services and how to guide their introduction for various target groups is much more important. The results obtained in Africa and China (Stigter et al. 2005d), the present pilot projects in China and India and those in preparation in Brazil, India, Vietnam and Cambodia (Stigter in prep.) confirm this.

Already in this recent publication (WMO 2006a) it is recognized that enhancement of the communication channels for the improved dissemination of agricultural meteorological information should take into account the literacy levels of users, socio-economic conditions, level of technological development and accessibility to improved technology and farming systems. The Chinese results reported on above explain details of this picture. They also improve and refine the idea that in the developing world, "lack of resources and skills are the basic limitation to enhanced web-based dissemination of information", but the emphasis asked for rural radio use is confirmed in China and many other developing countries (WMO

2006a). WMO (2006b) also recently recognized and illustrated that experience has shown that a major gap remains in the identification of clear and useful guidelines on the exact nature of agrometeorological products that must be provided.

In this way the studies reported on from China are examples of how to better understand the importance of services, also agrometeorological services, in rural areas. This includes information approaches such as the use of intermediaries in training farmers and information technologies fit for the target groups concerned. The five "Aceh/Sumatra" issues discussed in section 2. should guide us. Such studies, but now specifically made with respect to agrometeorological services, would help us even more in getting the right picture and being able to give the right guidance for important differentiation and scaling up operations (Stigter 2006e).

But as John Locke once said, "opinions are always suspected and usually opposed, without any other reason but because they are not already common".

References

Ahmed S, Turral H, Ahmad M-u-D, Masih I (2006) Limits and opportunities to improve farm level land and water productivity of major crops in Rechna Doab, Pakistan. Paper submitted to Agric Water Manage

Auta SJ, Abdullahi YM, Usman Y (2003) Evaluation of rural youth's participation in agriculture in Nigeria. A research report submitted to NAERLS, Ahmadu Bello University, Zaria

Bakheit NI, Stigter K, Abdalla AT (2001) Underground storage of sorghum as a banking alternative. Low Ext Input Sust Agric Mag 17(1):13

Bakheit NI, Stigter, K (2004) Improved matmuras: effective but underutilized. Low Ext Input Sust Agric Mag 20(3): 14

Bakheit NI, Ahmed MA, Stigter CJ, Mohamed HA, Mohammed AE, Abdalla AT (2005) Economic aspects of traditional underground pit storage (matmoras). The case of Jebel Muoya, Central Sudan. Sudan J Agric Res 5:89–96

Björnsen A, Gurung P (2001) Resilience in farm level food security. Low Ext Input Sust Agric Mag 17(1):6

Boulahya M, Cerda MS, Pratt M, Sponberg K (2005) Climate, communications, and innovative technologies: potential impacts and sustainability of new radio and internet linkages in rural African communities. Clim Chang 70:299–310

Bouma J, Koning NBJ, Struik PC, Wienk, JF (2006) Preface and Epilogue. NJAS-Wageningen J Life Sci 53:247–252 & 387–394

Brandt SA, Spring A, Hiebsch C, McCabe JT, Tabogie E, Diro M, Wolde-Michael G, Yntiso G, Shigeta M, Tesfaye S (2001) The "Tree Against Hunger": Enset-based agricultural systems in Ethiopia. http://www.aaas.org/international/ssa/enset

Chattopadhyay N, Samui RP, Wadekar SN (2003) Weather based operational plant protection of leaf spot disease of groundnut. Mausam 54:463–470

China Agricultural University (2002) Thematic research paper on China's rural poverty. College of Humanities and Development, CAU, Beijing

Connelly S, Wilson N (2001) Trees for semi-nomadic farmers: a key to resilience. Low Ext Input Sust Agric Mag 17(1):10–11

Das HP, Adamenko TI, Anaman KA, Gommes RG, Johnson G (2003). Agrometeorology related to extreme events. WMO Technical Note No. 201. WMO No. 943, Geneva

Gadgil S, Rao PRS, Rao KN, Savithri K (2000) Farming strategies for a variable climate. In Sivakumar MVK (ed) Climate prediction and agriculture. Int START Secr., Washington, pp 215–248

Gao Chunxin, Zeng Yan (2001) Research results on national agro-information required with emphasis on the analysis of information resource construction (in Chinese). J. Agro-books Inf 4:31–33

Heemskerk W, Wennink B (2006) Stakeholder-driven funding mechanisms for agricultural innovation. KIT Publishers, Amsterdam

Holt-Gimenez E (2001) Measuring farmers agroecological resistance to hurricane Mitch. Low Ext Input Sust Agric Mag 17(1):18

Hounkonnou D, Kossou DK, Kuyper TW, Leeuwis C, Richards P, Röling NG, Sakyi-Dawson O, Van Huis A (2006) Convergence of sciences: the management of agricultural research for small-scale farmers in Benin and Ghana. NJAS-Wageningen J Life Sci 53:343–368

Huntingford C, Gash J (2005) Climate equity for all. Editorial in Science, 16 September, 309:1789

Ibrahim AA, Stigter CJ, Adam HS, Adeeb HM (2002) Water use efficiency of sorghum and groundnut under traditional and current irrigation in the Gezira scheme, Sudan. Irrig Sc 21:115–125

ILEIA (2000) Communities combating desertification: livelihoods reborn. An Editorial. ILEIA Newsletter 16(1):4–5

Kaimowitz D (2005) Forests and floods: drowning in fiction or thriving on facts? Centre for International Forestry Research (CIFOR), Bogor, Indonesia

Lassa J (2006) New direction needed for coping with disasters. Jakarta Post (Indonesia), January 16, p 6

LEISA (2001) Lessons for scaling up LEISA. An Editorial. Low Ext Input Sust Agric Mag 17(3):4–5

LEISA (2002) Changing information flows. An Editorial. Low Ext Input Sust Agric Mag 18(2):4–5

Lomas J, Milford JR, Mukhala E (2000) Education and training in agricultural meteorology: current status and future needs. In: Sivakumar MVK, Stigter CJ, Rijks D (eds) Agrometeorology in the 21st Century: needs and perspectives. Agric For Meteorol 103:197–208

Ministry of Agriculture of China (2000) Thematic research paper on the needs of market information in China. Department of Market Information, MoA, Beijing (in Chinese)

Morrow K (2002a) Information villages: connecting rural communities in India. Low Ext Input Sust Agric Mag 18(2):28–30

Morrow K (2002b) Accessing the Internet in the South: some tools and techniques. Low Ext Input Sust Agric Mag 18(2):18–19

Onyewotu L, Stigter K, Abdullahi Y, Ariyo J (2003) Shelterbelts and farmers' needs. Low Ext Input Sust Agric Mag 19(4):28–29

Rathore LS, Stigter CJ (2007) Challenges to coping strategies with agrometeorological risks and uncertainties in Asian regions. In: Sivakumar MVK, Motha R (Eds.) Managing Weather and Climate Risks in Agriculture. Springer, Berlin Heidelberg, pp. 53–69

Reijntjes C (2001) Resilience to disaster. Low Ext Input Sust Agric Mag 17(1):4–5

Rogers EM (1983, 3rd Ed) Diffusion of innovations. The Free Press, New York and London

Röling NG, Hounkonnou D, Offei SK, Tossou R, Van Huis A (2004) Linking science and farmers' innovative capacity: diagnostic studies from Ghana and Benin. NJAS-Wageningen J Life Sci 52:211–235

Sahni P, Ariyabandu MM (2003) Disaster risk reduction in South Asia. Prentice-Hall of India, New Delhi

Severn WJ, Tankard Jr JW (1997 4th Ed) Communication theories: origins, methods and uses in the mass media. Longman, New York (Chinese edition published by Huaxia Publishing House, Beijing, in 2000)

Sikka K (2001) Role of media in disaster preparedness. In Sahni P, Dhameja A, Medury U (eds) Disaster mitigation: experiences and reflections. Prentice Hall of India, New Delhi, pp 124–131

Sivakumar MVK (Ed.) (2002) Improving Agrometeorological Bulletins. AGM-5, WMO/TD No 1108, WMO, Geneva

Stewart JI (1991) Principles and performance of response farming. In: Muchow RC, Bellamy JA (eds) Climatic risk in crop production: models and management for the semi-arid tropics and sub-tropics. CAB International, Wallingford, pp 361–382

Stigter K (2004) The future of education, training and extension in agricultural meteorology: a new approach. In: Zheng Dawei et al. (eds) The future of agrometeorological education in China, China Agricultural University, Beijing, China

Stigter K (2005a) Scientific research in Africa in the 21st century, in need of a change of approach. Invited lecture for the Open University of Sudan at the Institute for Studies of the Future, Khartoum, on 23 April 2005. Also available at the author's homepage on the website of his former employer (www.met.wau.nl under employees, affiliated staff). Appeared in 2006 in a slightly revised form in Afric J Agric Res 1:4–8

Stigter CJ (2005b) Report of a first "Agromet Vision" mission to Africa: Sudan, 16 April till 1 May, Appendix 1. Agromet Vision, Bruchem, 6pp

Stigter K (CJ) (2005c) Building stones of agrometeorological services: adaptation strategies based on farmer innovations, functionally selected contemporary science and understanding of prevailing policy environments. Paper presented as the opening keynote lecture at the International Symposium on Food Production and Environmental Conservation in the Face of Global Environmental Deterioration (FPEC, 2004), Fukuoka, Japan. J Agric Meteorol (Japan) 60:525–528

Stigter K (2005d) Equity climate for all. Discussion available at the INSAM website (www.agrometeorology.org) under the topic of the "Agromet Market Place"

Stigter K (2006a) Even the present situation in Aceh/Sumatra, Indonesia, bears similarity with problems encountered with introduction of agrometeorological services: another story of parallels. Pre-study available at the INSAM website (www.agrometeorology.org) under the topic of the "Agromet Market Place"

Stigter K (2006b) Agrometeorological services in various part of the world, under conditions of a changing climate. Austin Bourke Memorial Lecture presented in the Royal Irish Academy, Dublin, in the evening of 2 March. Extended Abstract available on the INSAM website under "Accounts of Operational Agrometeorology" and as Appendix 1 in "Report of a first fully "Agromet Vision" mission to Asia: China, Iran, Indonesia, 26 September 2005 till 27 April 2006", Bruchem, The Netherlands, pp 17 – 19

Stigter CJ (2006c) Agrometeorological services, climate change and a new countryside. Invited presentation as a guest speaker at the celebration of the 50th Anniversary of the Department of Agrometeorology, College of Environment and Natural Resources, China Agricultural University, Beijing

Stigter CJ (2006d) Report of a first fully "Agromet Vision" mission to Asia: China, Iran, Indonesia, 26 September 2005 till 27 April 2006. Agromet Vision, Bruchem, The Netherlands & Bondowoso, Indonesia

Stigter K (2006e) Agrometeorological (Advisory) Services and agrometeorological services. Available on the INSAM website (www.agrometeorology.org) under "Accounts of Operational Agrometeorology"

Stigter K (in prep) Applied agrometeorology. To be published in 2009 by Springer, New York

Stigter CJ, Das HP, Murthy VRK (2003) Beyond climate forecasting of flood disasters. Invited Lecture on the Opening Day of the Fifth Regional Training Course on Flood Risk Management (FRM-5). Asian Disaster Preparedness Center (Bangkok) and China Research Center on Flood and Drought Disaster Reduction (Beijing), Beijing, September. Available from ADPC (Bangkok) on CD-ROM. Also available at the INSAM website (www.agrometeorology.org) under the topic of the "Agromet Market Place"

Stigter K (Ed), with contributions from Barrie I, Chan A, Gommes R, Lomas J, Milford J, Ravelo A, Stigter K, Walker S, Wang S, Weiss A (2005a) Support systems in policy making for agrometeorological services: the role of intermediaries. Policy paper for a CAgM/MG meeting in Guaruja, Brazil. WMO, Geneva, Management Group meeting of 30 March – 2 April, document 7.1, 6pp + 1 App. Also available at the INSAM website (www.agrometeorology.org) under "Needs for Agrometeorological Solutions of Farming Problems"

Stigter CJ, Zheng Dawei, Onyewotu LOZ, Mei Xurong (2005b) Using traditional methods and indigenous technologies for coping with climate variability. Clim Change 70:255–271

Stigter CJ, Oteng'i SBB, Oluwasemire KO, Al-Amin NKN, Kinama JM, Onyewotu LOZ (2005c) Recent answers to farmland degradation illustrated by case studies from African farming systems. Ann Arid Zone, in print

Stigter K (CJ), Kinama J, Zhang Yingcui, Oluwasemire T (KO), Zheng Dawei, Al-Amin NKN, Abdalla AT (2005d). Agrometeorological services and information for decision-making: some examples from Africa and China. J Agric Meteorol (Japan) 60:327–330

Tai Huajie, Yao Kemin, Liu Wenze, Lou Xiurong (1997) An outline introduction to agrometeorological information in China. China Ocean Press, Beijing, 100 pp

Tan Ying, Stigter K, Xie Yongcai (in prep.) Information needs of various farmers with different income levels in poor areas of China. Dept. Media and Communication, China Agricultural University, Beijing, China, and Agromet Vision, Bruchem, The Netherlands & Bondowoso, Indonesia

UNESCO (2002) Radio browsing the Internet: an option for rural communities. Low Ext Input Sust Agric Mag 18(2):25

Van Noordwijk M, O'Connor T, Manurung G (2005) Why has transition from relief to rehabilitation been so slow? Jakarta Post (Indonesia), December 26, p 6

Venkatesh H, Guled MB (2004) Seasonal rainfall forecast and variability for agricultural application. In: Singh KK, Reddy R (eds) Proceedings of the National Workshop on seasonal climate prediction for sustainable agriculture, Hyderabad, pp 116–122

Verdin J, Funk C, Senay G, Choularton R (2005) Climate science and famine early warning. Phil Trans R Soc B 360:2155–2168

Viet NV (2002) Some measures to cope with the impacts of climate disasters (extreme climate events) on agriculture in Vietnam. Agrometeorological Centre, Institute for Meteorology and Hydrology, Ministry of Natural Resources and Environment (translation of internal reports)

Viet NV (2005) Impact of heavy rain on erosion in the northwest mountains of Vietnam and some measures to cope with it. Agrometeorological Research Centre, Institute for Meteorology and Hydrology, Ministry of Natural Resources and Environment (translation of internal reports)

Viet NV, Liem NV (2005) Micro-agroclimatological observations in migration areas in the northwest mountain areas of Dienbien. Provincial Government Project, Agrometeorological Research Centre, Institute for Meteorology and Hydrology, Ministry of Natural Resources and Environment (translation of internal reports)

Wason A (2002) "Farmer to Farmer": participatory radio for Dekhon farmers in Tadjikistan. Low Ext Input Sust Agric Mag 18(2):16–17

WMO/CAgM (2005) International Workshop on "Coping with agrometeorological risks and uncertainties: challenges and opportunities". WCP/AGM/WORISK, ANNEX 1, Geneva

WMO (with Baier W, Motha R, Stigter K) (2006a) Commission for Agricultural Meteorology (CAgM). The first fifty years. WMO No 999, Geneva

WMO (2006b) Weather, climate, water and sustainable development. WMO Annual Report 2005. WMO No 1000, Geneva

Ye Jingzhong (2002) Processes of enlightenment. Farmer initiatives in rural development in China. Ph.D.-thesis, Wageningen University, The Netherlands

Zhao Caixia, Zheng Dawei, Stigter CJ, He Wenqing, Tuo Debao, Zhao Peiyi (2005) An index guiding temporal planting policies for wind erosion reduction. Arid Land Res Managem 20:233–244

Zheng Dawei, Ju Zhao, Tuo Debao, Stigter CJ (2005) Reversing land degradation from wind erosion in Inner Mongolia: the choice between grass and bush restoration or conservation tillage of contour strip plantings depends on hill slopes and rainfall. J Agric Meteorol (Japan) 60:337–341

Zhu Fengli (2002) Information required in west China poverty regions and the progress of the poverty alleviation information project (in Chinese). Developm & Res 5:43–44

CHAPTER 12

Information Technology and Decision Support System for On-Farm Applications to cope effectively with Agrometeorological Risks and Uncertainties

Byong-Lyol Lee

12.1 Introduction

12.1.1 On-Farm Applications Against Risks

On-farm applications to cope with agrometeorological risks and uncertainties cannot be defined objectively without detailed description of all the external and internal driving forces, related events, direct and indirect impacts, consequential effects, available technology and resources, and farmer's implementation ability, governmental supporting system and national infrastructure. Nevertheless, it may be practiced through an ordinary farm management system when combined or linked together with an appropriate early warning system for natural hazards, if available. The creation of data archives and information bases are essential to decision making as well as research on hazards and warning systems. Components of an early warning system include: observation, detection, monitoring, assessment, forecasting, warning, projection and, valuation.

Common on-farm applications for decision making support in agriculture can be grouped into three categories as follows in a simple manner. This grouping can be also applied to decision making support system against agrometeorological risks and uncertainties as a starting prototype. The three categories are: 1) Production Management System, including yield, quality, and post-harvest; 2) Pest Management System, including insect, disease, and weeds; and 3) Resource Management System including soil, water, air, biome, and infrastructure.

In most cases, production management system will be main target in coping with agrometeorological risks, not only because it is the most susceptible area to direct impact by risks, but also because it can be relatively easily managed by farmers unless resources are limited. On-farm applications in terms of production management can be described in two ways. These include: 1) Structural applications such as irrigation, water-harvesting, windbreak, frost protection, artificial climate/weather, etc., and 2) Non-structural applications such as seasonal to interannual climate prediction, medium range forecasts and crop insurance.

12.1.1.1
Agricultural Management

Agricultural management requires diverse decision making processes by farmers not only during production period but also pre- and post- production period, especially under fragile and variable environmental and social conditions. This implies that farming is a process of continuous decision-making all through the year.

Decision-making by farmers varies widely depending on the individual's farming goals. They can be either short-term or longer-term goals. Most decision-making, ultimately aims at minimizing risks or to maximize productivities in terms of income optimization. The roles of agrometeorology in farm management can be defined as two distinctive subjects: by making better use of agrometeorological information 1) as a natural resource for higher farm production, and, 2) as an early warning for stable farm production.

Under fragile conditions, farmers tend to maintain average productivity with minimal additional inputs considering resource availability. To cope effectively with expected risks, farmers have to consider such diverse aspects as production, marketing, social and human aspects of agricultural managements for desirable decision making in farm management, which is a complex process.

To make an appropriate decision on a timely basis, farmers should have a wide range of timely information and knowledge on both risk sources and vulnerability of farming system to impending risks. Unfortunately certain agricultural risks are very difficult to detect by farmers themselves, especially those risks with very short-led time like torrential rain, violent hailstorms, tornado, etc. They have to rely on local or national meteorological services for warnings and advisories on these short-range risks. Even with these kinds of early warnings, farmers should take account of uncertainties existing intrinsically within early warnings at farmer's site.

In general, decision making in farm management, though an ubiquitous process, is largely involved in the processes of production management, pest management, and resource management including associated infrastructure management. In production management, major decision will be made in terms of yield, quality, and post-harvest, while in resource management important physical and biological environmental factors can be dealt with in terms of conservation of quantity and quality of resources. Although agrometeorology particularly deals with production risks and evaluation of possible production decisions, to solve local problems of farming systems the other risk factors have to be taken into account in that same process.

12.2
Risk & Uncertainty in Agriculture

12.2.1
Agrometeorological risks

Agrometeorological risk is the random environmental variability associated with the farming process due to weather, soils, diseases and pests. Thus weather and climate are one of the biggest production risks and uncertainty factors impacting on agricultural systems performance and management. Extreme climatic events such as severe droughts, floods, cyclonic systems or temperature and wind disturbances strongly impede sustainable agricultural development. Hence weather and climate variability is considered in evaluating all environmental risk factors and coping decisions.

Agrometeorological hazards and uncertainties may be classified, though somewhat arbitrary, as follows (United Nations 2006)
- Hydrometeorological hazards: floods, tropical cyclones, wind disturbance (severe storms), extreme temperature (cold, heat), drought, air pollution, haze and smoke, dust and sand storms, snow avalanche and winter weather hazards (permafrost), famine.
- Geological hazards: earthquakes, Tsunami, volcanoes, near-Earth objects, landslides.
- Biological hazards: epidemics, locust swarms, wild animals
- Environmental degradation: desertification, wildland fire, loss of biodiversity
- Climate changes (variability)
- Agricultural Policy (inconsistency)
- Socio-political Policy (circumstance)

12.2.2
Risk Management in Agrometeorology

To cope with agrometeorological risk and uncertainties effectively, they should be observed, detected, monitored, assessed, forecasted, warned by relevant authorities at national or local levels and then delivered to a farmer's site in a timely manner and with certain level of reliability. It should be based on systematic framework provided by government's risk management authorities including NMHS of each country for better risk management. Agricultural sectors or authorities also should establish response strategies to cope with agrometeorological risks and uncertainties at national and local levels, furthermore at regional and global level through international collaborations.

There are more challenges in newly emerging countries, not only because of lack of available resources and systems against risks, but also because of complexities in domestic structures while experiencing transit from traditional to modern society. The new social system may be more prone or vulnerable to even the normal degree of normal environmental variations primarily due to accelerated degradation of surrounding natural environment through industrialization and urbanization.

Degrading environmental conditions, when compounded by severe climatic events such as recurrent droughts, will cause more serious negative effects, making the drylands increasingly vulnerable, and furthermore desertification. In coping with risks systematically, especially when main risks are compounded with other factors, a reliable risk management system is essential to prevent or mitigated potential risks and uncertainties through appropriate preparedness and response strategies, which will be also a major challenge in decision making in farm managements.

Risk management system in agricultural meteorology can comprise of early warning systems provided by government authorities and agricultural management systems operated at farmer's site: The former will be mainly responsible for issuing warnings and advisories from authorities, while the latter for preparedness and response measures being made by farmers. It also needs proper communication mechanisms between two systems to share information in timely manner.

Early warning helps to reduce economic losses by allowing farmers to better protect their assets and livelihoods. It can guide farmers in selling livestock or selecting appropriate crops for a drought. It aims at reducing not only the immediate impact of a disaster but also the knock-on effects on assets that can reduce economic well being and increase poverty. Early warning information allows farmers to make decisions that contribute to their own economic self-sufficiency and their sustainability. If well integrated into a systematic framework of risk reduction, early warning systems can provide many development benefits to farmers.

12.3
Decision-making Support Against Risks

12.3.1
Emergency Response System

The establishment of early warning systems and associated preparedness and response systems in agricultural managements has been an important contributor to the progressive prevention and reduction of natural hazards in agricultural production. This is true for drought and famine-affected regions, as well as for developed countries where early warning systems, and preparedness, mitigation and risk transfer measures are generally well developed.

These two parts can be integrated into an Emergency Response System (ERS) in agricultural managements as an on-farm application for decision-making support system (DMSS) against agricultural hazards. At each stage of ERS, farmers should take appropriate actions for prevention and mitigation of risks by mobilizing available resources and applying strategies that are associated with relevant agricultural managements for optimal risk management.

Emergency response system requires scientific knowledge, including improved science and technology for information dissemination. They need the creation of data archives and information bases that are essential to decision making and to research on hazards and warning systems. ERS may enhance community capacities through participation processes, public-private partnerships, and recognition of indigenous knowledge and values of local farming community.

Key components of ERS against Agrometeorological risks and uncertainties can be summarized like any typical early warning system: (NEMA 2006). These key components of on-farm applications against risks include: preparedness/insurance, early warning, planning/vulnerability, coping strategy, response/action, counter-measures, recovery/relief, mitigation, outreach/education, and awareness.

- **Knowledge on Risks and Uncertainties:** awareness, recognition, monitoring: In coping effectively with agrometeorological risks and uncertainties, one of the most important strategies is how to make better use of knowledge on risks and uncertainties such as climate changes and variability, which includes observing, detecting, monitoring, assessing, projecting on earth system, then responding to current weather.
- **Preparedness: strategy, planning, implementation for prevention, prediction:** The importance of preparedness to cope with risks and uncertainties has been getting more recognition in its effectiveness and cost-benefit advantage, as compared to the reactive practices such as response, recovery, and relief actions, etc. In order to be well prepared in advance for prevailing risks, it is prerequisite to establish most suitable practices at farm level based on applicable strategies against agrometeorological risks.
  As a part of better preparedness to reduce the impacts of the variability (including extremes) of climate resources on crop production, both structural and non-structural measures can be used. The structural prepared measures can reduce direct intensity, duration, quantity of hazards in large, while the non-structural ones can contribute to minimize uncertainties of risks on a relatively long-term basis.
- **Response:** reduction or mitigation actions: How to respond to concurrent risks is also very critical on-spot action that farmers can take by themselves to reduce or mitigate hazards against normal farm management. For appropriate response actions to be taken farmers should get on time reliable, quantitative information about the environment within which they operate.
- **Recovery:** relief, insurance, alternatives, contingency production: In addition, the likely outcome of alternative or relief management options can reduce uncertainties in crop productivity when available to farmers. All through ERS, quantification is essential and computer simulations can be used to project feasibility of relief and recovery actions among alternative management and relief options. Contingency planning is an important part of such strategies, as ways must be found to avoid, reduce, or cope with risks.
- **Evaluation:** feedback afterwards, outreach: From knowledge and experiences through past risk management, farmers can learn invaluable lessons. To establish more promising ERS, any existing system needs to be implemented through feedbacks from farmers based on their experiences. This experience can also be shared with non-experienced neighboring farming communities through outreach programs for education and training, including general public concerned. It must be reflected to strategy developments during ERS implementation process.

12.3.1.1
Components of DMSS for Agrometeorological Risks

A complete and effective Decision-Making Support System for risks and uncertainties in Agrometeorology may comprises four inter-related elements: risk knowledge, monitoring and warning service, dissemination and communication, and response capability (United Nations 2006). A weakness or failure in any one part could result in failure of the whole system.

Risks arise from the combination of the hazards and the vulnerabilities to hazards that are present. Assessments of risk require systematic collection and analysis of data and should take into account the dynamics and variability of hazards and vulnerabilities that arise from processes such as environmental degradation and climate change. Risk assessments help to prioritize early warning system needs and guide preparations for response and disaster prevention activities.

Warning services lie at the core of the system. They must have a sound scientific basis for predicting and forecasting and must reliably operate twenty-four hours a day. Continuous monitoring of hazard parameters and precursors is necessary to generate accurate warnings in a timely fashion.

Warnings must get to those at risk. For farmers to understand warnings, they must contain clear, useful information that enables proper responses. Regional, national and community-level communication channels and tools must be pre-identified and one authoritative voice established. The use of multiple communication channels is necessary to ensure that everyone is reached and to avoid the failure of any one channel, as well as to reinforce the warning message.

Farming communities must also know how to react to warnings. This requires systematic education and preparedness programs led by disaster management authorities. It is essential that disaster management plans are in place and are well practiced and tested. The farming community should be well informed on means to avoid damage and loss of property.

12.3.1.2
Requirements of DMSS Components

Considering acceptable levels of risk, accurate risk scenarios should be generated that can show the potential impacts of hazards on vulnerable groups. It requires capabilities to analyze not only the hazards, but also the vulnerabilities to the hazards, and the consequential risks as well by risk assessment using computer resources and readily available software like geographic information systems. Risks are usually characterized through risk mapping, frequency distributions, scenario plans and exercises, annualized risk mapping and qualitative measures. Capabilities in science, technology and research, and the availability and sustainability of observation networks will decide how efficiently high-quality data can be based on the magnitude, duration, location and timing of hazard events and to extract information on hazard frequency and severity from observational data sets. To fill up data gaps, poor communication network and computer resources.

Requirements of the DMSS components include: On-going, systematic and consistent observations of hazard-relevant parameters; Quality assurance and proper archiving of the data into temporally and geographically referenced and consistently catalogued observational data sets; Capacities to locate and retrieve needed data and to freely disseminate data to public users; and Sufficient dedicated resources to support these activities.

Improvements in the quality, timeliness and lead time of hazard warnings are also essential in EWS. They have been enhanced markedly through scientific and technological advances, particularly in computer systems and communications technology. Continuous improvements in the accuracy and reliability of monitoring instrumentation, and in integrated observation networks particularly through the use of remote sensing techniques are enormous. In turn these have supported research on hazard phenomena, simulation modeling and forecasting methods and warning systems.

Dissemination and telecommunication mechanisms must be operational, robust, available every minute of every day, and tailored to the needs of a wide range of different threats and different user communities. The dissemination of the information must be based on clear protocols and procedures and supported by an adequate telecommunications infrastructure. At the national level, effective dissemination and alert mechanisms are required to ensure timely dissemination of information to authorities and farmers at risk in even the most remote areas of the country. In order to reach all those who need to take action, countries are becoming aware of the need to design warnings for particular groups of stakeholders, such as different language groups, people with disabilities, and tourists.

12.3.1.3
Major Gaps of DMSS Components

Although long historical records do exist in many cases, particularly for hazards, in others data is scarce and there are significant variations in data quality. Data may be inaccessible because of non-digital format.

At the national level, the main challenges include: establishing and maintaining observing systems and data management systems; maintaining archives, including quality control and digitization of historical data; obtaining systematic environmental data for vulnerability analysis; and securing institutional mandates for collection and analysis of vulnerability data.

It is difficult to collect accurate data. As a result of security and ownership, information is increasingly restricted in its efficient utilization.

There is a danger in losing societal memory of past hazards, particularly for infrequent hazards. In addition to losing knowledge of hazards, young communities face losing knowledge about how to reduce vulnerability and how to respond to warnings.

Despite significant progress having been made on monitoring and forecasting hazards, many gaps still exist, particularly in emerging countries.

Key issues include: inadequate distribution of monitoring systems for hydrometeorological hazards; inadequate level of operational capabilities (resources, expertise and operational warning services); lack of systems for many hazards such as dust and sand storms, severe storms, flash floods and storm surges; lack of procedures to share essential data in a timely fashion for the development of modeling and for operational forecasting and warning systems; inadequate access to information (forecasts and interpreted data); insufficient multi-disciplinary, multi-agency coordination and collaboration for improving forecasting tools; inadequate communication systems to provide timely, accurate and meaningful forecasting and early warning information down to the level of farming communities

Warning messages do not reach all at risk. In developing countries this is largely a result of the underdeveloped dissemination infrastructure and systems, while in developed countries it is the incomplete coverage of systems. The resource constraints also contribute to the lack of necessary redundancy in services for information in many countries.

Other factors and gaps to be considered include: 1) Telecommunication systems and technology – There is also a need to upgrade telecommunications facilities, including equipment, service provisioning and operation, to be based on internationally agreed standards for the timely delivery of warnings from authorities to the public. It should be noted that non-technological systems are in many cases necessary and adequate and are usually tailored to those who use them, ensuring their sustainability. An example is the case of traditional knowledge and information acquired through educational and awareness-raising programs; 2) Inadequate standards – There is need for development of standards, protocols and procedures for exchange of data, bulletins, alerts, etc. for some of the hazards, which traditionally have not been exchanged internationally among countries (e.g., tsunami). Protocols are critical, particularly when the lead time is short; 3) Poor public interest and concern – Perhaps the most important reason for people failing to heed warnings is that the warnings do not address their values, interests and needs. Messages are often not sufficiently targeted to the users and do not reflect an understanding of the decisions stakeholders need to make to respond to the warning. Lack of public interest in warnings also occurs because early warning systems only provide information on impending crises. They do not report on positive developments in the system that would engender public confidence and trust in future warnings, such as scientific advances that will enhance the warning services, or positive outcomes of responses to previous warnings. To overcome this obstacle, the public needs to be periodically informed about the hazards and the level of risk they pose, and how this may be changing. This information should not be technical and should remind the population of similar events; 4) Proliferation of communication technologies - The use of the new information and communication technologies, particularly the Internet, in disseminating warnings is a useful advance for expanding the coverage and reducing time lags in warning dissemination, yet it is also creating problems of untargeted messages inducing wrong responses due to misinterpretation. This problem is also related to the type of hazard under consideration; 5) Ineffective engagement of the media - Warning dissemination may be inadequate because of ineffective engagement of warning authorities with the media. The media is interested in reporting news and not necessarily in disseminating useful warnings. Thus, conflicts can arise when the media publish inaccurate or misleading information about potential events that contradict the official warning messages; 6) Ineffective integration of lessons - Finally, warning dissemination can be ineffective if there is a lack of feedback on the system and its performance. Serious hazard events are relatively rare at any one location, and experience of an event may be quickly forgotten. Formal feedback processes are needed to ensure that the system continually evolves and improves based on feedback and learning from previous experience.

12.4
Information Technology Required

12.4.1
Requirements for Agrometeorological Products

The common features of Agrometeorological products include general descriptions of Agrometeorological characteristics of specific regions in terms of agricultural production and resource management. Depending on the requirements and priorities of end-users, the description details or expertise levels of the contents vary to a great extent. In general, due to the shortage of expertise as well as the limited space of bulletins, they contain insufficient levels of quantity or quality of information.

Despite these limitations, the essential components for a successful product can be identified as follows:
- End-Users: Farmers, Associations, Extensions, Researchers, Policy-makers, General public.
- Contents: Types: General, Advisory, Warning, Recommendation, Suggestion
  Weather/Climate/Forecast/Prognosis/Diagnosis information
  Extremes, Special Weather Phenomena, Energy Balance (Flux)
  (Flood, Drought, Frost, Heat wave, Fire, Landslide, Cold injury, etc.)
  Crop, Fruit, Grass, Forest, Animal Husbandry, Fishery
  (Growth, Development, Yield, Population, Reproduction, etc.)
  Disease, Insect, Pest, Weeds
  Farm Management
  (Cropping, Irrigation, Sowing, Harvesting, Post-Harvest, Spraying)
  Resource Management (Water, Air, Soil, Biome, Infrastructure)
- Data: Form: Digital / Document based : Bulletin, Brochure, Letter, Note, Leaflet
  Format: Text, Numeric, Table, Chart, Figure, Image, Map, etc.
- Communication: Sharing, Dissemination, Feed-back
  Phone/Mobile, Fax, TV, Radio, PC-Network, Internet, Dedicated line, SMS, etc.
- Providers/Developers/Producers/Authors/Publishers/Editors
  Meteorologists, Agronomists, Entomologists, Ecologists, Agrometeorologists, Soil scientists, Virologists, Epidemiologists, etc.
- Raw Materials: Meteorological, Agronomical data, non-Agricultural data
  Observed, Processed, Derived, Estimated (inter-/extra-polated)
  NWP Model Outputs, Agricultural Model Outputs
  Domestic or Foreign Origin

- Tools: Statistical packages, Graphic tools, GIS, Simulation models,
- Institutions concerned
  Meteorological, Agricultural, Hydrological, Others
  Research Institute, Extension Office, University, Private Sector, Cooperation
  Local, Central(Federal), Regional, Global Organizations

12.4.2
Requirements for DMSS Infrastructure

Agrometeorological products require diverse computer resources all through the processes such as data collection, processing, archiving, dissemination, etc. in a systematic way for better on-site application by farmers. Basic requirements for integrated system infrastructure can be listed as follows:

- **Hardware Systems.** Servers for simulation models, databases, system analysis, high speed network framework, mass data storage and DBMS;
- **Information.** Existing DB: RS, agronomy, management, climate, etc.; Met Data Resources: synoptic data, forecasts(S,M,L), prognosis, adaption data; Development Tools: simulation models for climate, crop, resource management, root zone dynamics, farm management, etc.; Derived Products: climate change scenarios, seasonal and interannual forecasts, crop growth and development, regional food demand/production, etc.;
- **Interfaces.** TCP/IP based Internet Web interface with GUI; object oriented architectures: free of OSs, languages, platforms, networks; multi-directional communication networks between end-users and researchers;
- **Operations.** Facilities, equipment, space, man power, budget, hardware, software, evaluation, etc.

12.5
Resource Sharing System: Case of WAMIS

As the WMO Information System (WIS) evolves to provide a single entry point for any data request, the Commission for Agricultural Meteorology (CAgM) is trying to extend its service to member countries under WIS umbrella by implementing the World Agrometeorological Information Service (WAMIS) into a Grid portal to share computer resources, especially for emerging countries in which limited IT resources are most critical barriers in improving its operational services in Agro-Meteorology.

An inevitable use of advanced ICTs such as information network, database, simulation models, tools for GIS, RS for agrometeorology should be made in its implementation. In this regard, sharing of resources including IT and human resources available among countries will be a promising way to solve the above mentioned problems. WAMIS grid portal will be a promising solution in improving resource sharing among CAgM member countries by allowing them to make better use of remotely located resources for agrometeorological services at national/regional scale, especially when it provides interactive forecast-based agrometeorological services via simple Internet access.

12.5.1
WAMIS as a Web Portal

The main objective of WAMIS is to provide a dedicated web server for disseminating agrometeorological products issued by WMO members. By providing a central location for agrometeorological information, WAMIS will aid users to quickly and easily evaluate the various bulletins and gain insight into improving their own bulletins. The web site will also host training modules to further help Members improve the quality and presentation of their agrometeorological bulletins.

- Overview
  - Dedicated WMO Web server for AgroMeteorology in WMO
  - Three sites including two mirror sites: USDA(USA) / KMA (Korea)/BMC (Italy)
  - Demonstration of sharing Bulletins among member countries
  - Cyber tutorials are available, with continued development
- Status (Issues)
  - Diverse languages are being used
  - Large gaps among countries in contents and technologies employed
  - Rare standard format or style between bulletins from different members
  - Poor user-friendly interfaces
  - No request/reply functions
  - Limited information & materials available
  - No archival in DBMS
- Requirements
  - Extended elements, types, resolutions in time & space
  - Successful case studies & pilot projects
  - Cyber tutorials on applications
  - Technical support on IT, tools, models, etc.
  - Computer resources for DB, model operation
  - Training/Education on advanced technologies
  - Better communication frameworks

12.5.1.1
Implementation Strategy

As both future customer and information provider to WIS, WAMIS needs to be implemented as a grid portal to provide not only information but also computer resources that are critical for strengthening agrometeorological services in member countries, especially with limited computer resources for agrometeorological service. Under grid environment together with legacy technology for high performance computing, large-scale diverse data and analysis servers, WAMIS will provide an IT framework for end-users with interactive remote operation of their service development and deployment based on NWP forecasts as a grid portal. Specific interface will be provided for interactive operation on region-specific applications at operational level that requires and provides non-meteorological information from diverse sources, e.g. AMBER(DWD), DSSAT(USA). WAMIS grid portal can be used as an initial step to collaborate closely with GEOSS in the near future that aims at integrating all the observations available on the Earth.

- Web Portal: information sharing (current role of WAMIS)
  - Transit to XML-based service: standard schema development
  - Machine translation: multi-lingual interfaces needed
  - Operational applications based on Web service architecture
  - Tutorial interfaces for real practices
- GRID Portal: Resource sharing (extended role of WAMIS)
  - Forecast-based AgroMeteorological services for researcher/extension
  - Benchmarking on AMBER(DWD), expanding with DSSAT(USA)
  - NCAR (WRF, MM5), DWD (GME, LM), KMA(GDAS) as NWPs considered
  - Super ensemble of Long-range/Seasonal Forecasts (APCC/METGRID.)
  - GISC/vDCPC dedicated to WAMIS be required (NCAR, DWD)
  - LIS (NASA) as a framework for LSM (GDS/LAS+GRID in the future)
  - uCAgM project -> WAMIS grid portal -> WIS pilot project -GEOSS Hub
- IT Implementations
  - Computational Grid
    Korea Nat'l Supercomputing Center shall provide for Testbeds
    KMA will provide support for developmental and operational service
    MM5 under Globus, then extends to LM under Unicore environment
    AMBER(DWD) for AgMet Models, later extends to DSSAT (USA)
  - Data Grid
    Globus DataGrid functions will be used primarily
    Legacy servers with links to Data Grid using Data Broker
  - WIS umbrella
    Virtual DC/PC should be constructed among new and existing DC/PCs
    Specialized/dedicated GISC can be considered to accommodate non-meteorological data (Fig. 12.1)

12.5.1.2
Key Components

In principle, WAMIS grid portal will stick to WIS architecture and standards with the additional functions of providing computer resources and considering non-meteorological data/information sharing (Fig. 12.2). Thus it should provide specific user interfaces to accommodate its specific user demands both from grid and legacy technology using service-oriented architecture, including semantic grid technology in its long-term perspectives.

- System (Fig. 12.3)
  - GRID Servers for Simulation models, Databases, System Analysis
  - High speed network frame (APAN)
  - Web service interfaces for simulation models with near real time DB access

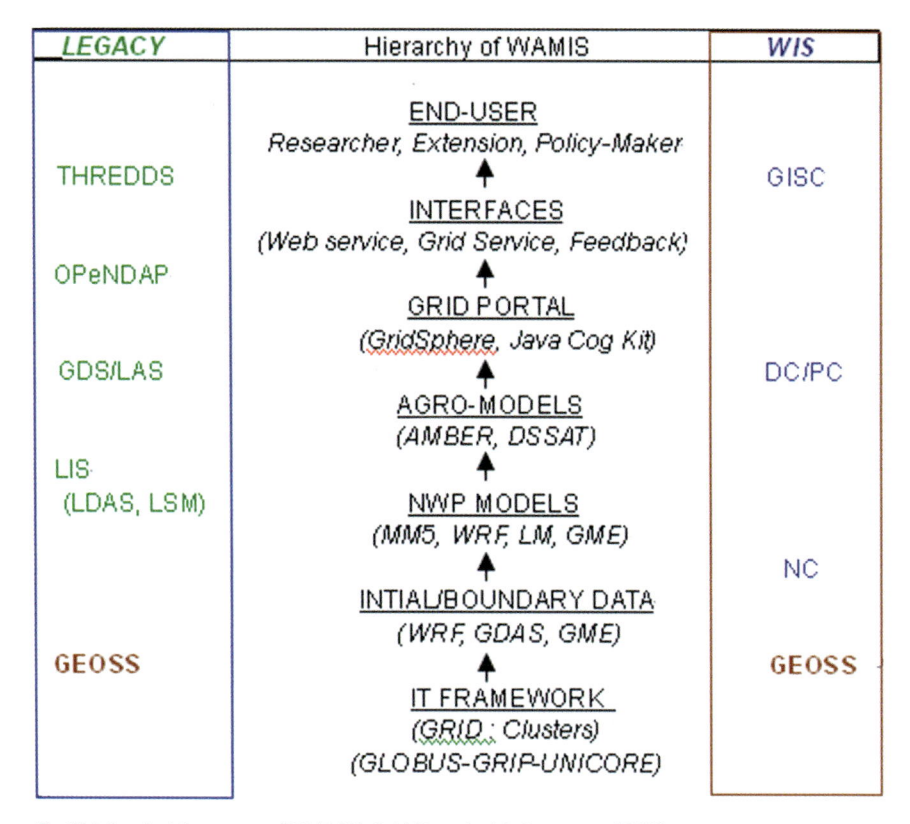

Chapter 12: Information Technology and Decision Support System

LEGACY	Hierarchy of WAMIS	WIS
THREDDS	END-USER Researcher, Extension, Policy-Maker ↑ INTERFACES (Web service, Grid Service, Feedback)	GISC
OPeNDAP	↑	
GDS/LAS	GRID PORTAL (GridSphere, Java Cog Kit) ↑ AGRO-MODELS (AMBER, DSSAT)	DC/PC
LIS (LDAS, LSM)	↑ NWP MODELS (MM5, WRF, LM, GME) ↑ INTIAL/BOUNDARY DATA (WRF, GDAS, GME)	NC
GEOSS	↑ IT FRAMEWORK (GRID ; Clusters) (GLOBUS-GRIP-UNICORE)	GEOSS

Fig. 12.1 Logical Structure of WAMIS Grid Portal with Legacy and WIS compo-nents

- Multi-tiered Interface Architecture under distributed computing environment
- Information
  - Existing DB: Remote Sensing, Agronomy, Management, Climate, etc.
  - Met Data resource: Synoptic data, Forecasts(S,M,L), Prognosis, Adaption data
  - Development tools: Simulation models for climate, crop, resource management, root zone dynamics, farm management, etc.
  - Derived Products: Climate change scenario, seasonal- and interannual-forecasts, crop growth and development, regional food demand/production
- Interfaces
  - TCP/IP based Internet Web service interface with GUI (JAVA)
  - Object Oriented Client/Server architectures with free of OSs, languages, platforms, networks. (SOAP/WSDL)
  - Multi-directional communication networks between end-users and researchers.

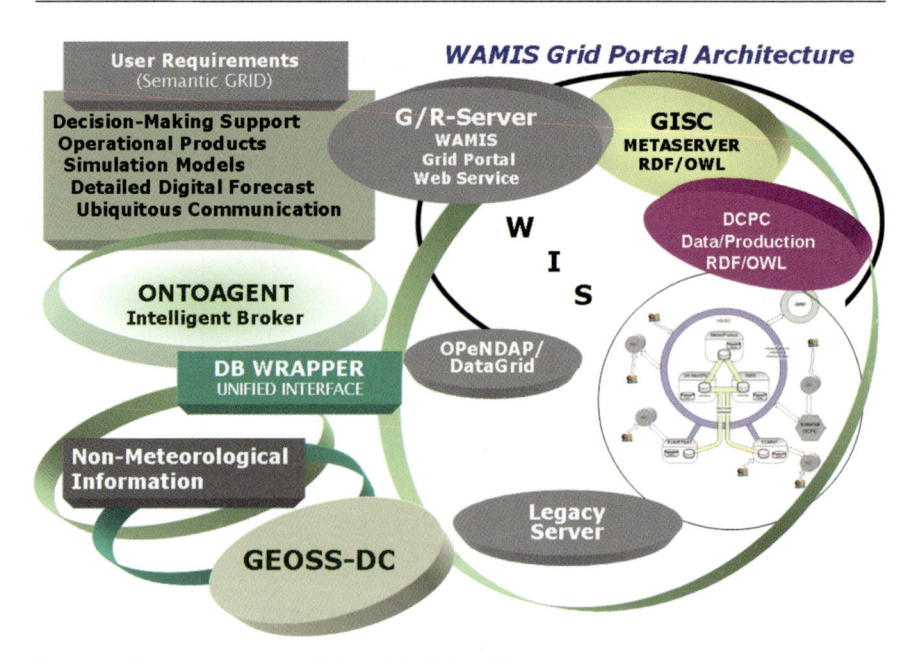

Fig. 12.2. Schematic Diagram of WAMIS Grid Portal Design

12.5.1.3
Service Architecture

WAMIS grid portal will try to make best use of service-oriented architecture while employing all the fundamental requirements under WIS framework. Key service layers consist of application, metadata, replica, resource and publishing services (Fig. 12.4). This service architecture will be able to meet the GISC requirements in metadata catalogues, Internet portal with local administration, data acquisition, discovery, distribution services, monitoring, and synchronization. In addition, computational grid technology is an important component of grid services implemented in WAMIS grid portal to provide computing resources for model operations via Internet through Web Service technology in the future.

- Application Service
    - Web service: Interactive user configuration on domain, applications
    - Grid service: Authorization/authentication, Brokers, Grid portal
- Metadata Service
    - Ontology broker: WMO Metadata Core Profile – extended with key words
    - Semantic Web: Protege3.0 based Ontology development (RDF/OWL)
- Replica Service
    - Catalogue: Globus with legacy catalogue (THREDDS)
    - Data Grid: Interface with legacy data servers (GDS, LAS)
- Resource Service
    - Model server: AgroMeteorological Models, NWP models

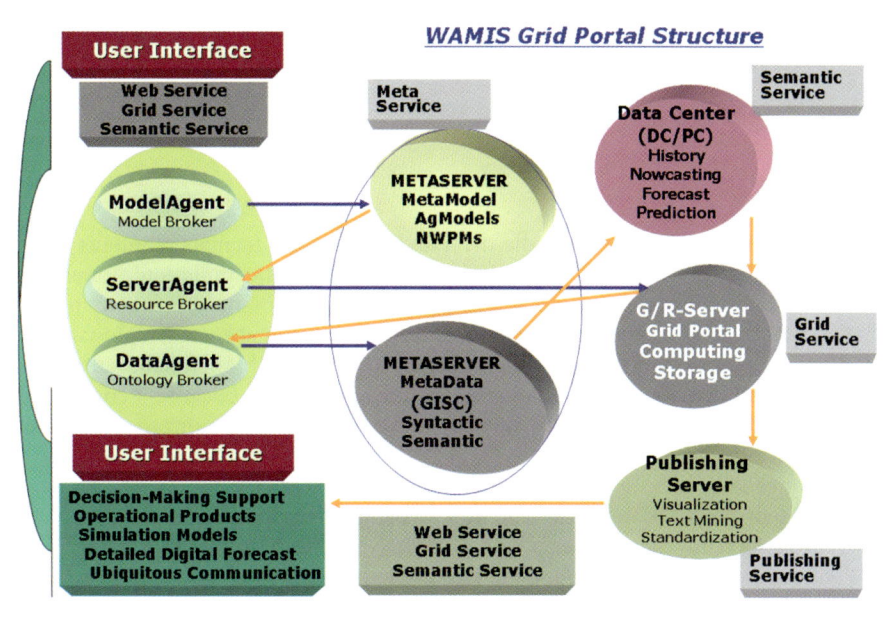

Chapter 12: Information Technology and Decision Support System

Fig. 12.3 System Components of WAMIS Grid Portal Structure

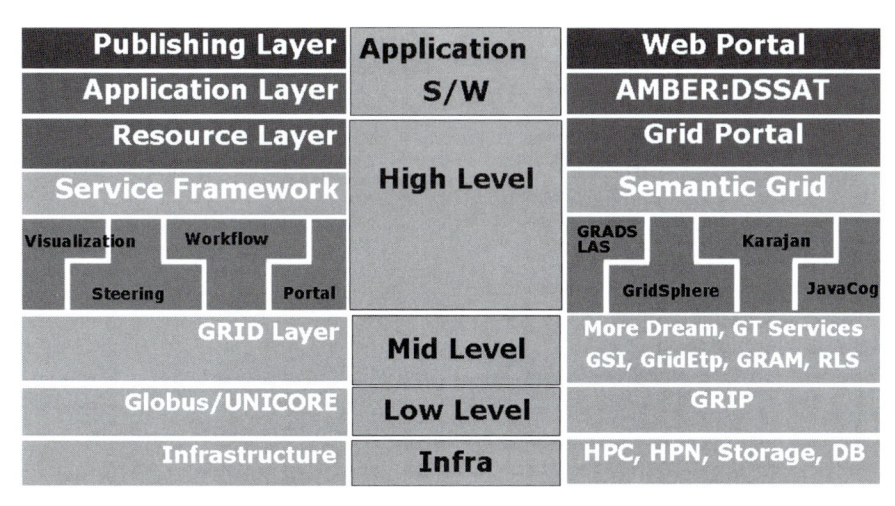

Fig. 12.4 Service Oriented Architecture of WAMIS Grid Portal

- Tool server: GIS, Graphic tools, Statistics, DBMSs
- Computational Grid: Globus, Unicore
- Publishing Service
  - GIS: Map display, spatial analysis, Web interface
  - Graphics: NCAR graphics, GrADS

- Interface: Web interface with feedback
- AccessGrid: Partial supplement for members available

12.5.1.4
DB requirements

WAMIS grid portal should handle diverse data sources, formats, contents from synoptic data, forecasts(S,M,L), prognosis, adaption data, simulation models for crops, resource management, root zone dynamics, farm management, etc. It also has to take care of derived products such as climate change scenario, regional food demand/production, etc. It indicates that WAMIS needs highly elaborated data handling and distribution mechanisms, including ontology, because it consists of various contents in different formats depending upon the origin or process of data manipulations.

- Data needed:
  - Meteorological data: historical, now-casting, forecast, prediction
  - Non-meteorological: energy flux, surface, vegetation, soil, moisture
- Time span: real-time, on-demand, fixed schedule
- Prerequisites:
  - Interface to legacy servers
  - Interface to non-meteorological information
  - Extended Metadata, relevant ontology
  - Metadata Catalogue, Replica service
- Measures
  - Dedicated GISC/DCPC to cover specific data/information
  - Supporting interface between legacy and DataGrid Servers
  - Computational Grid optional
  - AccessGrid may be required

12.6
Discussion & Conclusions

Coping actions with agrometeorological risks and uncertainties comprise appropriate decision making processes. Various strategic and response options can be taken by farmers as on-farm applications against risks. Site specific farm management systems can be used to prevent and mitigate potential risks. Those on-farm applications require systematic and consistent observations of hazard-relevant parameters, quality assurance and proper archiving of the data, catalogued observational data sets with capacities to locate and retrieve needed data and sufficient dedicated resources to support these activities.

Advanced ITs such as information network, database, simulation models, tools for GIS, Remote Sensing are invaluable in the implementation of decision makings against risk. In addition, sharing of resources between associated authorities and farming communities is necessary under limited resources.

The establishment of early warning systems and associated preparedness and response systems in agricultural managements is an important contributor to the progressive prevention and reduction of natural hazards in agricultural production. Emergency Response System (ERS) in agricultural managements can be considered as an on-farm application for decision-making support system (DMSS) against agricultural hazards.

Emergency response system requires scientific knowledge, including improved science and technology for information dissemination. They need the creation of data archives and information bases that are essential to decision making and to research on hazards and warning systems.

Data are scarce and there are variations in data quality. The main challenges include: 1) Establishing and maintaining observing systems and data management systems; 2) Maintaining archives, including quality control and digitization of historical data; 3) Obtaining systematic environmental data for vulnerability analysis; and 4) Securing institutional mandates for collection and analysis of vulnerability data.

Key issues on emergency response system include: 1) Inadequate distribution of monitoring systems for hazards, 2) Inadequate level of operational capabilities, 3) Lack of systems for many. 4) Lack of procedures to share essential data in a timely fashion for the development of modeling and for operational forecasting and warning systems, 5) Inadequate access to information, 6) Insufficient multi-disciplinary, multi-agency coordination and collaboration for improving forecasting tools, and 7) Inadequate communication systems to provide timely, accurate and meaningful forecasting and early warning information down to the level of farming communities

There is a need for development of standards, protocols and procedures for exchange of data, bulletins, alerts, etc. for some of the hazards. Protocols are critical, particularly when the lead time is short. The use of the new information and communication technologies, particularly the Internet, in disseminating warnings is a useful advance for expanding the coverage and reducing time lags in warning dissemination.

An inevitable use of advanced ICTs such as information network, database, simulation models, tools for GIS, RS for agrometeorology should be made in its implementation. In conclusion, WAMIS grid portal, as an example of IT sharing infrastructure will be a potential solution in improving resource sharing among CAgM member countries by allowing them to make better use of remotely located computer resources for early warning and risk management by providing both NWP outputs and Agricultural model outputs, especially when it provides interactive forecast-based agrometeorological services via simple Internet access.

References

NEMA (2006) Safe Korea (http://www.nema.go.kr)
United Nations (2006) Global Survey of Early Warning Systems, Pre-print version released at the 3rd International Conference on Early Warning, Bonn, Germany, 27-29 March 2006

CHAPTER 13

Coping Strategies with Agrometeorological Risks and Uncertainties for Crop Yield

Lourdes V. Tibig, Felino P. Lansigan

13.1
Challenges and opportunities

Farmers work within an environment characterized by highly variable biophysical, economic, political and institutional conditions. They are, thus, exposed to several types of risks which include production risk, yield risk, price or market risk, institutional risk, financial risk and human (or personal) risk (Belliveau et al. 2006). Hardaker et al. (1997) and Harwood et al. (1999) defined these agrometeorological risks as follows:
- Production risk spells the chance in losses in yield due to events beyond the control of the farmer and is often, related to weather, and/or related to technology.
- Price or market risk is the risk associated with changes in prices of outputs or inputs and may include market access.
- Institutional risk relates to changes in government policies which may impose unanticipated constraints on production practices.
- Financial risk results from the way farm's capital is obtained and is related to borrowing, interest rates, the ability to meet payment of debts and also, the willingness of lender to continue lending.
- Human or personal risk is associated with the farmers themselves, their families and any disruption of farm production and profitability in terms of labor, etc.

Any particular combination of risk-reducing measures can be defined as a risk management strategy. Risk perception is usually the first step in risk management. Individual farmers respond to these risks in highly variable ways, depending on the degree of their exposure and their coping abilities which are influenced by factors such as human and financial resources and networks, institutional support and most often, the quality of the natural resources available.

It is for these reasons that most developing countries, especially those in the tropics and sub-tropics, are almost always the ones who are less able to cope with agrometeorological risks and uncertainties. A majority of farmers from their burgeoning populations do not have access to resources, as they are mostly subsistence farmers. Their vulnerability to threatened food security, as a result of the external stresses, has become a very important challenge.

However, in some ways, the farmers' environment does not just present risks, but also opportunities to be exploited. For example, variations/changes in climatic conditions could have some beneficial effects. Or some areas in a farm could offer opportunities for cash crop production. Developments in technological innovations coupled with institutional support could lead to farmers using these innovations to not just improve agricultural yields, but also carry ancillary benefits to them and their families, and the environment.

13.2
Types of coping strategies with agrometeorological risks and uncertainties for crop yield

There is a wide variety of coping measures that producers/farmers employ in the field. These can be categorized into the following:
- optimal and sustainable use of resources;
- change in cultural practices/improved farming practices;
- modification of resource potential, including controlled micro-climates;
- local indigenous knowledge systems/networks;
- access to extension services;
- technological innovations such as new/ modified approaches; and
- others, such as resilience and divestment in natural capital.

13.2.1
Optimal and sustainable utilization of resources

The optional and sustainable utilization of resources involve any or all of the following:

13.2.1.1
Rational cropping patterns to fit resources available to farmers / producers

These rational cropping patterns could include crop improvement through crop diversification, altering crop mix and use of hybrids, use of cash crops that have secure markets, and practice of different cropping systems (double/multiple cropping, intercropping, mixed cropping, sequential cropping, ratoon cropping), and also rotation of crops.

Diversification of crops, cultivars and plot locations are most common means by which farmers attempt to stabilize agriculture income (Matlon 1991). Farm crop diversification basically refers to the diversification of crop species and the diversification of farmland ecosystems (Mengxiao 2000). There is, thus, a change in crops or a change of varieties, of cropping systems, as well as cropping intensities. Matlon and Fafchamps (1988) define the objectives of crop diversification as:
- to make more complete and efficient use of production factors by spreading their use across enterprises with different temporal profiles;
- to increase aggregate productivity by matching physiological requirements of crops to specific micro-environments;
- to meet domestic household consumption requirements in the context of multiple failures in both product and factor markets;

- to exploit morphological complementarities and compensatory behavior of crop components (as in the case of intercropping), and to improve and stabilize plot level productivity; and
- to improve aggregate production and reduce income risk.

In one of the Expert Consultations on Crop Diversification in the Asia-Pacific Region, it was recognized by all the country experts that crop diversification is a useful means to increase crop output under different conditions. The commonly understood mechanism is the addition of more crops to the existing cropping systems; and it is in effect, a broadening of the base of the system. This method of horizontal diversification has been responsible for production increases due to high cropping intensities. One of the experts acknowledged that the system of multiple cropping has been able to increase food production potential to over 30 tons/ha with an increase of the intensity by about 400–500 percent (FAO 2000).

The other type is vertical crop diversification which refers to the upstream and downstream activities of a particular crop or crops. There are tremendous opportunities for downstream activities such as minimally processed fruits, tropical fruit juices, natural food ingredients, functional food, frozen fruits, beverages and high fibre products (Yahya 2000).

In Bangladesh, the traditionally " rice-led " growth is now converted into a more diversified production base which includes several non-rice crops (Hoque 2000). There are programmes which promote crop diversification involving potatoes, oilseeds, pulses, spices and vegetables. A systematic arrangement of growing a variety of crops in rotation with rice has been launched and implemented. Diversified cropping systems are introduced to free upland areas in the winter season for non-rice crops; thereby, enabling the introduction of a third crop in the land under irrigated conditions (e.g., short-duration mustard or a sandwich crop of grain legume).

Meanwhile in China, attention is focused on guiding and encouraging farmers to adopt the market-oriented cropping structure using the agro-resource environment rationally. A substantial acreage has been changed to higher value and more profitable crops. Cultivation of three crops a year is widely practiced. Intercropping and multiple cropping are also extensively undertaken.

In India, the cropping patterns have occurred mainly from crops with declining demand and lower value potential to crops with an increasing demand and higher value potential (Hazra 2000). There is substantial diversification from coarse cereals to oilseeds, even if the current trend is a shift from coarse cereals like sorghum to superior cereals like rice and wheat.

Oil palm, rubber, cocoa and rice continued to be Malaysia's major crops (Yahya 2000). However, other crops such as coconut, tropical fruits, vegetables, flowers are also being grown. While in Nepal, to fit the variety of climatic conditions at any given time, a wide variety of crops is incorporated into the cropping system which is based on major staple food crops, but mostly rice-based cropping system in the lowlands and maize in upland areas (Sharma 2000).

Likewise in the Philippines, there are two perspectives of crop diversification. One is planting alternate crops after main crop (rice). The other is planting one or more crops in between a perennial crop (for example, coconut). Among the successful rice-based cropping patterns are rice-onion, rice-garlic, rice-peanut, and rice-mungbean (Espino 2000). While in the coconut-based system, the more successful are coconut + cacao, coconut + passion fruit, coconut + pineapple and many others.

In Sri Lanka, the marginal plantations of tea and rubber like the rice lands in the wet zone are diversified. The most important crops that are cultivated in crop diversification programmes are chillies, onion, shallots, vegetables, root and tuber crops and pulses (Weerasena 2000).

Crop diversification in Thailand aimed at improving socio-economic conditions of the farmers in the rainfed and irrigated areas centers on rice as the most important crop. Those areas not very congenial for rice production are being diverted to other crops, such as cassava, rubber, fruits and vegetable (Chainuvati 2000).

And in Vietnam, farmers in the Mekong Delta enjoy favourable conditions for practicing crop diversification integrating agriculture and the other related sectors like fisheries, forestry and livestock (Van Luat 2000). To address the annual flooding during the rainy season and droughts in dry seasons, farmers use the ditch and dike system of farming in their fields, in which the dikes are expected to be a preventive measure to avoid submergence of their crops. They lay pipes through the dike to take the water with silt and useful aquatic fauna and to facilitate drainage, thereby leaching decomposing matter. They plant many crops (with rice as main crop) on the dikes, and try to harvest the crops and fish from their fields before floods become imminent. After the floods that come during the rainy season recede, they plant their second crop. In all rice growing areas, advanced techniques of rice farming are being applied. Examples of these are the use of the row-seeding methods and the use of very short-duration rice varieties (80–90 days) with high grain quality and resistance to many pests and diseases which can yield 7–8 tons/ha.

Additionally, a number of researchers have indicated that an added advantage of intercropping is that it further improves stability at the plot level to the extent that crop mixtures yield better than sole equivalents in stress conditions, reduce the incidence and buildup of pests and diseases and manifest compensatory yield behavior due to the differences in crop structure, physiology or phenology.

On the other hand, rotation of crops spreads risks, improves soil tilth and other factors, and optimizes yields for the crops being rotated.

13.2.1.2
Crop improvement through varietal diversification, including use of cultivars/hybrids adapted to changed environment

Varietal diversification implies that farmers select and maintain a diversified set of varieties for their major crops to manage risks as follows:
- to spread the risk of loss due to period-specific stresses (e.g. brief periods of drought or insect population) as in the case of cultivating varieties with varying maturities permitting staggered planting;
- to reduce the risk of pest and disease losses since there is genetic variability in resistance or tolerance to biotic stresses;

- to cope with the changing environment stresses such as reductions in rainfall and growing periods; and
- to reduce the risk of crop losses due to the stresses associated with particular land types (as in the case of farmers matching crops to the micro-environments).

Gomez (2004) and Chalinor (2005) have emphasized the use of drought-tolerant varieties or new cultivars with improved response to altered climate conditions.

Although crop diversification is commonly practiced among the Asia-Pacific Region countries as an important strategy for economic growth, it is not however, without challenges. Foremost among these challenges are the possibilities that high crop intensity might cause degradation of ecology and natural resources; although, diversifying cropping patterns is being seen as a solution.

13.2.1.3
Crop production improvement through land-type diversification

When the character and properties of the soil vary with location on the toposequence, diversifying plot locations could reduce aggregate production and yield variability.

13.2.1.4
Change in intensification of production

A variety of crops is planted to fit the prevailing changes in the conditions in the fields (e.g., usually changes in rainfall, temperature or growing seasons). The intensification of production is based on the crop requirements and the changing conditions in the farm areas.

13.2.2
Change in cultural practices or improved farming practices

This coping measure consists of changes in farming practices to increase adaptive capacity through improved soil, crop and environmental quality. It includes adjustments in agricultural inputs such as the use of hybrids, flexible calendar of farming activities such as changes in timing of operations to address changes in temperature/moisture conditions, integrating crops with trees in a given production area to halt decline of land productivity and the use of farming systems such as organic and precision farming.

13.2.3
Modifications of resource potential including controlled micro-climates

These include changes in land topography, altered micro-climates through, for example, field afforestation to provide wind shelter/ windbreaks, and changes in location of crops to address hazard-prone areas and/or long-term loss of fertility due to salinity.

13.2.4
Local indigenous knowledge systems/networks

Farmers develop and enhance knowledge and technology from their previous experiences, the so-called local or indigenous knowledge. Examples of applications of local knowledge are in selection/improvement of seeds, choice of desirable and effective crops/varieties, cultural practices such as mulching, no tillage, etc.and many others. Farmer-developed technologies are actual manifestations of their coping mechanisms to various environmental challenges in their farms.

For example, in the tribal farms of the eastern part of Madha Pradesh, India, old agro-technologies are still carried on, with slight modifications where traditional varieties occupy more than 60 per cent of cultivated land and majority of the farmers still use indigenous techniques from seed selection to storage (Sharma et al. 2006).

In most poor, rural communities, the perception of the relationship between agriculture and weather/climate is quite different from that in the affluent farming areas. The farmers are able to identify specific and important weather patterns with the help of their perceived indicators (Valdivia et al. 2004). Farmers base their crop and other production decisions on their local knowledge systems, which have been developed from years of observations, experiences and experimentations. Predicting climate to them is an important cultural component, and the local knowledge systems provide them with the ability to make informed decisions, and in a way, prepare themselves psychologically.

In western cultures and also in countries where seasonal climate forecasting has already been developed to optimize the link of climate variation with agricultural production, there are now significant implications for improved agricultural yields. Perhaps, opportunities could now be open to link this scientific-based weather/climate forecast services with the local, indigenous knowledge, especially in rural communities.

13.2.5
Access to extension services

Agrometeorological extension services provide a very important input in farm productivity, whether it is in the form of inputs, like technical advice/ assistance to farmers who need technical information, or facilities like farmers' cooperatives.

13.2.6
Technological innovations

Climate is among the obvious sources of environmental risks in farming systems so that the development of technologies to assist farmers in managing the vagaries of weather has been an important focus of agricultural research. (Smithers and Blay-Palmer 2001). Technology innovations have permitted adapting cropping systems to a wider range of climatic regions, and also, farmers opting for varieties closely adapted to average and therefore, expected temperature regimes in their particular regions.

13.2.6.1
Direct-seeded rice (DSR) cropping system

In the northwest Bangladesh, a cropping system called direct-seeded rice (DSR) is emerging through the assistance of the International Rice Research Institute (IRRI) and the Bangladesh Rice Research Institute. This system has enabled early planting of crops which include T. aman rice, followed by potato and in some instances, even maize is planted in the furrows two weeks before the potato harvest. DSR is a system which enables optimal timing for different rice varieties. Even in the flood-prone regions in Bangladesh, like the Chalan Bil lowland region that can have only one rice crop a year (called the boro crop), DSR using boro rice results to much higher yields. The rice crops are also being grown at times different from the usual.

Rice is directly seeded either through dry (drilling the seed into a fine seedbed at a depth of 2–3 cm) or wet seeding (usually with the use of a drum seeder). However, weed management is a critical factor and timely application of herbicides, with one or two hand weeding can provide effective control.

This is also practiced in the Indo-Gangetic Plains in northern India (known as India's grain bowl, producing 50 per cent of the nation's rice and wheat). Farmers who tried the DSR cropping system during the previous wet season (kharif) attest to its advantages. Notable among these are that the farmers can sow earlier than when they used transplanting method, enabling them to take full advantage of the rains, have good yield, save labor and use less water (Singleton 2006).

One concern being addressed now by the scientists at the International Rice Research Institute is the possibility that there could be a buildup of weeds in DSR. A possible solution they are looking into is a rotation of crop (e.g., rice to sugar cane).

13.2.6.2
Raised bed cropping system

In most of the developing countries where irrigated cropping is widely practiced, a number of common problems emerge such as low yields, salinity of the soil, deteriorating soil structures, ground water depletion and water scarcity. In the rice-wheat belts of Pakistan and India, one innovative approach to address these issues is raised cropping beds (Page 2005). Beds are formed with narrow trenches between the beds serving as access and left in place for up to five years. Crops are planted into these beds.

In Pakistan, three types of beds are introduced for optimal production; namely wide beds, narrow beds and flat basin seed beds. With rabi wheat, the crops have been seen to have superior growth on raised beds than in traditional flat irrigation basins. There is also the advantage of having less salt in the crop root zones after a few planting seasons.

13.2.7
Others, including resilience and divestment of natural capital

In other farming areas, specifically in resource-poor regions, other options resorted to include resilience and a reduction in the land areas planted.

13.3
Some examples of coping strategies

Farmers and producers in both the developing and developed countries' face enormous challenges as a result of the multiple stresses that plague their day to day operations in the field. While the same risks and uncertainties could be found in both, production outcomes may not be the same. The main difference is a result of their access to resources (e.g. financial, technological innovations, external support, additional coping mechanisms such as crop insurances and others). While the options are there, when the farmer does not have the capacity to implement any because he does not have the needed resources and tools, he can only rely on resilience. What can make this worse is that in most developing countries, agriculture is not just about food production; it is survival for the resource-poor farmers and their families (Medina 2002).

This section looks at the various ways farmers/producers in the developed world differ from their counterparts in the developing regions in adjusting and coping with any set of constraints, risks and uncertainties, which could spell failures/shortfalls on one hand, and benefits on the other hand.

13.3.1
Canada

Agriculture remains a significant export commodity for Canada's economy. Furthermore, it is a mainstay of several regional economies. And, like in many parts of the world, weather and climate play an important role in Canadian agriculture. In a recent research on adaptation in Canadian agriculture to climate change and variability however, it was ascertained that the adaptive behaviour of Canadian farmers is prompted by uncertain variations from year to year, like for instance, rainfall intensity and duration, high temperatures and intensity of droughty conditions at critical periods, etc. (Bryant et al. 2000).

It is also noted, that recent research in farm-level adaptation in Canadian agriculture has identified the importance of the multiple sources of stress faced by farmers. It is these powerful forces that explain changes in cropping patterns involving crop substitution, new crops and new areas. Additionally, even though Canadian farmers may not explicitly acknowledge climate and weather conditions as an important element of their risk management strategy, the producers employ adaptation strategies to lessen the risk of negative impacts. These include crop and enterprise diversification and altering timing of planting in addition to the adoption of new technologies, such as improved land and water resource management, including changing the intensification of production to address the changing duration of growing seasons and associated changes in temperature and moisture, changing the location of crop production, using fallow and tillage practices, the use of irrigation systems, alterations to livestock management and those that are provided by external sources, like the use of seasonal climate forecasts and information, crop insurances and other income stabilization programs (C-CIARN 2004).

13.3.1.1
The Okanagan Valley, Canada

The Okanagan Valley is located in the southern interior of British Columbia. It is long and narrow with an area of 8,200 km^2. and is flanked by mountain ranges. Wine is its second highest grossing commodity.

Producers know that climate-related conditions are important for wine-growing operations. In good years, favorable conditions include a hot and dry summer with a long growing season and early spring. In bad years, the growing season experience lower temperatures and greater rainfall than normal. Additionally, they also identify risks associated with costs of production, pest and disease outbreaks, changing government policies, interest rates, failures in technology and risks associated with the market, such as loss of the tourist market.

Their efforts to manage climate-related risks, which to them are the ones they can readily adapt to, include measures such as fruit thinning or 'dropping crop' in cold and wet seasons to address poor weather for the grapes, and investing in risk-reducing technologies such as wind machines. By reducing the cropload, more energy is available to achieve a higher quality of grapes.

Some of their anticipatory strategies are site selection and changing topography, avoiding or removing frost pockets in the vineyard, choosing to plant varieties that mature earlier in the season, or having vineyards in different locations to have a greater chance that one region will sufficiently mature the grapes.

13.3.2
The United States

In the United States, sustainability in agriculture implies quite a number of differing farming systems. A few are mentioned here to emphasize the range of systems American farmers adopt depending on his access to resources.

Precision farming, also called site-specific management system of farming, is defined as a management strategy that employs detailed site-specific information to precisely manage production input. The philosophy behind precision agriculture is that production inputs should be applied only as needed for the most economic production (Searcy 1999). The farmers employ personal computers, telecommunications, global positioning systems (GPS), geographic information systems (GIS) and other advanced technical expertise. From knowledge of soil and crop characteristics unique to each part of the field, the production inputs are optimized.

On the other end of the spectrum of farming systems is organic farming. As defined by the USDA Study Team on Organic Farming, this system avoids/largely excludes use of synthetically compounded fertilizers, pesticides and livestock feed additives. It relies upon crop rotations, crop residues, animal manures, legumes, green manures, off-farm organic wastes, mechanical cultivation and aspects of biological pest control to maintain soil productivity and tilth, to supply nutrients and control weeds and other pests.

Perhaps, midway between the two extremes is low-input agriculture or low-input farming system. It seeks to optimize the management and use of internal production inputs or on-farm resources and to minimize use of production inputs or off-farm resources such as purchased fertilizers/pesticides whenever flexible and practicable to lower production costs, to avoid pollution of surface and ground water, to reduce pesticide residues in food, to reduce farmer's overall risk and to increase farm profitability (Parr et al. 1990).

13.3.2.1
Crop producers in Mississippi

Crop producers in Mississippi consider the multi-risk environment in their crop production as special challenges. They pay close attention to management strategies, whether what they manage are whole farms or otherwise. The following are tested techniques from which producers can choose from to sustain farm operations for whole farm management:
- Consider new crops and crop combinations,
- Diversify, match crops to soil capability and productivity, planting fields with low yield histories to alternative crops.
- Select proven varieties,
- Incorporate crop rotation and use herbicides that do not injure future crops in a rotation system,
- Carefully calibrate planting equipment to deliver a desired number of seeds at the optimum planting depth for a given crop,

- Carefully plan plant spacing as it directly affects the crop's ability to use light, water and fertilizer and ultimately affects crop yield, and
- Scout fields regularly throughout the growing season.

13.3.3
Latin America

13.3.3.1
The Andes of Peru and Bolivia

In the agropastoral system in the Andes of Peru and Bolivia (Valdivia et al. 2004), the following are the identified management strategies, to manage climatic risks, maximize use of resources and achieve multiple goals (e.g. food-security, etc.):
- diversification,
  - among crops, between crops and livestock, and non-agricultural
- access to forecasts and their use (even when there is limited use due to gaps/constraints),
- use of local knowledge systems,
- resilience /divestment in natural capital, and
  - shortening the fallow fields
  - not replenishing the soil
- livelihood strategies,
  - specializing in production that secure markets (ex. potato)
  - building buffers for bad years (ex. frieze-dried potato)
  - linkages to markets (ex. sales of value added crops)

13.3.4
Africa

In the African region, most adaptation strategies in agriculture are at the subsistence level and these include the use of new cultivars adapted to the environment, optimal use of crop calendar and crop diversification, alternating fallow land with strips of grain crops, strip cultivation in areas with intensive wind erosion and other management strategies like afforestation (windbreaks, etc.), change in land use in hazard prone areas, use of early warning systems (EWS) for natural disasters in agriculture in the fields, and institutional measures, like use of risk/hazard zoning maps.

13.3.4.1
Senegal

There are two important options that farmers/producers in Senegal employ, and these are the use of food production-related forecasting and early warning systems and the invention and diffusion of sustainable technology compatible with agricultural conditions.

13.3.4.2
Uganda

In Uganda, the farmers' coping toolkit for agriculture includes use of farmers' traditional knowledge of micro-environment diversification, intensification of vegetation on farmlands in order to reduce soil erosion and sedimentation rates near water catchment areas, integration of locally developed knowledge of soil, climate, biological resources and other physical factors with scientific assessment to maintain crop diversity. One example is the management and recycling of crop residues to improve and sustain productivity.

13.4
Case studies: Regional/national coping strategies

13.4.1
The Philippine experience:
Empowering farmers for rural development

The Philippines is an archipelago with more than 7,100 islands located in the tropics bounded by 4 ° and 21° latitudes and 116° and 127° longitudes. It is completely bounded by bodies of water. (See Figure 13.1). Its agroecological zone classification is warm humid tropics with four climate types based on rainfall patterns: Type 1 is described as two pronounced seasons, dry from November to April and wet during the rest of the year. Type 2 is characterized by the absence of a dry period with maximum rain period from November to January. Type 3 has a short dry period from November to February. Type 4 has more or less even rainfall distribution during the year.

Arable land per capita is 0.075, one of the lowest in developing Asia with agriculture contributing about one-sixth of total GDP. Rice constitutes about 30% of total crop harvested. As of 2000, about 67% of the 4 million-hectare rice area (but excluding the upland area planted to rice), is irrigated and the rest is rainfed. Upland rice is grown in both permanent and shifting cultivation systems scattered throughout the archipelago.

In the early 1960s, a technology innovation called Green Revolution changed the configuration of farming in the Philippine setting (Medina 2002). This technology which centered on the use of high-yielding varieties (HYVs) required high inputs of fertilizers, pesticides and irrigation. Together with support services like credit and extension personnel providing advice, yields had a dramatic increase. But, because Filipino farmers are mostly resource-poor, increase in yields did not compensate for the increase in external inputs. The burden of borrowed capital with associated interests and the uncertainties brought about by increased frequency of incidence of pests and diseases and also, the occurrence of natural hazards (typhoons, floods, drought events) had not improved the lives of most of the farmers.

Worse, other challenges had surfaced. Most of the varieties developed by the farmers were displaced by HYV monocrops. There were more frequent outbreaks

Fig. 1.1. The map of the Philippines.

of pests and diseases because most of the natural predators had vanished. A variety of available food sources like birds, fish, frogs and the like had been eliminated with the increasing use of pesticides. Also, farmers were exposed to health risks due to prolonged and frequent use of chemical inputs. More importantly, seeds commonly saved and exchanged among the farmers were gradually replaced by commercial seeds. And eventually, farmers lost control of their production assets like seeds, technology and in some cases, land.

To address these farmers' needs and concerns, a Farmer-Scientist Partnership for Development called Magsasaka at Siyentipiko sa Pag-unlad ng Agrikultura (MASIPAG) was formed in the mid-1980s (Medina 2002). This development approach

to address uncertainties and risks, and ultimately, improve the quality of life of the resource-poor farmers, had five strategies; namely,
- farmer-scientist partnership to combine the theories and technical knowledge of the scientists with that of the experience and practical knowledge of the farmers,
- bottom-up approach to prioritize farmers' needs,
- farmer-led research and training through the farmer-managed trials cum training center,
- farmer-to- farmer mode of technology transfer, and
- advocacy for sustainable/organic agriculture and other related issues.

The member-farmers either organized themselves or joined already organized units. They joined orientation workshops on local and global trends in agriculture, including alternatives like sustainable organic agriculture, which are organized by farmer-trainers from peoples' organizations (POs), or by the technical staff of the MASIPAG or partner non-government organizations (NGOs). Then they established trial farms where they planted from 50 to 100 traditional varieties and MASIPAG rice selections (those seeds which could not be defined technically as varieties because they do not meet the criteria for purity and uniformity) per trial farm. The seeds were to maintain more genetic variability, which gave wider possibilities to match selections to environmental conditions. The choice of starting traditional varieties was based on the less capital input required and those that were ecologically adapted to diverse agro-ecological conditions.

The farmers observed the characteristics of the different varieties and selections to assess them for suitability to the local environment conditions and pest resistance. The top ten performing locally adapted varieties were then chosen for planting. Farmers were given only between a hundred grams to one kilogram of seed per variety so that these farmers re-learned the skill of mass-producing their seeds. Result was the mosaic effect of the different neighboring varieties, creating a barrier to pests and diseases due to the differential resistance between the varieties. One positive effect of the farmer planting several varieties in his farm were the benefits of different rates of maturity, so that harvest was spread over a longer period allowing him to spread out the work, rather than hire labor.

The yields of MASIPAG_bred rice and some selected traditional varieties were in most cases higher or similar as those of HYVs. But because farmer partners did not use chemical fertilizers and pesticides, net income was significantly higher than the conventional technologies (See Table 13.1). Eliminating chemical inputs and focusing on the utilization of natural resources available in the farm for pest control not only minimized financial expense, but also provided ancillary benefits to the environment. Alternative pest management was focused on maintaining ecological balance in the farm. Pesticides were completely eliminated from the food chain and farmers were no longer exposed to toxic chemicals. This in return, had allowed the return of diverse food sources, which contributed to better nutrition of the farming family.

Moreover, the recovery and maintenance of 668 traditional rice varieties contributed to the conservation of the main staple food. Improvement of these varieties through a modified bulk selection breeding strategy had produced 539 MASI-

Table 13.1. Comparison of inputs used in organic MASIPAG farming versus conventional farming (Source: Medina, 2002)

Item	Conventional (Pesos)[1]	MASIPAG (Pesos)[1]
Straw application	0	225
Land preparation	1,500	1,500
Seeds	3,000	450
Uprooting/transplanting	0	1,500
Seed broadcast/seedbed	100	150
Weeding		375
Herbicide	542	0
Insecticide	1,829	0
Chemical Fertilizer	3,600	0
Harvesting/threshing	2,948	2,948
Total production cost	13,519	7,148
Yield/Gross income: 4560kg (PhP 7.40/kg) / 4620kg (PhP 7.40/kg)		
Net income	20,224	27,040
Net profit : cost ratio	1.49	3.78

Cost and return analysis per hectare of conventional farming (HVY) and organic MASIPAG RICE (Sinayawan, Valencia, Bukidnon, 1977)
[1] One US$=51.5 Pesos (as of Feb. 2002)

PAG selections, as of 2002. Also, for maize, a recent addition, 49 traditional varieties with 3 improved MASIPAG selections had been produced with two regional back-up farms for maize established.

One female farmer in Southern Philippines claimed that since she began shifting to organic farming her expenses on agricultural inputs, labor and land preparation in her 1.74-hectare farm cropping dropped from P15,000 to P2,480 ($ 300 to $ 50) per cropping. Another farmer, who used conventional farming for twenty years but has shifted to organic farming, not only saved in his expenses for farm inputs, but also experienced an increase in yield to as much as 68 cavans (or 4080 kgs) of rice per hectare.

The success of those who have used this approach has started to mobilize the other farmers in the rural communities in the southern Philippines (Zonio 2006).

13.4.1.1
The current practice

At the farm level, each MASIPAG rice farmer plants at least three rice varieties to ensure varietal diversity as an ecological design in preventing outbreak of pests as well as for genetic conservation. Farm diversification and integration of farming components and processes are also incorporated to avoid external chemical inputs and increase sustainability. Thus, vegetable and fruit trees are also conscientiously integrated in farm diversification. Without chemical inputs and with nutrient cycling in place, natural soil fertility has been improving. Farmers' experience has shown that it requires from three to five years of conversion from conventional to organic farming for soil nutrient recovery. Without chemical pesticides, ecological balance is re-established, with plant and animal diversity slowly but steadily increasing.

Advantages of this farming system are that there is improved access and control of seeds, enhanced capacity to develop and control technology and the availability of farmer-managed trial farms. The diverse variety of seeds maintained and readily available to them through their trial farms eliminates the cost of procuring expensive HYVs and has assured them of the adaptation to diverse agro-ecological conditions. The farmers are trained to do actual plant breeding and management, as well as evaluation and selection of plant cultivars so that they can develop seeds based on their resources, priorities or perceived needs.

13.4.1.2
Using farm wastes wisely

In one of the farmer communities in the country's premier granary (the Central Luzon plains), a group of farmers has been practicing some innovations such as the conversion of farm wastes (hay, straw and rice hull) into commercial products. Their farming approach consists of:
- use of organic fertilizers (3 bags of chicken manure substituted for a bag of chemical fertilizer),
- herbal spraying (fermented leaves and twigs of neem and "kakawate" trees and "makabuhay" plants),
- thorough land preparation, good water management and good crop establishment for control of weeds,
- small canal built near dikes for destroying mollusks of golden snails, and
- use of inbred rice varieties, instead of hybrids.

Their average yield is 110 cavans (6,600 kgs) per hectare and targeted net income is from $ 3,000 to $ 4,000 per hectare (Roque 2006).

13.4.1.3
Transforming the Cordillera (Philippines) rice terraces for organic food production

Organic farming is now being encouraged again in the Cordilleras (a range of mountains found in the northern Philippines which is the home of the Banawe rice terraces (Figure 13.2). This farming practice, indigenous to the tribes who live there, do not use synthetic pesticides, herbicides and chemical fertilizers; instead, time-tested principles of soil replenishment, biodiversity and ecological balance are applied. Some of the local farmers also use the inago, rice ratooning, rice-based cropping system, rice-vegetable farming system and rice-fish culture system, among others.

In the inago traditional cropping system, mulch mounds are established in the flooded rice paddies where vegetables, green onions and other condiments are grown. Elongating the inago mounds will enlarge the area for producing vegetables and condiments in the rice terraces. There are two benefits of this practice; one, it will augment farmers' income, and the other, it will correct the zinc deficiency associated with the continuous flooding of the rice terraces.

The practice of ratooning aims to increase rice production and cropping intensity. Ratooning produces a second crop without seeding, lengthy land preparation and replanting activities. It also enables shorter crop maturity and requires less fertilizer, water and labor.

Fig. 13.2. The Banaue rice terraces in the Cordilleras

Rice-vegetable (usually sweet potato) cropping system is practiced as an alternate to mono-rice. The dry cultivation of sweet potato aerates the soil and makes zinc available to rice plants. Integrating fish with rice in the rice terraces increase the cash and non-cash income (fish for home use readily available fingerlings and also control of weeds, insects and snails in the field).

This enhanced farming practice in the rice terraces is hoped to transform this resource into a vibrant producer of quality and safe rice, vegetables and condiments seasons after seasons. An added value to the Cordillera is the preservation of this cultural heritage for generations of Filipinos, not discounting the other benefits that organic agriculture offer the farmer practitioners, such as safeguarding public health by eliminating/minimizing use of toxic pesticides/fertilizers, renewed local economies since market opportunities are widened due to consumers demanding food safe for themselves and the environment, protection of natural resources (crops in accordance with local climate), preservation of biodiversity (through preservation of a greater number of strains), and ensured future of agriculture since organic farming produce food without depleting the system's ability to continue (grcenbiz.com).

13.4.1.3
Fruit production and the management of slopelands in the Philippines

High-value fruit trees such as mango, citrus, durian (Durio, Zibethinus, Murray), lanzones (*Lansium domesticum* Correa), pili (*Canarium ovatum* Engl.), banana and rambutan (*Nephelium lappaceum* L.) have become banner commodities of profitable fruit production enterprise in the Philippines. Production areas have expanded from flat rolling land onto hilly and marginal slopelands, so that the Sloping Agricultural Land Technology (SALT) has been developed and propagated.

Basically, there are four SALT models (Escaño and Tababa 1998):
- SALT 1: Ally cropping using leguminous tree or shrub species planted closely in a belt along contour lines. Annual and perennial crops are planted between the rows. These are a mixture of food and cash crops.
- SALT 2: Known as "simple agro-livestock technology", this recommends a land use of 40% for agriculture, 20% for forestry, and 40% for livestock, particularly goats. The cropping mix includes forage crops as well as cash and food crops.
- SALT 3: Known as "sustainable agroforest land technology", this promotes food-wood intercropping where trees are planted in slopes of more than 50%. Tree species are a mixture of fruit and timber crops.
- SALT 4 This is "small agrofruit livelihood technology", and recommends the planting of fruit trees on the upper two-thirds portion of a SALT farm.

Fig. 13.3. The West Africa semi-arid tropics (WASAT)

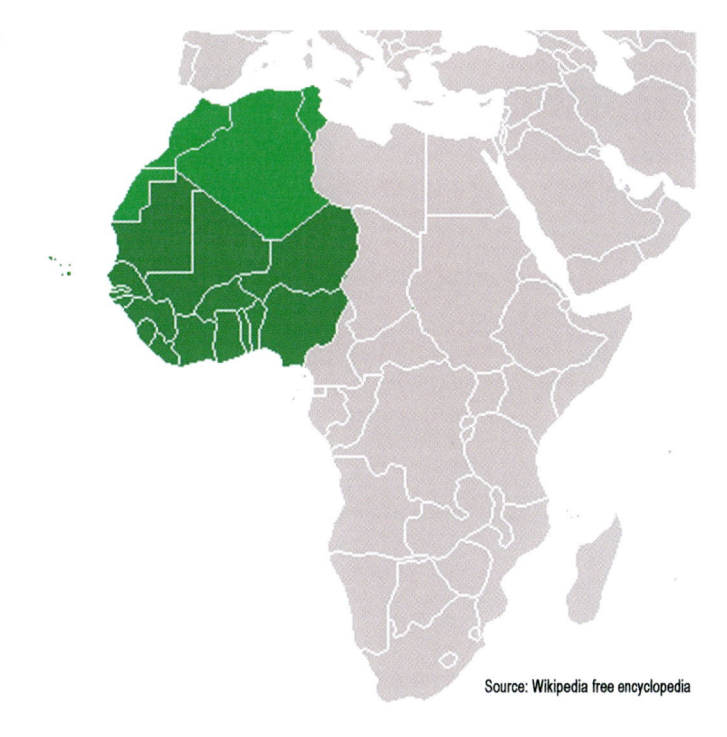

13.4.2
The West African semi-arid tropics (WASAT)

The West African semi-arid tropics (WASAT) are of three zones, the Sahel, Sudan savanna and North Guinea savanna (Figure 13.3). These three zones represent striking contrasts in production potential and risk. Climatic constraints (e.g., limited total precipitation, a short one-modal rainy season, high intraseasonal rainfall variability, high rainfall intensity and high evapotranspiration demands which peak at seedling and grain-filling stages, etc.) are most limiting in the Sahel and decline in importance in the Sudanian and Guinean zones (Matlon 1991). Its highly weathered soils reinforce these climatic constraints. The feature of the soils (loamy sands of the Sahel zone and sandy loams in the Sudanian and Guinean soils) result in low water-holding capacity, poor fertilizer use efficiency and high risk of periodic moisture stress.

Due to the close correlation of climatic and edaphic constraints across these agroclimatic zones, the lowest technical potential is located in the Sahel and the highest in the Sudan and North Guinean zones (see Table 13.2).

The WASAT farmers employ a combination of risk-reducing methods which can be defined as a management risk strategy. These tend to be interwoven as key elements of the farming systems in the region due to the particular constraints in the environment. Some of these methods are focused on reducing downside yield risk

Table 13.2. Zonal differences in cropping patterns and production potential in the WASAT (Source: Matlon, 1991)

	Sahel	Sudan Savanna	North Guinean Savanna
principal crops	millet, cowpea fonio, ground nut	millet, sorghum	cotton, maize, rice, cowpea, ground nut
secondary crops	sorghum	maize, ground nut, cowpea, cotton	millet
constraints / opportunities	– particularly harsh environment, – most limiting constraints for production – lowest water - holding capacity – short growing period	– low water - holding – capacity	– higher rainfall, – longer growing period allowing farmers to grow wider range of crops (solo, mixed or in relay cropping system)
productive potential	– lowest potential risk of highest failure	– higher production potential, – low risk	– higher production potential, – lowest risk
degree of crop diversification	– lowest diversification		– most diversified production pattern

directly, while the others are to generate compensatory income in the event that production shortfalls still occur (Matlon 1991).

These risk-reducing options include diversification of crops, cultivars and locations), varietal diversification in which farmers select and maintain a diversified set of varieties for the major crops in order to enable the spread of risk of loss due to period-specific stresses, diversify plot locations and land type diversification and match crops to the micro-environments to reduce the risk of crop losses due to stresses associated with particular land types, and interactive management which involves sequential decision-making in which cropping patterns and cultivation practices are sequentially adjusted to correspond to changes in conditions in the production area.

Additionally, Nyong in a 2005 study made on strategies of West African Sahel farmers for crop management noted that these include selection of varieties, crop field localization and ultimately, crop diversification. He has ranked these farming techniques used as management strategies according to preference of farmers and

Table 13.3. Ranking of farming techniques used in the West African Sahel as management tools according to order of preference and length of practice (Source:Nyong, 2005)

Farming Techniques	Preferred (ranking)	Length of practice (ranking)
Mixed cropping	2	1
Early planting	3	4.5
Wetland farming	1	2
Early maturing/drought-resistant crops	6	7
Increased spacing of crops	4	6
Change in crop type	7	4.5
Increase in farm area	5	3

number of years they have practiced them. The most preferred and practiced the longest are wetland farming and mixed cropping (see Table 13.3).

13.4.3
Improving rice-based cropping systems in the Indo-Gangetic Plains and in north-west Bangladesh

In rainfed agricultural areas, when the rice farmers wait for the monsoon rains to come, one crop management approach which has emerged as a promising part of the solution is simple: rather than transplanting rice seedlings into flooded field, rice seeds are sown directly into unflooded field.

The Indo-Gangetic Plains is a rich, fertile land encompassing most of northern and eastern India, the most populous parts of Pakistan, and virtually all of Bangladesh (see Figure 13.4). It is home to over 850 million people (Wikipedia). Rice-wheat is the principal cropping system occupying 13.5 million ha. Its sustainability is thus, vital to the livelihoods of the farmers of the region (Singh et al. 2005).

In India, rice and wheat are the staple food crops. Rice is traditionally transplanted at the end of the dry season (May/June) and wheat is sown after the rice harvest (November/December). Likewise in Bangladesh, rice is also its staple food; hence, it is of major importance due to the fact that its agriculture consists mainly of subsistence farming on very small farm areas. Yet, it is the single largest producing sector of the Bangladeshi national economy. (Wikipedia).

Bangladesh has annual monsoon floods and cyclones are frequent. But the High Barind Tract of NW Bangladesh is a drought-prone area. Its rainfed agriculture has a predominant cropping pattern of a single crop of transplanted rice (about 100,000 ha) grown during the monsoon 'aman' season from June to October dur-

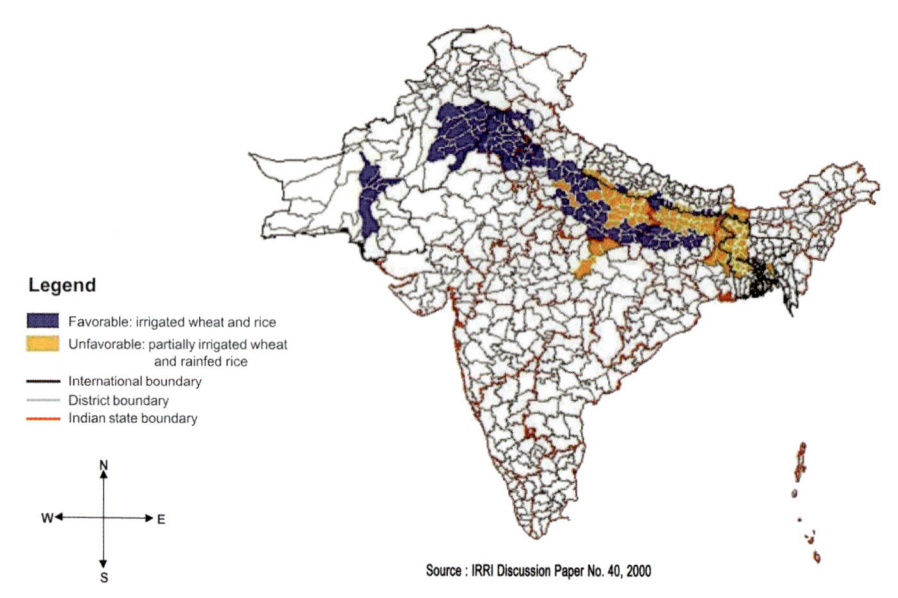

Fig. 13.4. Map of the Indo-Gangetic Plains showing the agroecological analysis of its rice-wheat area and productivity (Source: IRRI, 2000)

ing which 80 per cent of the 1200-1400 mm annual rainfall occurs (Mazid et al. 2006). After rice harvest, in the 'rabi' season, approximately 20,000 ha is sown to a range of dryland crops planted on the residual soil moisture. These include chickpea, linseed, mustard and/or wheat.

The late onset of the monsoon (as is being seen during most of the recent years) can delay rice transplanting, as a minimum of 400 mm cumulative rainfall is needed to complete land preparation for transplanted rice (Mazid et al. 2006). However, changing rice establishment from transplanting to direct seeding can result in yields (seasonally dependent and ranging from 2-4 tons/ha) similar to, or higher than those by conventional transplanting.

Direct seeding is either dry-seeding or wet-seeding. Dry seeded rice can be sown after land preparation with only 150 mm cumulative rainfall. On the other hand, wet-seeded (pre-germinated) rice sown by drum seeder on to puddle land removes the nursery bed requirement of transplanted rice and can advance the crop establishment by one month. Harvest time is advanced even when the cultivar used is not changed. Advancing crop establishment also reduces the risk of terminal drought and allows earlier planting to ensure more reliable establishment of a post-rice crop. A drawback though, is that the inherent advantage of weed suppression through puddling and transplanting rice into standing water is lost. The increased weed pressure after emergence of direct-seeded rice could, however, be overcome by the timely application of a pre-emergence herbicide after seeding and follow-up hand weeding.

In both India and Bangladesh, if soil moisture is adequate, pre-germinated rice seed may be either broadcast by hand or sown in rows with an inexpensive plastic drum seeder (Rice Today 2006). And in many northeastern Indian farm, farmers use tractor-mounted mechanical seeders that sow seeds at chosen rates and simultaneously apply fertilizer. Advantages cited by the scientists from different research centers assisting the farmers (the International Rice Research Institute, the UK-based Natural Resources Institute, the University of Liverpool in UK, the Bangladesh Rice Research Institute, and in India, the G.B. Pant University of Agriculture and Technology in Pantnagar, Narendra Deva University of Agriculture and Technology in Faizabad, C.S. Azad Agriculture University in Kanpur and Rajendra Agriculture University in Patna) include:

- relief in both the water and labor problems,
  At the eastern end of the Plains in Bangladesh, farmers need about 500 mm cumulative rainfall to establish a rice crop through transplanting. If farmers direct-seed, they can establish the crop from about one-quarter of this rainfall amount (Johnson and Mortimer 2005).
  In Barind, farmers are supposed to transplant by July. But if there is no rain, they can not transplant and the seedlings get older. Seedlings should be no older than 30 days to get the best yields. Additionally, labor requirements are less.
- avoidance of damages from early-season drought.
  Droughts during rice plants' flowering stage (prolonged monsoon breaks) can devastate the crop, causing yield losses of 50 per cent or more.
- Earlier establishment means earlier harvest; thus, increasing the chances of growing a dry-season crop (for instance, chickpea),
- avoidance of crop yield losses due to delay in planting , and
  In the rice-wheat belt of India, if rains arrive too late and there is no access to water (irrigation), the crop is compromised and the equally important wheat crop is also threatened. For every week beyond 1 November that wheat planting is delayed, the crop suffers a yield loss of 10 per cent in the most productive areas because of cold temperatures.
- cost savings
- Direct seeding is generally cheaper than transplanting, which incurs the expenses of nursery establishment and care and the labor that go with it. Additionally, even on larger farms, running tractors and machine seeders on dry, unpuddled fields is less expensive than on a flooded one. In Barind, the average crop establishment costs around US$120 ha^{-1} for transplanted rice, while for direct-seeded rice, there is a reduction of around 25 per cent in this cost with no yield disadvantage (Mazid et al. 2006).

Every crop management system has its own disadvantages, and for direct-seeded rice cropping, the constraint is in weed control. Because farmers can no longer rely on the flooding to suppress the weeds during the crucial initial period of crop establishment, there is a need for better management in the farms. The scientists assisting the farmers have already shown during the farmers' trials that successful weed management can be put in place in both rainfed and irrigated rice-cropping systems in India and Bangladesh (Rice Today 2006).

Fig. 13.5. Map of Indonesia.

13.4.4
Special Case: Small-holder rubber production in South Sumatra, Indonesia

The Indonesian archipelago extends from 6°N to 11°S latitude and from 95° to 141°E longitude (see Figure 13.5). There are more than 13,000 islands, with Sumatra as one of its largest. Indonesia has a moist climate with high temperatures and abundant rainfall.

Natural rubber is one of the most important agricultural industries in the Indonesian economy (Purnamasari et al. 1999). The industry is dominated by smallholders who have 85 percent of area planted and undertake 76 percent of production. A rubber producer's profit depends on the quality and quantity of latex yield and the costs involved in producing it. These factors depend largely on tree-management decisions such as used clone tree density, rotation length, tapping method and are also influenced by risks arising from climatic change and uncertainty of rubber prices.

To determine optimal management strategy for small-holder rubber production in South Sumatra, the use of a rubber agroforestry model was embedded in a dynamic economic model. The results of the model could be used as a guide in the tree-management options. For instance, the fact that the profitable cultivation period of the clonal tree (which contained improved genetic materials) is longer than for the wilding (unselected seedling), rotation length for the clone is longer than for the wilding and expected Net Present Value (NPV) is higher. Management practices recommended are earlier commencement of tapping from year 7 to year 6, higher density from 500 to 600 trees ha^{-1} and shorter cycles from 38 to 35 years for the clone and from 34 to 31 years for the wilding.

13.5
Conclusions

Coping strategies to address agrometeorological risks and uncertainties are many and varied. These are, however, largely dependent on the farmer's access to both on-farm and off-farm resources. There are multiple stresses, which at times are difficult challenges to surmount unless access to resources is assured. However, inasmuch as technology has leapfrogged, it is hoped globalization will begin to reduce the divide between the developed and developing world.

Yet, in many developing countries, persistence of poverty, particularly in the rural environments is most often due to the inability of the poor to gain access to support mechanisms in terms of technical expertise/technological innovations and others, including formal sources of credit and crop insurance.

References

Belliveau S, Smit B, Bradshaw B (2006) Multiple exposures and dynamic vulnerability: Evidence from the grape industry in Okanawagan Valley, Canada. Glob Environ Chang 16: 364–378.

Bradshaw B, Dollan H, Smit B (2004) Farm-level adaptation to climatic variability and change: Crop diversification in the Canadian prairies. Clim Chang 67:119–147.

Bryant CR, Smit B, Brklacich M, Johnston TR, Smithers J, Chiotti Q, Singh B (2000) Adaptation in Canadian agriculture to climatic variability and change. Clim Chang 45:181–201.

Chainuvati C (2000) Country Report, Thailand In Expert consultation on crop diversification in the Asia –Pacific region. FAO Corporate Documents Repository

Escaño CR, Tababa SP (1998) Fruit production and the management of slopelands in the Philippines. Food and Fertilizer Technology Center. FFTC 2006. In www.agnet.org/library/article/leb450.html

Espino RC (2000) Country Report, Philippines In Expert consultation on crop diversification in the Asia –Pacific region. FAO Corporate Documents Repository

FAO (2000) Expert consultation on crop diversification in the Asia –Pacific region. FAO Corporate Document Repository. In www.fao.org/docrep/003/x6906e/x6906c05.htm

Gomez BE (2004) Status of mitigation and adaptation strategies with respect to impacts of climate change/variability and natural disasters in agriculture-WMO RA I. In www.amis.org/agm/meetings/iccnd/Gomez_adapt

Guanghuo WA, Doberman C, Witt Q, Sun R, Fu G, Simbahan A (2006) Reducing the gap between attainable and potential yield in double rice cropping systems of Zhejing Province, China. Proceedings of the International Rice Research Conference, Rice Research for Food Security and Poverty alleviation. IRRI Copyright 2006

Hardaker JB, Huirne RBM, Anderson JR (1997) Coping with Risk in Agriculture. CAB Int., London, UK. pp 30–84

Hazra CR (2000) Country Report, India In Expert consultation on crop diversification in the Asia –Pacific region. FAO Corporate Documents Repository

Hoque ME (2000) Country Report, Bangladesh In Expert consultation on crop diversification in the Asia –Pacific region. FAO Corporate Documents Repository.

Harwood J, Heifner R, Coble K, Perry J, Somwaru A (1999) Managing Risks in Farming. USDA/ERS AER 774.

IRRI (2000) Improving the productivity and sustainability of rice-wheat systems of the Indo-Gangetic Plains: A Synthesis of NARS-IRR partnership research. In: Ladha JK, Fischer KS, Hossain M, Hobbs PR, Hardy B. (eds) Discussion Paper No. 40. International Rice Research Institute, Los Banos, Philippines.

Johnson DE, Mortimer AM (2005) Issues for integrated weed management and decision support in direct-seeded rice. In: Rice is life: scientific perspective for the 21st century. Proceedings of the World Research Conference held in Tokyo and Tsukub, Japan, November 4-7, 2004, Los Baños (Philippines): International Rice Research Institute, and Tsukuba (Japan): Japan International Research Center for Agricultural Research, CD 211–214

Lal Bose M, Abu Isa M, Bayes A, Sen B, Hossain M (2006) Impact of modern rice varieties on food security and cultivar diversity: the Bangladesh case. Proceedings of the International Rice Research Conference. Rice Research for Food Security and Poverty Alleviation. International Rice Research Institute, Los Banos, Philippines.

Maclean JL, Dawe DC, Hardy B, Hettel GP (Eds) (2000) Rice almanac. Los Baños (Philippines): International Rice Research Institute, Bouake (Cote d'l voire): West Africa Development Association, Cali (Columbia); International Center for Tropical Agriculture, Rome (Italy): Food and Agriculture Organization, 253 p.

Matlon PJ (1991) Farmer risk management strategies: The case of the West African semi-arid tropics. In: Holden D, Hazell P, Pritchard A (Eds) Risk in Agriculture. Proceedings of the Tenth Agriculture Symposium. The World Bank, Wash. D.C.

Matlon PJ, Fafchamps M (1989) Crop Budgets in Three Agro-Climatic Zones of Burkina Faso, ICRISAT Progress Report, International Crops Research Institute for the Semi-Arid Tropics, Hyderabad, India.

Mazid MA, Riches AM, Mortimer LJ, Wade M, Johnson DE (2006) Improving rice-based cropping systems in north-west Bangladesh. Proceedings of the Australian Weed Science Society Conference. Sept. 2006, Adelaide, Australia.

Medina CP (2002) Empowering farmers for rural development: the MASIPAG experience. Biotechnology and Development Monitor 49:15–18

Mengxiao Z (2000) Crop diversification in China. FAO Corporate Document Repository In: www.fao.org/docrep/0031x6906e/x6906e05.htm

Myers R ( ) Crop diversification opportunities in Missouri In:agebb.missouri.edu/sustain/crop divers/index.htm

Nyong A (2005) Adaptive capacity for food production in the West African Sahel. CCIARN Parallel Event of COP-11, Montreal, Canada, 02 December 2005

Page W (2005) Bedding down crops. Partners in Research for Development. October 2005. Australian Centre for International Agricultural Research, ACIAR ISSN-1031–1009

Pandey S, Behura D, Villano R, Naik D (2006) Drought risk, farmers' coping mechanisms and poverty: A study of the rainfed system in Eastern India. Proceedings of the International Rice Research Conference. Rice Research for Food Security and Alleviation. International Rice Research Institute, Los Banos, Philippines.

Parr J F, Papenclick RO, Youngberg IG, Meyer RE (1990) Sustainable agriculture in the United States. In:. Edwards CA, Rattan L, Madden P, Miller R , House G. (eds.) Sustainable Agricultural Systems, Soil and Water Conservation Society, Ankeny, IA. pp. 50–67.

Purnamasari RA, Cacho M, Simmons P (1999) Management strategies for Indonesian smallholder rubber plantation in South Sumatra: A bioeconomic analysis. Working Paper Series in Agricultural and Resource Economies. ISSN 1442 1909

Rice Today (2006) , April-June Vol. 5 No.2. International Rice Research Institute In www.irri.org/publications/today/pdfs/5-2/Rice Today_5-2.pdf

Ripple (2006) Northwest Bangladesh – Direct-seeded rice, drum seeders and misty mornings. Rice Research for Intensified Production and Prosperity in Lowland Ecosystems.

Risk Management (2000) Coordinated Access to the Research and Extension System. Mississippi State University Extension Service In http://msucares.com/pubs/misc/m1122.htm

Roque A (2006) Good for farmers' pockets, friendly to the environment. Philippine Daily Inquirer, June 25, 2006. PDI Philippines

Searcy SW (1999) Precision farming : A new approach to crop management. Texas Agricultural Extension Service, Publication L–5177

Sharma ML, Sharma PN, Khan MA, Tiwari RK, Hossain M (2006) Indigenous technology in the rice-based cropping systems among tribals of Eastern India. Proceedings of the International Rice Research Conference. Rice Research for Food Security and Alleviation. International Rice Research Institute, Los Banos, Philippines.

Sharma KC (2000) Country Report, Nepal In Expert consultation on crop diversification in the Asia –Pacific region. FAO Corporate Documents Repository

Smithers J, Blay-Palmer A (2001) Technology innovation as a strategy for climate adaptation in agriculture. Appl Geogr 21:175–197

Singh, G, Singh Y, Singh VP, Johnson DE, Mortimer M (2005) System-level effects in weed management in rice-wheat cropping in India. In: Proceedings of the BCPC International Congress on Crop Science and Technology-2005, SE CC Glasgow, UK, 1, pp. 545–550.

Singleton G (2006) Direct seeding of rice gets warm approval in the Indo-Gangetic Plain. In: Ripple (ed.) Northwest Bangladesh – Direct-seeded rice, drum seeders and misty mornings. Rice Research for Intensified Production and Prosperity in Lowland Ecosystems.

Sun Star (2004) Transforming the rice terraces for organic rice production In sunstar.com.ph/static/2004

Valdivia C, Gilles JL, Jette C, Quiroz R, Espejo R (2004) Coping and adapting to climate variability: The role of assets, networks, knowledge and institutions In sunstar.com.ph/static/2004

Van Luat N (2000) Country Report, Vietnam In Expert consultation on crop diversification in the Asia –Pacific region. FAO Corporate Documents Repository

Wall E, Smit B, Wandel J (2004) Canadian agri-food sector adaptation to risks. Position paper on climate change, impacts and adaptation in Canadian agriculture. C-CIARN Agriculture, University of Guelph, Ontario, Canada.

Weerasena LA (2000) Country Report, Sri Lanka In Expert consultation on crop diversification in the Asia –Pacific region. FAO Corporate Documents Repository

Yahya TMBT (2000) Country Report, Malaysia In Expert consultation on crop diversification in the Asia –Pacific region. FAO Corporate Documents Repository

Zonio A (2006) Poor village adopts organic farming. Inquirer Mindanao in the Philippine Daily Inquirer, PDI Philippines

Introduction to Sustainable Agriculture In www.grcenbiz.com

CHAPTER 14

Water management in a semi-arid region: an analogue algorithm approach for rainfall seasonal forecasting

Giampiero Maracchi, Massimiliano Pasqui and Francesco Piani

14.1
Introduction

Methods and results of this recent branch of atmospheric sciences must be the most simple and accessible as possible. For this reason, the Institute of Biometeorology, (part of the National Research Council, http://www.ibimet.cnr.it), has developed a physically – based statistical approach to obtain seasonal forecasts, regarding rainfall precipitation, over Sahel region.

The method is based on the "similarity" conditions of the sea surface temperature (SST) in three areas of the world defined as: Niño-3 (5S-5N;150W-90W), Guinea Gulf (10S-5N;20W-10E), Indian Ocean (5S-15N;60E-90E) which, in literature, are indicated as the most important areas to drive the precipitation patterns: Indian Ocean and Southern Atlantic with regard to the trends of precipitation and Niño-3 with regard to the interannual time scale (Giannini et al. 2003).

The importance of the sea surface temperature to force the long-term atmospheric anomalies, at least at a regional scale, is recognized with particular attention to the Pacific area affected by El-Niño Southern Oscillation (Dalu et al. 2006).

Many atmospheric Research Centers have developed their own methods to derive seasonal forecasts, based on the results of a large number of simulations of a Global (GCM) or a Regional (RCM) Circulation Model, namely "Ensemble Forecasts", or on statistical algorithms that relate the most important atmospheric variables, or both (see for example: http://www.ecmwf.int or http://iri.columbia.edu/).

This work describes a statistical method that relates the SST of three oceanic areas with the precipitation in the semi-arid region called Sahel. The chance of forecasting a reliable rainfall field is, in many parts of Africa, dependent on prevailing patterns of sea surface temperature, atmospheric circulations, the El Niño Southern Oscillation and regional climate fluctuations in the Indian and Atlantic Oceans. A brief summary of scientific background of the method is the following:
- West African Monsoon variability is strongly forced by the sea surface temperatures standardized anomaly (SSTAs) of the Gulf of Guinea. Warm Gulf of Guinea SSTAs generates a rainfall increase along the Guinean coast while the precipitation decreases over the Congo Basin. These features can be understood through the dynamical response of a Kelvin wave along the equator and a Rossby wave to the west of the SSTA. The first is associated with a weakening of the Walker circulation, while the latter tends to strengthen the West African Mon-

soon and the upward vertical velocity. The effects of Cold SSTAs are opposite, but weaker (Vizy and Cook 2001).
- The monsoon circulation influences the precipitation over the Sahel, in particular southern Sahel (10N-15N), in two main ways. The moisture is transported by the low-level southerly flow. The proximity of the monsoon circulation and circulation over Sahara generates a strong low-level convergence to force air parcels to rise vertically until the level of free convection (Vizy and Cook 2002).
- Positive SST anomalies in the Eastern Pacific and in the Indian Ocean, negative anomalies in the northern Atlantic and in the Gulf of Guinea are related with drought conditions over all the West Africa (Fontaine and Janicot 1996).
- Droughts limited to Sahel are due to a positive SST anomaly northward in the southern Atlantic and a negative pattern in the northern Atlantic. Floods along the West Africa are associated with positive anomalies in the northern Atlantic, while the floods limited to Sahel are related to different forcing: northward expansion of negative SST anomalies in the southern Atlantic, positive SST departures in the northern Atlantic, and development of negative SST anomalies in the eastern Pacific (Fontaine and Janicot 1996).
- The Principal Components Analysis (PCA) performed on the summer precipitation in the Sahel region, demonstrates that the two leading principal components (PCs) explain almost half of the variability of the precipitation. Moreover two main patterns are present: the first along the Gulf of Guinea coast, between the equator and 10°N, dominated the interannual variability, the second associated with the continental convergence in the Sahel (between 10°N and 20°N) affected by the interdecadal variability. The decomposition of these two leading PCs into high and low-frequency components shows the role of the SST of the Southern Atlantic and Indian Ocean for driving the long-term variability, while the interannual variations are driven by the ENSO (Giannini et al. 2003).

14.2
Methods and Dataset

In the method each "month" is defined by six variables: three are SSTAs while the other three take into account their respective tendencies (namely "Change Rates" or CRs). The CRs are defined as the difference between the current SSTAs and those of the previous month. The standardization is obtained with the subtraction of the 1979-2003 climatological mean and the division by the 1979-2003 climatological standard deviation. The "similarity" to the current SST conditions is evaluated by means of the minimization of the Euclidean distance to find the most similar year (namely analogue) and assign the values observed in that year to the forecast rainfall field. Due to the specific dynamical behavior of the West African Monsoon this simple analogue characterization is able to catch main features of rainfall precipitation patterns during the JJA period and a validation of this approach, through analysis of forecast skills, shows encouraging results.

SSTs used in the method are from three oceanic areas: the Niño-3 area (5S-5N; 150W-90W), the Guinea Gulf (10S-5N; 20 W-10E), the Indian Ocean (5S-15N; 60E-90E) as in Fig. 14.1.

The SST data have been obtained by the NCEP/NCAR Reanalysis dataset (2.5°x2.5° Lat-Lon, Kalnay et al. 1996; Kistler et al. 2001) while the precipitation data have been derived from the Global Precipitation Climatology Project (GPCP, Xie and Arkin 1996; Huffman et al. 1997; Xie et al. 2003) on a 2.5°x2.5° Lat-Long grid.

For each month and for each grid-point, the precipitation time series has been correlated to the SST time series, in order to have the relative weight of the three different oceanic areas with regards of the precipitation. Based on these weights and on the six variables defined above, the method searches for the most similar SST conditions in the past (the year obtained is called "Analogue Year"), assigning

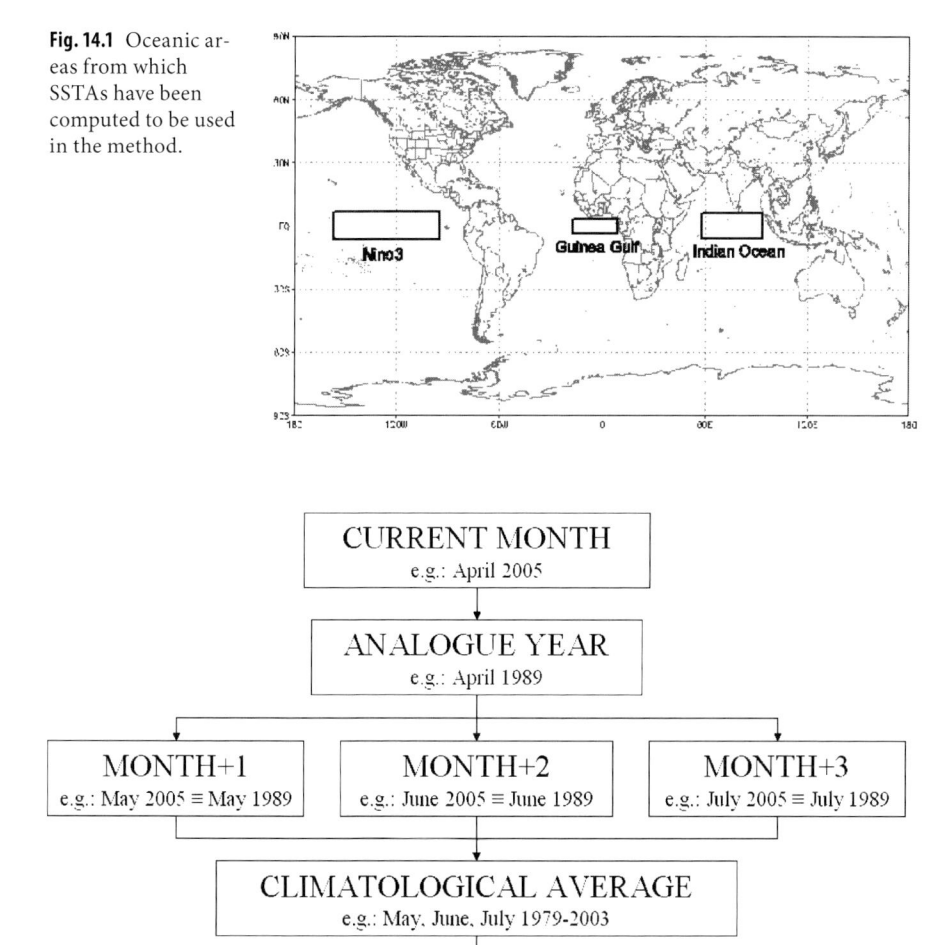

Fig. 14.1 Oceanic areas from which SSTAs have been computed to be used in the method.

Fig. 14.2 An example of the logical scheme applied for each precipitation grid point in order to derive the "Analogue" year.

the values observed in the closest year to the forecasts. Using 1979-2003 climatology, precipitation anomaly and percentage anomaly are then computed (Fig.14.2). The first is derived by means of the subtraction of the climatological mean from the forecast precipitation. The latter is obtained from the former, dividing for the climatological mean and it's represented as a percentage. The dry regions of the world, identified by a monthly cumulative precipitation under 30mm, are blanked to avoid large values of anomalies and percentage anomalies (Table 14.1).

Table 14.1. An example of forecast maps of percentage anomaly for the year 2003 with the correspondent observed values from CMAP (http://www.cdc.noaa.gov/cdc/data.cmap.html) Janowiak and Xie 1999: for each month, June, July and August, forecasted maps were shown with different time lags.

	June 2003	July 2003	August 2003
OBS CMAP			
1 month ahead			
2 months ahead			
3 months ahead			

14.3
Skill evaluation

A true validation strategy should be based only on data collected prior to the target month to be forecasted. Such a calculation should then be repeated for all available years. But the resulting skill depends on the amount of data used for each calculation. Another possible strategy is the adoption of a cross validation calculation. Each prediction was estimated using only data before and after that specific year. The cross evaluation hindcast method is able to represent a good forecast skill measure if two conditions were satisfied: the climate statistics do not change among the period considered and there is a weak autocorrelation between neighboring years data. In order to perform the cross – evaluation analysis we select all summer time forecasts (June – July – August) in the 1979 – 2005 period. For each month the rainfall anomaly has been computed based on the 1979 – 2005 mean value for the region defined as Lat: 15°.0 – 17.5°, Long: -10° – 10°. For this area the entire monthly anomaly ensemble was divided into three categories below 33% percentile (Below hereafter), above 66% percentile (Above hereafter) and between them (Normal hereafter). Each monthly anomaly was aggregated in order to form three ensembles. The same computation was performed for different forecast ranges: 3 months ahead, 2 months ahead and 1 month ahead respectively. For each observed anomaly group, by means Below – Normal – Above, all the forecasted anomalies were computed and showed as "chocolate wheels" graphs in Fig.14.3 to Fig.14.5 (Hayman 2000). As expected, using SSTA, for forecasting precipitation introduces a lag time in the peak performances of the method, by means: best performanc-

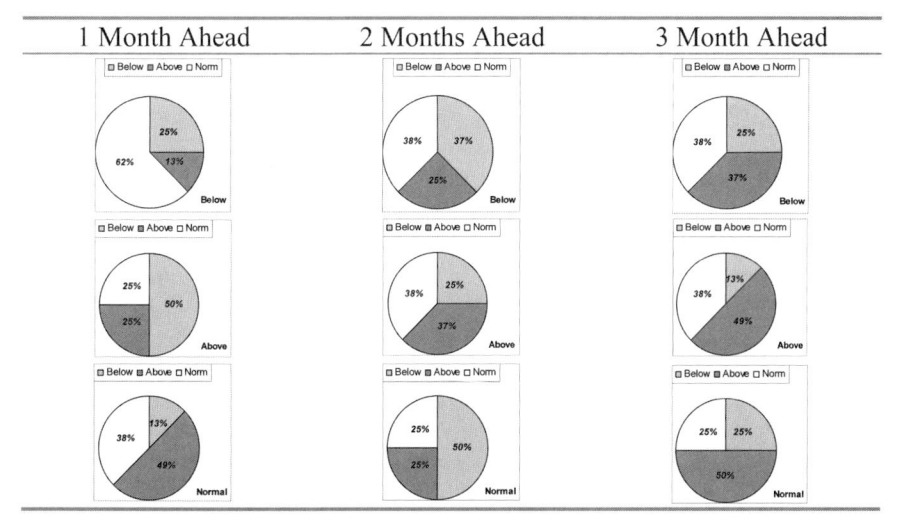

Fig. 14.3 Chocolate wheels for Sahel area evaluation. Numbers represent the percentage of historical cases in the lower (light grey), middle (white) and upper (dark grey) terciles for June. Table columns represent time lags and table rows represent different observed classes (Below – Above – Normal from top to bottom).

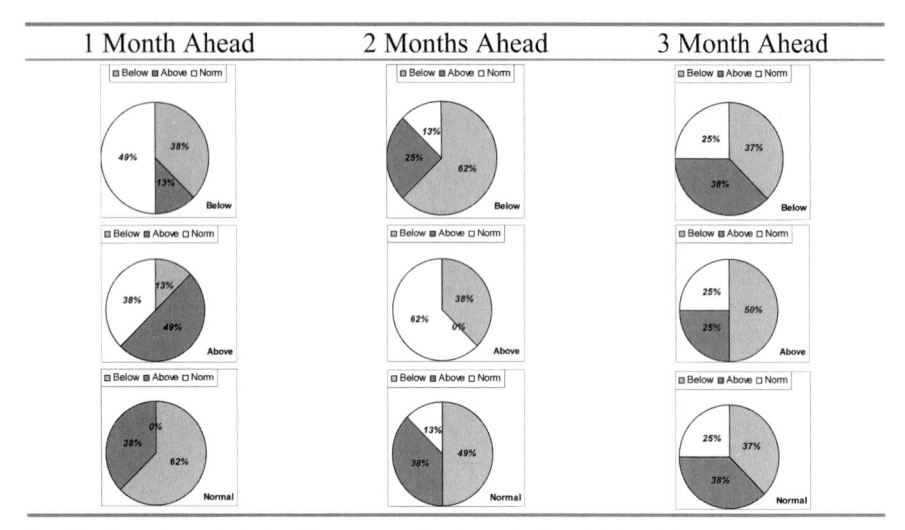

Fig. 14.4 Chocolate wheels for Sahel area evaluation as in Fig. 14.3 for July.

es were obtained 2 o 3 months in advance. Same behaviors are present for all target months. Normal months were forecasted worse than Above or Below months, probably this is a link to choice of SSTA as predictors emphasizing strong rainfall anomalies.

One limitation of the evaluation strategy is the short period of time used: just 26 years. One single forecasted event can alter the statistics greatly. Further analysis will be focused on extending this period back in time increasing the statistical ensemble.

14.4
Conclusions

The weather predictions today are established on solid theoretical and practical bases, their reliability and accuracy are steadily increasing and their usefulness is widely recognized in a variety of fields and applications, often in the frame of automatic integrated prediction systems.

The current state of the art of the weather forecasts allows the short-term prediction of rare and dangerous local events such as rainstorms, frost with high reliability and accuracy and very high spatial resolution, as well as the accurate medium range prediction. The role of conventional weather forecasts is precisely defined as a strategic one, and as such, is considered the national as well as the regional weather services (Soderman et al. 2003). Having information on the future trend of precipitation three months or more in advance could be of extreme importance in many fields of large economic, social, environmental and strategic relevance: agriculture and forestry, land and landscape management (to forecast droughts or heat waves for example), international cooperation and catastrophy management (i.e. food shortages, droughts, production and distribution of energy).

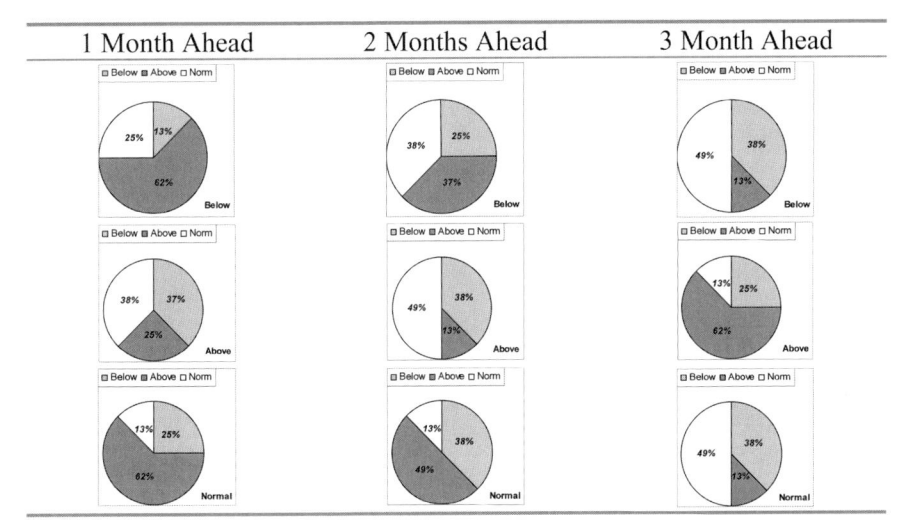

Fig. 14.5 Chocolate wheels for Sahel area evaluation as in Fig. 14.3 for August.

Seasonal forecasts could answer these questions but not with the same efficiency, accuracy and reliability of the meteorological forecasts. Although many enhancements have interested this branch of atmospheric science in the last decade, large errors and uncertainties still affect this type of products. At the same time, seasonal forecasts have assumed a relevant role for planning and decision making.

Authors would like to underline the experimental character of the results of the method. They must be used just as an indication of the possible future trend of the precipitation.

References

Dalu GA, Gaetani M, Pielke RA Sr., Baldi M, and Maracchi G (2006) Regional variability of the ITCZ and of the Hadley cell. Geophy Res Abst 6:10-2–2004.
Fontaine B, Janicot S (1996) Sea surface temperature fields associated with West African rainfall anomaly types. J Clim 9:2935–2940.
Giannini A, Saravanan R, Chang P (2003) Oceanic forcing of Sahel rainfall on interannual to interdecadal time scales. Science 302:1027–1030.
Hayman PT (2000). Communicating probabilities to farmers: pie charts and chocolate wheels. In Petheram RJ. (ed.) Tools for participatory R&D in dryland cropping areas. p133–136 RIRDC00/132. Rural Industries Research and Development Corporation, Canberra.
Huffman G J and co-authors (1997) The Global Precipitation Climatology Project (GPCP) combined data set. Bull. Am. Met. Soc.. 78:5–20.
Janowiak JE, Xie P (1999) CAMS_OPI: a global satellite-raingauge merged product for real-time precipitation monitoring applications. J Clim 12:3335–3342.
Kalnay E and coauthors (1996) The NCEP/NCAR 40-Year Reanalysis Project. Bull Am Met Soc 77:437–471.
Kistler R and coauthors (2001) The NCEP–NCAR 50-Year Reanalysis: Monthly Means CD-ROM and Documentation. Bull Am Met Soc 82: 247–268.

Soderman D, Meneguzzo F, Gozzini B, Grifoni D, Messeri G, Rossi M, Montagnani S, Pasqui M, Orlandi A, Ortolani A, Todini E, Menduni G, Levizzani V (2003) Very high resolution precipitation forecasting on low cost high performance computer systems in support of hydrological modeling. Prepr. 17th Conf. on Hydrology, AMS, Long Beach.

Xie P, Arkin PA (1996) Global precipitation: a 17-year monthly analysis based on gauge observations, satellite estimates, and numerical model outputs. Bull Am Met Soc 78:2539–2558.

Xie P, Janowiak JE, Arkin PA, Adler R, Gruber A, Ferraro R, Huffman GJ, Curtis S (2003) GPCP pentad precipitation analyses: an experimental dataset based on gauge observations and satellite estimates. J Clim 16:2197–2214.

Vizy EK, Cook KH (2001) Mechanics by which Gulf of Guinea and Eastern North Atlantic Sea Surface Temperature Anomalies can influence African Rainfall. J Clim 14, 795–821.

Vizy EK, Cook KH (2002) Development and application of a mesoscale climate model for the tropics: Influence of sea surface temperature anomalies on the West African Monsoon. J Geophys Res 107(D3), 10.1029/2001JD000686.

CHAPTER 15

Water Management – Water Use in Rainfed Regions of India[1]

YS Ramakrishna[3], GGSN Rao, VUM Rao, AVMS Rao and KV Rao[2]

Central Research Institute for Dryland Agriculture, Hyderabad

Abstract

Large investments of about Rs.800 billion since Independence has gone into development of surface irrigation projects and the gross irrigated area increased from 22.56 m ha to 75.14 m ha by 2000-01 in India. In spite of large-scale developments in irrigation sector, the agricultural production remains static at 212 mt, a cause of great concern, which is mainly attributed to the inefficient water management practices, poor maintenance of structures and water conveyance systems. In this review article, various issues, perspectives and strategies in water management research programs were highlighted. The impact of climate change on water resources at global level and at national level has also been discussed. A few case studies on improving the water use efficiencies through watershed programs carried out at CRIDA, Hyderabad are mentioned. Social problems in implementing water management strategies have been indicated.

15.1 Introduction

Huge investments amount to Rs.790.55 billions have gone in development of surface waters by the Government of India during the period 1947-2001 (Parthasarathy, 2006). As a result, the gross irrigated area increased from 22.56 m ha in 1950-51 to 75.14 m ha in 2000-01, thus creating largest irrigated area in the world. It is reported that about 4400 (large, medium and small) dams have been constructed

---

[1] Paper presented at International Workshop on Agro meteorological Risk Management, New Delhi during 25-27 October 2006
[2] Director, Project Coordinator (Ag. Met.) , Principal Scientist (Ag. Met.) and Scientist Senior Scale (Ag. Met.) and Senior Scientist
[3] Corresponding author – Y.S.Ramakrishna, Director, CRIDA, Santosh nagar, Hyderabad – 500059,AP. India. E mail: Ramakrishna.ys@crida.ernet.in, Phone: 91-040-24530177, FAX : 91-040-24531802

Table 15.1 Crop wise status of irrigation facility across the country – India

Crop	Irrigated area (1000 ha)			
	North	South	West	East
Barley	125	92	158	NA
Cotton	1749	466	1671	NA
Fruits	444	278	417	250
Groundnut	NA	600	224	72
Maize	413	303	523	138
Millet	363	266	459	121
Potatoes	147	92	138	83
Pulses	1309	141	1839	248
Rape seed	154		302	56
Rice	8,788	7,004	1,970	6,129
Sorghum	305	224	387	102
Soybean	286	179	268	161
Sugarcane	1650	809	777	NA
Vegetables	394	246	369	222
Wheat	6526	204	9,994	3,671
All irrigated crops	20,651	10,905	19,496	11,251
Equipped for irrigation	16,032	10,020	15,030	9,018
Cropping intensity	129	109	130	125

so far in India. The expansion in irrigated area in the country is mainly due to developments in ground water exploitation and nearly 60 percent of the irrigation in the country is met from ground water resources. These have contributed to the increased agricultural production from 50 mt to 212 mt, thus becoming self-sufficient in meeting the foodgrian requirements of the increased population to above 1.1 billion. The status of cropwise irrigation facilities across the country is given in the Table 1.

It is inferred from the table that the irrigated area is high under rice in north, south, west and eastern India followed by wheat in north, west and eastern India. Sugarcane, cotton and pulses occupy the third place. In recent years, the agricultural food production remains static, which is hovering around 210 mt, a cause of great worry to the administration in meeting the future food requirements. Per-

Table 15.2 Estimates of Water Need for India (M ha m)

Activity	1990	2000	2025
Irrigation	46.0	63.0	77.0
Domestic	2.5	3.3	5.2
Industrial	1.5	2.7	12.0
Energy	1.9	2.7	7.1
Others	3.3	3.5	3.7
Total	55.2	75.2	105.0

haps, the expected agricultural production levels could not be achieved due to mismanagement and over-exploitation of irrigated waters and little has been done to improve the rainwater use efficiency in the rainfed regions of the country. Stagnation or fall in agricultural production was also noticed especially in Indo-Gangetic Plains due to practicing of mono-cropping system (Rice-Wheat).
The future total water need estimated for India by 2025 is shown in Table 2.
- The entire water potential of 1122 BCM need to be developed by all means by 2025 through surface and ground water development.

It is seen from this table that irrigation water requirement is increasing drastically along with Industry requirement. It is hard task to achieve this unless better water management practices are adopted to enhance efficient use of water. Lack of understanding of the importance of the cost involved in providing the irrigation water absolutely free of charge among the farmers, non-practicing of modern irrigation techniques in saving water, faulty structures and water transport system, absence of strict legislation in controlling the abuse of water are some of the important reasons in recording low water use efficiency of water resources in the country. This paper is aimed at reviewing the different water management practices for improving and sustaining the food grain productivity from rainfed regions (61% of net sown area) which contribute to about 44 percent of the total food production in the country.

15.2
Water Resources of the Country

The total surface water resource of the country is estimated at 1869 km^3 at 50 percent dependability and approximately 1500 km^3 at 75 percent dependability. It has been estimated that due to extreme variability in precipitation which disallows storage of flash and peak flows, and due to non-availability of suitable storage sites in hills and plains, only about 690 km^3 of surface water can be stored for beneficial use. In addition, on yearly recharge basis, about 430 km³ of ground water is available for different uses. Thus the total estimated utilizable water from surface and ground water sources becomes 1120 km³

Ministry of Water Resources of Government of India has estimated the per capita annual availability of water for the population based on 1991 census as 2208 m3. It has been estimated that per capita withdrawal of water in India during 1990 was 611 m3 as against 1870 and 665 m3 in the USA and France, respectively (Seckler, 1999). Therefore, there is a dire necessity for adopting efficient water management practices in all sectors of its use and for agriculture, in particular.

15.3
Rainwater Management

Rainwater, a crucial natural resource, is the key input in Indian agriculture. It is the prime mover in agricultural development in general and in rainfed agriculture in particular. In India, 65 percent of the total cropped land is rain dependent and hence subjected to vagaries of monsoon. In view of the fact that about 40 percent of total annual precipitation goes as runoff, efforts should be made to capture this precious rainwater for crop production. Storage of water in the soil and in natural or man-made structures and efficient utilization of given quantity of water are important aspects of water conservation (Singh, R.P., 2000).

In situ water conservation is a more feasible and practical proposition under most situations. The strategy for *in situ* moisture conservation lies in soil management, which aims to maximizing the use of rainfall by increasing infiltration and storage. Soil cover management (mulching / Canopy), tillage and land configurations (ridge and furrow, BBF, etc.) are practices aimed at increasing infiltration and soil moisture storage.

15.4
Issues and Perspective in Water Management

Some of issues, perspectives and strategies in water management under various systems as envisaged by the National Commission for Farmers (Swaminathan, 2006) are reported below.
a) Creation of new infrastructure
- Conflicting interests of participating States
- Land acquisition a long process
- Relief and rehabilitation process is very slow
- River linking has raised political and hydrological concerns.

b) **Constraints in existing infrastructure**
- Wide disparity between design and delivery
- Inadequate command area development
- Water logging and salinity at head reach and deficit water at tail end
- Heavy conveyance losses and the irrigation efficiency is only 40 percent

- Poor water delivery and reliability
- Budgetary constraints adversely affect the maintenance
- Under pricing of water led to cultivation of water intensive crops
- Decline in irrigation by tanks in the southern peninsular region due to decrease in carrying capacity
- Less inflows into the rivers with increased watershed activities.

c) Ground Water Exploitation
- Abundant and timely supply of ground water improved agricultural production
- Ground water utilization reduced water logging and salinity
- Over-exploitation of ground water recorded at many locations, results in deepening of water table and reduced recharge capacity of dug wells in hard rock regions
- Least attention given to ground water recharge.

d) Problems in Rainfed Areas
- Extreme variability in rainfall both in spatial and quantum dimension
- Evaporative demand higher than rainfall during greater part of the year
- Expansion of deep tube wells in hard rock regions aggravated water crisis
- Poor ground water quality (saline / brakish water)
- Deterioration in soil health in the intensively cropped rice systems
- Low input farming practices and low fertility status and drought conditions at different crop growth stages
- Less adaptation of *in situ* water conservation techniques by the farmers
- Implementation of Watershed Development Programs by various agencies
- Poor maintenance of watersheds and community involvement after development has been negligible
- Limited availability of drought tolerant crop species.

15.5
Strategies for Improving the Water Management and Water Use Efficiencies

a) Surface Irrigation (Major Irrigation):
- Priority to allocate resources across the projects on the basis of additional irrigation from a given investment and time
- Quick disposal of inter-state river water disputes by the Government through River Board Authority
- Assured supply of required budget grants
- Water availability in the new projects should be assessed properly in view of Watershed Development Programs
- The National Water Policy of 2002 proposed should be implemented to avoid financial burden on the State Governments
- For long-term sustainability of irrigation system pricing water should be encouraged

- Capacity building of water users as well the staff managing irrigation system needs to be taken up for improving the water use efficiency
- Proper repairing and maintenance of canals to be taken up for effective implementation of participatory irrigation management
- Water release into the command area should be remotely regulated based on the demand from various sectors.

b) Minor Irrigation (Tanks and Small Reservoirs)
- Community surface water storage facilities be strengthened in supplementing the drinking water needs supplied by PHED.
- Water harvesting storage structures without any sluice gates such as kohlis in Maharashtra, tanka, nadis, khadings in Rajasthan need to be improved and strengthened through local voluntary organizations.
- Promotion of ground water recharge through construction of low cost check dams in drought prone areas be given priority.
- Renovation of local community based irrigation system for increasing the carrying capacity, which may improve the rural employment.

c) Excessive Ground Water Utilization
- Assured power supply would reduce the risk of over irrigation.
- Water supplied through field channels must continue for sufficient time to reduce loses due to absorption.
- Ground water recharge should be encouraged through well-planned and maintained Watershed Program.
- A water literacy movement among the stakeholders should launch and regulations be developed for long-term use of ground water on a sustainable basis.
- Constitution of Pani Panchayat at each village for maintaining equity distribution of water.
- The farmers should be encouraged to tap good ground water resources within the irrigation schemes.
- Allocation of surface irrigation waters should be based on good quality ground water.
- Sub-surface drainage system and efficient irrigation methods be planned to prevent further salinization.
- Wherever possible, surface irrigation systems should be used for ground water recharge.

d) Rainfed Areas
- *In situ* water conservation techniques such as compartment bunding, ridges and furrows, tide ridges, double cropping, strip cropping, mulching and vegetative barriers for improving soil moisture needs to be further strengthened for the benefit of small and marginal farmers.
- Watershed development is the key to success for sustainable and improved agricultural output from rainfed areas. Hence, liberal funding is essential for Watershed Programs.
- Integrated Development of Watershed at macro-level "Watershed Plus" which not only focuses on soil and water conservation but should deal and integrates

measures that increases productivity and provide value addition to the community living.
- The proposed National Rainfed Area Authority should be given responsibility to manage entire Watershed Program, thus enabling the farming community to achieve "Jal Swaraj" in relation to drinking and irrigation water.
- Rainwater harvesting through farm ponds for supplemental irrigations and recharging dead open dug wells be given top priority for enabling ground water recharge as well as enhancing productivity.
- A Million Well Recharge Program proposed by NCF indicated that to create awareness among the farmers about the importance of ground water recharge for achieving future foodgrian demands by providing a rebate in the ratio of internet under enhanced Agricultural Credit Program on priority basis.
- Getting more crops per drop through efficient irrigation methods such as drip, sprinkler, need to be promoted vigorously to conserve the water resources, food security and enhance income.
- To improve the soil moisture in the black cotton soils, chiseling at 1 m interval should be undertaken at few selected watershed areas on a pilot scale to assess the economic viability and its impact on environment.

15.6
Water Management through Watershed Program

To overcome the uncertainty in production from rainfed regions due to frequent failure of monsoon rains, the solution lies with the development of watersheds on a large scale to achieve the second green revolution. The Watershed Programs taken up by agencies that were funded through different Ministries of Government via its various developmental programs are given in the following Table-3.

a) Crop Diversification for Efficient Water Use

The most farmers' choice is paddy when water is available and many farmers keep their land fallow both in kharif and *rabi* in anticipation of good rains or ground water, but both are uncertain. Paddy is considered to be the poor water user and requires 1200mm compared to 300 to 400mm by other irrigated dry (ID) crops (Table 15.4). A two-pronged strategy (direct and indirect interventions) has been launched in Mahaboobnagar, a drought prone district of Andhra Pradesh in a cluster comprising of 4 villages in a project implemented by CRIDA and BAIF Institute of Rural Development, Karnataka (BIRD-K). Farmers have been convinced to move away from paddy particularly during *rabi* by educating them that some crops like chickpea, maize, ragi, etc., requires less water and hence can be cultivated in more area using the some quantity of water as required for paddy.

Alternatively crops namely chickpea, maize and ragi which are essentially dryland crops but are able to produce substantially high yields with limited irrigation have been introduced in the cultivators fields through supply of seed. These crops could give substantially higher returns to the cultivators because of their higher water use efficiency (Table 5).

Table 15.3 Watershed programs details since inception on area coverage and expenditure

Scheme name	Area treated (million ha.)	Expenditure (Rs. in billion)
Agriculture Ministry		
National Watershed Development Program for Rainfed Areas	8.56	26.71
River Valley Project & Flood Prone area	6.25	20.38
Watershed Development Program for Shifting cultivation Areas	0.35	2.56
Reclamation of Alkali Soils	0.69	1.06
Watershed Development Fund	0.04	0.21
Externally Aided programs	2.80	49.80
Total	18.69	100.72
Rural Development Ministry		
Drought Prone Area Program	6.57	50.61
Drought Development Program	3.53	19.61
Integrated Watershed Development Program	8.45	22.28
Externally Aided programs	0.36	2.13
Total	18.92	94.62
Environment and Forests Ministry		
National Afforstation Eco-development Program	0.88	8.52
Grand Total	38.49	203.87

b) Other Rainfed technologies developed / tested by Central Research Institute for Dryland Agriculture for efficient rainwater management
   1. In-situ measures for rainwater management in rainfall areas.
- *Off season land treatment:*
  - Reduces weed growth and retains more moisture
  - Summer tillage for alluvial, red and other light soils
  - Compartment bund for heavy black soils for assured *rabi crops*.
- *Conservation furrows*
  - Retains about 37% additional soil moisture compared to farmer's practice.
  - Better crop growth and higher yields by about 17%
- *Ridges and furrows system in cotton*
  - Additional yield over farmer's practice

Table 15.4 Water requirement of paddy and ID crops

Crop	Water requirement (mm)	Area equivalent of paddy (ha)
Paddy	1200	–
Groundwater	400	3.0
Maize	400	3.0
Chickpea	250	4.8
Ragi	400	3.0

Table 15.5 Net returns and water use efficiency of different cropping systems based on water requirements during kharif and rabi for a 3 acre farm.

Practice /Intervention	Cropping system	Total net returns (Rs)	Water use efficiency Rs. per mm
Farmers' practice 1*	Paddy (3) – Paddy (3)	36000	5.00
Farmers' practice 2**	Paddy (3) – Paddy (1)	24400	5.08
Farmers' practice 3**	Paddy (3) – fallow (3)	18300	5.08
Intervention 1	Paddy (3) – groundwater (3)	37560	7.83
Intervention 2	Paddy (3) – Maize(3)	31854	6.64
Intervention 3	Paddy (3) – Chickpea (3)	27936	6.42
Intervention 4	Paddy (3) – Paddy (1), Chickpea (2)	30824	5.82
Intervention 5	Paddy (3) – Paddy (1), Maize (2)	33436	5.97

Note: Figures in parenthesis indicate acreage
* When sufficient water is available for cultivating all the area in both the seasons.
** When water is available for cultivating all area in *kharif* season and less area in *rabi* season

- *Cover cropping*
  – Improves soil quality with on farm generation of organic matter in off-season.
- *Micro catchments*
  – Improves the perennial plant establishment even on steep slopes.
2. Medium term measures rain water management in rainfed areas.
   – Stone and vegetative field bunds for soil and water conservation
   – Graded line bund helps in efficient drainage.
   – Trench cum bund for soil and water conservation.
3. Long term measures for rain water management in rainfed areas

- *Water harvesting*
  - Contour trenching for runoff collection.
  - On-farm reservoirs
  - CRIDA developed low-cost water harvesting structures
  - Ground water recharge structure (percolation tanks).
  - Recharge through defunct wells.

4. Strategies for improving water use efficiencies
- Irrigation –furrows improves the efficiency of stored water
- Micro irrigation techniques
  - Drip irrigation
  - Sprinkler irrigation
  - Supplemented irrigation with harvested runoff.
  - Crop diversification.

5. Alternate land use system
    - Bush farming in arable and non-arable lands
    - Agric silviculture
    - Agric horticulture
    - Participating ground water evaluation for efficient alternate land use pattern.

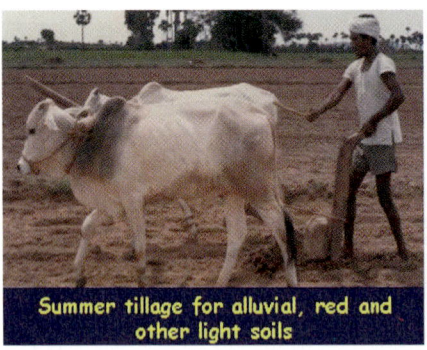

Fig. 15.1 Off-season land treatment.

Fig. 15.2 Conservation furrow in castor and groundnut.

6. Climate Change and its Impact of Water

Climate Change can affect the regional atmospheric circulation patterns, which is important for taking decisions about water and land use planning and management. The information available from GCMs focuses on how climate changes will affect the water balance. Considerable efforts have gone into study the effect

Fig. 15.3 Ridges and furrows system.

Fig. 15.4 Cover cropping.

Fig. 15.5 Medium term measures rain water management in rainfed areas.

Fig. 15.6 Contour trenching for runoff collection.

of global warming on water systems both space and in time. The reports of IPCC, 1996a & b have indicated the following:

- GCMs indicate that there will be some changes in the timing and regional patterns of precipitation (very high confidence), but researchers have low confidence in projections for specific regions because different models produce different detailed regional changes.

Fig. 15.7. CRIDA developed low-cost water harvesting structures.

- GCMs consistently show that average precipitation will increase in higher latitudes, particularly in winter (high confidence). Models are inconsistent in other estimates of how the seasonality of precipitation will change.
- Research results consistently show that temperature increases in mountainous areas with seasonal snow pack will lead to increases in the ratio of rain to snow and decreases in the length of the snow storage season (very high confidence). It is likely that reductions in snowfall and earlier snowmelt and runoff would increase the probability of flooding early in the year and reduce the runoff of water during late spring and summer.
- Increases in annual average runoff in the high latitudes caused by higher precipitation are likely to occur (high confidence).
- Research results suggest that flood frequencies in some areas are likely to change. In northern latitudes and snowmelt-driven basins, research results suggest that flood frequencies will increase (medium confidence), although the amount of increase for any given climate scenario is uncertain and impacts will vary among basins.
- Models project that the frequency and severity of droughts in some areas could increase as a result of regional decreases in total rainfall, more frequent dry spells, and higher evaporation (medium confidence). Models suggest with equal confidence that the frequency and severity of droughts in some regions would decrease as a result of region increases in total rainfall and less frequent dry spells.
- Higher sea levels associated with thermal expansion of the oceans and increased melting of glaciers will push salt water further inland in rivers, deltas, and coastal aquifers (very high confidence). It is well understood that such advances would adversely affect the quality and quantity of freshwater supplies in many coastal areas.
- Water-quality problems will worsen where rising temperatures are the predominant climate change (high confidence). Where there are changes in flow, complex positive and negative changes in water quality will occur. Water quality may improve if higher flows are available for diluting contaminants. Specific regional projects are not well established at this time because of uncertainties in how regional flows will change.
- A large number of studies suggest that climate changes will increase the frequency and intensity of the heaviest precipitation events, but there is little agree-

Fig. 15.8 Micro irrigation techniques.

Fig. 15.9 Bush farming in arable and non-arable lands.

Table 15.6 Rainfall and fiver flows and their projections in two major river systems in India

River Basin	Baseline (1961-1990)		Future (2071-2100)	
	Annual Rainfall (cm)	Annual Flow (km^2)	Annual Rainfall (cm)	Annual Flow (km^2)
Krishna	91	60	112	67
Godavari	166	98	201	116
Ganga	134	482	150	543

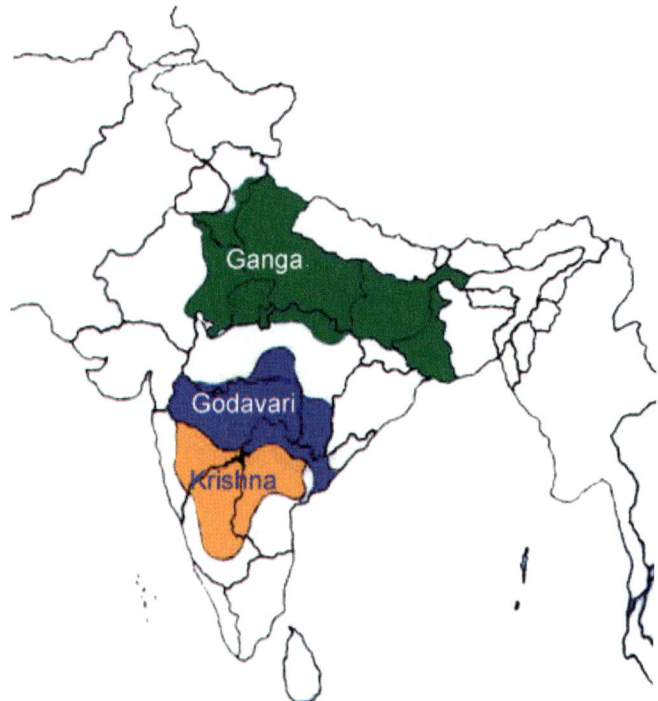

Fig. 15.10 Rainfall intensity at three major river basins.

ment on detailed regional changes in storminess that might occur in a warmed world. Contradictory results from models support the need for more research, especially to address the mismatch between the resolution of models and the scales at which extreme events can occur.

The Indian water resources under climate change scenario studied by Indian Institute of Tropical Meteorology suggest the following:
- The hydrological cycle is predicted to be more intense, with higher annual average rainfall as well increased drought.
- There is a predicted increase in rainfall in all three river basins towards the end of the 21st Century (Fig.15.4). The Godavari basin is projected to have higher precipitation than the other two given in the following Table 6.
- The intensity of daily rainfall is also predicted to increase in these basins (Fig. 15.11).
- Changes in the number of rainy days when examined, with results indicating decreases in the western parts of the Ganga basin, but with increases over most parts of the Godavari and Krishna basins.
- Thus surface water availability showed a general increase over all 3 basins (though future populations projections would need to be considered to project per capital water availability).
8. Weather-based Agro-advisories and Crop Water Management Strategies

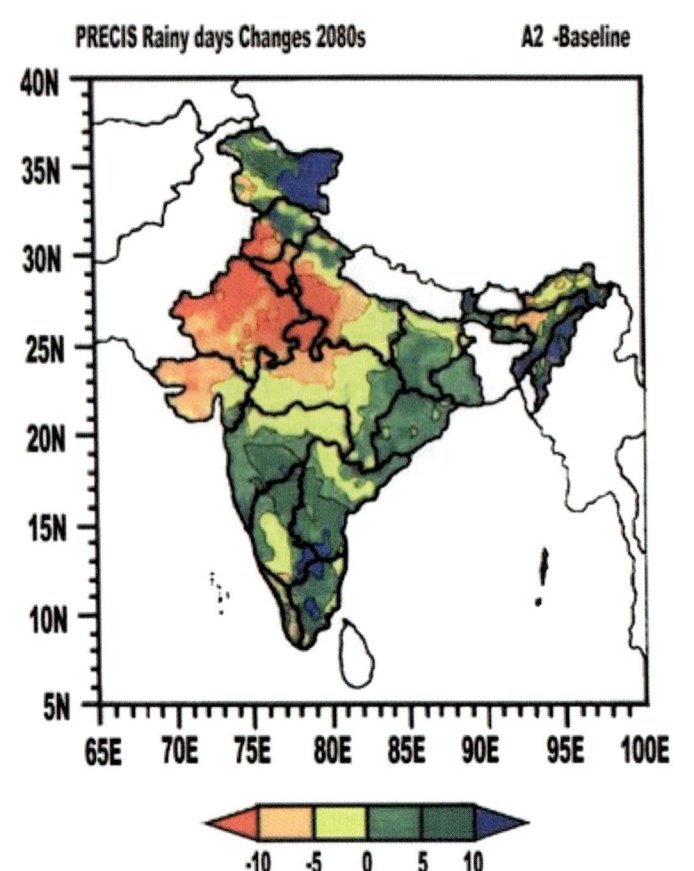

Fig. 15.11 Changes in annual number of rainy days (A2 scenario).

Efficient irrigation water management plays a key role in improving agricultural productivity and also protects the soil environment. Proper and timely dissemination of agro-advisories related to irrigation fertilizer and pesticide management helps the farmers for better planning of agricultural operations. The present system of weather-based agro-advisories is issued by 107 Agro meteorological Field Units (AMFU) operating at State Agricultural Universities (SAUs) and Indian Council of Agricultural Research (ICAR) Institutes. Using the information on latest crop condition at the region concerned and the medium range forecast issued by NCMRWF for that region, the agro-advisories are prepared by a group of experts and the same is disseminated to the farmers through different mass communication network systems such as Radio, TV, Newspapers. The information that generally provided in Agro Advisory Services are status of crop condition, current and expected weather and the advisory, which consists of agronomic measures and plant protection measures to be followed for each crop in the next few days to come.

Fig. 15.12 Agro Advisory Network Group.

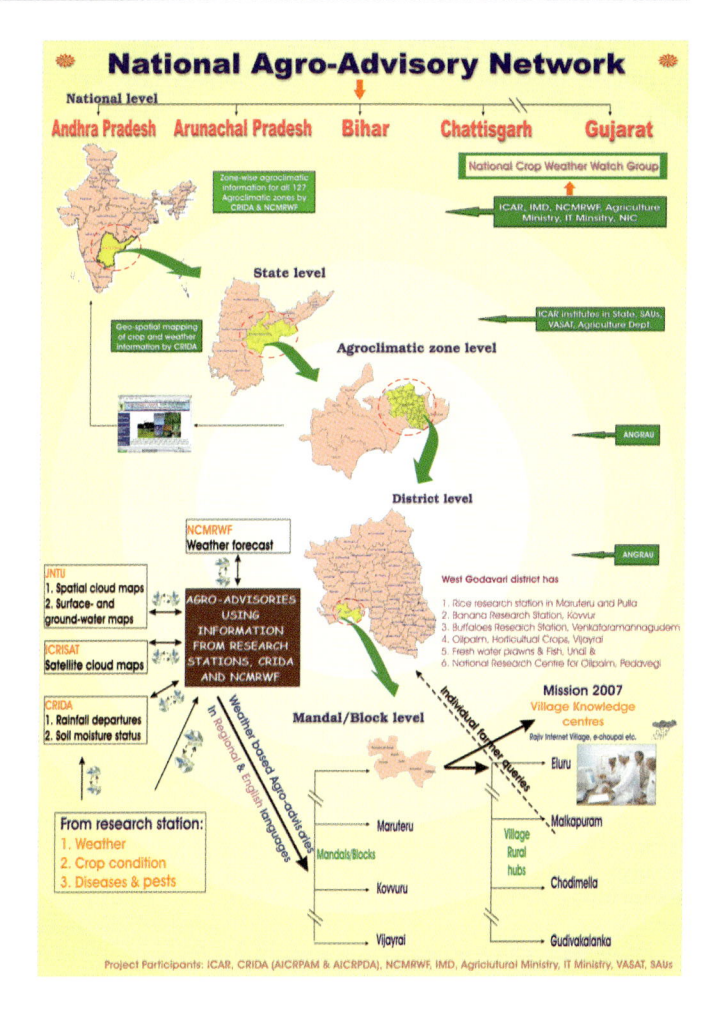

With the expansion of IT Network in the country, the information is also made available through website which is expected reach each village shortly through the efforts of Government of India and NGOs. The agro-advisories issued by 25 centers of AICRP on Agrometeorology (AICRPAM) located in various agro-climatic zones spread across the country are made available through a website www.cropweatheroutlook.org. On a trail basis, the Acharya N.G. Ranga Agricultural University at Hyderabad in collaboration with various organizations such as NCMRWF, IMD, CRIDA, JNTU AND ICRISAT is issuing agro-advisories on district basis based on input provided by agencies located in each district. Similarly, efforts are on at different States to promote more regional websites.

Though the present websites are providing necessary advisories to the farming community, it is feared that it may not be reaching the needy farmers well in time. Therefore, All India Coordinated Research Project on Agrometeorology

(AICRPAM) Unit at CRIDA has planned a National level Agro Advisory Network for efficient dissemination of information both to the stakeholders and to the planners. A sketch diagram of the proposal is shown below:

The aim of such national network is to disseminate information upto village level and the farmers will be able to interact with the Agro Advisory Network Group located at district level.

Conclusions

The review suggests that there is scope to improve the water management efficiency for sustainable agricultural productivity by adopting Integrated Watershed Development Program through participatory approach method. A Million Well Recharge Program proposed by NCF by providing some incentives may be given priority to educate and create awareness about the importance of fresh water availability in the coming years. The impacts of climate change on regional water resources need more attention. Expansion of the current Agro-advisory Network System at National level shall play an important role in improving the water use through efficient water management practices.

References

Central Research Institute for Dryland Agriculture, 2006. "Enabling Rural Poor for better Livelihoods through Improved Natural Resource Management in SAT India". Final Technical Report 2002-2005, DFID-NRSP (UK) Project R8192; Hyderabad, India: Central Research Institute for Dryland Agriculture; Bangalore, Karnataka, India: University of Agril. Sciences; Hyderbad, India: ANG Ranga Agril. University; Tiptur, Karnataka, India: BIRD-K and Hyderabad, India: ICRISAT.

FAO Statistics database (www.FAO.org)

Intergovernmental Panel on Climate Change (IPCC). (1996a) Climate Change 1995: The Science of Climate Change: Contribution of Working Group I to the Second Assessment Report of the Intergovernmental Panel on Climate Change. Cambridge University Press, New York.

Intergovernmental Panel on Climate Change (IPCC). (1996b) Climate Change 1995: Impacts, Adaptations and Mitigation of Climate Change: Scientific-Technical Analyses: Contribution of Working Group II to the Second Assessment Report of the Intergovernmental Panel on Climate Change, Cambridge University Press, New York.

Mark W. Rosegrant, Ximing Cai, and Sarah A.Cline (2002) Global Water Outlook to 2025: Averting an Impending Crisis , Jointly published by IFPRI, Washington, D.C., U.S.A. and International Water Management Institute, Colombo, Sri Lanka

Pant, GB., (2005) Climate Change Impacts on Water Resources in India. Key Sheet 5. Submitted to Ministry of Environment & Forests, Government of India.

Parthasarathy, S. (2006) From Hariyali to Neeranchal. A Report of the technical committee on watershed programmes in India submitted to Ministry of Rural Development, Government of India, pp 222.

Seckler, D., (1999) World Water Scarcity and the Challenge of Increasing Water Use Efficiency and Productivity. In "Sustainable Agricultural Solutions". The Sustainable Agricultural Initiative Action Report, Novello Press, Ltd., London, 116-126

Singh, RP., (2000) Rainwater Conservation, Recycling and Utilization. In Intl. Cong. On Managing Natural Resources for Sustainable Agricultural Prouduction in the 21st Century. Invited Papers, pp 202-208

Swaminathan MS., (2006) A draft national policy for farmers. Fourth report submitted to the Indian Union Minister of Agriculture on 13 April 2006. pp: 148-186.

CHAPTER 16

Examples of coping strategies with agrometeorological risks and uncertainties for Integrated Pest Management

A.K.S. Huda, T. Hind-Lanoiselet, C. Derry,
G. Murray and R.N. Spooner-Hart

16.1
Introduction

Some risks in the agricultural sector are unavoidable while others can be managed. Agrometeorological risks in the farming sector include the temporal and spatial variability of rainfall, temperature, evaporation and, in climate change scenarios, atmospheric carbon dioxide levels. While such factors may impact directly on plant growth and development they can also exert an important indirect effect by influencing the life cycles of plant diseases and pests. In addition they may have a profound influence on attempts to control such pests, as is seen when an unexpected rainfall event causes dilution or early hydrolysis of a surface pesticide, or when hail damage opens the way for mould, bacterial or insect attack. Integrated pest management (IPM) must take into account such risks if crop damage is to be minimized. The implications of agrometeorological risk studies in countries such as Australia offer not only local perspectives on IPM but also provide information for improved crop profitability, natural resource usage and agricultural sustainability in other countries, where a critical relationship between crop success, regional food security and human survival may exist.

The capacity of an individual farming enterprise to carry out IPM depends largely on its given financial and economic situation. A business with a high level of debt may only have capacity for low cost management options, such as the planting of disease-resistant varieties and routine pre-crop disease control, while a business that is leveraging and expanding its asset base may be able to cope with higher cost management options, such as the introduction of new crops or rotation of crops to preserve the long-term status of the land (Lloyd Kingham, NSW DPI, personal communication, 2006). Given these differences in ability to cope, perception of risk may vary considerably with level, type and location of enterprise.

Global climate change will inevitably present a challenge to those engaged in agroclimatic risk modeling in the interests of IPM. Agribusiness units most at risk are likely to be those already stressed as a result of factors such as land degradation, salinization and ecological change. Local economic setting must also be taken into account when estimating possible impacts. In countries with a low level of agricultural industrialization, many units may be based on low capital investment resulting in short-term land use policies, while in industrialized countries units may exist at the other extreme, having over-capitalized on items such as dedicated irrigation systems, slow-growing cultivars and on-site processing facilities. Units at both ends of this capitalization spectrum may, however, be economically marginal in terms of climate change, with increased reliance on state subsidy, off-farm income, or secondary industry support. Such situations do not offer much leeway for farming sustainability through IPM in areas where climate change may be accompanied by increased disease occurrence or pest invasion.

One way to cope with risk has traditionally been through the use of insurance, although agricultural economists are becoming increasingly skeptical about insurance as a regionally-sustainable risk management strategy. Some crop insurance, however, may enable an enhanced ability to apply IPM in special situations. For example, an agrometeorological risk that can be insured against in Australia is hail. Hail can cause wound sites which allow pathogens to breach external defenses and gain access to plant tissues, resulting in exacerbation of initially superficial damage. In such cases insurance indirectly allows for a measure of protection against microorganic degradation and pest damage. Only farmers in a stable or growing financial situation may, however, be able to afford this luxury (Lloyd Kingham, NSW DPI, personal communication, 2006).

Australian crops of wheat and canola have a local advantage in that the full potential spectrum of destructive pathogens has not yet been established through assiduous quarantine control and an integrated agricultural management system. In some cases, pathogens found in Australia offer less virulent or aggressive forms than those found elsewhere. Pathogenicity generally relies on a subtle interplay between genetics and external biotic and abiotic factors operating within local ecosystems. Abiotic factors may include farming practice, soil differences, seasonal characteristics or climatic conditions.

The expenditure on integrated control is supported by studies which show that when incursion of an exotic pest occurs or a new variety evolves locally, the result is considerable loss to the industries concerned, with reduction of both quantity and quality of a crop. Furthermore, there is an indirect cost associated with environmental damage resulting from the need to apply additional pesticide (White 1983; Zadoks and Schein 1979). Where epidemic threats are anticipated, contingency planning can enable the use of proactive or less extreme intervention, resulting in reduced pest damage, pesticide use, and ecological impact (Murray and Brennan 2001).

Two climate-sensitive diseases of Australian field crops are stripe rust of wheat and Sclerotinia rot of canola, both having a high risk-ranking in the list of Australian crop diseases (Murray and Brennan 2001). Stripe rust of wheat is estimated to cause on average a loss of about US$ 142 million per annum (Brennan and Murray 1998), while stem rot has been reported as causing losses up to US$ 37 million, (Hind-Lanoiselet 2006).

In terms of the gross economic production value, wheat is the most important crop in Australia, attracting a large share of public funds for research and development. A substantial part of those funds is raised from production levies that are matched by government funds then disbursed by bodies such as the Grains Research and Development Corporation (Brennan and Murray 1998). Research carried out in terms of this and other funding has already identified the importance of agrometeorological risk assessment (Huda et al. 2004; Wallace and Huda 2005).

This chapter discusses some of the Australian research into climate sensitive diseases, in particular wheat and canola, to present some thoughts on the approaches needed for coping with the risks and uncertainties associated with IPM. A perspective on future considerations in this area is also given.

16.1.1
Crop Diseases - Stripe rust in wheat and Sclerotinia rot in canola

Stripe rust in wheat is caused by *Puccinia striiformis* f.sp. *tritici* (Figure 16.1). Infection and growth is favored at temperatures between 12 to 15°C, with longer time required at lower and higher temperatures. At ideal temperatures, the cycle from infection to new spore production takes about 12 to 14 days, given susceptible plants and sufficient humidity. In the warmer months up to two cycles of the disease can occur per month (Murray *et al.* 2005).

Sclerotinia rot is caused by *Sclerotinia sclerotiorum* (Lib.) de Bary (Figure 16.2). Sclerotia (asexual resting propagules) remain viable for many years in the soil. When weather conditions are favorable, the sclerotia germinate to produce apothecia (sexual fruiting bodies) (Le Tourneau 1979; Morrall and Thomson 1991). Apothecia produce thousands of air-borne ascospores that can be carried several kilometres by the wind (Brown and Butler 1936; Schwartz and Steadman 1978). Spores that land on canola petals may lodge in the lower canopy of the crop during senescence at the end of flowering. Germinating spores use the petal as a source of nutrient, producing a fungal mycelium that grows and invades the canola plant. Germination and infection are enhanced by wet weather (McLean 1958; Rimmer and Buchwaldt 1995).

Sclerotinia is a monocyclic disease in Australia in keeping with the flowering periodicity of canola, although in some parts of New South Wales (NSW) where a

Fig. 16.1. Stripe rust on wheat (photo, Paul Lavis).

Fig. 16.2. Sclerotinia stem rot on canola (photo, Tamrika Hind-Lanoiselet, NSW DPI)

summer-irrigated crop is grown, two cycles may occur in one year (Hind-Lanoiselet, unpublished data, 2006).

A potentially sustainable way of controlling crop diseases is the breeding of resistant plants, although to optimise control other management strategies such as fungicide use and good land management practices have to also be used. The latter can include crop rotation and general crop hygiene (Murray and Brown 1987). The incorporation of multiple disease management strategies reduces the chance and hence severity of attack and limits fungicide use which in turn reduces the risk of the pathogen acquiring resistance to the fungicide. Conventional breeding has, however, not always been successful at producing resistant plant varieties, as can be seen until recently with attempts to breed resistance for *S. sclerotiorum* into canola (Buchwalt et al. 2003). Even when resistance for a pathogen such as stripe rust has been successfully bred into a variety for several decades this can be overcome by the introduction of a new stripe rust race as occurred in Western Australia in 2002, with spread of rust to the eastern Australian states by 2003 (Murray et al. 2005).

When plant resistance cannot be relied on for the level of pathogen management required, other management strategies such as fungicide application are generally used (Sansford et al. 1995). Routine seasonal application of fungicides is, however, not profitable as procurement and application costs are high, and disease incidence varies greatly with year, region and locality (Sansford et al. 1995; Twengstrom et al. 1998). A major consideration is the potential removal or dilution of fungicides by early or unexpected rains, and in this regard short-term climatic modeling is highly desirable. The result of climate-pest modelling only take on meaning when interpreted in terms of broader risk management considerations.

Models can range in complexity from a simple set of anecdotal rules applied by the subsistence farmer to complex, computer-based models such as those constructed by researchers in collaboration with state departments. The simplest models are likely to be based on relatively simple causal or "push-pull" relationships (deterministic) whereas complex models are likely to be based on webs of such relationships involving a large number of agrometeorological factors and confounders, with ability to take into consideration the chance of each causal factor potentiating with time, or in terms of some spatial distribution (probabilistic).

All models offer a predictive dimension, giving opportunities for anticipating and hence limiting crop damage. Some use early warning systems, such as changing weather patterns, to allow for early, corrective action (proactive), whereas others rely on the onset of disease as an action trigger (reactive). The latter models are likely to be of limited effectiveness in controlling damage, hence the need to explore new data and methodologies which can be effectively used in modelling for proactive disease management (Gugel and Morrall 1986; Zadoks 1984). A cornerstone of epidemiological modelling is the collection of relevant local data for disease occurrence and related risk factors, from which risk management models can be developed to allow for a range of actions based on the excedence of limit values within a predetermined data range (Abawi and Grogan 1975; Last 2001).

16.1.2
Implications for technology transfer

A preliminary step to breaking the epidemic cycle of disease in plant populations is to identify strategic intervention points in the life cycle of the agent. This requires a thorough knowledge of characteristics relating to the crop or plant population itself (host), the pathogen (disease-causing agent) and the place in which the disease occurs (environment). While epidemiological study based on these characteristics can yield great insights as to the establishment and continuation of disease in plant populations, observations may be very specific to time and place and for this reason great circumspection is required when transferring conclusions and hence control strategies from one region to another.

Agrometeorological factors may vary considerably, as evidenced when early European farming traditions were first imported to Australia. The intensive cropping system used on European farms and originally imported into Australia and other Asia-Pacific countries has been found in many cases to be unsuitable in terms of available area and soil characteristics. Disease control approaches for intensive horticulture may also be unsuitable for broad acre field crops, for example the practice of liming soil to pH 7.0-7.5 to control clubroot (*Plasmodiophora brassicae*) of Brassica *spp.* (Donald et al. 2003).

Large scale migration from Europe to Australia from the early 19th century brought farmers into contact with semi-arid and arid environments for the first time. The response was to perceive drought as a symbolic national enemy, and to attempt technological solutions to solve the "drought problem" with extensive economic support. A better approach might have been to accept that certain areas of the land were simply unsuitable for certain types of agriculture (Royal Geographical Society of Queensland 2001).

Relevance of specific technologies also changes with time. Tillage was traditionally used in Australia to reduce the incidence of soil borne diseases including *S. sclerotiorum*, but improved conservation practice suggests that zero tillage is desirable in conserving soil moisture, reducing erosion and limiting costs (Kharbanda and Tewari 1996; Paulitz 2006). Furthermore, a number of studies on conventional and no-till systems have not found significantly different levels of disease incidence (Paulitz 2006). To support place-sensitive technology transfer throughout the Asia-Pacific region, Australia is developing a range of generic modeling and intervention strategies which will be validated at selected sites in the region. An important component of this ongoing research thrust, however, is the securing of funding from regional agencies to augment support which the Australian government is prepared to commit to such a project.

Australia offers a range of research experience relating to an integrated monitoring system which sees regularly updated fact sheets for disease control distributed to farmers throughout State Agriculture Departments (such as the NSW Department of Primary Industries, http://www.dpi.nsw.gov.au), based on modelling studies supported in government and universities by funding bodies such as Grains Research Development Corporation (http://www.grdc.com.au).

16.1.3
Resource allocation for risks

A rational allocation of resources for the control of plant diseases is based on the potential economic losses which they may cause. This applies both at individual level, when a grower decides whether or not control of a particular disease is financially warranted, and at the national level, when funds are apportioned to research, risk communication and disease control. As the disease spectrum and economic environment change with time, estimates of disease losses need to be based on current data, if resource allocation is to be optimized (Brennan and Murray 1998).

In countries where primary industries are in an early stage of development, farmers may have little income and may rely on loans at high interest rates for input investments, and for crop protection. When there is crop failure due to high climatic variability, as may result in droughts, farmers with low financial capacity may loose their entire investment. Ultimate outcomes are not only economic; farmers in India have been reported to perceive this as personal failure and widespread anguish with high rates of suicide has been recorded (Sivakumar 2000). In

Australia the suicide rate in male farm owners is about twice the crude national average for males in all sectors, despite the fact that in Australia several organizations provide farmers with financial assistance when extreme weather conditions have resulted in severe challenges, such as drought and floods.

Such assistance includes income support from Centerlink, interest subsidies from the Rural Assistance Authority, advice and funding for developing a business plan and succession planning from the Farm Help Program, and funding for establishment of a farm Environmental Management System (EMS) from various state agencies. In order to further the mental health and wellbeing of farmers in NSW, Australia, a blueprint has been developed to improve access to mental health support, including counselling, crisis lines and the teaching of coping skills (NSW Farmers Association 2005).

While applied IPM is likely to remain a technological area, resource allocation within a risk management framework needs to involve interdisciplinary collaboration if the very real threats to the mental and physical health of those engaged in farming as primary industry are to be addressed.

16.1.4
Supportive Decision-Making Tools

A decision support tool called RustMan was developed for stripe rust of wheat in the 1990s. RustMan estimates the likely impact that stripe rust will have on wheat yield and the benefits from spraying to control the disease. RustMan uses results derived from field experiments at Wagga Wagga and Yanco from 1984-1987, with the addition of current information on the reaction of wheat varieties to the races of stripe rust in Southern and Central NSW. Estimates require the input of average weather conditions occurring over one agricultural season (Gordon Murray, personal communication, 2006).

Sufficient macroclimatic data have now been collected for the development of a similar tool for Sclerotinia rot on canola although the higher impact of post-treatment climatic variation demands a longer-term forecast record.

16.1.5
Effectiveness of decision-making tools

The effectiveness of decision making tools depends on their ability to predict and to facilitate risk management or mitigation, with subsequent assessment of outcome (Meinke and Stone 2005). Some points relating to this effectiveness are:
- Farmers are only able to respond and adapt to climatic conditions, they cannot expect the model to assist them to manage or mitigate the climatic event itself,
- Adaptation or 'responsive adjustment' as risk ameliorating strategy, must be targeted and may be complex,
- The proactive dimension in risk amelioration is important if damage is to be minimised,

- Outcomes need to be seen in practical terms if individual and societal benefits through improved risk management practices and better targeted policies are to be optimised.

16.1.6
Importance of Experimental Observation

Hind-Lanoiselet et al. (2004, 2006) demonstrated the importance of rainfall distribution in relation to disease development, having observed that in years with low rainfall and high temperatures Sclerotinia rot is not a limiting factor for crop development. In such years the application of fungicide had impact on early disease but no ultimate impact on yield (Figure 16.3). Such findings suggest that in similar years finances can be directed away from this problem to be used more productively in other areas, with the proviso that this practice is not rigidified so as to prevent successful financial re-targeting in subsequent growing seasons.

16.1.7
Desirable level of complexity

A management tool should only include easily obtained information and results should be relatively simple for the user to interpret. Complex tools based on advanced modeling need to clearly communicate the data for general stakeholder use. The economic component needs to be included as part of a risk-benefit framework which only encourages the use of fungicide when it is likely to enhance profitability.

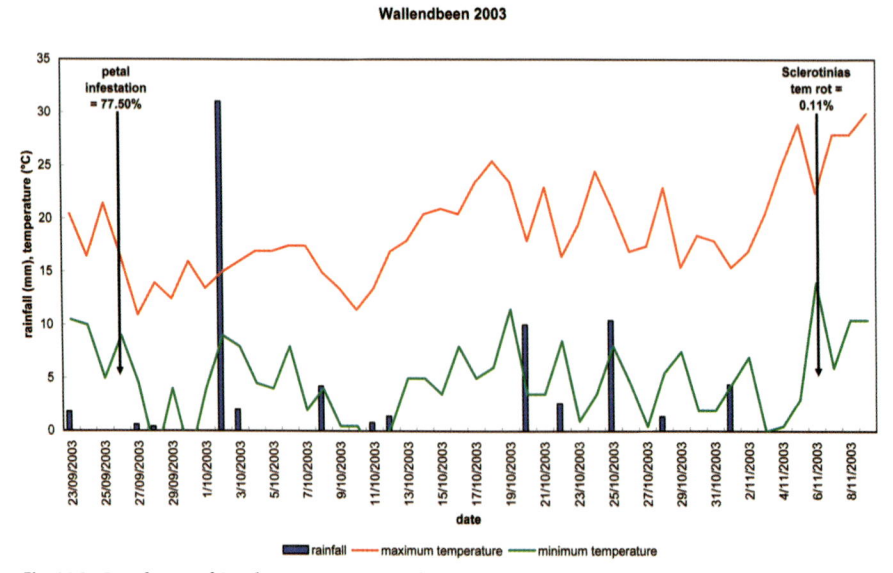

Fig. 16.3. Incidence of *S. sclerotiorum* at 20% flowering on canola petals and disease incidence before harvesting after a dry finish at a trial in Wallendbeen in 2003.

16.1.8
Economic balance in control

Due to the sporadic nature of stem rot it is uneconomical to apply fungicides routinely, although to be effective they need to be applied before the plant becomes infected. In Australia, growers are advised to consider the current price of both chemical and canola to determine the viability of Sclerotinia control before applying a fungicide (Hind-Lanoiselet and Lewington 2004, and Hind-Lanoiselet et al. 2005). A table is used to help determine the level of Sclerotinia infection that would justify a fungicide application (Table 16.1).

Note: Net returns from Sclerotinia control for each fungicide are based on a 2 t/ha potential yield and chemical and application costs of $82/ha for Rovral, and a rule of thumb that yield loss = 0.5 (disease incidence).

The data in the table show that:
- A yield loss of 10% to 15% would be required to break even and justify using the fungicide
- A 10% yield loss would represent 20% stem rot in the crop
- A 15% yield loss would represent 30% stem rot in the crop, a high disease level

The RustMan support tool can be profitably used before the rust is seen so that farmers can make early decisions (Gordon Murray, personal communication, 2006). In reality the software is typically not used until the disease has taken hold at which point effective management may not be possible. This delay is exacerbated by the limited stock and thus appreciable waiting time for fungicide. To improve this situation RustMan needs to be used early on in the assessment and to facilitate this, relevant output information needs to be effectively communicated through one of the public access web sites or by electronic mail.

16.1.9
Towards the Future

Agriculture in Australia has shown considerable capacity to meet challenges through farm management practice, appropriate crops and cultivar selection, technologies to increase water use efficiency, and pest control. Global warming, however, poses a much greater and broader challenge than those previously experienced and current financial resources may be inadequate. Dissention about the potential outcomes of global warming is problematic. While the agricultural impacts of drought and floods of specific duration and intensity can be estimated, some parts of Australia may, in response to global warming, experience improved conditions as a result of longer growing seasons, fewer frosts, higher rainfall (northern Australia) and increased atmospheric carbon dioxide (Australian Greenhouse Office 2006).

A future trend in climate prediction is likely to be in the area of macroclimate forecasting, with medium range weather forecasts (3-10 days) being increasingly

Table 16.1. Returns from the use of Rovral fungicide for Sclerotinia control

Rovral % yield loss	yield loss (t/ha) at 2t/ha potential	On Farn Price Canola ($/tonne)													
		$270/t	$280/t	$290/t	$300/t	$310/t	$320/t	$330/t	$340/t	$350/t	$360/t	$370/t	$380/t	$390/t	$400/t
5	0.1	-$55	-$54	-$53	-$52	-$51	-$50	-$49	-$48	-$47	-$46	-$45	-$44	-$43	-$42
10	0.2	-$28	-$26	-$24	-$22	-$20	-$18	-$16	-$14	-$12	-$10	-$8	-$6	-$4	-$2
15	0:3	-$1	$2	$5	$8	$11	$14	$17	$20	$23	$26	$29	$32	$35	$38
20	0.4	$26	$30	$34	$38	$42	$46	$50	$54	$58	$62	$66	$70	$74	$78
25	0.5	$53	$58	$63	$68	$73	$78	$83	$88	$93	$98	$103	$108	$113	$118
30	0.6	$80	$86	$92	$98	$104	$110	$116	$122	$128	$134	$140	$146	$152	$158

used in operational farm management decisions. There is increasing capacity to integrate seasonal climate forecasts, medium range weather forecasts and historical climate information, to enhance the availability and accuracy of data to be included in proactive decision making. In crop protection there have been simultaneous advances in describing the relationships between plant diseases and crop microclimate, such as those relating to field temperature and leaf wetness.

Further research is, however, urgently required to explore relationships between macroclimate (climate of a region) and microclimate (climate immediately within and surrounding a plant canopy), and this will require improved collection of microclimate and local disease incidence data. Recent work in Australia suggests the value of such information in risk and opportunity-management decision making (Wallace and Huda 2005; Huda et al. 2004).

Epidemiology and risk assessment will undoubtedly play an increasing role in anticipating the complex interaction between climate and disease. In its broadest sense, epidemiology is "the study of the distribution and determinants of health-related states or events in specified populations, and the application of this study to the control of health problems" (Last 2001). While originally developed as the science of disease control in human populations (demos being Greek for "the people"), epidemiological approaches are today fundamental to disease control in the agricultural sector.

A basic concept in traditional epidemiology is the Host-Agent-Environment (HAE) disease model which in its simplest form is represented by a triangle as shown in Figure 16.4.

The model proposes that for a disease to exist, all three co-factors must be present. In the case of the fungal diseases discussed in this paper, these factors include:

- Host (crop) factors: plant species and variety, time of planting, crop rotation, general crop hygiene, coexistence of other pathologies.
- Agent factors: mould types present, load and viability of infectious forms, state of spore activation.
- Environment factors: climate (including macroclimate, microclimate and seasonal change, weather immediately following fungicide or pesticide application,

Fig. 16.4. The Host-Agent-Environment disease triangle.

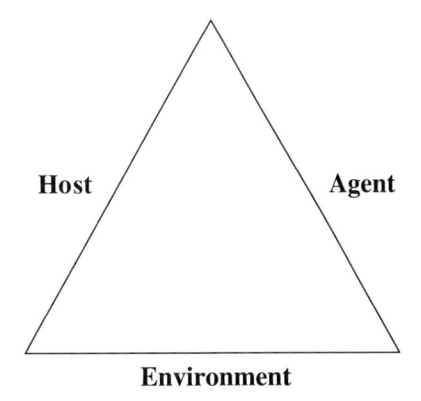

hail damage and relative humidity), insect damage, sunlight duration, presence of atmospheric gases including carbon dioxide, and acid-forming gases, soil factors (macronutrients, micronutrients, salinity and iron sulfides), irrigation factors (volumetric and qualitative, application technique, leaf runoff, soil pooling, nutrient residue on leaf), availability, type and application of fungicide, and agricultural practice.

Some plant epidemiologists have suggested the addition of a fourth factor, time, to the model (forming a "disease pyramid") to take into account the temporal process of disease development (Stevens 1960; Van der Plank 1975). The authors, however, view time not as a single, independent variable but as integral to each of the three base variables, because of the need to consider distinct and complex time-series when developing probabilistic risk-factor distributions for a range of polycyclic processes in risk modeling (Zadoks and Schein 1979).

Epidemiology is not only a study system but one committed to the management of problems. The term "coping strategies" in this chapter title relate to an Australian commitment to view crop disease control not only as a theoretical field but as a field of endeavor aimed at securing sustainable regional economic, social, ecological and health outcomes. The centrality of integrated plant production and pest control in achieving food security with limited environmental impacts has been clearly identified by major international organizations (Food and Agriculture Organisation 2005; Unnevehr and Hirschhorn 2000).

Good risk assessment alone is powerless to bring about change unless operating within a framework for sound and intersectoral risk management. This is particularly true where a project must bring together a number of countries in collaborative effort to ensure effective regional risk management.

At a recent workshop in Hyderabad it was proposed that an integrative model proposed by Derry et al. (2006) be used to guide an Asia Pacific Network research project into Asia-Pacific regional climate and disease risk management (Figure 16.5).

The model facilitates early identification of climatic hazard or change likely to impact on agricultural security in terms of epidemiological realities (stage 1). Proactive risk assessment (stage 2) incorporates the consideration of existing climate/crop-disease models and the possible development of downscaled models on the basis of local epidemiological records and observed agricultural practice. In practical terms this stage is already under development in Australia, with models relating to disease frequency and impact being investigated. Risk assessment information communicated to government and farming organizations enables the fine-tuning of policy (stage 3), to encourage proactive and cost-effective epidemiological and economic interventions (stage 4). Examples are the application of fungicide during a period of expected high humidity with suitable temperature range for mycotic growth, or the avoidance of fungicidal leaf treatments prior to a period of predicted rainfall, when wash-off can occur. Developing systems for monitoring changes in crop health status following intervention (stage 5) provides feedback for the further fine-tuning of policy and interventions.

It should be noted that the overall process is a cyclic one, with a potential starting point at any one of the five loci. Thus there is no "correct" place to start, and

Fig. 16.5. Risk management model (Derry et al. 2006).

all meaningful work can be "banked" at a relevant point within the conceptual framework. Effective intersectoral and multi-staged communication of risk lies at the hub of the model, which will involve the development of communication pathways and a common dialogue between scientists, managers and communities. The Hyderabad workshop was seen to provide opportunities for such collaboration on a regional level.

In terms of the model some envisaged policy-related strategies are:
- The assistance of agricultural development by anticipating short-term climatic variations, in order to improve economic yield, and hence security relating to food supply with positive outcomes on socioeconomic conditions and population health
- The provision of a suitable framework for policy modification in the anticipation of important, short-term climatic change, enabling the incorporation of proactive intervention in agricultural practice
- The exploration of new approaches to managing crop diseases and the application of pesticides and herbicides to ensure economic use, and prevent overuse, as an important component in human health and aquatic ecosystem protection
- The encouragement of multilateral agricultural risk communication and dialogue between all stakeholders in the agrometeorological process

16.2
Conclusions

In addressing risks and uncertainties for integrated pest management, Australian researchers have concluded that more needs to be known about the complex relationships between climate and pest cycles relevant to local place. In this regard, collaborative activity is required between scientists, risk managers, government and local farmers to determine best practice approaches for addressing pest management, with the aim of achieving economically-sound and ecologically-sustainable outcomes.

Research results relating to Sclerotinia rot in Australian canola and stripe rust in wheat offer useful practical findings for the development of pest management systems elsewhere. A major focus of Australian research is the optimization of natural controls relating to informed planting strategies, and the minimization of pesticide application through the prediction of climatic influences, which can in turn lead to optimal effectiveness in the control of disease agents. Technology transfer is, however, a highly specialized area which has resulted in errors in the past, and which must therefore be treated with circumspection.

The relationship between macro- and microclimate, and the effects on the cycles of disease agents, needs special attention if quantity of applied pesticide is to be minimised, while optimising disease control outcomes.

While improvements in meteorological and crop-pest monitoring and modeling will remain important, a sound understanding of local economic, ecological and social realities is essential if the effectiveness and accountability of interventions is to be assured.

Acknowledgements

Support from the World Meteorological Organization (WMO), Asia-Pacific Network (APN), Australia-India Council, University of Western Sydney and New South Wales Department of Primary Industries (NSW DPI) in carrying out this research is appreciated.

References

Abawi GS, Grogan RG (1975) Source of primary inoculum and effects of temperature and moisture on infection of beans by Whetzelinia sclerotiorum. Phytopathology 65:300-309.
Australian Greenhouse Office (2006) http://www.greenhouse.gov.au/ impacts/publications/risk-vulnerability.html.
Brennan JP, Murray GM (1998) Economic Importance of Wheat Diseases in Australia, NSW Agriculture, Wagga Wagga
Brown JG, Butler KD (1936) Sclerotiniose of lettuce in Arizona. Arizona Agricultural Experimental Station Technical Bulletin 63:475-506
Buchwaldt L, Qun Yu F, Rimmer R, Hegedus DH (2003) Resistance to Sclerotinia sclerotiorum in a Chinese Brassica napus cultivar 8th International Congress of Plant Pathology, incorporating 14th Biennial Australasian Pathology Conference, 2 - 7 February 2003, Christchurch, New Zealand
Derry C, Attwater R, Booth S (2006) Rapid health-risk assessment of effluent irrigation on an Australian university campus. International Journal Hygiene Environmental Health, 209, 159-171.

Donald C, Porter I, Edwards J (2003) Clubroot (Plasmodiophora brassicae) an imminent threat to the Australian canola industry. In: Edwards J (ed). NSW Agriculture, Tamworth, New South Wales, Australia, pp 114-118

Food and Agriculture Organisation (2005) The State of Food and Agriculture 2005, FAO, Rome, Italy.

Gugel RK, Morrall RAA (1986) Inoculum-disease relationships in Sclerotinia stem rot of rapeseed in Saskatchewan. Canadian Journal of Plant Pathology 8:89-96

Hind-Lanoiselet T (2006) Forecasting Sclerotinia stem rot in Australia. PhD, Charles Sturt University

Hind-Lanoiselet T, Lewington F (2004) Canola Concepts: Managing Sclerotinia. Report No. Agnote 490, NSW Department of Primary Industries

Hind-Lanoiselet T, Lewington F, Wratten K, Murray GM (2006) NSW Department of Primary Industry results for Sclerotinia trials in 2004 and 2005

Hind-Lanoiselet TL, Lewington F, Murray GM, Hamblin P, Castleman L (2005) Economics and timing of fungicide application to control Sclerotinia stem rot in canola. In: Potter T (ed) 14th Australian Research Assembly on Brassicas. SARDI, Primary Industries and Resources South Australia, Port Lincoln, South Australia, 3 - 7 October 2005, pp 29-33

Hind-Lanoiselet TL, Lewington F, Wratten K, Murray GM (2004) NSW Agriculture results for Sclerotinia trials in 2003 http://www.canolaaustralia.com/information/pest_and_disease

Huda AKS, Selvaraju R, Balasubramanian TN, Geethalakshmi V, George DA, Clewett JF (2004) Experiences of using seasonal climate information with farmers in Tamil Nadu, India. In: Huda, A.K.S. and Packham, R.G. (editors) Using seasonal climate forecasting in agriculture: a participatory decision-making approach. ACIAR Technical Reports No. 59, (ISBN 1 86320 475 X print, 1 86320 476 8 electronic), Australian Centre for International Agricultural Research, GPO Box 1571, Canberra, ACT 2601. pp 22-30 (http://aciar.gov.au/web.nsf/doc/ACIA-6797SA)

Kharbanda PD, Tewari JP (1996) Integrated management of canola diseases using cultural methods. Canadian Journal of Plant Pathology 18:168-175

Last JM (2001) Dictionary of Epidemiology, Oxford, Melbourne.

Le Tourneau D (1979) Morphology, cytology and physiology of Sclerotinia species in culture. Phytopathology 69:887-890

McLean DM (1958) Role of dead flower parts in infection of certain crucifers by Sclerotinia sclerotiorum (Lib.) de Bary. Plant Disease Reporter 42:663-666

Meinke H, Stone RC (2005) Seasonal and inter-annual climate forecasting: the new tool for increasing preparedness to climate variability and change in agricultural planning and operations. Climate Change 70:221-253.

Morrall RAA, Thomson JR (1991) Petal test manual for sclerotinia in canola, University of Saskatchewan, Canada

Murray GM, Brennan JP (2001) Prioritising Threats to the Grains Industry within Disease Regions of Australia, NSW Department of Primary Industries. Report for Grains Research Development Corporation, Wagga Wagga

Murray GM, Brown JF (1987) The incidence and relative importance of wheat diseases in Australia. Australas Plant Path 16:34-37

Murray GM, Wellings C, Simpfendorfer S, Cole C (2005) Stripe rust: Understanding the disease in wheat. Report No. Publication number 5617 - NSW Department of Primary Industries, NSW Department of Primary Industries, Australia

NSW Farmers Association (2005) The mental health network blueprint, http://www.aghealth.org.au/blueprint/index.html

Paulitz TC (2006) Low input no-till cereal production in the Pacific Northwest of the U.S.: the challenges of root diseases. European Journal Plant Pathology 115:271-281

Rimmer SR, Buchwaldt L (1995) Brassica Oilseeds. In: Kimber DS, McGregor DI (eds) Diseases. CABI International, New York, pp 111-140.

Royal Geographical Society of Queensland (2001) http://www.rgsq.gil.com.au/heath12c.htm.

Sansford CE, Fitt BDL, Gladders P, Sutherland KG (1995) Oilseed rape disease development and yield loss relationships. GCIRC Bulletin 11:101-104

Schwartz HF, Steadman JR (1978) Factors affecting sclerotium populations of, and apothecium production by, Sclerotinia *sclerotiorum*. Phytopath 68:383-388

Sivakumar MVK (ed.) (2000) Climate prediction and agriculture. Proceedings of the START/WMO International Workshop held in Geneva, 27-29 September 1999, International START Secretariat, Washington DC, USA.

Stevens, RB (1960) Cultural practices in disease control. In: Plant Pathology, An Advanced Treatise (Eds: Horsfall, J.G. and Dimond, A.E.), vol. 3, Academic Press, New York. Switzerland, International START Secretariat, Washington, DC, USA, 322 pp.

Twengstrom E, Sigvald R, Svensson C, Yuen J (1998) Forecasting Sclerotinia stem rot in spring sown oilseed rape. Crop Prot 17:405-411

Unnevehr, L, Hirschhorn, N (2000) Food Safety Issues in the Developing World, World Bank, Washington.

Van der Plank (1975) Principles of Plant Infection, Academic Press, New York.

Wallace GE, Huda AKS (2005) Using climate information to approximate the value at risk of a forward contracted canola crop. Australian Farm Business Journal (AFBM) vol 2 (1) : 75-83. Also available on web http://www.afbmnetwork.orange.usyd.edu.au/afbmjournal

White GB (1983) Economics of plant disease control. In: Kommedahl T, Williams PH (eds) Challenging Problems in Plant Health. American Phytopathology Society, Minnesota, pp 477-486

Zadoks JC (1984) A quarter century of disease warning, 1958-1983. Plant Dis 68:352-355

Zadoks JC, Schein RD (1979) Epidemiology and Plant Disease Management, Vol. Oxford University Press, New York

CHAPTER 17

Coping Strategies with Agrometeorological Risks and Uncertainties for Drought Examples in Brasil

O. Brunini, Y. M. T. da Anunciação, L. T.G. Fortes, P. L. Abramides,
G. C. Blain, A. P. C. Brunini, J. P. de Carvalho

17.1 Introduction

The 1997–1998 El-Niño caused an extreme drought in the northeastern region with considerable losses for agriculture, livestock, water resources and society. Regionally, the impact of these anomalies can be striking. In the southeastern region, for example, in the State of São Paulo in the El Niño period, the effects caused by this phenomenon were quite different with above average rainfall in months like May and June. This situation can be observed, as indicated by the rainfall anomalies represented by the monthly Standardized Precipitation Index (SPI) for the month of May in 1998 (Figure 17.1). The occurrence of these anomalies lead the State Government to create a task force involving the various sectors of society, such as, research institutes, universities and the civil defense, to propose mitigation measures.

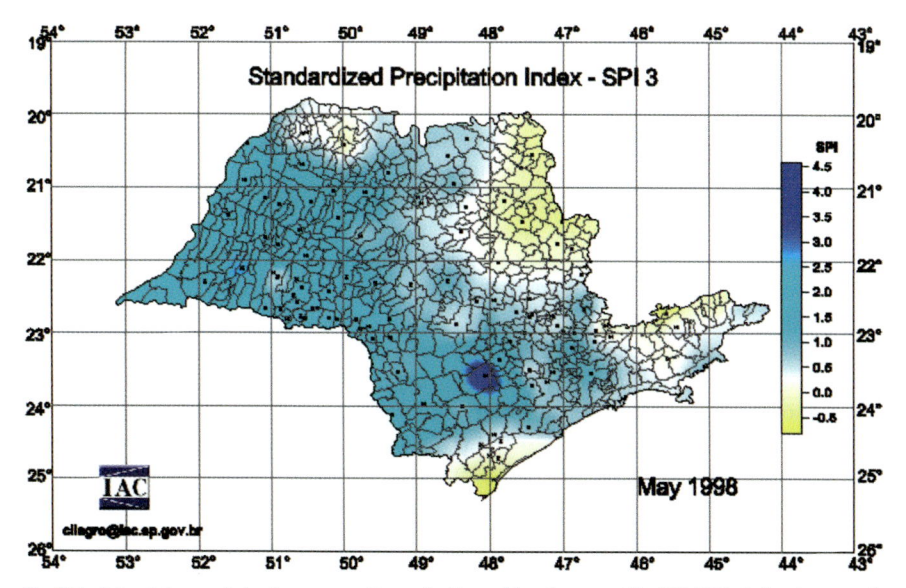

Fig. 17.1. Monthly precipitation anomalies as indicated by the monthly SPI (SPI-1) for the month of May 1998 in the State of São Paulo.

The National Meteorological Institute (INMET) determines the occurrence of droughts by means of the SPI, and also in deciles and the monthly deviation in precipitation compared to the climatological standard from 1961 to 1990. Studies have shown that 18 to 20 years of drought occurs every 100 years. The frequency of the drought occurrence in the Brazilian northeast is associated with the frequency of the El-Niño and of the Atlantic Ocean dipole; and the frequency of the drought occurrence in the southern region is associated with the frequency of the La-Niña. The areas affected by the drought vary in intensity, extension and time duration.

When a drought situation is confirmed through precipitation anomaly indices, technical material is prepared containing the precipitation monitoring for the affected region with a climate prognosis for the following quarter and this material is forwarded to the federal authorities in order to support the Brazilian government emergency actions. In the northeastern region of Brazil, there are several institutions and technical and technological infrastructure to detect drought. A limiting factor to ease detection of drought and corresponding mitigating actions is the lack of training and capacity to define the applicable methodologies.

In addition to the National Meteorological Institute, some States of the Federation developed specific studies for droughts to support not only agriculture, but also the civil defense activities and water resources planning and studies. An example is the State of São Paulo, through its Integrated Agrometeorological Information Center (CIIAGRO), and the Drought and Hydrometeorological Adversities Mitigation and Monitoring Center (INFOSECA). In this aspect, the assessments of the drought conditions and prognosis are prepared and distributed to farmers, rural cooperatives and other sectors of society.

The Ministry of Agriculture at Federal level and the Agricultural Secretariat of the São Paulo State Government, at State level, apply the reports and bulletins of drought monitoring in the Agricultural Activity Assurance Program (PROAGRO) and as a subsidy to the Agricultural/Livestock Expansion and Agricultural Insurance (FEAP) for the federal, and for São Paulo State government, respectively.

The immediate results of these actions are a reduction in the request for coverage for climatic events and the reduction of risks in Meteorological Adversities upon agriculture, in addition to the monitoring of the insurance operations and the agrometeorological management of PROAGRO and FEAP (sources: www.agricultura.gov.br; www.agricultura.sp.gov.br).

Regionally, there are programs that involve research institutions and the community in order to minimize risks for agriculture during drought situations. In the northeastern region, the state governments have mechanisms of their own to aid the population, such as distribution of water and foodstuff. In the drought areas, communities are supported by the federal and state governments and NGOs that orient the population. In order to improve health and reduce infant mortality, efficient methods of collecting and storing water by means of rural cisterns, underwater reservoirs and desalinization units are being applied.

In the northeastern and southern regions, regional forums for the quarterly climate prognosis for the rainy seasons are held. There is no specific forecast for drought, but the climate prognosis indicates beforehand if there is a probability of precipitation remaining below or above the normal. The State of Ceará, through its Secretariat of Rural Development and the Ceará Meteorological Foundation indicates the beginning of the sowing time by means of the climate prognosis, and the drought probability studies using real-time monitoring of precipitation and soil moisture content.

In the southeastern region, the government of the State of São Paulo implemented the Drought and Hydrometeorological Adversities Mitigation and Monitoring Center (INFOSECA), which is subordinated to the Instituto Agronômico (Agronomy Institute). The work performed by INFOSECA along with the activities carried out at CIIAGRO is pioneering in Brazil the agrometeorological monitoring of drought and its effect on agricultural activities (source: http://ciiagro.iac.sp.gov.br – www.infoseca.sp.gov.br).With its major territorial portion restricted to the equatorial humid or tropical areas, the effects of meteorological adversities on the Brazilian territories, and most notably drought, are very distinct. An assessment of the Humidity Index as proposed by Thornthwaite and Mather (1955) is presented on Figure 17.2, involving some states in the southern, northeastern and midwestern states.

In general, the macroclimatic characteristics indicate humid climate conditions for the states in the southeastern and midwestern regions. Nevertheless, even for humid regions, the climatic oscillations cause, in specific years, a drought condition that is highly unfavorable to crops. This statement is supported by the monthly variation of the SPI for the areas of Campinas and Ribeirão Preto in the State of São Paulo for the month of January (Figure 17.3). Even though the month of January normally presents high rainfall indices, on specific years a meteorological drought occurs. This phenomenon has an elevated consistency with values that are highly unfavorable and prejudicial to crops. The same aspect can be observed by the monthly variation of the Palmer Drought Severity Index for the locations of Votuporanga and Assis during the month of October (Figure 17.4).

Figure 17.4 further indicates an incisive factor which is the higher incidence of dry periods in the month of October in the last 15 years, shifting the beginning of planting of the summer crops to early November. This relationship with the PDSI, as well as with the SPI oscillations support the importance of monitoring and prognosis of drought in Brazil from the meteorological, hydrological and agronomic standpoints, with greater focus on the socio-economic effects of this meteorological adversity.

The methodologies and parameters used at federal level by the National Meteorological Institute (INMET) and at a state level by the Integrated Agrometeorological Information Center – CIIAGRO, and by the Drought and Hydrometeorological Adversities Mitigation and Monitoring Center (INFOSECA), of the Agricultural Secretariat of the State of São Paulo are described below.

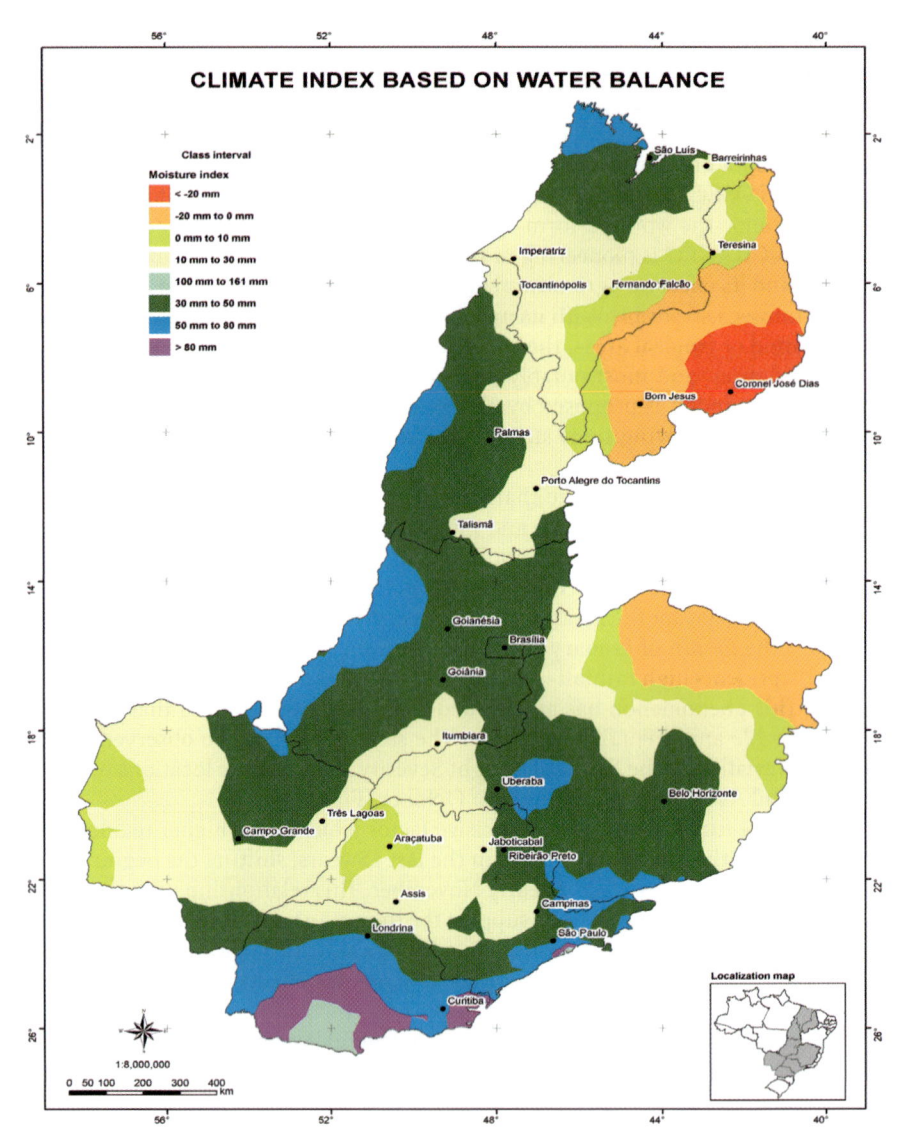

Fig. 17.2. Macroclimatic characteristics of some states in the southern, southeastern, midwestern and northeastern states based on the climatic classification proposed by Thornthwaite-Mather (1955).

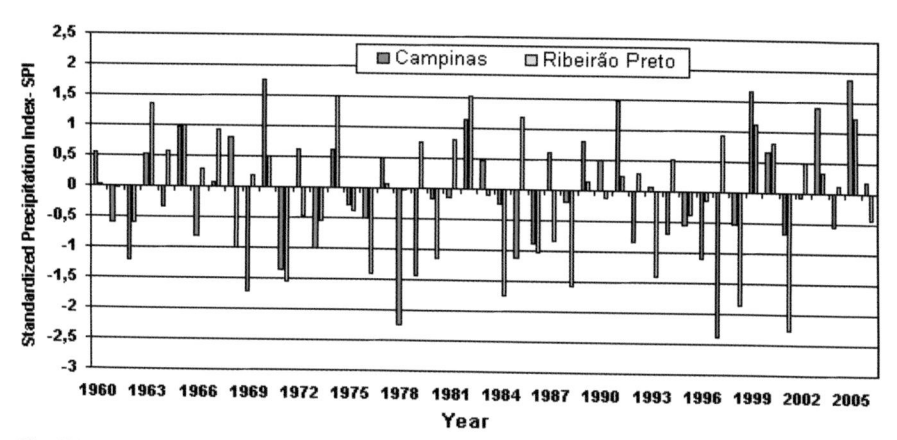

Chapter 17: Coping Strategies with Agrometeorological Risks 285

Fig. 17.3. Seasonal variation of the Standardized Precipitation Index (SPI) on a monthly scale (SPI-1) for the month of January in the regions of Campinas and Ribeirão Preto in the State of São Paulo.

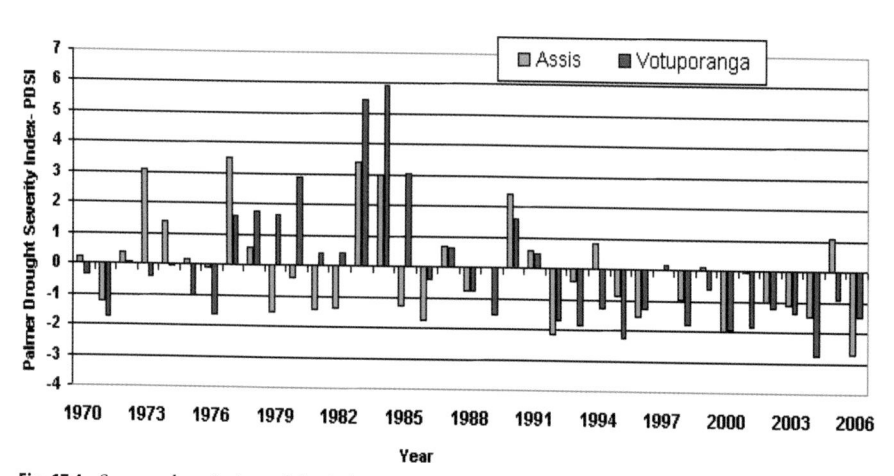

Fig. 17.4. Seasonal variation of the Palmer Drought Severity Index (PDSI) for the month of October in the regions of Assis and Votuporanga in the State of São Paulo.

17.2
Methodologies to Assess Precipitation Anomaly and Drought

17.2.1
Meteorological Indices

17.2.1.1
SPI Standardized Precipitation Index

The Standardized Precipitation Index (SPI), proposed by McKee et al. (1993), corresponds to the number of standard deviations that the observed accumulated precipitation deviates from the climatological average, for a determined period of time. The State of São Paulo (Brunini et al. 2000, INFOSECA 2005), Pernambuco (Santos and Anjos 2001) as well as INMET have been monitoring droughts through the SPI, presenting results that enable the use of the information to anticipate and mitigate adverse effects.

It is common to see in literature an association between a range of values for the SPI and the qualitative assessment of precipitation observed during the corresponding period. The most frequent association is suggested by IRI (2005), as per Table 17.1.

Calculation of the index begins with the adjustment of the gamma probability density function to the monthly rainfall series. After this phase, the accumulated probability of the occurrence for each monthly total observed is estimated. The normal inverse function (Gaussian) is applied to this probability and the result is the SPI.

In this method, precipitation can be totalized in several scales (1 to 72 months). When the time scale used is small (1, 2 or 3 months, for example), the SPI moves frequently above or below zero, observing the meteorological drought regime. As the assessment scale increases (12 or 24 months, for example) the SPI responds slower to changes in precipitation observing the hydrological drought regime.

Table 17.1. Arbitrary correspondence between the SPI values and the climate categories (adapted by Mckee et al., 1993)

SPI Values	Categories
SPI >+2	Extremely Wet
+1.50 a +1.99	Very Wet
+1.00 a +1.49	Moderately Wet
-0.99 a +0.99	Near Normal
-1.00 a -1.49	Moderately Dry
-1.50 a -1.99	Severely Dry
<-2.00	Extremely Dry

The gamma probability density function (GPDF) assumes distinct forms, according to the variation of α. Values for this parameter inferior to 1 indicate a strong asymmetric distribution (exponential form) with g(x) tending to infinite when x tends to 0. In the case of α = 1 the function intercepts the vertical axis in β for x=0. The increase in the magnitude of this parameter reduces the asymmetric degree (deviation from the mode) of the distribution (the probability density is displaced to the right). Values for α greater than 1 result in a GPDF with the maximum point (mode) in β*(α-1). An increase in the β parameter stretches the GPDF to the right, lowering its height and reducing the probability of the occurrence of the mode value. Similarly, as the density is compressed to the left (reduction of the β magnitude) and the height of the function becomes greater, the probability of the event increases.

Thus, the spatial variation of α and β in a state or country, indicate which are the regions with greatest degree of asymmetry in the temporal distribution of precipitation (rainfall irregularity). Considering the phenomenon of drought, anomalies in relation to environmental conditions of each area, these regions are at a greater risk of being subjected to meteorological droughts.

17.2.1.2
Palmer Drought Severity Index Adapted to the State of São Paulo – Pdsi Adap

The most important step of the PDSI is the calculation of precipitation, "Climatologically Appropriate Existing Conditions" (**P**) which can be understood as the amount of monthly precipitation necessary for a given area to remain under normal climatic conditions. This parameter is calculated as described by Palmer (1965). For the calculation of the monthly water anomaly (d), the precipitation observed in the month (Pi) is compared to **P** in the same period.

$$d = Pi - \mathbf{P} \tag{1}$$

As Palmer (1965) developed a standardized index compared to different locations at any period of time, it needs to be standardized (weighted) on a regional basis (Karl 1986). Thus, Palmer (1965) developed the climatic characterization factor designated by the letter K.

$$K = 17.67 * K' / \sum_{i=1}^{12} DK' \tag{2}$$

Where,

$$K' = 1.5 \log_{10}[(T + 2.8)/(D)] + 0.5 \tag{3}$$

T - the ratio between the demand and supply of water in a region, and
D - the monthly average of the absolute values for d.

Table 17.2. Arbitrary correspondence between the PDSIadap and drought categories

PDSI adap	Categories
≥ 3.00	Extremely Wet
2.00 a 2.99	Severe Wet
1.00 a 1.99	Moderately Wet
0.51 a 0.99	Slightly Wet
0.50 a -0.50	Near Normal
-0.51 a -0.99	Slightly Dry
-1.00 a -1.99	Moderately Dry
-2.00 a -2.99	Severely Dry
≤ -3.00	Extremely Dry

According to Blain (2005) adaptation of the PDSI to the State of São Paulo, had its major focus on the K factor of climatic characterization. The other elements of the original methodology, such as precipitation, "Climatologically Appropriate Existing Conditions" and the d index were calculated as described in the original paper by Palmer (1965). Drought categories, according to the PDSIadap are presented in Table 17.2. The final expression for K in State of São Paulo is:

$$K = (22.8K')/\Sigma DK' \qquad (4)$$

and the final equation adapted to the State of São Paulo is:

$$PDSIadapi = (Z_i/0.94) + 0.15 * PDSIadapi\text{-}1 \qquad (5)$$

17.2.1.3
Decile Method

The method consists of, initially, the organization in ascending order and subsequent classification of the historic precipitation data accumulated during the period of interest (normally 1, 3, 6, 12 or more months) in 10 intervals of equal frequency (10 percent probability of occurrence in each class). These intervals are denominated deciles and are normally numbered 1 to 10. N being the number of historic observations registered, the first decile will contain the n1 smallest values for precipitation, where n1 corresponds to the integer part of (N/10), the second decile will contain the following values (n2 & n1), where n2 = (N/20), and so on.

Subsequently, a category will correspond to each decile, in other words, a descriptive concept of the rainfall intensity, in which deciles may be grouped, this means more than one decile may be associated with the same category. If we asso-

Table 17.3. Alternative classifications used with the decile method

Decile	Originally Proposed Classification	Classification Currently Adopted by the Australien Office of Meterology	Classification Adopted by INMET	
	Category	Category	Category	Index
1	Much Below Normal	Lowest on Record	Extremely Below Normal	−3
		Very Much Below Average		
2	Below Normal	Below Average	Below Normal	−2
3			Slightly Below Normal	−1
4				0
5	Near Normal	Average	Normal	1
6				2
7	Above Normal			3
8		Above Average	Slightly Above Normal	1
9	Much Above Normal		Above Normal	2
10		Very Much Above Average	Extremely Above Normal	3
		Highest on Record		

ciate a color coding to each category, for example, we can plot precipitation behavior maps verifying, for each point, a class corresponding to the rainfall value observed during the period of interest, painting the point on the map with the color associated with this category.

Originally, the proponents of this method suggested using a rainfall classification per deciles as defined in the first part of Table 17.3. More recently, the Australian Bureau of Meteorology adopted the classification defined in the second part of Table 17.3. On the other hand, INMET adopted the convention defined in the second part of Table 17.3. On the other hand, INMET adopted the convention defined on the third part of Table 17.3 and further associating, at each concept, a numerical index between −3 and +3.

17.2.1.4
Quantile Method

In summary, the Quantile method, consists in the classification of the accumulated precipitation values during the period of interest (timescale), X, in five categories as defined below:

Table 17.4. Rainfall anomaly classification based on Quantile methodology

Preciptation level	Associated Probability	Categories (Observed Precipitation)
Quantile 1	15%	Very Dry
Quantile 2	20%	Dry
Quantile 3	30%	Normal
Quantile 4	20%	Wet
Quantile 5	15%	Very Wet

- First Quantile, $0 \leq X \leq Q_1$, where Q_1 is such that the Probability $(X \leq Q_1) = 0.15$
- Second Quantile, $Q_1 < X \leq Q_2$, where Q_2 is such that the Probability $(X \leq Q_2) = 0.35$
- Third Quantile, $Q_2 < X \leq Q_3$, where Q_3 is such that the Probability $(X \leq Q_3) = 0.65$
- Fourth Quantile, $Q_3 < X \leq Q_4$, where Q_4 is such that the Probability $(X \leq Q_4) = 0.85$
- Fifth Quantile, $X > Q_4$

Similarly to the SPI, to determine the Qi, i=1,...5, values, a probability model is adjusted (normally a Gamma distribution) to the historic data observed. X being the precipitation for the period and F(x) the Accumulated Density Function adjusted to the historic values for X, and F1 to the inverse F function, thus:

$$Q_1 = F^1 (0.15), Q_2 = F^1 (0.35), Q_3 = F^1 (0.65) \text{ e } Q_4 = F^1 (0.85) \qquad (6)$$

Each of the five quantiles defined above is associated with a qualitative classification as indicated on Table 17.4. As with the previous methods, the period of interest is normally 1, 3, 6, 12 or more months.

17.2.1.5
Comparison between methods

With the exception of the Palmer Index, the intrinsic principle of the various methods discussed above is the same and their results will differ only in the distinct conventions adopted for the classification of precipitation in categories and by the treatment, parametric or not, applied to the historic data. This comparison is discussed by Fortes et al (2006), which presents the chart reproduced on Figure 17.5.

17.2.2
Agrometeorological Indices

The understanding of the effect of the meteorological variables and their effect on crops is vital to determine the indices that adequately reflect the climate-plant in-

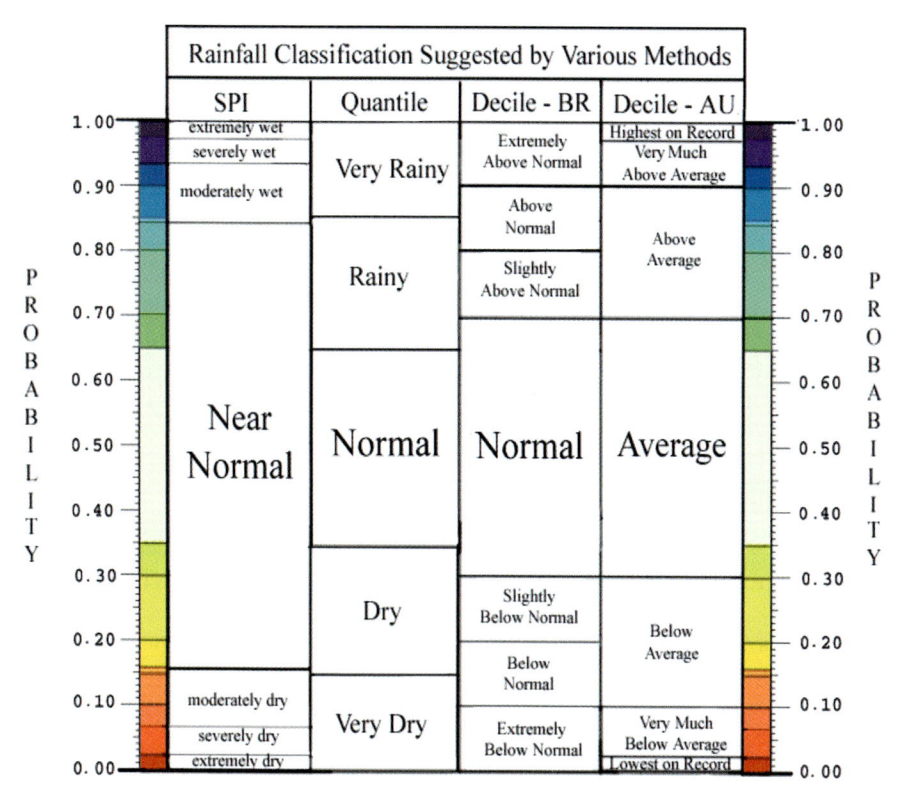

Rainfall Classification Suggested by Various Methods				
PROBABILITY	SPI	Quantile	Decile - BR	Decile - AU
1.00 – 0.90	extremely wet / severely wet / moderately wet	Very Rainy	Extremely Above Normal	Highest on Record / Very Much Above Average
0.90 – 0.70		Rainy	Above Normal / Slightly Above Normal	Above Average
0.70 – 0.30	Near Normal	Normal	Normal	Average
0.30 – 0.10		Dry	Slightly Below Normal / Below Normal	Below Average
0.10 – 0.00	moderately dry / severely dry / extremely dry	Very Dry	Extremely Below Normal	Very Much Below Average / Lowest on Record

Fig. 17.5. Numerical scale indicating the estimated probability through the historical values of precipitation, in order to verify if a specific recorded rainfall value is smaller, equal to or larger than the historical case.

teraction and crop yield and that can be used in a constant, dynamic and easily handled manner.

It is worth mentioning that the drought phenomenon can be assessed or monitored, with emphasis on the meteorological, hydrological, agronomic and socialeconomic aspects, however, from an agronomic standpoint, this monitoring and prognosis must be evaluated with tools that involve agronomy and agrometeorological knowledge and that integrate them in the process for weather and climate forecast.

In this aspect, the *Instituto Agronômico do Estado de São Paulo* (Agronomy Institute) has been developing the pioneering work with the implementation of CIIAGRO in 1988, and subsequently with the Drought and Hydrometeorological Adversities Mitigation and Monitoring Center (INFOSECA) in 2005. The agrometeorological indices used on a routine and continuous basis by CIIAGRO and INFOSECA are given below.

17.2.2.1
Actual Evapotranspiration Standardized Index (IPER)

Developed by Blain and Brunini (2006), and based on the SPI methodology, this index begins with the adjustment of the beta probability density function to the series of water balance in a ten days step. After this phase, the cumulative probability of a given estimated value for ETR is calculated. The normal inverse function (Gaussian), with a zero average and unit variance is applied to the accumulated probability. The result is the value of the new index, named "Actual Evapotranspiration Standardized Index" (IPER).

Considering that the beta distribution is defined in the interval [0 and 1] and that the de average temperatures in the State of São Paulo do not allow decendial ETR values above 100mm, the following variable transformation was chosen:

$$ETR" = ETR/100 \tag{7}$$

Where,

ETR" actual evapotranspiration variable transformed so that $0 < ETR" < 1$

G(ETR") is then transformed into a normal variable (final value for the IPER) through the equations developed by Abramowitz and Stegun (1965)

$$IPER = -\left(t - \frac{c_0 + c_1 t + c_2 t^2}{1 + d_1 t + d_2 t^2 + d_3 t^3}\right) \text{ para} 0 < H(x) \le 0.5$$

$$IPER = +\left(t - \frac{c_0 + c_1 t + c_2 t^2}{1 + d_1 t + d_2 t^2 + d_3 t^3}\right) \text{ para} 0.5 < H(x) < 0.5 \tag{8}$$

Where:

$$t = \sqrt{\ln\left(\frac{1}{(G(ETR))^2}\right)} \tag{9}$$

$$t = \sqrt{\ln\left(\frac{1}{(1 - G(ETR))^2}\right)} \tag{10}$$

And the values of the constants are defined as:
$c_0 = 2.515517$; $c_1 = 0.802853$; $c_2 = 0.010328$; $d_1 = 1.432788$; $d_2 = 0.189269$; $d_3 = 0.001308$

IPER values close to or greater than 0 indicate that the accumulated ETR in a 10-day period is close to or greater than the climatologically expected value of this parameter in this period. Negative values for the index indicate that the actual evapotranspiration in a given 10-day period is below the expected level for this given period. Variation of this index is directly related to the number of standard deviations that a given value of the ETR is below the climatologically expected value for

Table 17.5. Arbitrary correspondence between the IPER values and the drought categories to address the crop water requirements

IPER Values	Categories
IPER > -0.5	Near Normal
-0.5 to -1.0	Moderately Dry
-1.1 to 1.99	Severely Dry
IPER < -2.0	Extremely Dry

a given period of time and location. Table 17.5 offers an arbitrary correspondence between the IPER value and the water conditions for the soil to address the needs of crops.

17.2.2.2
Crop Moisture Index (CMI)

Palmer (1968) developed the Crop Moisture Index (CMI) in order to perform weekly monitoring of crop conditions on a climatological scale, based on the average temperature and the total precipitation for the current week. According to this author, in simple terms, agricultural drought is an "evapotranspiration deficit". However, if the potential evapotranspiration is used as the maximum estimated moisture required by plants, sub-humid and semi-arid areas will have a evapotranspiration deficit during summertime. It is suggested that the actual evapotranspiration anomaly be used, in other words, an estimating the total, the actual evapotranspiration dropped in relation to the expected actual evapotranspiration for that week. The CMI quickly responds to changes in climatic conditions for a region, being as such, appropriate for monitoring in small time scales (weeks or 10-day periods). The index is not adequate for a larger time scales, such as months, quarters and others.

17.2.2.3
Crop Development as a Function of Soil Moisture

Developed by Brunini (2005), this index seeks to relate the current soil moisture conditions and the development of the crop, aiming at quantifying and qualifying the water conditions in the soil which are favorable or unfavorable to plant development. In this case, the crop water development factor (CWDF) is the function between the ratio between the amount of water available in the soil (DAAS) and the maximum available water (DISPMAX).

$$CWDF = DAAS/DISPMAX \tag{11}$$

Where:

$0 < \equiv \text{CWDF} <= 1$
$\text{DAAS} = 0$ implies in $\text{CWDF} = 0$
$\text{DAAS} = \text{DISPMAX}$ indicates that $\text{CWDF} = 1$

Based on the agronomic, pedologic and agrometeorological aspects, the following relationship is established as seen on Table 17.6.

Considering that the soil moisture factor (CWDF) can be observed as a punctual value, as well as an average assessment, or an averaged value for soil moisture and characteristics of the crop development above or below the median value, a index was introduced to the crop development (CWDI), which considers the average

Table 17.6. Arbitrary relationship between the average soil ratio (CWDF) and the plant agrometeorological development conditions.

Average Soil Water Ratio	Plant Development Conditions
$0.8 <= \text{CWDF} <= 1$	Very Good
$0.6 <= \text{CWDF} < 0.8$	Favorable
$0.4 <= \text{CWDF} < 0.6$	Reasonable
$0.3 <= \text{CWDF} < 0.4$	Not Favorable
$0.2 <= \text{CWDF} < 0.3$	Harmfully
$0.1 <= \text{CWDF} < 0.2$	Severe
$0.0 <= \text{CWDF} < 0.1$	Critical

Table 17.7. Arbitrary relationship between the average soil moisture index and the conditions related to the plant water satisfaction index

Soil Moisture Index	Conditions Related to the Plant Water Satisfaction Index
$1.0 <= \text{CWDI} <= 1.5$	Good
$0.5 <= \text{CWDI} < 1.0$	Favorable
$0.0 <= \text{CWDI} < 0.5$	Reasonable
$-0.25 <= \text{CWDI} < 0.0$	Not Favorable
$-0.5 <= \text{CWDI} < -0.25$	Harmfully
$-0.75 <= \text{CWDI} < -0.5$	Severe
$\text{CWDI} < -0.75$	Critical

characteristics of soil moisture and the crop. This relationship is indicated by the formula below and in Table 17.7.

$$CWDI = [\{(CWDF/0.40)-1\}] \qquad (12)$$

These parameters and indices enable the monitoring of a culture, considering the periods of crop development, type of soil and culture, as well as date of sowing and the phenological phases.

17.2.2.4
Soil Water Supply Conditions and Water Stress on a Crop

Many drought indices consider either rainfall only or, in some cases, the interaction with the water available in the soil as passive. As such, Brunini (2005) introduced the Crop Water Stress Index (CWS), which is based on the relationship between actual evapotranspiration, potential evapotranspiration and the water available in the soil. In addition that to the water availability follows the evolution of the root system.

In this case, values are estimated for general crops, in which the crop coefficient **Kc** is not employed. However, assessments are made involving specific groups of plants defined by Z1, Z2, Z3, Z4, which corresponds to the depth of the root system, as shown below:

Z_1 (25 cm) = potato, onion, garlic, rice, garden produce, beans
Z_2 (50 cm) = beans, peanuts, corn, sorghum
Z_3 (75 cm) = soybean, citrus, coffee, sugarcane, cotton
Z_4 (100 cm) = coffee, citrus, sugarcane

This diversity in depth aims at differentiating crops, as well as to the different water retention capability of the soil, which can reflect in a larger or smaller exploration volume of the roots.

The water stress concept, based on the ETR.ETP relationship was developed as a result of works from Brunini (1981, 1987); Camargo and Hubbard (1994), Camargo and Hubbard (1999), in which the reduction in crop yield or plant development is based on the sum or product of the (ETR/ETP) in the period.

In this case, we analyzed only the response of a plant and the average (ETR/ETP) values during this period indicating the relationship between these two parameters. We then have a combination of **Z** for each value of the DAAS, which is a double entry table. In other words:, for each value of water available in the soil and for each potential evapotranspiration at the same period there is a unique value of **Z**; having in mind that:

i) CWDF = (DAAS/ DISPMAX) (13)
and37. Mund- und Rachentherapeutika
Judith Günther
  ii) $Z = f [(CWDF) (ETR/ETP)]$ (14)

iii) CWS=1− Z (15)

Table 17.8 indicates the relationship between the Crop Water Stress Index (CWS) and the plant water supply, while Table 17.9 represents the relationship between the average value of stress for a given culture in a given time interval (ACWS), which is determined by the relationship:

$$ACWS = (\Sigma(CWS)/n \quad (16)$$
$$i = 1$$

n being the number of intervals used

17.3 Results and Analysis

17.3.1 Meteorological Aspects of Drought Monitoring and Prediction

Considering the Brazilian territory, the INMET performed the follow-up of the drought in the 2005/2006 period, monitoring the monthly rainfall values and

Table 17.8. Arbitrary relationship between the Crop Water Stress Index (CWSI) and the plant agrometeorological development conditions

Crop Water Stress Index	Plant Development Conditions
$0 = CWS < 0.1$	Good
$0.1 \leq CWS \leq 0.2$	Favorable
$0.2 \leq CWS < 0.4$	Ordinary
$0.4 \leq CWS < 0.6$	Reasonable
$0.6 \leq CWS < 0.8$	Not Favorable
$0.8 \leq CWS \leq 1.0$	Critical

Table 17.9. Arbitrary relationship between the average Crop Water Stress Index (CWS) and the average development conditions of the plant during the period

Average Crop Water Stress Index	Average Development Conditions of the Plant
$0.8 \leq ACWS \leq 1$	Good
$0.6 \leq ACWS \leq 0.8$	Favorable
$0.4 \leq ACWS < 0.6$	Ordinary
$0.2 \leq ACWS < 0.4$	Reasonable
$0.1 \leq ACWS < 0.2$	Not Favorable
$ACWS = 0.1$	Critical

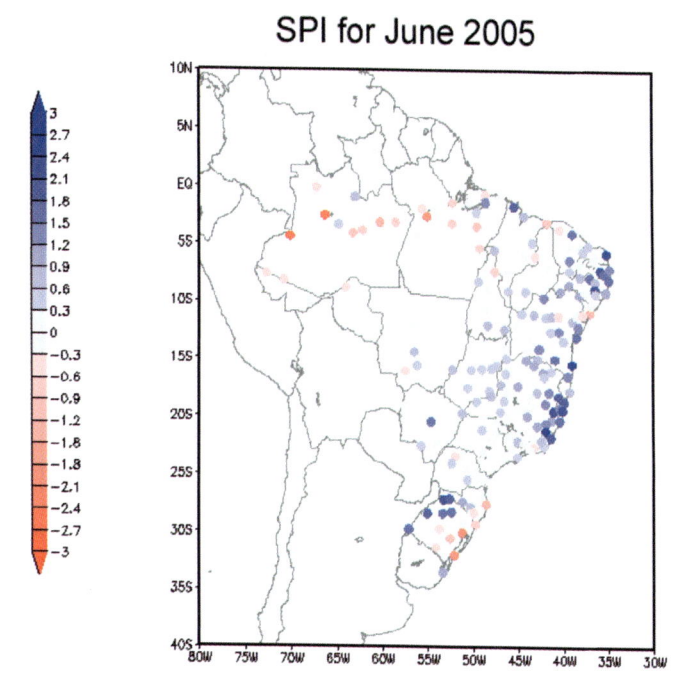

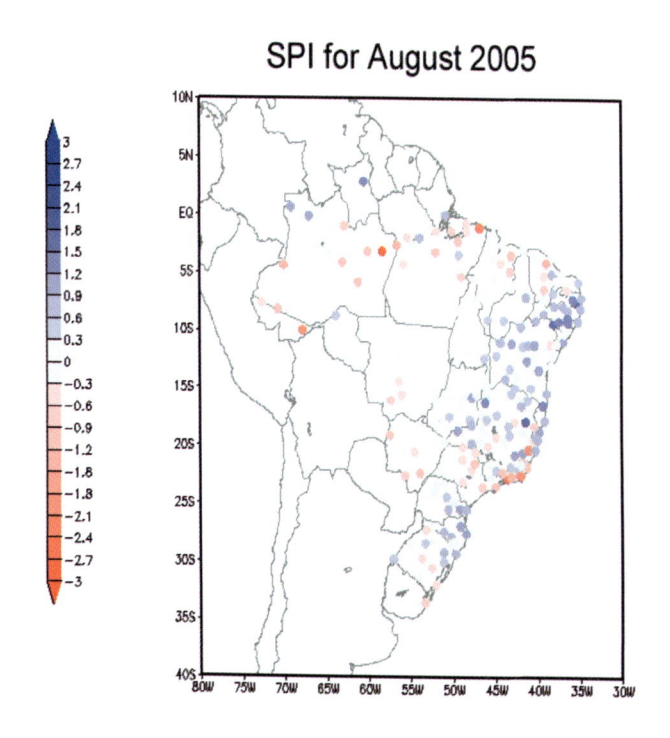

Chapter 17: Coping Strategies with Agrometeorological Risks 297

Fig. 17.6 a. Monthly monitoring of rainfall anomaly in the Brazilian territory as indicated by the SPI for June/05 to August/05.

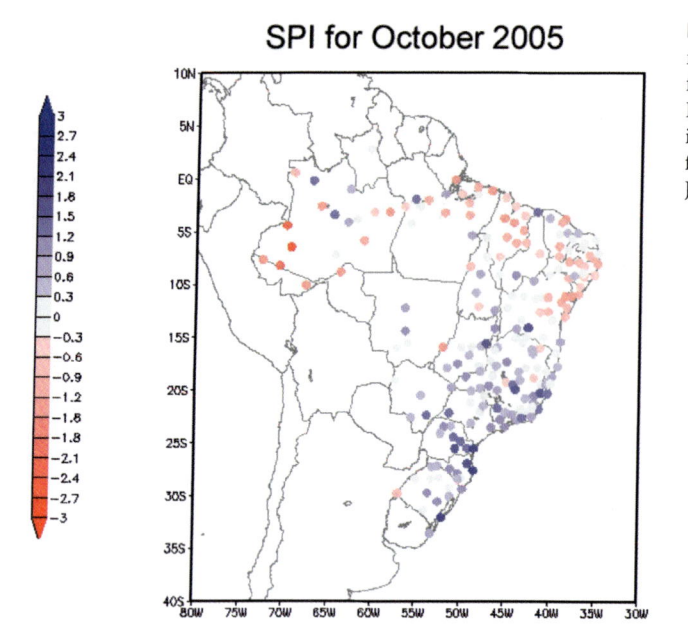

Fig. 17.6 b. Monthly monitoring of rainfall anomaly in the Brazilian territory as indicated by the SPI for October/05 to January/06.

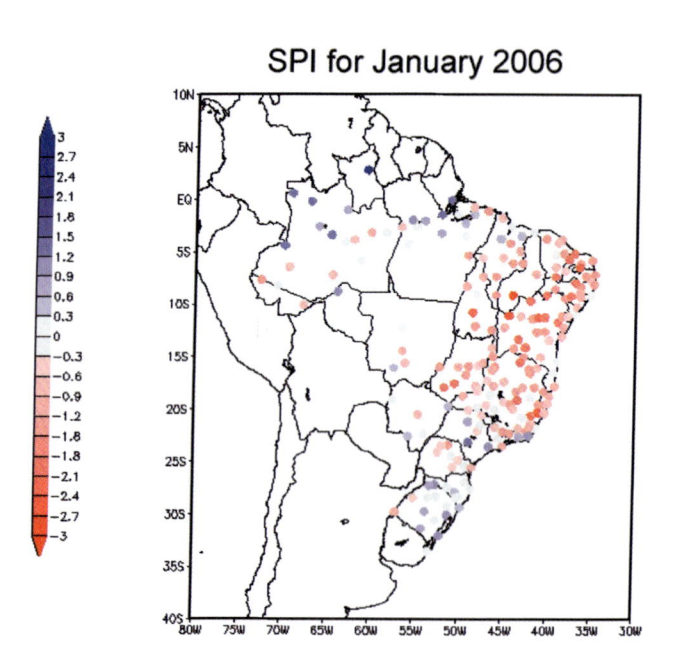

their duration (Figure 17.6 a; and Figure 17.6 b), which is presented below in the sequence of figures corresponding the months from June 2005 to January 2006. Extreme precipitation negative indices as indicated by the SPI value were observed during this period up to the month of November 2005.

The severe drought experienced in the southern regions of Brazil during the summer of 2005 caused important economic losses to the region, which has its key agricultural cycle from October to March. This drought is well represented in Figure 17.7 and shows the SPI-3 index calculated for February 2005 (which considers the accumulated precipitation from December 2005 to February 2006).

As a result of this adverse phenomenon observed in 2005, during the 2006 crop growing season, INMET adopted a special procedure for follow-up, based on the monitoring of the total accumulated precipitation measured on a daily, monthly, quarterly, semesterly timescale in the stations that the Institute maintains in that region (Anunciação and Fortes 2006). This enabled keeping the authorities responsible for agriculture in the country alert to the repetition trends for the phenomenon and provided subsidies for decision making concerning preventive and mitigating measures.

The southeastern region of Brazil presents a severe history of critical drought periods that often caused serious economic and social losses. Even though the values observed in Figures 17.2 and 17.3 indicate a punctual oscillation of the SPI and PDSI for some locations, they did not quantify the magnitude of this adverse event on a spatial scale.

As example, the 1963 and 1964 periods were extremely dry in this region, as indicated in Figures 17.8 and 17.9 where we have the SPI on a monthly scale (SPI-1) for the month of December 1963 and on an annual scale (SPI-12) for March 1964. It can be noted that the SPI values indicate very severe conditions with negative effects on agriculture, and very serious on water resources, indicating that the socioeconomic losses were extremely severe. The figures obtained, inferior to -2.0, reflect a likelihood of occurrence of 2 to 3 times every 100 years. This phenomenon occurred again in the 2004/2005 period, and this rainfall anomaly based on the SPI values in a month analysis is presented in Figure 17.10. Such results have shown for the State authorities the importance of drought monitoring and the proposed mitigation aspects to reduce the social impact of this phenomenon.

Perspectives for the summer of 2005/2006, based on precipitation data observed and the climatic prognosis for the January/March quarter 2006 were analyzed in greater detail by INMET. Long dry periods in the southern region of the country were observed, most notably in the State of Rio Grande do Sul, severely penalizing the agricultural crops and resulting in economic difficulties to a large number of producers.

The assessment presented below is based on the monitoring of the total accumulated precipitation measured on a daily, monthly, quarterly, 6-month basis in the INMET weather stations located respectively in Passo Fundo northern region of the state; Santa Maria, Central Region, and Bagé, southern region of the state as pointed on the map in (Figure 17.11) defines, by black circles, the three stations considered. The deviation of the precipitation throughout 1, 3 and 6 months is assessed by means of the Standardized Precipitation Index (SPI). This monitoring enabled forecasting of meteorological droughts and the issuance of alerts to deci-

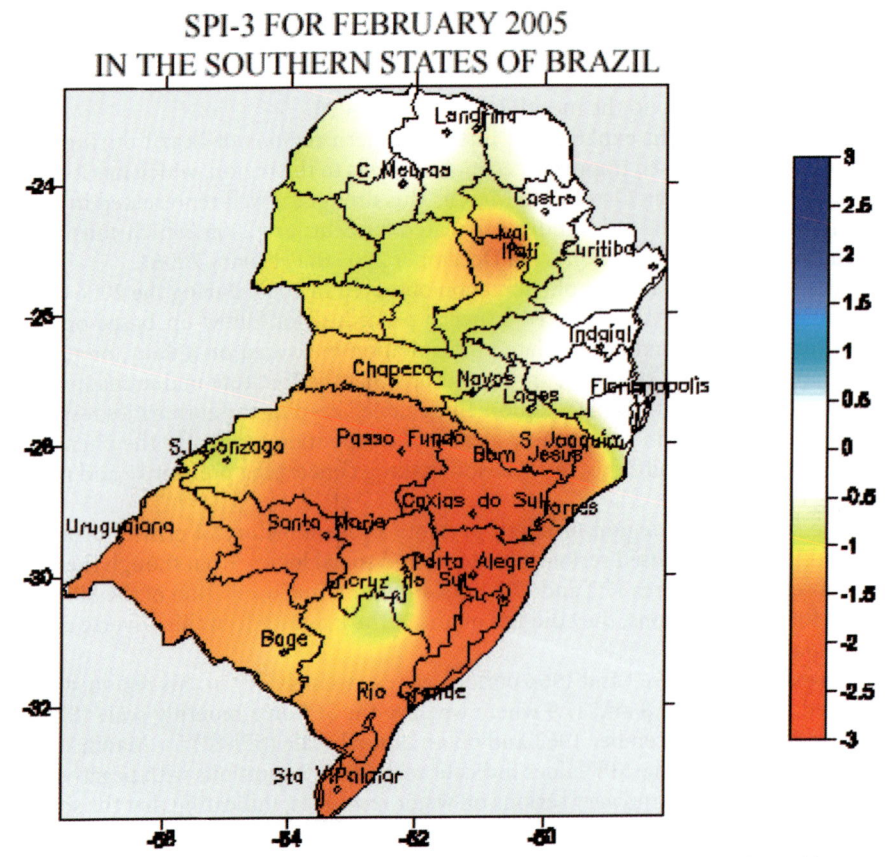

Fig. 17.7. Rainfall anomaly for the period December 2005 to February 2006 as indicated by SPI, in a quarter analysis (SPI-3).

sion makers. Monitoring of the accumulated precipitation in the 1 and 6-month periods facilitated the comparison of the current behavior with that of previous years, as well as the formulation of a prognosis for the coming years. Figure 17.11 indicates that in the September-November quarter in 2005, the precipitation index was situated slightly above the average for Passo Fundo and Santa Maria and within the average for Bagé. The trend for the subsequent months was subsequently analyzed for each station separately. In this sense, the concern was using the available data in order to provide subsidies for the decision making process within the Ministry of Agriculture. As a result, the handling of each case was not homogenous; in fact, there is more information available for Santa Maria, which has been studied for longer period of time. In the future updating of this work, the treatment will be uniform.

Considering climate prediction and prognoses of the rainy season, the National Meteorological Institute (INMET), and the National Space and Research Institute

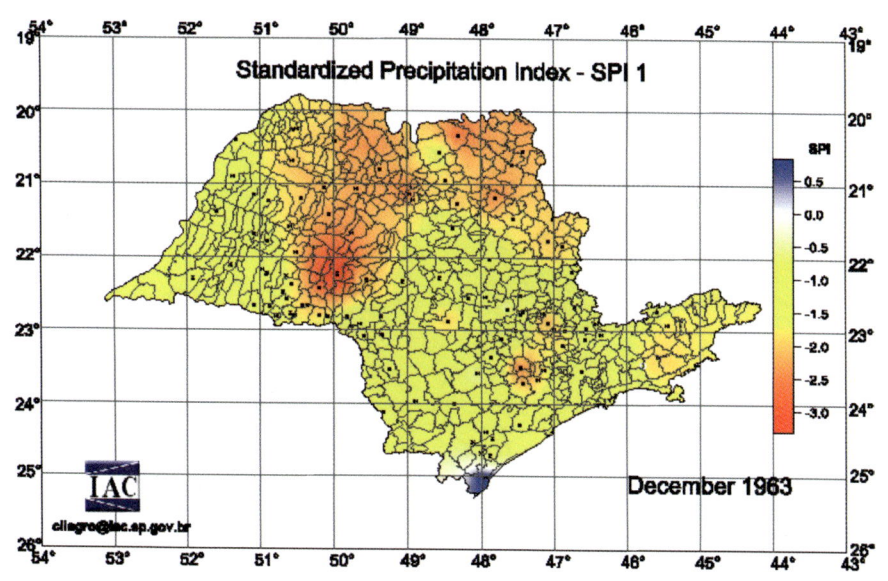

Chapter 17: Coping Strategies with Agrometeorological Risks 301

Fig. 17.8. Variation of SPI on a monthly scale (SPI-1) for the month of December 1963 in the State of São Paulo.

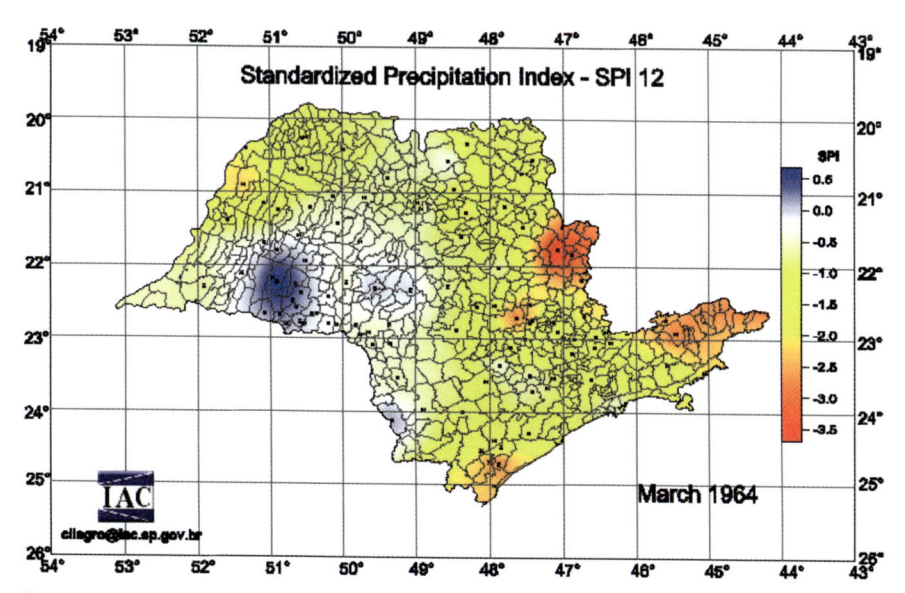

Fig. 17.9. Variation of SPI on and yearly scale (SPI-12) for the month of March 1964 in the State of São Paulo.

302   O. Brunini et al.

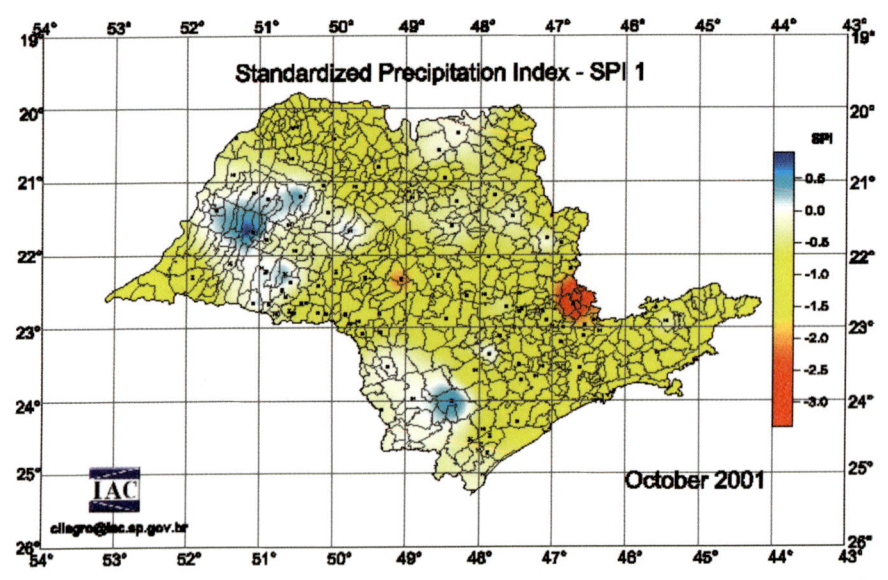

Fig. 17.10. Rainfall anomaly for the period October 2001 in São Paulo State based on the SPI values (SPI-1).

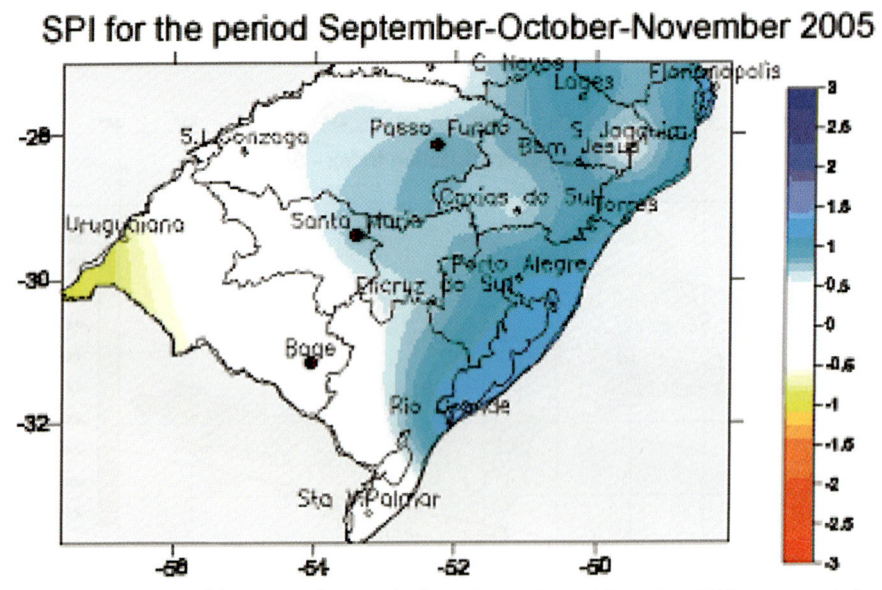

Fig. 17.11. Variation of the SPI for the months from September to November 2005, on a quarterly scale (SPI-3) for the State of Rio Grande do Sul, highlighting the location of the Auto-mated Meteorological Stations.

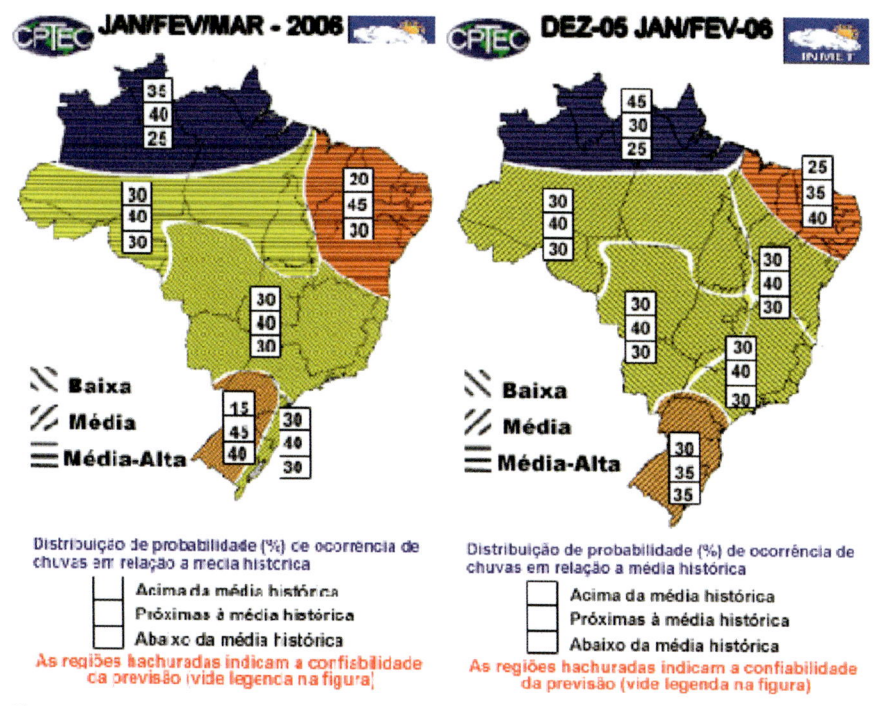

Fig. 17.12. Summary of the climatic prognosis corresponding to the December 2005 to February 2006, and January to March 2006 quarters, prepared by CPTEC/INPE and INMET.

(INPE) periodically carry out a climatic prognosis indicating if the meteorological conditions, especially precipitation, will be above or below the historic average (Figure 17.12). This trend is used by several institutions especially those dedicated to agricultural planning.

Figure 17.13 presents the prognosis for precipitation prepared for the State of Rio Grande do Sul by the 8th Meteorological District of INMET, together with the Meteorology Department of the Pelotas Federal University (CPPMet/UFPEL, 2005), for the months of January, February and March 2006.

For CIIAGRO and INFOSECA in the State of São Paulo, climatic prognosis is not carried out, but the INMET/INPE prognosis is used for agricultural purposes and planning. One of the uses is to establish a monthly prognosis of the SPI and its effect on agriculture. Since the climatic likelihood for the coming months was of precipitation below the average the SPI trend was projected as a function of the possibility of this event for the July-August quarter (Table 17.10). Since results indicate the persistency of the meteorological drought at least to the end of September in the regions comprising the states of Paraná, São Paulo and Minas Gerais, projecting unfavorable conditions for the sugarcane, coffee and citrus crop, and a delay for summer crops planting.

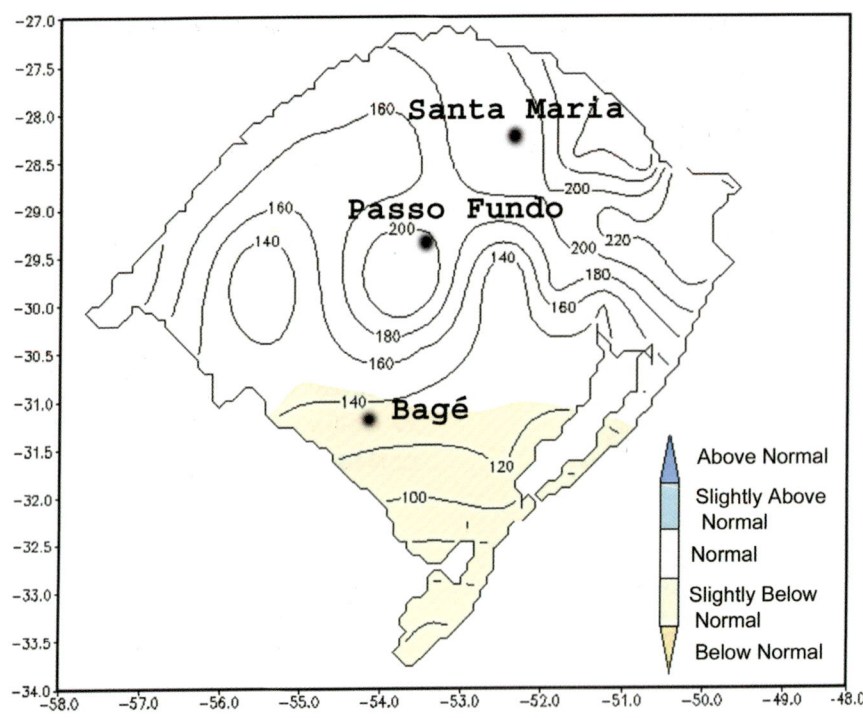

Fig. 17.13. Prognosis for precipitation in the State of Rio Grande do Sul in the months from January to March 2006. The areas represented in white indicate rainfall in the climatological average, yellow below average and blue above-average (source: www.inmet.gov.br/climatologia/cond-clima/bol-dez2005.pdf

17.3.2
Agrometeorological Aspects of Drought

Several institutions in Brazil try to make quantification and the monitoring of drought from a meteorological and agronomic standpoint. Some examples are the Ceará Meteorological Foundation (FUNCEME), National Meteorology Institute (INMET) and the National Space Research Institute (INPE). Nevertheless, few of these institutions routinely consider the assessment and characteristics of this phenomenon towards agriculture and civil defense, embracing agronomy soil characteristics and crop behavior.

With regards to the SPI, several assessments have been made specifically for the southern region of Brazil, the INMET has tried to compare the behavior of the crops with the SPI values. Figure 17.14 shows the behavior of soybean yield and the SPI index for six months (SPI-6) calculation based on data from the Passo Fundo (soybean producing region) and Santa Maria, for the period ranging from October to March. The estimation for the index for the October 2005 and March 2006 semester, and the forecast for the next crop, performed by CONAB, are indicat-

Table 17.10. Estimated monthly values for the Standardized Precipitation Index (SPI-1) in relation to the prognosis of rainfall

Estimated Monthly Values for the SPI (SPI-1)			
Locality	July	August	September
Araçatuba -SP	−0.89	−0.72	−0.55
Catanduva -SP	−0.86	−0.69	−0.52
Jaú -SP	−0.84	−0.66	−0.49
Piracicaba -SP	−0.81	−0.63	−0.46
Ribeirão Preto -SP	−0.78	−0.60	−0.43
São José do Rio Preto -SP	−0.75	−0.58	−0.40
Cambará-SP	−1.05	−0.93	−0.81
Joaquim Távora -PR	−1.02	−0.9	−0.78
Maringá -PR	−0.99	−0.87	−0.75
Paranavaí-PR	−0.96	−0.84	−0.72
Uberaba-MG	−0.69	−0.66	−0.63

ed with distinct colors and highlighted by the oval (source production – CONAB 2005). The results highlight the importance of analyzing crop yield and rainfall patterns.

Concerning the State of São Paulo, the adverse effect of these precipitation anomalies has been assessed for some crops. For example, the assessment of sugarcane yield in the Ribeirão Preto region, demonstrated a good correlation between the SPI values in 9-month scales (Figure 17.15) and sugar yield. Normally the growing period for the crop is from September to May, in which accumulation and the increment of dry matter is directly influenced by the climate and in such a case. The SPI for May with the 9-month recurrence (SPI-9) adequately reflects the water conditions in this soil for this crop. But for maize in the off-season cropping in the Assis region when planting is performed between January and March, it is observed that the averaged SPI on a monthly scale adequately reflects the water conditions for this crop. This relationship is presented on Figure 17.16, and a good relationship between the SPI and the productivity levels can be observed.

Another parameter that is adequately related to the agricultural production is the Palmer Drought Severity Index (PDSI). The relationship between the average PDSI adap values and maize yield in the State of São Paulo is presented in Figure 17.17, indicating the potential of this easily used index. These results are quite will correlated to overall maize grain production in the State, and the same figures were observed in the 2005/2006 crop growing season.

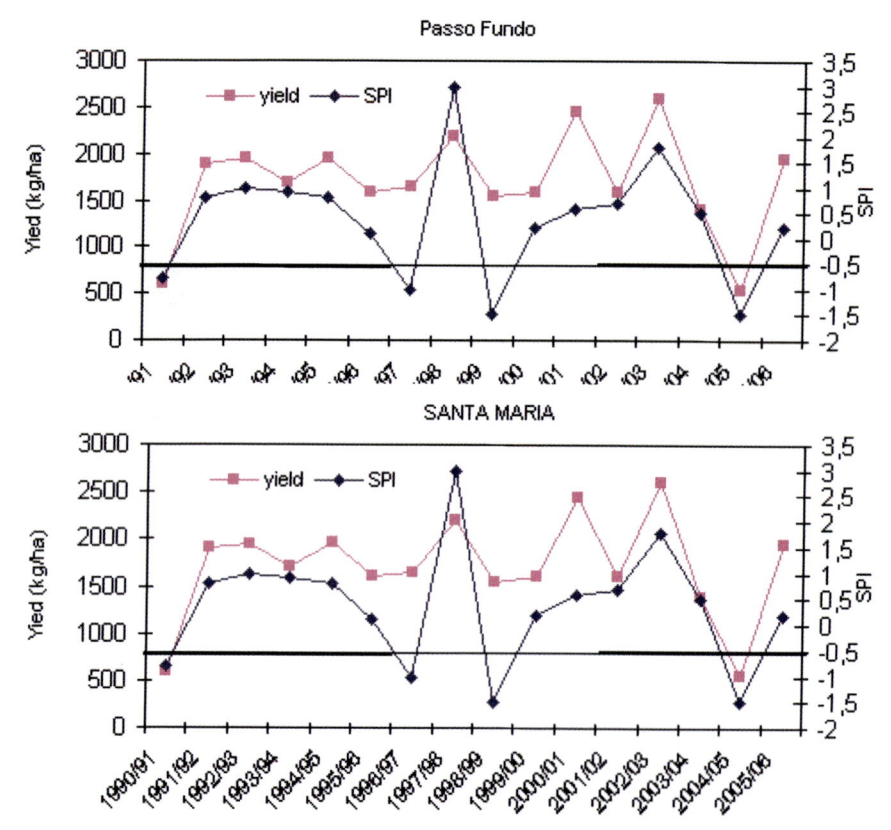

Fig. 17.14. Comparison between soybean yield and the values for the SPI on a six-month scale (SPI-6) for the State of Rio Grande do Sul, considering the period from October/05 to March/06.

17.3.3
Drought Monitoring and Mitigation Center

The State of São Paulo, through the Agronomic Institute (IAC) in a partnership with the State Extension Service Agency CATI created the INFOSECA (Drought and Hydrometeorological Adversities Mitigation and Monitoring Center), an operational system that brings immediate reports of the actions and effects of meteorological adversities upon agriculture and proposes ways of monitoring and mitigating the negative impact of these adversities, most notably, drought.

The work of INFOSECA allows systematically following up on the evolution of drought conditions in the State, proposing mitigating and relief measures, as well as physical and agronomic processes to bypass the problem. These processes may include future prognosis of the drought conditions, and is available at the site: www.infoseca.sp.gov.br.

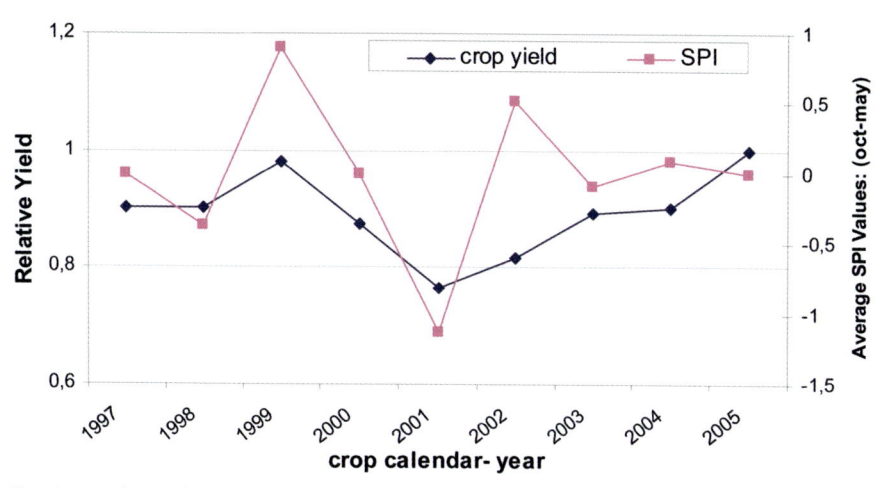

Fig. 17.15. Relationship between the decreasing productivity for sugarcane and the SPI values on the nine-month scale (SPI-9) for the Ribeirão Preto – SP, region.

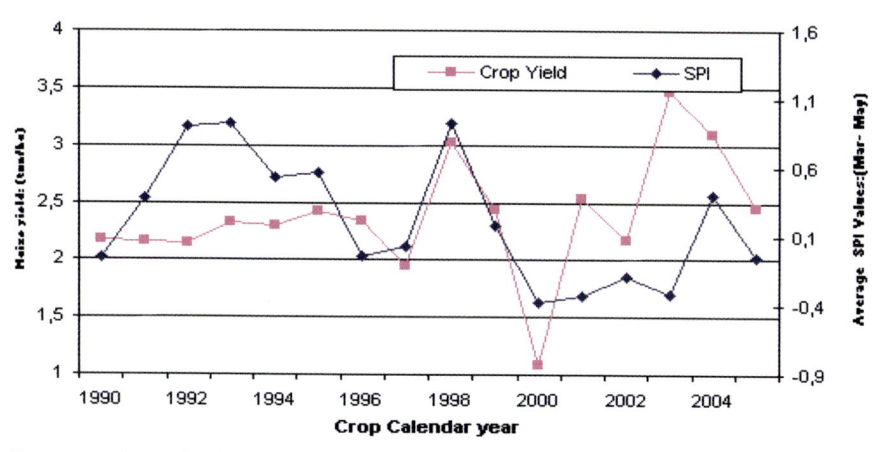

Fig. 17.16. Relationship between the yield for the off season maize and the average monthly values for the (SPI-1).

The users have two basic lines of work, in other words, a user may analyze the effect of the drought from a fully meteorological as well as, an agrometeorological standpoint. The INFOSECA system has the purpose of processing and making available the agrometeorological information related to drought indices, and communicates agrometeorological warning and outlook of these adversities to the agribusiness. This system is based on agrometeorological parameters and relies on a management model and the direct data input via web from the meteorological stations. Furthermore, it has a module to provide information and counseling and real-time consulting via Internet. Meteorological data are collected (mainly, precipitation, maximum and minimum air temperatures) from 130 locations in dif-

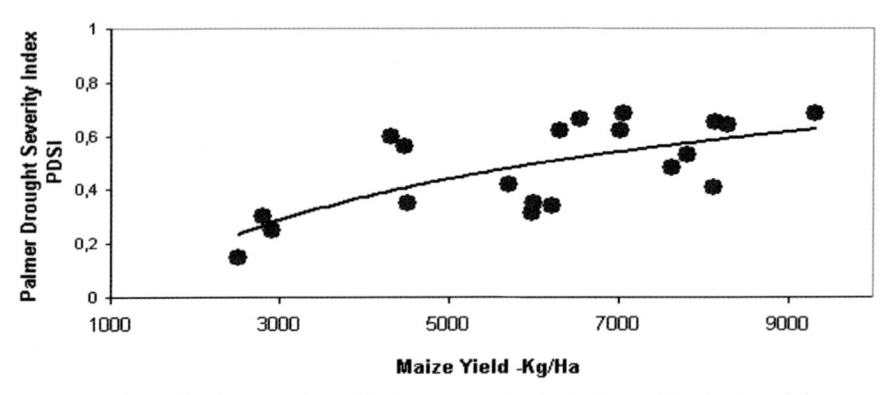

Fig. 17.17. Relationship between the yield of summer maize in the State of São Paulo and the average values for the Palmer Drought Severity Index (PDSI) values during crop growing sea-son.

ferent regions of the State of São Paulo, that are recorded in to the CIIAGRO system. Data are consisted, assessed and transformed into agronomic parameters and displayed in the form of tables and maps of indices (SPI, Palmer, ETM/ETP and DI) and the agrometeorological indices (CMI, CWS, CWSI, IPER and Crop Development Index). A daily bulletin containing drought prognosis is supplied. The system was developed using the Sis Plant technology and is based on the HTML, ASP, VbScript and SQL languages. Communication of Web data and database server is performed via ODBC, using the MySQL database. Information is provided at municipal level and consolidated by Administrative Region, Regional Development Offices – EDR/CATI, Water Resource Management Units – UGRH and Regional Research Centers.

The study allows the analysis of the meteorological conditions and drought through the use of the universally adopted indices and introducing new analysis that take into account soil characteristics, crop evapotranspiration and the relationship between potential evapotranspiration and water availability in the soil and the development of the root system. In this aspect, results referring to the different depths of root systems are also presented, as for crop with superficial root systems and consequently more sensitive water storage such as rice, beans, onions and deeper root systems, such as citrus, coffee and fruit.

Table 17.11 presents the average water stress conditions for the off-season maize crop with a root system at 50cm in the period ranging from March 1st to April 30th, 2006, as well as for the sugarcane crops during the same period, however with a the root system of 1 m deep. It can be noted that for sugarcane, the agrometeorological conditions were not considered critical, due to larger soil volume exploration by the sugar cane rooting system, however, for the maize crop, the situation was highly prejudicial.

Table 17.11. Average conditions of water stress for the off-season corn crop (Z=50cm) and for the sugarcane crop (Z=100cm) in the period ranging from March 1, 2006 to April 30, 2006

Locality	Rooting Depth (cm)	ACWDI	Condition
Guariba	(Maize-early stage) 25	0.03	Extreme Severe
Guariba	(Maize- tasseling period stage) 50	0.33	Not Favorable
Guariba	(Sugar cane-full development stage) 100	0.65	Good
Jaboticabal	(Maize -early stage) 25	-0.27	Extremely Severe
Jaboticabal	(Maize -tasseling period) 50	-0.31	Extremely Severe
Jaboticabal	(Sugar cane-full development stage)100	0.23	Harmfully

17.3.4
Climatic Risk Zoning

One of the most important aspects of agrometeorology is to define the timeframe and location with probability of occurrence of drought and other adverse phenomena for specific crop development stage, or the climatic risk assessment for agriculture exploitation. Specifically considering drought, this assessment of water shortage probability is made by comparing the crop water demand and the water availability in the ecosystem imposed by the rainfall precipitation regime. Crop Water Requirement Index (CWRI), can be defined as:

$$CWRI = (ETR/ETM) \tag{25}$$

where:
    ETR – actual crop evapotranspiration ; and
    ETM – maximum crop evapotranspiration, as defined by

$$ETM = Kc \cdot ETo \tag{26}$$

where:
    Kc – crop coefficient
    ETo – reference crop evapotranspiration

The studies that sought to quantify the climate-plant relationship and the risks of meteorological adversities are one of the basic tools used in the agricultural financing programs. As examples we can name the PROAGRO at Federal Government level and the FEAP at the State of São Paulo Government level.

Figure 17.18 presents the climatic risk zoning for the summer maize crop in the State of São Paulo (Brunini et al. 2001) used in the PROAGRO Agricultural Insurance Program.

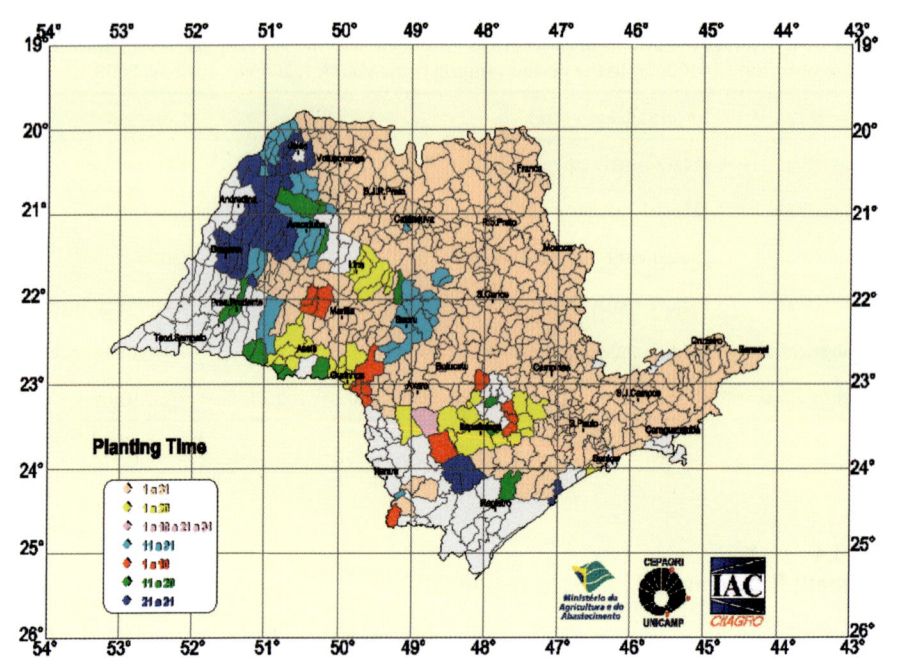

Fig. 17.18: Probability of water supply during the tasseling period of the maize crop in the State of São Paulo. Between the 1st and 10th of October (source : Brunini et al 2001).

One further step was taken by the government of the State of São Paulo in this system for the risk characterization, with the introduction of the *"Sistema de Avaliação de Riscos Climáticos e Monitoramento Agrometeorológico de Culturas"* (Climatic Risk Assessment System and Agrometeorological Monitoring of Crops).

In this process, climatic risks related to drought are assessed as well as the probability of addressing the water demand for any crop, be it annual or perennial. The likelihood of addressing the water requirement is made on the beta distribution (β), that the best represents the agro-system being analyzed, since the ETR/ETM ratio has values between 0 and 1. Furthermore, following up on the evolution of the agrometeorological parameters and behavior is allowed. The study can be made for all critical phenological phases of the crop, and a subroutine allows that the soil volume for each the crop inferred by the root system is also inferred (climate Risk Evaluation and Crop Agrometeorology System).Information on the probability for meeting crop water requirements for each planting scheduling for each critical phase of the crop is automatically inserted into the CIIAGRO, enabling the on-line assessment of climatic risks.

Table 17.12 indicates the probability of attending crop water requirements water for the off-season maize crop in the region of Palmital – SP, as well as the risk of occurrence of frost or agricultural drought in the tasseling period.

Even though tables and charts allow the indication or the results of the occurrence of adverse phenomena, and the response of a crop in a given region, they do

Table 17.12. Probability of attending crop water demands during specific phenological phases of the corn crop planted between 1–5 January, and the risk of high or low air temperature

Day/Month	Phenological Stage	Prob ETR/ETM 0.7<p<0.8	Prob ETR/ETM p>=0.8	Frost risk %	Dry Spell Min-Days	Dry Spell Max-Days
1/1	Sowing	8.03	91.62	0	0	6
6/1	Sowing	0.03	99.96	0	0	6
½	Fast growing	0.01	99.98	0	0	13
6/2	Fast growing	0.01	99.98	0	0	13
16/2	Tasselling	0.01	99.98	0	0	21
21/2	Tasselling	0.01	99.98	0	0	21
26/2	Tasselling	0.01	99.98	0	0	21
1/3	Tasselling	0.01	99.98	0	0	21
6/3	Tasselling	0.03	99.96	0	0	21
¼	Ripening	0.01	99.98	0	0	21
6/4	Ripening	0.01	99.98	0	0	21
11/4	Ripening	0.01	99.98	0	0	21

Prob – Probability function

not provide the spatial visualization of these parameters or their degree of occurrence in different time frames.

In order to make this information more readily understood by the general users and by the decision makers, these data are transformed into agrometeorological maps. Two basic tools were used for this– SURFER and ARG-GIS.

Figure 17.19 shows the water stress conditions for maize crop using the Surfer methodology. Note the differences as a function of the spatial variability and the topography of the state when the different types of soil are included.

On the other hand, with the use of the ARC-GIS, this information is more detailed, enabling the overlapping of other variables. Figure 17.20 presents the same map with the water stress conditions for the maize in the ARC GIS system. In this case, minimum and maximum air temperatures lower than 16°C and higher than 32°C were superimposed on the map indicating restrictive areas due to thermal insufficiency or elevated temperatures, as well as the water supply.

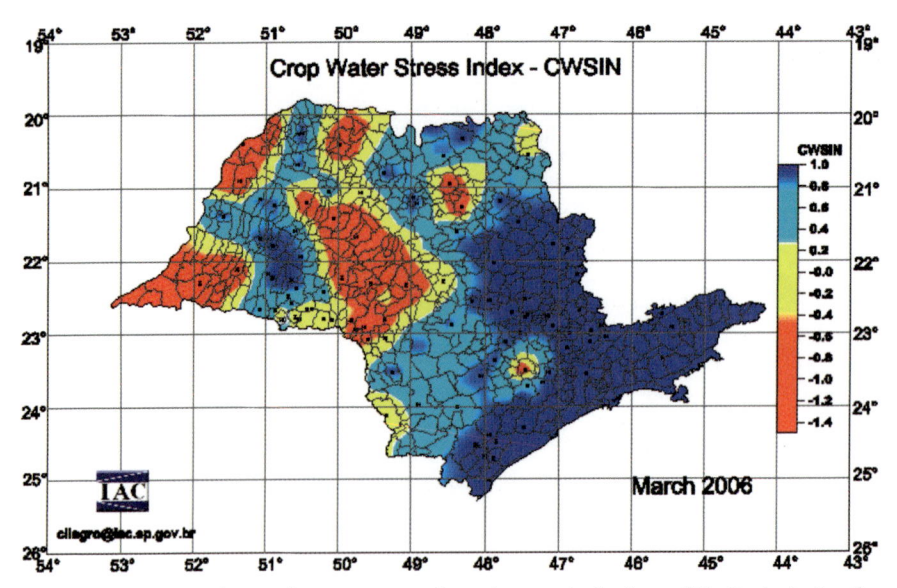

Fig. 17.19. Average condition of water stress on the maize crop in the State of São Paulo during the month of March 2006, by the Surfer system.

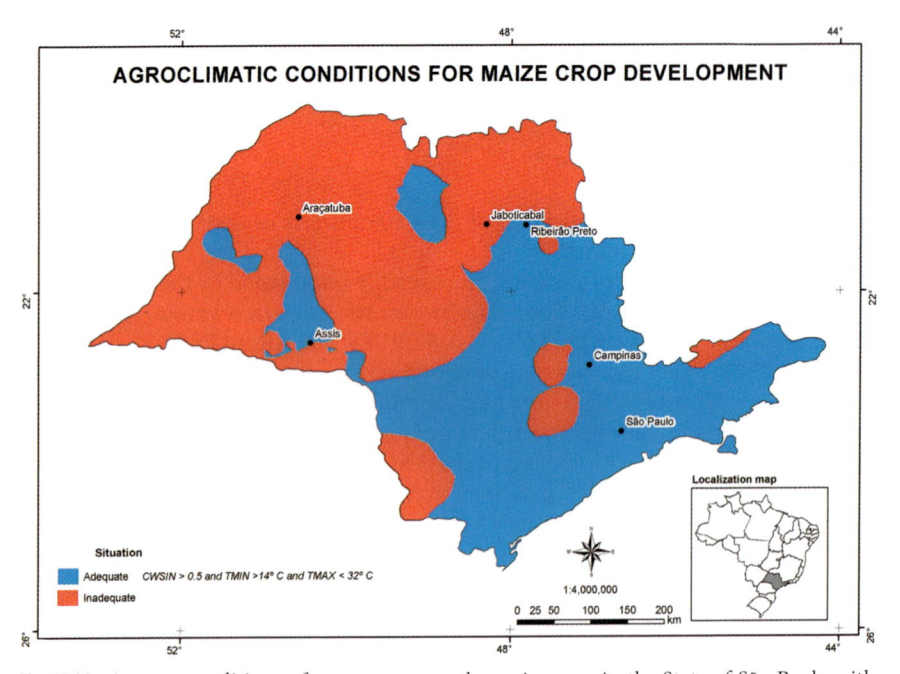

Fig. 17.20. Average conditions of water stress on the maize crop in the State of São Paulo with overlapping of the areas with minimum air temperature below 14°C, by the ARC-GIS system.

17.4 Conclusions

The assessment of the aspects presented and discussed enabled the following premises:

Drought is a constant phenomenon in agriculture in Brazil, thus requiring continuous prediction and monitoring to provide valuable mitigating measures.

As a result of the territorial extension, the mitigating measures are not necessarily identical and must take into consideration the cultural aspects of the population the climate regime, and the agricultural exploitation.

The various indices presented have proven to be adequate for monitoring and mitigating the effects of drought, nevertheless, adjustments are necessary for the use of these indices for each region and crop. For the PDSI, the parameters of the equations should be estimated for each region in Brazil.

Every state should create a Drought Monitoring and Mitigation Center, subordinated to the State's Agricultural Secretariat. It should be the responsibility of the INMET, in association with state agencies, to propose norms and to define standards and policy for the monitoring and mitigation of drought on a regional and nationwide scale.

It should be understood that the drought phenomenon cannot be assessed and interpreted by only one field of expertise, but rather by a set of specialists and institutions. Furthermore, it is extremely important that researchers and specialist in the areas of agronomy, agrometeorology, meteorology, civil defense, agro extension service, and others, should be involved in the study of the drought phenomenon.

References

Abramowitz M Stegun IA (1965). Handbook of mathematical function. Dover, p.1046.

Anunciação YMT, Fortes LTG (2006) Monitoramento da precipitação no rio grande do sul com vistas à previsão para o verão 2005/2006. Brasília: INMET, 2006. 25p. (INMET/cdp/rt-001-2006)

Assad E, Macedo MA, Zullo Junior J, Pinto HS, Brunini O (2003) Avaliação de metodos geoestatiticos na espacialização de indices agrometeorológicos para definir riscos climaticos. pesquisa agropecuária brasileira, Brasilia 38:161–171.

Blain GC (2005) Avaliação e adaptação do índice de severidade de seca de palmer (pdsi) e do índice padronizado de precipitação (spi) às condições climáticas do estado de são paulo. campinas: IAC, 2005. 120f. dissertação (mestrado em agricultura tropical e subtropical) – Instituto Agronômico, Campinas – sp, 2005.

Blain GC, Brunini O (2006) Quantificação da seca agrícola pelo índice padronizado de evapotranspiração real (iper) no estado de São Paulo. Bragantia 65:517–525.

Blain GC, Brunini O (2005) Avaliação e adaptação do índice de severidade de seca de Palmer (PDSI) e do indice padronizado de precipitação (spi) às condições climáticas do estado de São Paulo. Bragantia 64:695–705

Brunini O (2005) Centro de monitoramento e mitigação de seca e adversidades hidrometeorológicas. disponível em: www.infoseca.sp.gov.br. acesso em julho 2005.

Brunini O (2005). Infoseca. Campinas: Instituto Agronômico (fôlder)

Brunini O (1983). Tese de modelo agrometeorológico para estimativa da produtividade de cultura de soja ciclo precoce. In: Congresso Brasileiro de Agrometeorologia, 3, 1983, Campinas, SP. Anais, Campinas: SBA, 1983. pp 245-253.

Brunini O, Abramides PLG, Brunini APC, Blain GC, Pedro Junior MJ, Camargo MBP, Tremocoldi WA, Pezzopane Jr, Rolim GS (2005) O infoseca no monitoramento e mitigação de seca e adversidades hidrometeorológicas. In: Congresso Brasileiro de Agroinformática, 5, Londrina. Anais, Londrina-Paraná: Sociedade Brasileira de Agroinformática, 1:1-15.

Brunini O, Abramides PLG, Rolim GS, Brunini APC, Ernandes ES, Blain GC (2006) O infoseca no alerta e monitoramento de risco de incêndio. In: Congresso Brasileiro de Biometeorologia, 4, Ribeirão Preto, SP. Anais, Ribeirão Preto: Instituto de Zootecnia/Sociedade Brasileira de Biometeorologia, 1:98-104. CD-ROM.

Brunini O, Blain GC, Brunini APC, Santos RL, Brigante RS, Almeida EL (2002) Monitoramento das condições de seca no estado de são paulo utilizando o índice padronizado de precipitação. in: congressobrasileiro de meteorologia, Foz do Iguaçú. Proceedings, Sociedade Brasileira de Meteorologia, 1:1135-1139. CD-ROM.

Brunini O, Grohmann F, Benincasa M (1977). Variação estacional do armazenamento e das perdas de água em unidades de solo do estado de são paulo. Cientifica, Jaboticabal, V. Especial 5:119-123

Brunini O, Grohmann F, Santos JM (1981) Balanço hídrico em condições de campo para duas cultivares de arroz sob duas densidades de plantio. Revista Brasileira de Ciência do Solo, 5: 1-6

Brunini O, Pinto HS, Zullo Junior J, Barbano MT, Camargo MBP (2000) Drought and desertification and preparedness in Brazil: The example of São Paulo State. In: Wilhite D, Sivakumar MVK (eds.) Early Warning Systems for Drought Preparadness and Drought Monitoring, Genebra, Suiça: Organização Meteorológica Mundial, pp. 89-103.

Brunini O, Sawazaki E, Duarte AP, Tsunechiro A, Kanthack RAD, Blain GC, Brunini APC, Zullo J (2003) Milho safrinha: zoneamento e riscos climáticos no estado de São Paulo. In: Congresso Brasileiro de Agrometeorologia, 13, Santa Maria-RS. Anais, Sociedade Brasileira de Agrometeorologia, 1:629-630.

Brunini O, Zullo Junior J, Pinto HS, Assad E, Sawasaki E, Paterniani MEZ (2001) Riscos climáticos para a cultura do milho no estado de são paulo. Revista Brasileira de Agrometeorologia, Santa Maria, 9:519-526.

Camargo MBP, Brunini O, Miranda MAC (1986) Modelo agrometeorológico para estimativa da produtividade para cultura de soja no estado de são paulo. Bragantia 45 :231-286.

Camargo MBP, Hubbard KG (1999). Drought sensitivity indices for a sorghum crop. J Prod Agric 12:312-316

Camargo MBP, Hubbard KG, Flores-Mendoza F (1994) Test of a soil water assessment model for a sorghum crop under different irrigation treatments. Bragantia 53:95-105

Companhia nacional de abastecimentos (conab). produtividade de soja. safras de 1990/1991 a 2004/2005: séries históricas. disponível em: http://www.conab.gov.br/ download/safra/ sojaseriehist.xls). acesso em nov. de 2005.

Fortes LTG, Anunciação YMT, Lucio PS (2006) Comparação entre métodos alternativos para informar anomalias de precipitação. Brasília: INMET, Trabalho submetido ao XIV Congresso Brasileiro de Meteorologia.

Infoseca (2005) Disponível em http://www.infoseca.sp.gov.br. acesso em dez. de 2005.

Instituto Nacional de Meteorologia (2005) Oitavo distrito meteorológicxo (INMET/8 DISME). Universidade Federal de Pelotas. Centro de Pesquisas e Previsões Meteorológicas. (UFPEL. CPPMET). Prognóstico climático para o estado do rio grande do sul. Disponível em: http://www.inmet.gov.br/climatologia/cond_clima/bol_dez2005.pdf. acesso em dez. de 2005.

Instituto Nacional de Pesquisas Espaciais. Centro de Previsão de Tempo e Estudos Climáticos (INPE, CPTEC). Instituto Nacional de Meteorologia (INMET) (2005). Prognóstico climático de consenso (CPTEC/INPE e INMET para o trimestre janeiro, fevereiro e março de 2006).

Disponível em: http://www.inmet.gov.br prev_clima_tempo/prognostico_climatico_ trimestral/pc/pc0512_a.pdf.acesso em dez. de 2005.

Internacional Research Institute for Climate and Society (IRI) (2005) Global spi analyses. disponivel em: http://www.ingrid.1deo.columbia.edu/maproom/.global/.precipitation/spi. html). acesso em dez. de 2005.

Karl T (1986) The sensitivity of the palmer drought severity index and palmer's z-index to their calibration coefficients including potencial evapotranspiration. J Clim Appl Meteorol 25: 77–86.

Mckee TB, Doesken NJ, Kleist J (1993) The relationship of drought frequency and duration to time scale. In: Proceedings of the Conference on Applied Climatology, Boston: AMS, pp 179–184.

Normais Climatológicas 1961–1990 (1992). Brasília: Departamento nacional de meteorologia, 84 pP.

Palmer WC (1968) Keeping track of crop moisture conditions, nationwide: the new crop moisture index. Weatherwise 21:156–161

Palmer WC (1965) Meteorological drought. US Weather Bureau Res. No. 45, 58 pp

Santos FAS, Anjos RJ (2001) Utilização do índice de precipitação padronizada (spi) no monitoramento da seca no estado de pernambuco. In: Congresso Brasileiro de Agrometeorologia, Fortaleza, Anais, SBA/FUNCEME, pp 121–122.

Secretaria de Agricultura e Abastecimento do Estado de São Paulo (2006) Sistema de avaliação e monitoramento de riscos climáticos e agrometeorológicos. São Paulo, CD-ROM.

Thornthwaite CW, Mather JR (1955) The Water Balance. Publications in Climatology, Vol. 3, No..1, 104 pp.

CHAPTER 18

Coping Strategies with Desertification in China

Wang Shili, Ma Yuping, HouQiong, Wang Yinshun

18.1 Introduction

Desertification was defined as "land degradation in arid, semi-arid or sub-humid dry areas resulting from various factors including climatic variations and human activities" in the United Nations Convention to Combat Desertification (UNCCD). About two-thirds of the countries of the world, one-fifth of the global population and one-fourth of the land of the earth are now affected by desertification with a direct economic loss about US$ 42.3 billion every year. Desertification has become a source of poverty and a constraint to socioeconomic sustainable development. Combating desertification, ecological improvement and sustainable development is an imperative hard task for the world.

China is one of the countries with a large area, wide coverage and heavy losses from desertification and sand encroachment in the world. With rapid increase of population and driven by economic benefits in socioeconomic development, various human activities that deteriorate vegetation in sandy areas had prevailed, such as over-grazing, wasteland cultivation, excessive firewood gathering, excessive gathering of arenicolous plants and irrational use of water resources. The serious land desertification and sand encroachment have been threatening China's ecological security and sustainable socio-economic development.

Upon the entry of the twenty-first century, Chinese government raised its attention to desertification combating and incorporated desertification management oriented ecological improvement into the economic and social development plan. Consequently several significant actions were taken, including promulgation and execution of the *Desertification Combating Law*; release of the *State Council Decision About Further Accelerating the Desertification and Sadification Combating Work*, formulation of the *National Desertification and Sadification Combating Plan*. A number of key programs on sand encroachment prevention and control have been launched and implemented, such as the six major forestry programs, grassland protection and improvement program, small watershed integrated program.

In order to master the status and dynamic changes of desertification and sand encroachment land nationwide, national monitoring survey for desertification was carried out three times in 1994, 1999, and 2004, respectively. The results of the 3rd national monitoring survey for desertification and sand encroachment show that the expansion tendency of desertification and sand encroachment in China has been primarily brought under control and the 'stalemate of rehabilitation and destruction' has been realized (State Forestry Administration, 2005).

18.2
Status of Desertification in China

18.2.1
Status of Desertified Land

The land suffering desertification nationwide in 2004 was 2.6362 million km^2, taking up 27.46% of the territory, located in 498 counties of 18 provinces including Beijing, Tianjin, Hebei, Shanxi, Inner Mongolia, Liaoning, Jilin, Shandong, Henan, Hainan, Sichuan, Yunnan, Tibet, Shaanxi, Gansu, Qinghai, Ningxia and Xinjiang.

18.2.1.1
Distribution of desertification in different Climatic zones

The area of desertified land in arid, semi-arid and sub-humid arid region is 1.15 million km^2, 971,800 km^2 and 514,400 km^2, taking up 43.62%, 36.86% and 19.52% of total desertified land area respectively.

18.2.1.2
Desertification types

Wind-eroded desertification, water-eroded desertification, salinization and freeze-thawing desertification lands cover 1.8394 million km^2, 259,300 km^2, 173,800 km^2 and 363,700 km^2 respectively.

18.2.1.3
Degree of desertification

Lightly, moderately, severely and extremely severely desertified land respectively covers 631,100 km^2, 985,300 km^2, 433,400 km^2 and 586, 400 km^2.

18.2.1.4
Distribution of desertification in various provinces (Autonomous Regions)

Desertification in the country is mainly located in Xinjiang, Inner Mongolia, Tibet, Gansu, Qinghai, Shaanxi, Ningxia and Hebei provinces (autonomous regions). The area of desertified land in these 8 provinces occupies 98.45% of the total desertified land area.

18.2.2
Status of land most vulnerable to sand encroachment

Land most vulnerable to sand encroachment is a kind of degraded land between sand encroachment and non-sand encroachment due to over utilization of land or shortage of water resource. The area of land most vulnerable to sand encroachment is 318 600 km^2, accounting for 3.32% of the total country's territory, which is mainly distributed in 4 provinces (autonomous region) including Inner Mongolia, Xinjiang, Qinghai and Gansu.

18.2.3
Dynamic Changes of Desertification

Compared with 1999, the national desertified land area decreased by 37,924 km^2, representing an annual drop of 7585 km^2. In terms of different desertification types, the area of desertified land caused by wind erosion decreased by 33,673 km^2, by water erosion decreased by 5525 km^2, while that of desertified land caused by salinization increased by 930 km^2. In terms of desertification degree, the area of lightly desertified land increased by 90,700 km^2, moderately desertified land increased by 117,300 km^2, while that of severely desertified and and extremely severely desertified lands decreased by 131,700 km^2 and 114, 200 km^2 respectively. As for major provinces, the area of desertified land in 16 provinces has decreased.

18.3
Development and Causes of Desertification in North China

18.3.1
Development and Cause Analysis

18.3.1.1
Historical situation

Before the term desertification was defined, Chinese scientists were more concentrated on sandy desertification caused by wind erosion. Researches showed that in the Pleistocene about 10 thousand years ago, several kinds of severe and frequent climatic oscillations happened in China, the occurrence and reversion of desert were mainly controlled by global climate change driven by earth orbital parameters. Since the Holocene, especially recent 2000 years, due to the increase of population, innovation of productive tools, and increase of the extension and intensity in farming, human activities gradually became another important factor. However, sandy desertification was still primarily caused by climate change. In the last one hundred years, the process of sandy desertification was affected by not only climate change, but also the intensified human activities (Dong et al. 1998). For exemple in Gonghe Basin of Qinghai province in China, during the 50s to 80s of 20th century, the accumulated afforestation and conservation area was 1,700 ha, but the destructed forest area was five times this area, and the expansion area of desertification even up to 32 times this area (Dong et al. 1998).

18.3.1.2
Type of human activities

In North China, where wind erosion is the main factor affecting desertification, various human activities operated, including overgrazing (30.1%), excess reclamation (26.9%), excess firewood gathering (32.7%), irrational use of water resource (9.6%) and lack of environmental protection in building factories, mines and transportations (0.7%). In the southeast of Horqin grassland, which once was sparse forest grassland, 1,333 square kilometer ground were cultivated once, some of them even two or three times during the last one hundred years. The percentage of shifting sand area occupying total land increased from 14% in late 1950s to 32% in 1970s. In late 1980s and 1990s, it was up to 41.2% and 54% (Zhu 1998).

18.3.1.3
Key vulnerable region

The most severe area of desertification in Northern China is farm-pastoral transition zone and rainfed agriculture area with annual precipitation of 200 mm- 400 mm and annual evaporation of about 2000 to 2500 mm. Since the region is at the edge of East Asia monsoon, there is vulnerable environment and sensitivity to climate change, resulting from low rainfall, arid climate, loose soil, and frequent high wind. During early 1950s to late 1970s, as a result of excess reclamation and firewood gathering, desertification developed further including acceleration of wind erosion of sandy grasslands and increase in the area of shifting sands. Until late 20th century, it began to reverse in most area according to the monitoring results from remote sensing. An example is Xijingzi village in semi-arid area which belonged to Shangdu County in Inner Mongolia where sandy desertification area made up about 41.3% of whole region in the early 1960s, and then jumped to 57.8% in 1978. Owing to adjustments in land use and countermeasures, the percentage dropped to 22.7% in late 1980s.

The climate shifted to warm and dry in these areas. However, great changes in land use have been taken place since 1980s. So it could be found that environmental changes resulting from climate change provided a basis for desertification in farm-pastoral zone; but improper land use leads and accelerates its formation and development.(Xue et al. 2005; Li and Lu 2002). It is obvious that desertification in semi-arid and semi-humid areas in China during modern time is due to complicated interactions among climate change, human activities and desertified land under vulnerable ecosystems. (Dong et al. 1998).

18.3.2
Possible influence of climate change on desertification

Future climate change would continue to influence the development and adverse processes of desertification. On the basis of meteorological data at national level, the influences of climate change on desertification in China were estimated by HADCM2 model (Ci et al. 2002; Ci 1994). The results illustrated that if CO_2 doubled and temperature increased by 1.5°C in 2030, the desertification area would expand 184,023 km^2. In 2056, another 175,024 km^2 would be desertificated. According to estimates from statistical regression prediction model (Dong et al. 1997; Shang et al. 2001), rising temperature and decreasing precipitation would lead to further desertification expansion in the future 80 years in northern desertification land. Another research also indicated that the development process of desertification in North China would be rapid and severe during the first ten years in 21st century with climate change (Shang et al. 2001).

Moreover, recent studies made clear that during the future 10 to 50 years, temperatures might be higher by 1.9 to 2.3 °C, precipitation would increase by 19% in Northwest China (Ding 2002). It will be advantageous to control natural expansion and accelerate the adverse process of desertification. However, there are uncertainties in long-term climate change, and regional climate is also likely to change to warm and dry. Thus, the postive influence of increasing temperatures and precipitation on desertification should not be overestimated.

18.4
Desertification Monitoring in China

18.4.1
Indicator system for desertification monitoring and evaluation

In order to make clear the status and damage degree of desertification and to operate early warning, it is essential to establish a scientific indicator system for desertification monitoring and evaluation. In 1977 at the United Nation Conference on Desertification (UNCOD) held in Nairobi, a "Map of the World Distribution of arid Regions" at 1:25 million scale was prepared by FAO, UNESCO, WMO and UNDP together, and the global desertification was evaluated (Lu et al. 2000). Aimed at making desertification maps and evaluating desertification status, indicator system was paid more attention by many scholars and international organizations (Berry and Ford 1977; Reining 1978; FAO and UNEP 1984; Oldeman 1998; Marbutt 1986; Hunsaker and Carpenter 1990; Hammond 1995; Rubio and Bochet 1998; and Dregne 1999).

A series of indicators for sandy desertification evaluation was determined in China in the past (Zhu and Liu 1984). In recent ten years, more progresses were made in the studies of indicator system for desertification monitoring and evaluation (Wang 2003). According to the definition of desertification and its characteristics, an indicator system was developed consisting of four components: driving forces indicators, state indicators, impact indicators and control indicators (Wang

2003). It was pointed out that for national or regional level the first step is to develop a frame of indicator system at macroscopic level. On the basis of it, executable indicator systems could be established at different areas. Then relative data are gathered according to the indicator system to monitor and evaluate desertification (Wang 2003; Wu et al. 2005).

18.4.2
Desertification monitoring in China

18.4.2.1
National desertification surveys in the history of China

From 1994 to 1996, the State Forestry Administration (former the Ministry of Forestry) of China organized experts and technicians to investigate desert, gobi and sandy desertification land in whole country. China national maps of desertification land distribution at 1:1 million and 1:2.5 million scales were made in 1996 respectively, and China Country Paper to Combat Desertification was finished (Wang 2003).

The second nationwide desertification survey was carried out in 1999. The frame of indicator technical rules for desertification evaluation consists of 3 climate zones, six types of land use and 4 desertification land types.

The 3rd National Monitoring Survey for Desertification and Sand Encroachment was completed in 2004 under the auspices of the State Forestry Administration and with the involvement of such sectors as agriculture, water conservancy, environmental protection, meteorology and the Chinese Academy of Sciences. The focused ground survey is combined with interpretation of remote sensing data, as well as application of GIS and GPS. About 156 million units of information was obtained based on investigation in 5.02 million sites. A National Geography Information Management System for Desertification and Sand Encroachment was established.

18.4.2.2
Content and mission of China's desertification monitoring

18.4.2.2.1
Scope of monitoring
It consists of arid, semi-arid and sub-humid area, distributed generally in ten provinces belonging to North and West China and involving 270 counties. Key monitoring area was in farm-pastoral transitional zone occupying 82 counties of 9 provinces of North China.

18.4.2.2.2
Contents of monitoring
The status of desertification land distribution and dynamic macroscopic data in arid areas for the state, provinces and typical regions are requested to be provided at definded times. And maps of desertification land distribution should be pre-

pared in a timely when needed. Countermeasures and suggestions for desertification combating are put forward based on surveys and analysis.

18.4.2.2.3
Monitoring classification system
It consists of determination of land use type, desertification type and degree.

18.4.2.3
China's Desertification monitoring system

There are three levels of desertification monitoring.

18.4.2.3.1
The China National Desertification Monitoring Centre
Its main task is to provide timely data (number, figure and image) to central government in making strategic policy and specific measurements for desertification combating and land protection, to provide dynamic decision-making and consultation on some key development regions and sensitive eco-regions.

18.4.2.3.2
Provincial Sub-center
It is in charge of province monitoring.

18.4.2.3.3
Desertification monitoring station
Sequential investigations and records of desertification area, type and degree are operated in the stations with representative zone, and then reported to sub-center and national center.

18.4.2.4
Monitoring techniques and construction of information management system

18.4.2.4.1
Main techniques of macroscopic desertification monitoring
The ground observations are taken at fixed sample lands at appointed times. The sampling method, combined with ground investigation and 3S (remote sensing, GIS,GPS) techniques, is used to estimate desertification land area. Systematic sampling and percentage sampling method are used in estimation of desertification land area status and dynamic macroscopic monitoring. And aerial photograph and remote sensing monitoring techniques are more and more applied to large area desertification monitoring.

18.4.2.4.2
Construction of monitoring information management system

A uniform information management system was established. It consists of database and application analysis models. There are various applied models such as of dynamic prediction, environment evaluation and desertification loss evaluation. The desertification monitoring information system for nationwide and provinces were built respectively.

18.5
China's Key Forestry Programs on Combating Desertification

The essential strategy in combating desertification in China is to control structure and function of agro-forest complex ecosystem, rationally use water and land resources, so as to promote a virtuous cycle within ecosystem. The emphasis is put on prevention and control of further desertification extension, especially for vulnerable eco-regions. At the national level, a number of key programs on sand encroachment prevention and control have been launched and implemented, such as the six major forestry programs, grassland protection and improvement program and small watershed integrated program (State Forestry Administration 2005).

18.5.1
Program for converting cropland to forest/shrubbery

18.5.1.1
Background

The first objective is to halt cultivation in the area with severe soil and water loss, desertification, salinization and Karst rocky desertification, or low and unsteady yields, and the second is to seed up tree and grass under specific local conditions and restore vegetation.

18.5.1.2
Scope of Program Construction

In this program, 25 provinces and Xinjiang Production and Construction Group were involved covering 1897 counties. The emphasis was put on West and Middle China. And the important arable land with high ecological function in riverhead region, steep slope land surrounding lakes, severe soil and water loss land, and severe windstorm areas are given priority. In all, 856 counties were determined to be key counties for program construction, occupying 29.9% of total national counties, and 45.1% of total counties involved in the program region.

18.5.1.3
Objectives and Assignment of Program Construction

The objectives and assignments by 2010 are fixed to achieve 14.7 million ha afforestation from arable land and 17.3 million ha afforestation from barren hills, to finish converting cropland to forest in slope land, and increase vegetation fraction by 4.5 percent.

18.5.2
Programme of Combating desertification in the wind sand sources areas affecting Beijing and Tianjin city

18.5.2.1
Background

In order to mitigate the damage of wind and sand storm and build an ecological defense for North China, central government made decision to implement a programme of combating desertification in the wind sand sources areas affecting Beijing and Tianjin.

18.5.2.2
Scope of Program Construction

This project region involves five provinces of Beijing, Tianjin, Hebei, Shanxi and Inner Mongolia with 75 counties in total. The whole area of this project is 458 thousand km^2, in which the area of sand encroachment is 101.2 thousand km^2.

18.5.2.3
Countermeasures and objectives

This project took comprehensive control measures mainly on forest and grass vegetation restore. It included conversion of cropland to forest, afforestation, agriculturalmeasure includeing manual grass planting, aerial seeding, enclosure, grassland construction, seed base, grazing prohibition, warm shed building, as well as hydrlogical measure, such as riverhead project, water-saving irrigation project, small watershed comprehensive treatment. There will be eco–immigration with 180 thousand people.

18.5.3
Three-North Shelterbelt Programme and Shelterbelt Programme in upper and middle reaches of the Yangtze River

18.5.3.1
Three North Shelterbelt Program

18.5.3.1.1
Background
In order to change the status of wind-sand damage and soil and water loss in Northwest, North and Northeast China, Three North Shelterbelt Program started to be implemented in November 1978.

18.5.3.1.2
Scope of program construction
This program covers 551 counties in northeast, north and northwest China, with a total area of 4.069 million km^2, i.e., 42.4% of China's total land area.

18.5.3.1.3
Construction contents and size
The program started in 1978, and will completed in 2050 with a projected life span of 73 years, including three phrases and eight period projects. A total of 35.08 million ha of plantation is expected to be established under the program. When the program is completed, forest fraction in Three North area would increase from 5.05% to 14.95%, and sand-storm damage and soil-water loss would be controlled effectively. The fourth period started in 2001.

18.5.3.2
Five Shelterbelt Programs for Upper and Middle Reaches of Yangtze River and other areas

Five shelterbelt programs started successively. It contains Shelterbelt Programs for Upper and Middle Reaches of Yangtze River, Coastal Shelterbelt Program, Ecological virescence project of plain, Ecological virescence project of Taihang Mountain, and Shelterbelt Program for Pearl River.

18.6
Practical Strategies and Countermeasures to Combat Desertification

The strategies and countermeasures to combat desertification are different in various regions due to variations in natural conditions. The people in sandy areas obtained large numbers of experiences in combating desertification during a long period of time, and some of them are successful in preventing, alleviating, and reversing.

The major measures involve stabilizing sands, afforestation, small watershed management, rational use of water resources, and agro-forest ecological system (Combating Desertification Management Center 2001; Wang 2003).

18.6.1
Stabilizing sands techniques system

The aim of stabilizing sands is to transform shifting sandy land into semi-fixed sandy land or fixed sandy land so as to establish stable sandy ecosystem.

18.6.1.1
Biological stabilizing sands techniques

Compared with other stabilizing sands techniques biological techniques have more benefits: (1) lower cost and stable function; (2) improvement of soil properties; (3) improvement of ecosystem of sandy areas; (4) stabilizing sands plants may be used as feed, fuel, wood, fertilizer etc. Conventional biological stabilizing sands techniques are difficult to practice, which include man-made forest/shrubbery, aerial seeding, so some operational and easy techniques are adopted in China.

A sands enclosure to restore natural vegetation is a typical and major measure of biological stabilizing sands technique. It refers to fencing in vegetation-destroyed land in arid and semi-arid areas so as to prevent human activities and animal use, and gradually restore natural vegetation. The ways of enclosure are fully enclosure, half- enclosure and alternate enclosure. Full enclosure means to forbid all the hu-

Fig. 18.1. Sands enclosure in pasture

man activities destroying plants growth. The duration of enclosure is 3-5 years, sometimes 8-10 years. Half- enclosure is implemented mainly during the season. Alternate enclosure refers to be implemented around the divided zones in turn. Sands enclosure is proving to be the low cost alternative and highly efficient (Fig. 18.1).

18.6.1.2
Engineering stabilizing sands techniques (stabilizing sands with sands barriers)

Sand barriers, made by straw, brushwood or branch, are used to control the direction and speed of wind sand so as to prevent deposition of sand carried by wind. Sands barriers are also the precondition and necessary condition for biological stabilizing sand (Fig. 18.2). There are vertical and horizontal sand barriers. The vertical sand barriers are 50-100 cm high above sand surface, horizontal barriers are 20-50cm.

Height of vertical sand barriers: The sand particles move at the lower layer of ground (bellow 10 cm), so the height of sand barriers is normally 15-20 cm, and sometimes 30-40 cm.

Lacuna of sand barrier: The accumulated sand scope in front of barrier is 2-3 times of height of sand barriers, with 25% lacuna. The smaller the lacuna, the shorter is the reach distance of the sand. Normally 25%–50% is adopted.

Shape of sand barrier: The benefit of preventing and stabilizing sands is determined by shape of sand barrier. The sand barrier should be vertical to dominant wind direction with a check- board of 1m × 1m.

Fig. 18.2. Sands barriers.

18.6.1.3
Chemical stabilizing sands techniques

Some chemical materials are sprinkled on the sandy surface, forming a concretion layer so as to fix the sands. The materials include oil-products, macromolecule polymer etc.

18.6.2
Shelterbelt techniques system

18.6.2.1
Shelterbelt system in oasis

Shelterbelt system in oasis consists of stabilizing sands belt with sands enclosure, shelterbelt on the edge of oasis and cropland shelterbelt inside of the oasis (Fig. 18.3). In order to prevent springing up wind and head off shifting sand, there is a need to build a wide shrub shelterbelt with a width of 200 m at least outside of oasis. As the second defense line a shelterbelt on the edge of oasis is located between shrub stabilizing sands belt and cropland. Cropland shelterbelt inside of the oasis plays the important role in adjusting microclimate on cropland, sustaining oasis and improving agricultural production.

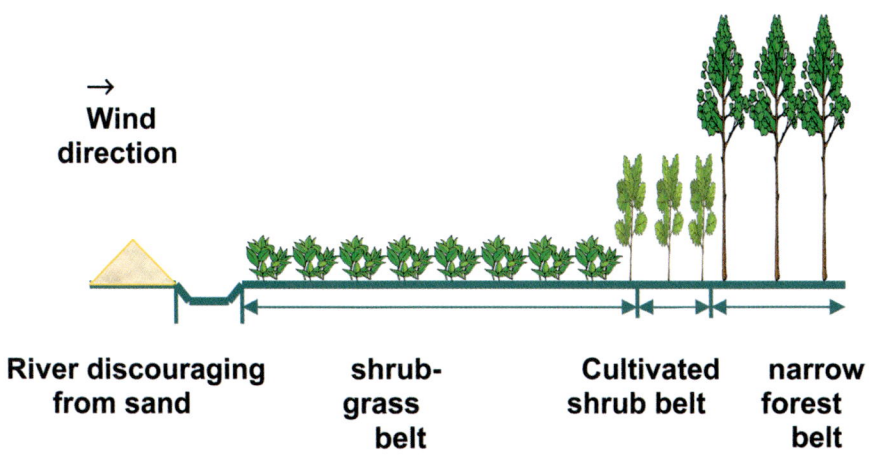

Fig. 18.3. Shelterbelt system in oasis. (Source: Combating Desertification Management Center of State Forestry Administration)

18.6.2.2
Shelterbelt system in pasture of sandy land

It is meaningful in combating desertification to develop shelterbelt in pasture with annual precipitation of 350-400 mm. The main shelterbelt should be vertical to dominant wind direction and accessory shelterbelt is orthogonal with main shelterbelt. The density of forest network has a direct effects benefit on combating wind sands.

18.6.2.3
Shelterbelt for railway in sandy land areas

Building shelterbelt for railway in desertification and semi-desertification is the most difficult projection. Shapotou railway Shelterbelt in Ningxia of China is a successful example of combating desertification (discussed details bellow).

18.6.3
Typical models in combating desertification in China

18.6.3.1
Typical models for combating desertification in oasis of arid areas– Shelterbelt system in oasis in Hetian, XinJiang Autonomous Regions

18.6.3.1.1
Natural conditions
Hetian county is located north of the Kunlun Mountain, southwest of the Taklimakan Desert with an annual precipitation of 34.8 mm and an annual evaporation of 2564 mm. The Taklimakan Desert is approaching Hetian county at a speed of 3-5 m per year.

18.6.3.1.2
Strategies for combating desertification
Sands enclosure, protecting and constructing shelterbelt were implemented by making use of local fine water condition. By selecting appropriate economic trees and fruits trees, an ecological –economic shelterbelt was build up.

18.6.3.1.3
Practical countermeasures
Sands enclosure: To protect and restore natural desert vegetation such as *Populus enphratica* forest and *Tamarix ramosissima* so as to maintain and expand oasis.

Shelterbelt system in oasis: To build shelterbelt system with tree-shrub-grass and multi-trees varieties in the edge of oasis and desert. The length of main forest belt is 600 km while a desert vegetation protection belt with 10-15 km built outside main forest belt.

Cropland shelterbelt network: To build a cropland shelterbelt network with narrow forest belt and small network in oasis, such as economic trees or grape aisle.

18.6.3.2
Typical model for combating desertification for railway in arid areas – Shapotou, Ningxia railway sands stabilization

18.6.3.2.1
Natural conditions
Shapotou District, Zhongning Couty, Ningxia is located in south-east of Tengger Desert and bordered upon irrigated Plain in Zhongwei. There are lots of high shifting dunes threatening the Bao-lan railway. Mean annual precipitation is 186.2 mm with ground water at 80 m. The vegetation fraction is less than 1%.

18.6.3.2.2
Strategies for combating desertification
In order to prevent the middle segment of Bao-lan railway from the invasion of forwarding sand dunes, a forefront stopping sands belt along with railway was built.

18.6.3.2.3
Practical countermeasures
The shelterbelt system consists of five belts outward of the railway with the length 300 meters of windward and 200 meters of leeward and a straw barrier belt is the core part.

Sands stabilization and fireproof belt: Stones, loess or slag are laid along with railway roadbed, 20 m windward and 10 m leeward.

Afforestation belt with irrigation: The aim is to reclaim terraces for farming and build irrigation channel for afforestation in the scope of 60 m of windward outside of the first belt and 40 m of leeward outside. The trees are planted mixed with shrub.

Straw barrier belt: In the scope of 240 m of windward outside and 160 m of leeward outside, straw check-board barriers of 1m × 1m are set up to stabilize sands in autumn, then shrub seeds are planted in the center in the season with more rainfall and application of irrigation. With the benefits of vegetation shelterbelt in a long time the soil macrobiotic crust gradually formed on the sands surface and become thick, which adjusted soil water and nutrition, and consequently with some vegetation survival (Bai et al. 2003; Long and Li 2003).

Forefront stopping sands belt: The vertical sand barriers made from wattle are set up at the top of dunes, and the barriers with height of 1 m are buried 30 cm underground.

Sands enclosure belt: The windward slope of dunes in forefront stopping sands belt, is enclosed with sand barriers and planting shrubs so as to promote growth and reproduction of natural vegetation.

The combating techniques in Shapotou received an award of special prize in Science and Technology in China, and also by UNEP and UNDP.

18.6.3.3
Typical model for combating desertification in farm-pastoral transition zone in semi-arid region

18.6.3.3.1
Model 1: Integrated measures in sand area in Yulin county Shaanxi Province

18.6.3.3.1.1
Natural conditions

The Yulin Sand Area (Shaanxi Province) is located in south of Ordos Plateau, south-east of Mu Us Sand Land. The annual precipitation is 316-450 mm and annual evaporation is 2092-2506 mm, with a mean annual wind speed of 2.0-2.5 m/s. The high wind days total is 14-33 on the average, with maximum of 77 days. The annual dust days totaled 81 with dust storm days of 11.5 per year.

18.6.3.3.1.2
Strategies of combating desertification

Firstly the shifting sandy land is stabilized and followed by adjusting the structure and ways of land use.

18.6.3.3.1.3
Practical countermeasures

To build shelterbelt system with different structures: The shelterbelt systems with different structures are built according to topography, water resources and features of sandy plants.

To develop integrated exploitation models: Agricultural production is developed giving priority to farm land with irrigation. Integrated models with agro-forest system are carried out including fruits trees, medicinal materials, economic crops, trees nursery, and animal husbandry.

18.6.3.3.2
Model 2: Converting cropland to forest shrub land in the north of Yinshan Mountain in Inner Mongolia

18.6.3.3.2.1
Natural conditions

It is located in the north Yinshan Mountain of Inner Mongolia with an annual precipitation of 220–350 mm, most of which was received during July to September. Average wind speed is 4-6 m/s and there are 50-80 days with wind force scale beyond 8.

18.6.3.3.2.2
Strategies of combating desertification
The converting cropland to forest and shrub is mostly emphasized.

18.6.3.3.2.3
Practical countermeasures
Bio-measures are given priority: The wide shelterbelt is modified to narrow shelterbelt. Shrubs are dominant with tree-shrub-grass combination.

Hills as unit for rehabilitation: The top of upland hills with severe wind erosion and sloping cropping field with gradient greater than 15° are converted to grass and shrubs. Sloping cropping fields along with slope at the middle and bottom of the hills are changed as contour ploughing so as to control water erosion. The bottomland is improved as cropland with high productivity.

18.6.3.3.3
Model 3: Integrated ecological-economic exploitation in family courtyard in Wongniute county in Inner Mongolia

18.6.3.3.3.1
Natural conditions
Wongniute county is located in middle of Chifeng city in Inner Mongolia, west of Horqin Sandy land with a mean annual precipitation of 340 mm, mean annual evaporation of 2233.7 mm, and an average annual wind speed of 4m/s and 73.9 days of wind force scale beyond 8.

18.6.3.3.3.2
Strategies of combating desertification
An integrated exploitation model in family courtyard was adopted. It consists of utilizing solar energy, developing biogas, growing vegetables in plastic green house and feeding pigs in heating shed.

18.6.3.3.3.3
Practical countermeasures
The key technique is to transform solar energy to heat in green houses and shed for vegetable, grape tree and animal husbandry in winter, and also to use biogas produced from dejecta of pigs to heat and fuel for family.

18.6.3.3.4
Model 4: Combating desertification and sandification in Naiman county of Inner Mongolia

18.6.3.3.4.1
Natural conditions
It is located in central of Horqin sand land interlaced by dunes and interdunes. The desertified soil acreage increased from 20% in 1950s to 77.6 % in beginning of 1990s.

18.6.3.3.4.2
Strategies of combating desertification
The basic principle is to modify current land use status with unitary agriculture, to implement the sand enclosure, and set up plant barriers to stabilize sands in the surface of dunes.

18.6.3.3.4.3
Practical measures
In undulating sandy lands, converting farmland to forest/shrubbery is given a priority, and shelterbelt forest network is set up. As for the sand encroached area with interdunes, different countermeasures are taken according to local natural conditions.

18.6.3.4
Typical model for combating desertification in steppe in semi-arid region – Fenced grassland development in Erdos of Inner Mongolia

18.6.3.4.1
Natural conditions
In Erdos, the mean annual precipitation is 266-412 mm, with an average annual wind speed of 3.5 m/s, and sand storm days totalling 11–24.

18.6.3.4.2
Strategies of combating desertification
"Ku lun" refers to fenced grassland in Mongolia language. It is the basis of pasture development, and its establishment is focused on "water, grass, forest, feed, and machine."

18.6.3.4.3
Operational countermeasures
The combination of tree-shrub-grass is adopted to control sand damage. The objective is to promote integrated development with agriculture-forestry-animal husbandry.

18.7
Services for combating desertification in Chinese Meteorological Offices

18.7.1
Research on desertification development and combating in terms of meteorological conditions

A large number of observations and research on the development and causes of desertification as well as climatic effects of combating desertification measures were carried out in China Meteorological Administration (Lu et al. 2006; Bai et al. 2006;

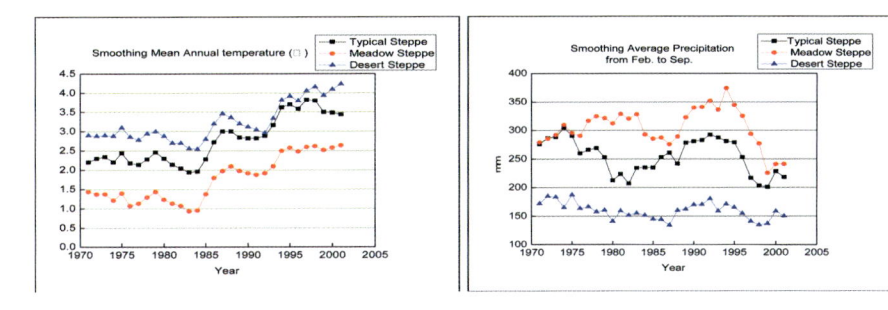

Fig. 18.4. Changes in average annual temperature and precipitation Apr. to Sep.

Li and Lu 2002; Zhou et al. 2002). A case study on causes of desertification in Xilinguole Steppe made by Xilinhaote Institute of Animal Husbandry-Meteorology is demonstrated in paper.

18.7.1.1
Effects of climate change on desertification in Xilinguole Steppe

Figure 18.4 shows the smoothing average of annual temperature and precipitation in Apr.-Sep. in meadow steppe, typical steppe and desert steppe. It is clear taht the temperature increases and precipitation falls during growing period of pasturage.

According to calculations, the decrease of every 1 mm precipitation would reduce grass yield by 100-200 kg/ha. So drought results in sand encroachment and degradation.

18.7.1.2
Effects of man-made interference on wind erosion and soil degradation in Xilinguole Steppe

The effects of man-made interference on capacity of anti-wind erosion in typical steppe was carried out by means of wind tunnel and taking soil under original status (Xu et al. 2005). Several man-made interferences includes free grazing, fenced grassland in 1987 and 2003, and cropland. The results indicate that when the wind speed is lower than 12 m/s the wind erosion increases slowly, while it rapidly increases at wind speed of 12-16 m/s. It also can be seen that at any wind speed the wind erosion rate in free grazing land is higher than fenced grassland, and wind erosion rate in earlier fenced grassland (1987) is lower (Fig. 18.5). It is proved that fencing is an effective measure to control degradation of typical steppe.

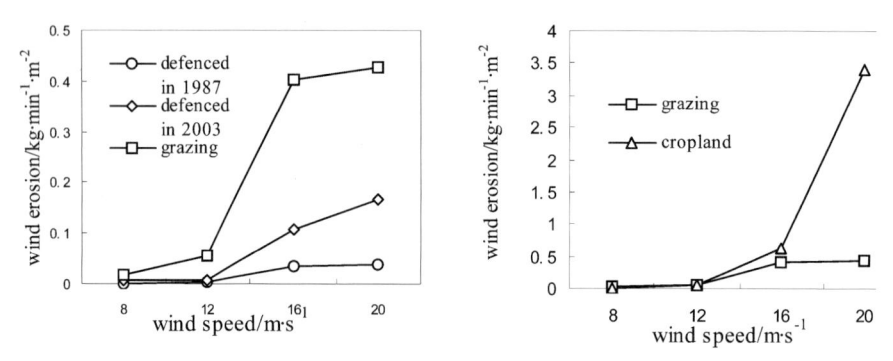

Fig. 18.5. Wind erosion rate under different wind speed in different steppes. (Source Yu et al.)

18.7.1.3
Effects of man-made interference on biomass in Xilinguole Steppe

The biomass above the ground was measured in free grazing land, fenced grassland (in 1987, 1997, and 2003) and cropland respectively in April, July, August and October. The results show that the order of biomass value is: fenced grassland in 1987> fenced grassland in 1997> fenced grassland in 2003>free grazing>cropland. It indicates that reclaimation and grazing might reduce the biomass of ecosystem, while fencing is in favor of increasing biomass.

18.7.2
Monitoring and assessing services to combating desertification of grassland

18.7.2.1
Estimation of carrying capacity in animals in grassland

In Xilinhaote Steppe feed of animals is mainly from reaped hay in autumn and remainder on grassland in winter, therefore the carrying capacity for animals in cold season depends on growth status of natural grass and snow recovery on grassland in winter. A model to calculate carrying capacity for animals in cold season was developed (Yang et al. 2001) as follows:

$$M = \sum_{i=1}^{n} \frac{Y_{ic}}{e \cdot t} + \sum_{i=1}^{n} \frac{(g_i \cdot Y_{ci} + Y_{di}) a_i b_i}{e \cdot t} (1 - \frac{f(L_i)}{P_i f(h_i)})$$

The rational carrying capacity for animals in cold season in 1996 and 1997 were calculated and compared with actual values (Table 18.1). It can be seen there are imbalances in carrying capacity with excess or shortage. This information is useful for local government to rationally use resources in grassland and alleviate desertification.

Table 18.1. Comparison between rational and actual carrying capacity for animals in cool season in Xilinguole (sheep unit/km^2)

steppe Type	Harvest year		Rational values		Actual values		Excess −) shortage (+)%	
	1996	1997	1996	1997	1996	1997	1996	1997
desert	bumper	poor	42	28	50	61	−19	−118
typical	bumper	bumper	72	50	107	83	−49	−66
meadow	bumper	bumper	154	113	76	101	+51	+11

Source: Yang et al.

18.7.2.2
Monitoring and forecasting system of animal husbandry in steppe

In order to provide meteorological information services for farmers and government a monitoring and forecasting system of animal husbandry for northern steppes in China was established (Wei et al. 2005). The system includes 5 sub-models: data base, climatic monitoring and diagnosing sub-model, forecasting sub-model and decision-making as well as service sub-model. In climatic monitoring and diagnosing sub-model the effects of climate conditions on growth, development and yield formation of grass are assessed. The forecasting sub-model includes functions of predicting date of green grass return and date of enough grass for animals, prediction of grass output, death rate of animals in spring and carrying capacity for animals in cold season.

18.7.2.3
Meteorological information services to animal husbandry of steppe

The animal husbandry-meteorology station and forest- meteorology station in the agrometeorological station network in China Meterological Administration (CMA) play a role in combating desertification and restoring ecosystem. For example, the Xilinhaote Institute of Animal Husbandry-Meteorology provides the information services of monitoring and forecasting of animal husbandry and consultation services in ecosystem restoration and combating desertification. Various bulletins are reported for the government and the department of ecosystem protecting and combating desertification, which include climate review, effects of climate on ecological environment, outlook of climate-environment, and suggestions on ecosystem protecting and combating desertification. The forecasting of date of green grass return and date of enough grass for animals is adopted by government as the basis for determining beginning and ending dates of seasonal sand enclosure, and the estimation of carrying capacity for animals in cold season is used to determine proper number of full-grown animals.

18.7.3
Monitoring and predicting of dust storms in China

The occurrence and development of dust storm are related to ecosystem and general circulation, so the monitoring and predicting of dust storm is not only a part of preventing from dust storm, but also a service task of NMHSs to combating desertification. CMA set up monitoring network of sand storm in China, and developed numerical forecast system of sand storm.

The sand storm monitoring network in China includes:

18.7.3.1
Observation network of CMA

It consists of 2456 meteorological observation stations. Apart from conventional observation data in CMA, some data of adjacent countries is made available by WMO data exchange are used in dust storm monitoring.

18.7.3.2
Sand storm monitoring network in CMA

There are 24 observation stations located in the areas suffered by sand storms in northern region of China, important data such as PM10 (particle matter with diameter less than 10 μm), visibility measured by instruments and dust fallout are obtained in real time.

18.7.3.3
Sand storm monitoring network in China Environment Protect Administration

There are 45 observation stations located in the 11 provinces in northern region of China. PM10 and TSP (total suspended particle) are measured.

18.7.3.4
Monitoring using meteorological satellite

The data from meteorological satellite -China Fengyun 2C are retrieved to obtain information on the distribution and extension of east Asia sand storms.

18.7.3.2
Present operational services of Monitoring and numerical predicting of dust storms in China

18.7.3.2.1
China dust storm website and information services

China dust storm website (http://www.duststorm.com.cn) issues monitoring and forecasting information on dust storms provides knowledge and research progress in dust storm to public. The contents of website are dust storm monitoring, dust storm forecast, dust storm assessment, dust storm yearbook, yearbook knowledge as well as dust storm research results.

18.7.3.2.2
Numerical forecasting of dust storms

Center for Atmosphere Watch and Services (CAWAS) in CMA developed an advanced numerical forecast system of dust storms, which describe the concentration distribution of sand dust in Asia, and has been put into operation in CMA. It predicted well the dust storm weather process in 2004, 2005 and 2006. The numerical forecast information is issued on the official website of CMA (http://www.cma.gov.cn) and website of WWRP, International Sand storm Research Programme.

18.8
Conclusions and Discussion

Desertification has been threatening human survival and society development. It is the result of interaction between irrational human activities and vulnerable environment especially atrocious weather and extreme climate. The occurrence of desertification is accompanied with extreme climate events, such as drought, high winds and sand storms etc.

Much progresses in combating desertification were made around the world. However we are still faced with serious challenges in coping with desertification. The global climate warming and frequent and severe droughts are existing facts, and future climate change will continue to influence desertification. At the same time there are still various human-driving factors leading to deteriorating vegetation in sandy areas and extension of sand encroachment. Combating desertification is a big problem with a long time horizon and large scale, which requires more dense monitoring networks making observations more frequently. The research on relationship between climate and occurrence, development of desertification as well as combating countermeasures should be carried out furthermore. And it will take a long time to rehabilitate the plant community to a stable status and to restore ecological system of sandy lands.

There are lots of opportunities to cope with desertification. In view of characteristics of desertification, some structural measures are needed. It is important to promote the measures of prohibiting reclamation, grazing and firewood collection. Successful practical technologies and models for desertification prevention and control should be popularized and applied. The principle contains conforming to natural laws, and adopting biological, agronomical and engineering measures and combining artificial rehabilitation with human-promoted natural restoration according to local conditions. In addition, techniques for training and demonstration are useful to make farmers master and apply countermeasure to cope with desertification.

Non-structural measures could be taken by NMHSs. The desertification monitoring in terms of meteorological conditions in desertified region should be strengthened. Moreover, further research should be carried out to clarify the relationship between climate and desertification, and combating countermeasures. The agrometeorological information services are useful for governments and farmers to prioritize and adopt countermeasures in combating desertification.

Acknowledgements

This research was funded by China Natural Sciences Foundation Committee 40675071, Study on Assessment of Climate and Agro-ecological Effect for Converting Farmland into Forest or Grass land Program in Shan-Gan-Ning Region nd by China Ministry of Science and Technology (2006BAD04B01). We would like to thank the colleagues in Inner Mongolia Meteorological Bureau, Ningxia Meteorological Bureau and Combating Desertification Management Center of State Forestry Administration for help in investigation, thank Dr. Zhang Li for help in editing the paper

References

Bai ML, Hao Q (2006) Impact of clamtic variation on ecological environmental evolution in Hunshandake Sand land, J Des Res 26:484-488 (in Chinese with English abstract)

Bai XL, Wang Y, Xu J (2003) Characteristics of reproduction and growth of mosses in the soil crust of fixed dunes in Shapotou area, J Des Res 23:171-176 (in Chinese with English abstract)

Berry L, Ford RB (1977) Recommendations for a system to monitor critical indicators in areas as prone to desertification. In: Worcester M. Program for international development. Clark University.

Ci LJ, Yang XH, Chen ZX (2002) Potential impact of climatic change and human activities on desertification in China, Earth Sci Front 9:287-294

Ci LJ (1994) The impact of global change on desertification in China. J Nat Res 9:289-303

Combating Desertification Management Center (2001) Operational countermeasures and models for combating desertification in China, China Environment Science Press, Beijing, pp 222 (in Chinese)

Ding YH (2002) Prediction of environmental variety in West China. Beijing: Science press, pp 17-43

Dong GR, Jin HL, Chen HZ, Zhang CL (1998) Geneses of desertification in semiarid and subhumid regions of northern China, Quarter Sci 2:136-144

Dong GR, Shang KZ, Wang SG (1997) The possible development tendency of contemporary natural desertification processes in Northern China, In: Studies on climate change and influences in China. Beijing: Meteorological Press. pp 416-425.

Dregne HE (1999) Desertification assessment and control. In: New Technologies to Combat Desertification, Proceedings of the International Symposium in Iran in 1998, The United Nations University, pp 95-102.
FAO and UNEP (1984) Provisional methodology for assessment and mapping of desertification. Rome: FAO
Hammond A (1995) Environmental indicators. Washington: World Resources Institute.
Hunsaker CT, Carpenter DE (1990) Ecological indicators for the environmental monitoring and assessment program. US EPA Office Research and Development, Research Triangle Park, NC.
Li DL, Lu LZ (2002) Climate characters and evolution of agricultural and pasturing interlaced zones in China. J Des Res 22:483-488
Long LQ, Li XR (2003) Effects of soil macrobiotic crust on seedling survival and seedling growth of two annual pants, J Des Res 23:657-660 (in Chinese with English abstract)
Lu JT, Zheng XJ, Li XL (2006) Climatic changes of the desertified regions in China over the past five decades, In: Dynamics of desertification and sand encroachment in China. Beijing: China Agricultural Press. pp 20-49.
Lu Q, Yang YL, Jia JD (2000) China's desertification. Beijing: Kaiming Press. pp 21-28
Marbutt JA (1986) Desertification indicators. Clim Chang 9:113-122.
Oldeman LR (1988) Guidelines for general assessment to the status of human induced soil degradation. Wageningen: ISRIC.
Reining PH (1978) Handbook on desertification indicators. Washington: AAAS Publication Number 78-7
Rubio JL, Bochet E (1998) Desertification indicators as diagnosis criteria for desertification risk assessment in Europe. J Arid Env 39:113-120.
Shang KZ, Dong GR, Wang SG (2001) Response of Climatic Change in North China Deserted Region to the Warming of the Earth. J Des Res 21:387-392
State Forestry Administration (2005) A bulletin of Status Quo of Desertification and Sand Encroachment in China.
Wang CH (2003) Climate change and desertification. Beijing: China Meteorological Press. pp 206 (in Chinese)
Wei YR, Hao L, He JJ (2005) Development and application of monitoring and forecasting system of animal husbandry on grassland in North China. Pratac Sci 22:59-65
Wu B, Su ZZ, Yang XH (2005) A frame work of indicator system for desertification monitoring and evaluation. For Res 18:490-496 (in Chinese with English abstract)
Xu ZQ, Li WH, Min QW (2005) Experimental research on the anti-wind erosion of typical grasslands. Env Sci 26:164-168
Xue X, Wang T, Wu W, Sun QW, Zhao CY (2005) Desertification development and its cause of agro-pastoral mixed regions in North China, J Des Res 25:320-328
Yang YW, Wang Y, He JJ (2001) Research of Establishing the Carrying Capacity Model on the Basis of Remote Sensing (RS) Information in Cool season Grassland. Chin J Agromet 22:39-42
Zhou XD, Zhu QJ, Sun ZP (2002) Preliminary study on regionalization desertification climate in Region. J Nat Disast 11:125-131 (in Chinese with English abstract)
Zhu ZD, Liu S (1984) The Concept of Desertification and the Differentiation of Its Development. J Des Res 4:2-8 (in Chinese with English abstract)
Zhu ZD (1998) Concept cause and control of desertification in China. Quart Sci 5:146-155

CHAPTER 19

Coping strategies with agrometeorological risks and uncertainties for water erosion, runoff and soil loss

P.C. Doraiswamy, E.R. Hunt, Jr., V.R.K. Murthy

19.1
Introduction

The pressure of increasing world population demands for higher crops yields from the finite area of productive agricultural lands. Meeting the needs especially in developing countries through more intensive use of existing agricultural lands and expansion into more marginal lands will substantially increase erosion. There is an urgent need to take preventive and control measures to mitigate the threat to global food security. These concerns are supported by a report by El Swaify (1994) that the annual rates of soil erosion can often range between 20 to over 100 t ha^{-1}, which results in about 15-30 per cent annual decline in the soil productivity. An estimated loss of about 6 million ha annually is estimated as a result of degradation by erosion and other causes (Pimental et al. 1993). The data for these estimates are often selective from small scale studies conducted over short time periods, however, this does draws attention to the increasing problem of soil loss. In the western world the loss in productivity from erosion may be masked or compensated by increased costly and efficient management practices such as improved crop varieties, fertilizer, pesticides, and irrigation. Even under these management practices, soil erosion has continued and sediment loss has become a very costly factor in the overall picture.

The processes of water erosion are closely linked to the pathways taken by water in its movement through vegetation cover and over the ground surface. During a rainstorm, part of the water falls directly on the soil, because there is no vegetation or because it passes through gaps in the plant canopy. The rain falling directly on the soil surface can potentially produce rain splash erosion. Rain intercepted by the vegetation may either evaporate or drip down the plant to the soil surface. The rain and intercepted water reaching the soil may infiltrate contributing to the soil moisture storage. However, when the soil is either saturated with water or surface conditions prevent infiltration, the excess contributes to runoff on the surface, resulting in erosion by surface flow causing rills and gullies. The infiltration rates are controlled by the soil characteristics such as the water holding capacity and hydraulic conductivity, surface dryness, the rate of rainfall, land slope and soil management practices. Higher rates of runoff from eroded surfaces wastes valuable moisture—the principal factor limiting productivity in arid lands.

The agrometeorological coping strategies would first require the study and evaluation of the causes of water erosion, runoff and soil loss for the area of interest. The erosion process begins at the local level, but requires evaluation at the landscape and regional levels to adapt effective measures to minimize and control the erosion and soil loss. Soil runoff models are used to study the temporal and spatial extent of erosion and soil loss. The EPIC model (Williams 1995) is one example used extensively in the U.S. and internationally to study the process and extent of erosion at local and watershed levels. Data from satellite remote sensing can help in defining the channel flow of surface water based on the digital elevations imagery maps and also other surface characteristics such as landuse and surface roughness.

Soil erosion is a three stage process: (1) detachment, (2) transport, and (3) deposition of soil. Different energy source agents determine different types of erosion. There are four principal sources of energy: physical, such as wind and water, gravity, chemical reactions and anthropogenic, such as tillage. Soil erosion begins with detachment, which is caused by break down of aggregates by raindrop impact, sheering or drag force of water and wind. Detached particles are transported by flowing water (over-land flow and inter-flow) and wind, and deposited when the velocity of water or wind decreases by the effect of slope or ground cover. Three processes – dispersion, compaction and crusting – accelerate the natural rate of soil erosion. These processes decrease structural stability, reduce soil strength, exacerbate erodibility and accentuate susceptibility to transport by overland flow, interflow, wind or gravity. These processes are accentuated by soil disturbance (by tillage, vehicular traffic), lack of ground cover (bare fallow, residue removal or burning) and harsh climate (high rainfall intensity and wind velocity).

The effects of erosion and soil loss on soil properties vary by soil series, management, landscape position and climate. In general, soil erosion affects the chemical properties by loss of organic matter, plant nutrients, and exposure of subsoil materials with low fertility or high acidity (Olson et al. 1999). The changes in physical properties of soil, include structure, texture, bulk density, infiltration rate, rooting depth, and available water-holding capacity (Frye et al., 1982). The mineralogical properties of soils are also affected by the thinning of the plow layer (Ap horizon) and subsequent mixing of the subsoil (B horizon) into the Ap horizon by tillage. Eroded soils are subject to higher temperatures, have lower porosity and microbial activity.

19.2
Agrometeorological coping strategies

The soil erosion process is modified by biophysical environment comprising soil, climate, terrain and ground cover and interactions between them (Figure 19.1). Soil erodibility – susceptibility of soil to agent of erosion - is determined by inherent soil properties e.g., texture, structure, soil organic matter content, clay minerals, exchangeable cations and water retention and transmission properties (Lal 2001). Climatic erosivity includes drop size distribution and intensity of rain, amount and frequency of rainfall, run-off amount and velocity, and wind velocity. Important terrain characteristics for studying soil erosion are slope gradient, length, aspect and shape. Ground cover strongly reduces the impact of the eroding energy before it has a chance to reach the soil; hence, most strategies to limit erosion begin with having some sort of vegetation, live or dead, covering the ground.

The risk of soil erosion begins when natural vegetation, grasslands and forested areas, are either cleared for cultivation or used for grazing. The problem is accelerated by attempting to farm slopes that are too steep, cultivating up-and-down hills, continuous use of land for the same crop without rotation or fallow, inadequate use of fertilizers and organic manures, compaction of soil through of heavy machinery and pulverizing of the soil when creating seed-beds. Soil conservation strategies are also aimed at protecting the soil from direct exposure to natural elements such as establishing and maintaining good ground cover. Of particular importance in areas of the world where the first rains of a wet season area highly erosive, is the selection of crops that can establish ground cover rapidly. Where climatic conditions permit, early-season and between-season vegetation cover can be provided by off season crops that can be disked in or destroyed by pre-planting applications of herbicide.

Soil conservation relies upon good management of soil combined with agronomic practices and the use of mechanical measures playing a supportive role (Figure 19.1). The basics of soil conservation practices to minimize soil erosion and enhance its prevention are shown in Figure 19.1. The initial task in adaptation of conservation practices to minimize the risk of water erosion, run off and soil loss would be to study the specific conditions in the study area or region that is con-

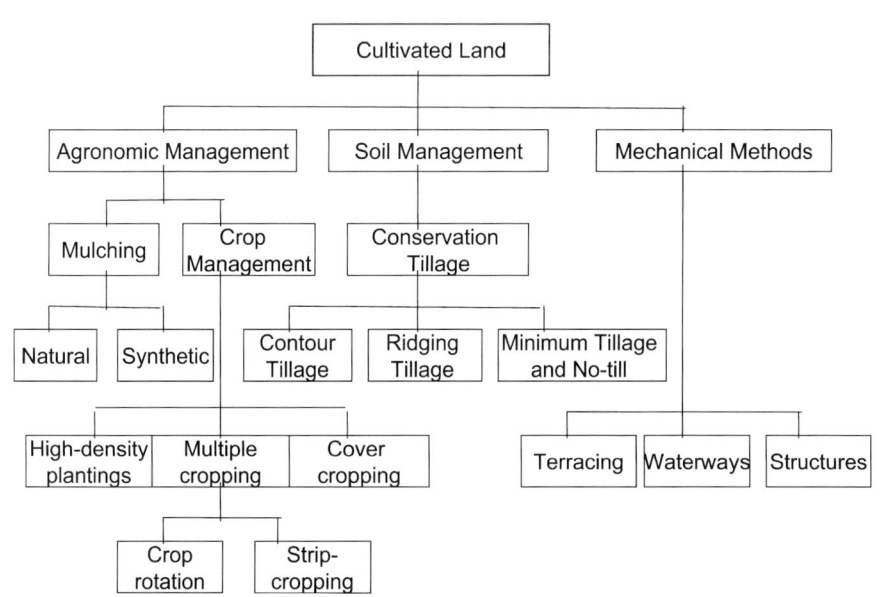

Soil conservation strategies for cultivated land (El-Swaifly et al, 1982)

Fig. 19.1. Soil conservation strategies for cultivated land (El-Swaify et al. 1982)

tributing to the degradation. Some very obvious examples is the use of crop residue to increase soil organic matter which may be impractical in some countries as the residue are used for animal feed. Introduction of no-till for crop cultivation is another example that maybe impractical in countries where there is no mechanization and most cultivation is done by human labor. Particularly in developing countries, limitation in resources and traditional methods of farming can hinder the adoption of a particular management practice.

19.3
Soil Management Strategies

19.3.1
Organic Matter

The ultimate goals of sound soil management are to maintain the fertility and structure of the soil. Highly fertile soils produce higher crop yields while maintaining good plant cover and minimizing the erosive effects of rainfall, runoff and wind. One of the ways to achieve and maintain soil fertility is to apply organic matter which improves the cohesiveness of the soil of the soil, increases its water retention capacity and promotes a stable aggregate structure. Organic matter may be added as green manures or residue. Under no-till management, the crop residue and the roots are left as added organic matter. Green manures, which area normally leguminous crops ploughed in, have a high rate of fermentation and yield a rapid increase in soil stability (Kolenbrander 1974). However, the long-term use of the land for cropping reduces the organic carbon content. Soil erosion accounts for only part of the reduction. Biological mineralization of the carbon from the soil *in situ* is the most important process, particularly in semi-arid areas where bare soil summer fallows are used to conserve moisture (Rasmussen et al. 1998).

19.3.2
Tillage Practices

Tillage is an essential management technique that provides a suitable seed bed for plant growth and in some areas of the world, the only non-chemical method to control weeds. The tillage tools pulled upwards by a tractor are designed to apply an upward force to cut and loosen the compacted soil, sometimes to invert it and mix it. The other main negative effect of driving a tractor across a field is compaction of the soil. Compaction may result in an increasing shear strength through an increase in bulk density, followed by low infiltration and increased runoff and erosion. The compaction generally extends to the depth of the previous tillage, up to 300 mm for deep ploughing, 180 mm for normal ploughing and 60 mm with zero tillage (Pidgeon and Soane 1978). Spoor et al. (2003) completed a study summarizing the risk of soils to compaction in relation to texture and wetness.

Conventional Tillage is the standard system of tillage practice involving ploughing, secondary cultivation, with disc harrowing, suitable for planting for a wide range of soils. The mouldboard plough inverts the soil in the plough furrow and moves all the soil in the plough layer to a depth of 100-200 mm. Secondary disc cultivation helps form seed beds and remove weeds by breaking up the cloddy surface produced by ploughing. The increased surface roughness due to tillage can be a successful deterrent to water erosion (Cogo et al. 1984).

Contour Tillage for planting and cultivation can reduce soil loss from sloping land compared with standard cultivation up-and-down the slope. It is inadequate as the only conservation practice to reduce soil loss for lengths greater than 180 m at 1° steepness (Troeh et al. 1980). This technique may be effective against soil loss and water erosion only for low rainfall intensity. Protection against more extreme storms is enhanced by supplementing contour farming with strip cropping, discussed later in this paper. Contour ridging and connecting the ridges with cross-ties over the intervening furrows, thereby forming a series of rectangular depressions that fill with water during rain can be very effective in controlling soil erosion along slopes. This practice known as tied ridging, should be used only in well drained soils to minimize water logging and damage to crops. Tied ridging on sandy soils in Zimbabwe with no till over a period of three years gave soil losses of less that 0.5 t ha^{-1} compared with up to 9.5 t ha^{-1} for conventional ploughing with mouldboard under maize cultivation (Vogel 1994).

Conservation Tillage can be defined as any practice that leaves at least 30% cover on the soil surface after planting. Numerous studies have examined the effects of different types of conservation tillage as outlined in table 1 (Natural Resources Conservation Service, USDA 1999). The success of various systems is highly soil specific, landscape, climatic pattern and also dependent how well weeds, pests and disease are controlled. The main barriers to adoption are the expense of specialized equipment for managing cultivation in crop residues, problems of weed control and increases in pest, particularly rodents.

Practice	Description
Conventional	Standard practice of ploughing with disc or mouldboard plough, one or more disc harrowing, a spike-tooth harrowing and surface planting.
No Tillage	Soil undisturbed prior to planting, which takes place in a nar row, 25-27 mm-wide seed bed. Crop residue covers of 50-100% retained on surface. Weed control by herbicides.
Strip Tillage	Soil undisturbed prior to planting, which is done in narrow strips using rotary tiller or in-row chisel, plough-plant, wheel track planting. Intervening areas of soil untilled. Weed control by herbicides and cultivation.
Mulch Tillage	Soil surface disturbed by tillage prior to planting using chisels, field cultivators, discs, or sweeps. At least 30% residue cover left on surface. Weed control by herbicides and cultivation.

In general the better drained, course- and medium textured soils with low organic content respond best and the systems that are not successful occur on poorly drained soils with high organic content or heavy soils, where the use of the mouldboard plough is essential. The effectiveness of the tillage practice is very dependent on the types of crops planted and the use of crop rotations.

No tillage system usually restricts tillage only to that necessary for plantings and carried out by drilling directly into the stubble of the previous crop. Weed control by herbicide application is an essential part of this system. This conservation technique has been found to increase the water-stable aggregates in the soil compared to disc cultivation and ploughing (Suwardji and Eberbach 1998; Mrabet et al. 2001). It is not suitable in soils that easily compact and seal because it can lead to lower crop yields and greater runoff.

In the Corn Belt of the USA, where maize, soybean and wheat are cultivated in rotation, no tillage is one of leading technologies to control erosion in areas. Moldenhauer (1985) reviewed the various tillage systems in this area and showed that annual soil loss under no till on a range of erodible soils was 5-15% of that from conventional tillage. In the southern regions of the USA, no till is recommended on ultisol soils which have become severely eroded after 150 years of continuous cropping. In this area, Langdale et al. (1992) studied three no tillage systems under soybean production as an alternative to the conventional system of growing soybean with a bare fallow over winter: Soybean with winter cover using in-row chisel; soybean with barley as winder cover and using fluted coulters; and soybean with rye as a green manure and using fluted coulters. The respective mean annual soil losses for the four systems (starting with conventional tillage) were 26.2, 0.1, 0.1 and 3.4 t ha^{-1}, respectively. One other environmental benefit to no tillage operations is that there is no exposure of the sub-layer soils to oxidation and release of carbon dioxide to the atmosphere.

In *strip tillage*, the soil is prepared for planning along narrow strips, with the intervening areas left undisturbed. In a single plough-planting operation, typically up to one-third of the soil is tilled. The plough-plant systems caused the least soil compaction, conserved most soil moisture and reduced losses of organic matter, nitrogen, phosphorus and potassium (Quansah and Bonnie 1981). In studies conducted in Ghana, the technique reduced soil loss in maize plots from multiple rain storms totaling 452 mm to 0.2 t ha^{-1} compared with 1.4 t ha^{-1} for traditional tillage using hoe and cutlass (Baffoe-Bonnie and Quansah 1985).

Mulch Tillage system in general has been successful to reduce water and wind erosion and to promote the conservation of soil moisture in drier wheat-growing areas (Fenster and McCalla 1970). The soil is prepared in such a way that at least 30% of the surface is covered with plant residues, or other mulching materials, are specifically left on or near the surface. Mulch tillage is a broad term and includes practices such as no-till, disk plant systems, chisel plant systems, and strip tillage systems. Sometimes a cover crop, usually a legume, is specifically grown within the cropping cycle to produce mulch material. Another variant of planted fallowing, practiced in North America, is referred to as summer fallow or ecofallow. The latter is a system of fallowing in which weed growth is restricted by shallow cultivation or by using herbicides to conserve soil moisture. Crops are grown every other year or once in 3 years. This type of "cropless" fallow is mostly used in arid climates to conserve soil moisture, without having to resort to irrigation.

Minimum Tillage or reduced tillage is a practice using chiseling or disking to prepare the soil while retaining a 15-25 % residue cover. Minimum tillage is not an easy option; it demands commitment, time and patience. This option is selected based on the soil and cropping needs of a particular region. Drier and more stable structured soils are best suited to minimum tillage. Chisci and Zanchi (1981) found that minimum till was suitable for silt clay soils in northern Italy under continuous wheat production. Soil moisture was maintained and in turn reduced the cracking while promoting high infiltration rates. Despite higher runoff, annual soil loss was almost one third of that under conventional tillage. In the U.S. Corn Belt, farmers use chiseling in the autumn to produce a rough surface but retain residue cover, followed by disc cultivation in the spring to smooth the seed bed and cover the residue. This can reduce soil loss by an order of magnitude compared to conventional tillage (Siemens and Oschwald 1978).

19.3.3
Crop Management Strategies

Agronomic measure and strategies for soil conservation use the protective effect of plant cover to reduce erosion and soil loss. In considering effectiveness of crop management strategies, there are many factors that need to be considered such as type of crop or vegetation used in a crop rotation, tillage practices, landscape characteristics, rainfall magnitude and intensity. In general, row crops are less effective and may cause some erosion. In a 100 year study, land under continuous maize has only 44% as much of the top soil compared to land kept under grass, whereas the land under a rotation of maize and pasture had 70 % of the top soil (Gantzer et al. 1990). Legumes and grasses used on a rotation scheme with row crops can provide good ground cover and also help maintain the organic content of the soil, contributing to the soil fertility and stability of aggregate structure. This type of rotation can also potentially increase the yields of the main crop.

Maize grown on 2-4° slopes as a row crop with conventional tillage and clean weeding, can result in soil loss between 10 and 120 t ha^{-1}, taking data from India (Singh et al. 1979) and Zimbabwe (Hudson 1981). Soybeans are often used in the rotation with maize because of their apparent ability to reduce soil loss by intercepting a higher percentage of the rainfall and that it requires minimal nitrogen fertilizer. However, studies in the U.S. Midwest indicate that soybean cultivation can result in as much if not more erosion than maize (Laflen and Moldenhauer 1979). The mean soil loss over a seven year period for maize in a silt-loam soil with a 4% slope was 7 t ha^{-1} for continuous maize, 6.5 t ha^{-1} for maize followed by soybean. Lal (1976) found that when using sequential cropping near Ibadan, Nigeria, on a slope of 6°, maize crop followed by cowpeas with no tillage produced a soil loss of only 0.2 t ha^{-1} but when growing cowpeas followed by maize but with tillage gave a loss of 6.2 t ha^{-1}. Although the effects of tillage and no till are difficult to isolate, it seems that a maize-cowpeas sequence produces less erosion than a cowpeas-maize because maize is a soil-depleting crop and when grown second, is planted into an already partially exhausted soil.

Fallow crops can be useful only for grazing or fodder, but have no immediate value to the farmer who does not have a cattle farm. Thus this type of management strategy is not followed by the main cereal-growing areas and is not a practical to control soil erosion. An alternative approach in the temperate regions is to minimize the period of bare ground, for example, by leaving crop residue on the land after harvest and delaying ploughing until the following spring. Another practice is to have a winter cover crop that is planted late in the fall and ploughed in to form a green manure prior to sowing the main crop. The cover crops are rye, oats, mustard, sweet clover and other similar crops. Although the cover crop grows rapidly and retains nutrients in the soil that would be otherwise lost to leaching, the cost of growing a cover crop may outweigh the benefits an individual farmer receives. This is especially true on small holdings where farmers do not have sufficient cash reserves. Depending on the seasonal rainfall pattern, the cover crop could complete for the available soil moisture and, in dry areas, adversely affect the growth of the main crop.

An area that is receiving significant attention now as a coping strategy for soil erosion and also carbon sequestration is mulching or covering the soil with crop residue and sometimes standing stubble. The cover protects the soil from raindrop impact and reduces the velocity of runoff and wind. This is a useful alternative form of cover crop in dry areas where insufficient rain prevents the establishment of ground cover before the onset of heavy rains or strong winds or where a cover crop completes for soil moisture with the main crop. In semi-humid tropics, one of the benefits of mulching is to lower the soil temperature, and prevents surface evaporation to maintain the soil moisture which may increase crop yields. However, in cool climates, a reduction in soil temperature may shorten the germination and the growing season, whereas in wet areas, higher soil moisture maintained in mulch soils may induce anaerobic conditions. Furthermore, crop yields can be reduced because fungal and bacterial decomposition of mulch competes for nitrogen with the main crop so mulching generally requires added nitrogen fertilizer for compensation. Therefore one has to be selective in applying mulches or maintaining crop residue and aware of the potential changes in soil environment while reducing the rate of soil erosion. Numerous studies have successfully demonstrated the benefits of mulches and crop residue in minimizing soil loss under different landscape and slope conditions (Lal 1976; Khybri 1989; Sherchan et al. 1990).

19.3.4
Mechanical Control Strategies

Mechanical methods are usually adapted to support agronomic and soil management strategies. The methods adapted depend on whether the objectives for controlling movement of water over soil is to reduce the velocity of runoff, increase the surface water storage capacity or safely dispose excess water. We will briefly review only control strategies that help minimize runoff and soil loss. Construction of contour bunds and terraces are the most common strategies of mechanical methods for erosion control. Contour bunds are soil banks 1.5 to 2 m wide constructed across the landscapes with slopes of 1-7° as a barrier for runoff and form water storage areas and to break up the slope into shorter segments. The banks are placed at 10 to 20 m intervals depending on the slope and extent of the landscape.

Soil characteristics such as texture can be considered in the design of the counter bunds and there are no precise specifications for their design.

Numerous studies have demonstrated a range of effectiveness in the construction of contour bunds (Hurni 1987; Clark et al. 1999).

Terraces are an earth embankment, or a combination ridge and channel, constructed across the field slope that intercepts, detains and safely conveys runoff to an outlet. Terraces are used to shorten the length of long slopes and serve as small dams to catch water and guide it to an inlet. They also serve as a guide for a contour row pattern in the field and help to improve water and soil quality. Terraces can be classified into three main types: diversion, retention and bench. The primary purpose of a diversion terrace is to intercept runoff and channel it across the slope to a suitable outlet. These diversion terraces can be as narrow as 3-4 m wide and cannot be cultivated with machinery. Also for cultivation to be possible, the banks should not exceed 14° slope if small machinery is used or 8° if larger farming equipment is used. In general diversion terraces are not suitable for agriculture if ground slopes exceed 7° because of the expense of construction and close spacing required.

Retention terraces are used where water storage is necessary on the hillside and generally designed with a capacity to store the expected runoff volume. They are recommended only for permeable soils on slopes that do not exceed 4.5°. Bench terraces are generally used where landscape with slopes up to 30° needs to be cultivated. A series of shelves and risers are constructed and should be properly constructed and maintained to prevent erosion. Vegetation cover or stones are usually used to minimize erosion and unprotected risers can be the source of most of the erosion in terraced systems (Critchley and Bruijnzeel 1995). Bench terraces are constructed in various modifications to suit the landscape, soil properties and rainfall patterns or water supply.

19.4
Conclusions

The risk farmers take in cultivating crops and uncertainties that follow due to natural disasters is always present. Preparing for uncertainties has to always be part of the management strategies even in the case of natural disasters. Unfortunately such preparations are not feasibly in most small farmers in developing countries. In more advanced agricultural systems, country level policies are established to mitigate the impact of natural disasters and provide assistance to farmers. The problems of soil erosion and decline in soil quality are confounded by high demographic pressure, low per capita arable land area, resource-poor farmers who cannot afford to invest in soil erosion control measures. The grain production is directly linked with soil quality and productivity which are declining along with the per capita arable land area, especially in densely populated countries of Asia and Africa. Soil erosion and erosion-induced decline in productivity in these regions are an increasing problem.

The discussions presented here are primarily on methods to reduce the risks and uncertainties in crop production through intervention management techniques for the farming systems to help minimize soil erosion and loss of top soil to excessive water runoff. The entire farming systems should be considered for developing a comprehensive solution in reduction of soil erosion, soil loss and runoff. This system should include the soil management practices such as tillage, the crops cultivated, sequence of cropping with cover crops planted where possible, the use of crop residue, and finally development of mechanical barriers to slow down the runoff over sloping landscapes. The specific strategy that is suitable for a particular site depends on many factors such as landscape characteristics, soil properties, rainfall patterns and intensity, adaptability of soil and crop management practices in the region. Developing a proper combination of these strategies based on a good assessment of the problem is critical for a successful implementation of coping risks and uncertainties in crop production. Finally, we need further research to develop and refine new methods for application in Asia and Africa that account for the soils, crops and resources available for these areas.

References

Baffoe-Bonnie E, Quansah C (1975) The effect of tillage on soil and water loss. Ghana J Agric Sci 8:191-195.

Chisci G and Zanchi C (1981) The influences of different tillage systems and different crops on soil losses on hilly silty-clayey soil, In Morgan RPC (ed) Soil conservation: problems and prospects. Willey Chichester, pp. 211-217.

Clark R, Duron G, Quispe G, Stocking MA (1999) Boundary bunds or piles of stones. Using farmers' in Bolivia to aid soil conservation. Mount Res Dev 19:235-240.

Cogo NP, Moldenhauer WC, Foster GR (1984) Soil loss reductions from conservation tillage practices. Soil Sci Soc Am J 48:368-373.

Critchley WRS, Bruijnzeel LA (1995) Terrace risers: erosion control or sediment source? In: Singh RB, Haigh MJ (eds) Sustainable reconstruction of highland and headwater regions, Oxford and IBH Publishing, New Delhi, pp. 529-541.

El-Swaify SA (1994) State of the art for assessing soil and water conservation needs and technologies. In: Napier TL, Comboni SM, and El-Swaify SA (eds). Adopting conservation on the farm. An international perspectives on the socioeconomics of soil and water conservation. Soil and Water Conservation Society, Ankeny, IA, pp.13-27.

El-Swaify SA, Dangler EW, Armstrong CL (1982) Soil erosion by water in the tropics. College of Tropical Agriculture and Human Resources University of Hawaii.

Fenster CR, McCalla TM (1970) Tillage practices in western Nebraska. Agricultural Experimental Station Lincoln Bulletin 597.

Frye WW, Ebelhar SA, Murdock LW, Blevins RL (1982) Soil erosion effects on properties and productivity of two Kentucky soils. Soil Sci Soc Amer J 46:1051-1055.

Gantzer CJ, Anderson SH, Thompson AL, Brown JR (1990) Estimating soil erosion after 100 years of cropping on Sanborn field. J Soil Wat Cons 45:641-644.

Hudson NW (1981) Soil conservation. 2nd Edn Bradford, London.

Hurni H (1987) Erosion-productivity-conservation systems in Ethiopia. In Plasentis I (ed) Soil conservation and productivity, Sociedad Venezolana de la Ciencia del Suelo Maracay, pp. 654-674.

Kolenbrander GJ (1974) Efficiency of organic manure in increasing soil organic matter content. Trans 10th Int Congr Soil Sci 2:129-136.

Khybri ML (1989) Mulch effects on soil and water loss in Maize in India. In: Moldenhauer WC, Hudon NW, Sheng TC, and Lee SW (eds). Development of conservation farming on hillslopes. Soil Wat Cons Soc, Ankeny, IA, pp. 195-198.

Lal R (1976) Soil erosion problems on an alfisol in western Nigeria and their control. IITA Monograph 1.

Lal R (2001) Soil degradation by erosion. Land Degradation & Development 12: 519–539.

Langdale GW, Mills WC, Thomas AW (1992) Conservation tillage development for soil erosion control in the southern Piedmont. In: Hurni H and Kebdo Tato (eds).

Laflen JM, Moldenhauer WC (1979) Soil and water losses from corn-soybean rotations. Soil Sci Soc Amer J 43:1213-1215.

Moldenhauer WC (1985) A comparison of conservation tillage systems for reducing erosion. In D'Itri FM (ed). A systems approach to conservation tillage. Lewis Publishers, Chelsea, MI: 111- 120.

Mrabet R, Saber N, El-Brahli A, Lahlou S, Bessam E (2001) Total particulate organic matter and structural stability of a Calcixeroll soil under different wheat rotations and tillage systems in semiarid areas of Morocco. Soil Till Res 57:225-235.

Natural Resources Conservation Service (1999) CORE4 conservation practices training guide. The common sense approach to natural resources conservation. USDA-NRCS (http://wwwnrcsusdagov/technical/ECS/agronomy/core4pdf).

Olson KR, Mokma DL, Lal R, Schumacher TE, and Lindstrom MJ (1999) Erosion impacts on crop yields for selected soils of North Central United States. In: R Lal (ed.) Soil Quality and Soil Erosion , Soil and Water Conservation Society, Ankeny Iowa, CRC Press, pp 259-283.

Pidgeon JD, Soane BD (1978) Soil structure and strength relations following tillage zero tillage and wheel traffic in Scotland. In: Emerson WW, Bond RD and Dexter AR (eds) Modification of soil structure, Wiley Chichester, pp 371-378.

Pimental D, Allen J, Beers A, Guinand L, Hawkins A, Linder R, McLaughlin P, Meer B, Musonda D, Perdue D, Poisson S, Salazar R, Siebert S, and Stoner K (1993) Soil erosion and agricultural productivity. In: Pimental D (ed) World soil erosion and conservation. Cambridge University Press Cambridge, pp. 277-292.

Quansah C (1985) Rate of soil detachment by overland flow with and without rain and its relationship with discharge slope steepness and soil type. In: El-Swaifly SA, Moldenhauer WC, Lo A (eds) Soil erosion and conservation. Soil Conservation Society of America Ankeny IA, pp. 406-423.

Quansah C, Baffoe-Bonnie E (1981) The effect of soil management systems on soil loss runoff and fertility erosion in Ghana. In: Tingsanchali T and Eggers H (eds) Southeast Asian regional symposium on problems of soil erosion and sedimentation. Asian Institute of Technology Bangkok, pp. 207-217.

Rasmussen PE, Albrecht SL, Smiley RW (1998) Soil C and N changes under tillage and cropping systems in semi-arid Pacific Northwest agriculture. Soil Till Res 47:197-205.

Sherchan DP, Chand SP, Thapa YB, Tiwari TP, and Gurung GB (1990) Soil and nutrient losses in runoff on selected crop husbandry practices on hill slope soil of the eastern Nepal. In: Proceedings International symposium on water erosion sedimentation and resource conservation. Central Soil and Water Conservation Research and Training Institute. Dehra Dun, India, pp 188-198.

Siemens JC and Oschwald WR (1978) Corn-soybean tillage systems: erosion control effects on crop production costs. Trans Amer Soc Agric Eng 21:293-302.

Singh G, Bhardwaj SP, Singh BP (1979) Effects of row cropping of maize and soybean on erosion losses. Ind J Soil Cons 7:43-46.

Spoor G, Tijink FGJ, Weisskopf P (2003) Subsurface compaction: risk avoidance identification and alleviation. Soil Till Res 73:175-182.

Suwardji P, Eberbach PL (1998) Seasonal changes of physical properties of an Oxic Paleustalf (Red Kandosol) after 16 years of direct drilling or conventional cultivation. Soil Till Res 49:65-77.

Troeh FR, Hobbs JA, Donohue RL (1980) Soil and water conservation for productivity and environmental protection. Prentice Hall Englewood Cliffs NJ.

Vogel H (1994) Conservation tillage in Zimbabwe. Evaluation of several techniques for the development of sustainable crop production systems in smallholder farming Geographica Bernensia African Studies Series A11.

Williams JR (1995) The EPIC Model In: Computer Models of Watershed Hydrology Water Resources Publications, Highlands Ranch CO, Chapter 25, pp 909-1000.

CHAPTER 20

Developing a global early warning system for wildland fire

Michael A. Brady, William J. de Groot, Johann G. Goldammer, Tom Keenan, Tim J. Lynham, Christopher O. Justice, Ivan A. Csiszar, Kevin O'Loughlin

20.1
Introduction

Fire is a prevalent disturbance on the global landscape with several hundred million hectares of vegetation burning every year. Land and forest fires (collectively referred to as wildland fires) occur annually on every continent except Antarctica, and most global fire is unmonitored and undocumented (Figure 20.1). Increasing trends in wildland fire activity have been reported in many global regions. Wildland fires have many serious negative impacts on human safety, health, regional economies and global climate change. Developed countries spend billions every year in an attempt to limit the impact of wildland fires. In contrast, developing countries spend little, if any, money to control fire, yet they are often the most susceptible to the damaging impacts of fire because of increased vulnerability of human life and property (due to limited fire suppression capability), increased risk due to high fire frequency (often caused by the cultural use of fire), and sensitive economies (tourism, transport).

To mitigate these fire-related problems, forest and land management agencies, as well as land owners and communities, require an early warning system to identify critical time periods of extreme fire danger in advance of their occurrence. Early warning of these conditions with high spatial and temporal resolution incorporating

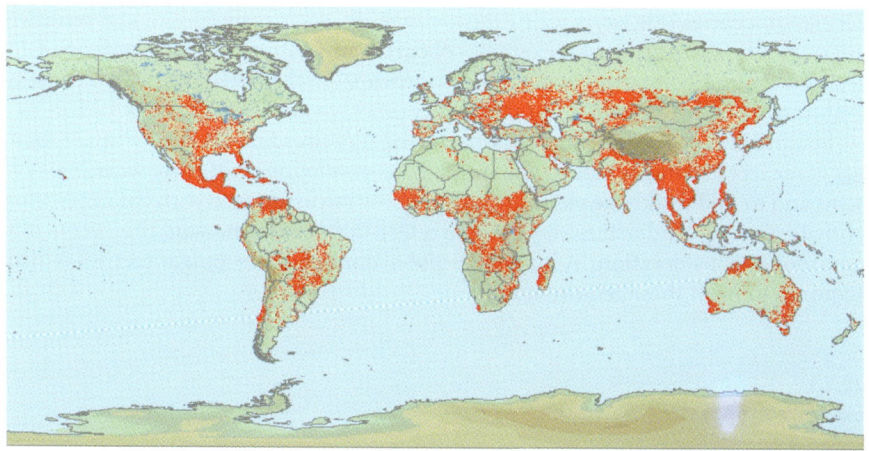

Fig. 20.1. Global vegetation fires detected in April 2005.

measures of uncertainty and the likelihood of extreme conditions allow fire managers to implement fire prevention, detection and pre-suppression plans before fire problems begin. Considering the fact that the majority of uncontrolled and destructive wildfires are caused by humans as a consequence of inappropriate use of fire in agriculture, pastoralism and forestry, it is crucial that international wildland fire early warning systems are developed to complement relevant national fire danger warning systems where they exist, to provide early warning where national systems do not exist, and to enhance warnings applied or generated at the local (community) level (people-centered early warning systems – as requested by the UN Secretary General and as laid down in the Hyogo Framework for Action 2005–2015: "Building the Resilience of Nations and Communities to Disasters"). This will ensure delivery of targeted information reflecting specific local conditions and allowing the involvement of local communities in wildland fire prevention.

Fire danger rating is a mature science and has long been used as a tool to provide early warning of the potential for serious wildfires. Fire danger rating systems (FDRS) utilize basic daily weather data to calculate wildfire potential. FDRS early warning information is often enhanced with satellite data such as hot spots for early fire detection, and spectral data on land cover and fuel conditions. Normally, these systems provide a 4-6 hour early warning of the highest fire danger for any particular day that the weather data is supplied. However, by using forecasted weather data, as much as 2 weeks of early warning can be provided, depending on the length of the forecast. Ensemble weather prediction systems through multiple realisations of forecasts provide distributions of weather forecasts and capture their inherent predictability and uncertainty associated with such forecasts. As well, FDRS indices can be calibrated with local data to provide longer term early warning, such as a 30-day early warning tool developed for Southeast Asia to indicate the potential for disaster-level haze events from peatland fires (see section 20.8).

FDRS tools for early warning are highly adaptable and have demonstrated their application to a wide range of users, from independent remote field stations (for making local fire suppression and preparedness decisions) to global and regional fire information centres (for large-scale decision making, such as multi-national resource sharing). There are numerous examples of current operational systems utilizing GIS technology and computer modelling of landscape level fire danger, which process and transfer early warning information very quickly via the World Wide Web.

Long-term knowledge of conditions during wildfires and the utility of fire danger forecasts are important to the immediate development of early warning systems and to undertake the planning and preparation associated with the impacts of climate change. Understanding the characteristics of extreme wild fire events is a paramount consideration. A long-term global dataset of fire danger metrics is also required to meet these requirements.

20.1.1
EWS-Fire Proposal

Following the recommendations of the UN World Conference on Disaster Reduction (WCDR) in Kobe, Japan, January 2005, and the proposal of the UN Secretary General to develop a Global Multi-Hazard Early Warning System (GEWS), a call for project proposals for building a GEWS was issued in preparation for the 3rd International Conference on Early Warning (EWC-III) (27–29 March 2006, Bonn, Germany), sponsored by the United Nations International Strategy for Disaster Reduction and the German Foreign Office (www.ewc3.org/). An international consortium of institutions cooperating in wildland fire early warning research and development submitted a proposal for the EWS-Fire, and it was selected for presentation at EWC-III. The outcomes of the discussions, which are documented on the GFMC Early Warning Portal (www.fire.uni-freiburg.de/fwf/EWS.htm), reveal the high interest in and endorsement by government and international institutions.

The early warning system is also included as a yet-to-be-funded task in the 2007–2009 work plan of the Global Earth Observation System of Systems (GEOSS), an international initiative involving more than 69 countries and 46 international organizations. Regionally, it was presented at the Conference on Promoting Partnerships for the Implementation of the ASEAN Agreement on Transboundary Haze Pollution, 11–13 May 2006 in Hanoi, Vietnam. Within the international fire science community the scientific and technical aspects of the EWS-Fire were presented by de Groot et al. (2006) at the Fifth International Conference on Forest Fire Research (24 November to 1 December 2006, Figueira da Foz, Portugal).

Objectives and development stages of the EWS-Fire, and the international consortium of institutions to be involved are described below. To illustrate the anticipated sustainability of the EWS-Fire, a case study is presented on the transfer and ongoing operation of a regional FDRS in Southeast Asia.

20.2
Objectives and Expected Impact of EWS-Fire

The goal of the proposed Global Early Warning System for Wildland Fire (EWS-Fire) is to provide a scientifically supported, systematic procedure for predicting and assessing international fire danger that can be applied from local to global scales. The objectives of the EWS-Fire are to develop and implement:

- a global early warning system for a 0-10 day prediction of wildland fire danger based on existing and demonstrated science and technologies;
- an information network to quickly disseminate early warning of wildland fire danger that reaches global to local communities;
- an historical record of regional and global fire danger information for early warning product enhancement, validation and strategic planning purposes; and
- a technology transfer program to provide the following training for global, regional, national, and local community applications:
  - early warning system operation;
  - methods for local to global calibration of the System; and
  - utility of the products and System for prevention, preparedness, detection, and, where appropriate, fire response decision-making.

The EWS-fire is not intended to replace the many different national FDRS currently in use, but rather to support existing national fire management programs by providing:
- new longer term predictions of fire danger based on advanced numerical weather models; and
- a common international metric for implementing international resource sharing agreements during times of fire disaster.

An additional benefit is that the EWS-Fire provides a FDRS for the many countries that do not have the financial or institutional capacity to develop their own system. Because the system can be used at the local level, it can support local capacity-building by providing a foundation for community-based fire management programs.

There are several expected impacts of developing the EWS-fire. Firstly, early warning (0-10 day) of wildland fire danger, on a regional and global basis, will provide international agencies, governments and local communities the opportunity to mitigate fire damage by assessing threat likelihood and possibility of extreme behaviour enabling implementation of appropriate fire prevention, detection, preparedness, and fire response plans before wildfire problems begin. Secondly, a globally robust operational early warning framework with an applied system will provide the foundation with which to build resource-sharing agreements between nations during times of extreme fire danger. Thirdly, technology transfer and training will develop local expertise and capacity building in wildland fire management for system sustainability.

20.3
Planned System Development

Development of an operational EWS-Fire is proposed through three phases. The overall system structure, including information technology and fire and weather science, is prepared during the system design phase. Establishment of standard operating procedures to provide and distribute daily early warning products will occur during the operational implementation phase. Finally, a technology transfer phase will facilitate sustainability of system operations and ensure applicability of the system to practical fire management through training programs.

20.3.1
Warning System Design

The EWS-Fire needs to address two distinct issues: 1) to establish a methodology to use forecasted weather to provide predicted fire danger, and 2) it must also provide a means of interpreting the fire danger in practical and locally relevant fire management terms. These criteria imply that the FDRS used for the EWS-Fire must have fairly simple and reasonably predictable weather inputs over the forecasted range, and it must be possible to locally calibrate the fire danger indices. The following steps are envisaged, with additional detail provided by de Groot et al (2006):

- Review and summarize literature and data on global fire activity to assess risk to global communities and areas of priority.
- Adapt a current risk monitoring system for global application, using the Canadian Forest Fire Weather Index System (FWI) in a prototype (Figure 20.2).
- Develop protocols for utilizing state of the art (0-10 day) global weather forecasting models for fire danger prediction.
- With latest numerical weather prediction ensemble prediction techniques, adapt FWI System to operate in a forecasting mode providing probability of event characteristics.
- Utilize historical satellite detected hotspot and archived numerical analysis of FWI to provide: further calibration of the FWI system for early warning purposes; a fire status product (where current fire problems are and basis to assess severity of forecasted fire danger conditions); and historical records of fire danger and behaviour regionally and globally.
- Studies to assess form and utility of products with end users and their social and economic impacts.

20.3.2
Operational Implementation

Daily (actual) FWI products are already being generated for many countries and some global regions on an operational basis (Fig. 20.2). Linking or networking these agencies would provide a significant start to compiling a global product of current fire danger. An advantage of building on existing national and regional systems is that it will ensure direct connection and applicability of the EWS-Fire from local to global scales. Provision of a complete global product of current fire danger will require a coordinating agency or facility to compile existing spatial data, to produce fire danger maps for regions where there currently are none, and to integrate the current fire danger maps with hot spot data (and possibly other remotely sensed data in the future, such as vegetation and rainfall). This will be a large task and may need the assistance of a number of regional agencies. The overall structure and procedures for this will depend on many factors, including the range and scale of products, availability of data, facilities, etc.

The following steps are envisaged:
- Undertake concept evaluation through operational trials with users.
- Support the institutional arrangements required for relevant agencies in the region to consider the new operational system.
- Develop procedures within the robust framework of the World Weather Watch (global network of operational meteorological services) to run the EWS-Fire on a daily operational basis, which includes analysis and production of current and forecast fire danger assessment, and dissemination of early warning information through multiple channels.
- Establish procedures with operational meteorological services to maintain and update the System as new tools and products are developed.

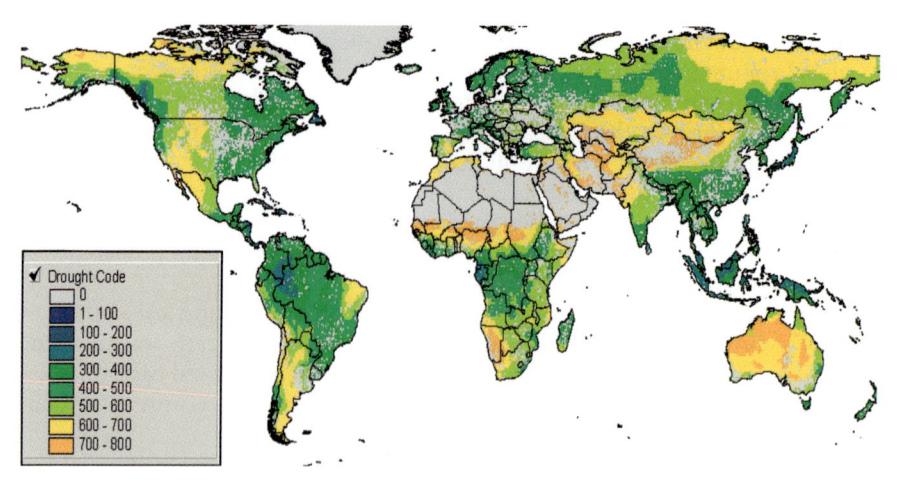

Fig. 20.2. Example of potential EWS-Fire product using the Drought Code (DC) component of the FWI System.

20.3.3
Technology Transfer

Training on the operational and application aspects of the system will occur through established institutions specializing in technology and information transfer. Through the WMO framework and the United Nations University, training and workshops will be provided in:
- EWS-Fire operations
- Basic understanding of fire danger and early warning
- Calculating FWI components
- Provision of FWI algorithms
- Developing and implementing decision-aids based on early warning to mitigate the impacts of fire through prevention, preparedness, detection, and fire response
- Involvement of local communities in the application of early warning information in wildland fire management (Community-Based Fire Management – CB-FiM), especially in wildfire prevention, and preparedness for coping with wildland fire disasters (including smoke pollution and public health)

As well, the EWS-Fire project will be promoted through presentations to land and forest fire managers at conferences, professional meetings, etc., and documents will be published on the EWS-Fire.

20.4
Implementing Organizations and Division of Tasks

Development of the EWS-Fire requires an integrated effort by organizations with different areas of specialization. The six organizations below include the required areas of expertise, resources and experience. Other organizations will be asked to participate as needed, particularly during operational implementation and technology transfer.

20.4.1
Natural Resources Canada - Canadian Forest Service (CFS)

Provide scientific, technical and fire management systems expertise to:
- Expand current international fire danger (FWI) monitoring system to full global coverage (presently monitoring approx. 1/3 global area for current daily conditions).
- Develop criteria to interpret FWI output in terms of fire danger levels for early warning purposes (e.g., conditions that define Low, Moderate, High, and Extreme fire danger).
- Develop practical decision-aid tools based on early warning for fire prevention, preparedness, detection, and fire response (e.g., when fire danger is High, open burning restrictions are imposed, tower detection is implemented, fire fighting staff and equipment are ready for immediate dispatch, etc.).
- Prepare and assist in delivering a technology transfer program to train early warning system users.

20.4.2
Bureau of Meteorology Research Centre (BMRC)

Australia with UK Met. Office, in cooperation with WWRP/WMO:
- Adapt Canadian Forest Fire Weather Index System as a Numerical Weather Prediction (NWP) Suite Module to provide medium range (up to two weeks) forecasts of FWI. Approach includes an ensemble of current operational deterministic global model outputs e.g. Canadian Meteorological Service, UK Met. Office, National Center for Environmental Prediction, Japanese Meteorological Agency, European Centre for Medium-Range Weather Forecasts, Australian Bureau of Meteorology, Deutscher Wetterdienst, etc. This activity will provide an initial globally-based FWI employing current global weather prediction systems in a manner that optimizes forecast skill.
- Validation and evaluation of initial Ensemble Prediction System (EPS)-based FWI. Employ satellite based hotspot data and ground-truthing where available.
- Provide FWI module to WMO members as required for global, regional, national and local implementation within the World Weather Watch operational framework.

20.4.3
Bushfire Cooperative Research Centre, Australia

- Validation and intercomparison of FWI with other wild fire indices. Employ satellite based hotspot data and ground truthing where available.
- User evaluation and optimisation of EPS Global FWI products.

20.4.4
University of Maryland (UMD), USA, acting on behalf of GOFC-GOLD

- Provision of global MODIS hotspot data from the daytime and night-time orbits of the Terra and Aqua satellites.
- Provision of global MODIS Normalized Difference Vegetation Index (NDVI) data, aimed at the assessment of vegetation condition, based on deviations from optimal and average conditions.

20.4.5
Global Fire Monitoring Center (GFMC), Germany on behalf of the UNISDR Wildland Fire Advisory Group / Global Wildland Fire Network and the United Nations University (UNU)

- Collection and global dissemination of FWI and associated early warning products.
- Technology transfer – through the United Nations University, to facilitate local-level implementation of a people-centered Early Warning System for wildland fire / CBFiM.

20.4.6
Global Observation of Forest and Land Cover Dynamics (GOFC-GOLD) Secretariat, Edmonton, Canada

- Assist the coordination and outreach mechanism for observations and products.
- Support the utilization and evaluation of products and assist in system implementation through EWS-Fire application within the GOFC-GOLD and UN regional networks.

20.5
Sustainability

The sustainability of the EWS-Fire will be secured through the long-term scope of the research, operational and extension agendas of the participating organizations. All organizations have demonstrated and will continue to have an interest in a coordinated and collective approach in dedicated research and capacity building,

notably through the existing and expanding networking activities. The UNISDR Global Wildland Fire Network and the GOFC-GOLD Fire Implementation Team and associated Regional Networks will jointly work in:
- technology transfer and training for development of local expertise and capacity building in wildland fire management; and
- involvement of the United Nations University through the new partnership agreement signed on 7 October 2005 (GFMC is now functioning as an UNU Associated Institute).

The anticipated sustainability of EWS-Fire can be illustrated through the case study below, which describes the development and operational implementation of new fire early warning systems at the regional, national and local levels in the fire prone areas of Southeast Asia.

20.6
Case Study in EWS-Fire Development

Forest and land fires in Southeast Asia have many social, economic, and environmental impacts. Tropical peatland fires affect global carbon dynamics, and haze from peat fires has serious negative impacts on the regional economy and human health. To mitigate these fire-related problems, forest and land management agencies require an early warning system to assist them in implementing fire prevention and management plans before fire problems begin. From 1999 to 2004 the Canadian Forest Service, in collaboration with partner agencies in Southeast Asia, developed an FDRS for the region, as well in Indonesia and Malaysia (de Groot et al. 2007). The FDRS' provide early warning of the potential for serious fire and haze events. In particular, they identify time periods when fires can readily start and spread to become uncontrolled fires and time periods when smoke from smouldering fires will cause an unacceptably high level of haze.

20.6.1
System Development

The FDRS were developed by adapting components of the Canadian Forest Fire Danger Rating System, including the Canadian Forest Fire Weather Index System, to local vegetation, climate, and fire regime conditions. A smoke potential indicator was developed using the Drought Code (DC) of the FWI System. An ignition potential indicator was developed using the Fine Fuel Moisture Code (FFMC) of the FWI System. Historical hot spot analysis, grass moisture, and grass ignition studies were used to calibrate the FFMC to track the ability for grass fires to start and spread. The Initial Spread Index (ISI) of the FWI System was used to develop a difficulty of control indicator for grassland fires, a fuel type that can exhibit high rates of spread and fire intensity. To provide early warning, the FDRS identifies classes of increasing fire danger as the FFMC, DC, and ISI approach key threshold values, which were defined through several calibration and validation studies.

20.6.2
Operational FDRS

In 2002 the Indonesian FDRS commenced operations nationally at the Indonesian Meteorological and Geophysical Agency (http://meteo.bmg.go.id/fdrs/index.html). In 2003 the Malaysian Meteorological Service began operating the Malaysian FDRS (www.kjc.gov.my/english/service/climate/fdrs1.html) and displaying regional outputs for the Association of Southeast Asian Nations (www.kjc.gov.my/english/service/climate/fdrs1.html) (Figure 20.3). The FDRS are being used by forestry, agriculture, environment, and fire and rescue agencies to develop and implement fire prevention, detection, and suppression plans.

Sustained use of newly transferred technology is often a challenge. To ensure ongoing FDRS operations after the Southeast Asia FDRS Project, a summary of FDRS technical information, reference material, guides for interpretation, and practical applications for user groups were compiled in a manual. As well, FDRS operating agencies in ASEAN, and in Indonesia and Malaysia prepared manuals of standard operating procedures (Figure 20.4). Also, a regional team of training specialists prepared a train-the-trainer course curriculum. By 2004, 20 fire management specialists from Indonesia, Malaysia, and Brunei completed the one-week course to become agency trainers. Finally, a Southeast Asia Fire Science Network was formed in 2001 to continue development of fire science expertise within the region. The network includes scientists from local universities and management agencies that share common interests and the goal to further advance knowledge in areas such as fire weather, fuel models, and fire behaviour.

20.6.3
Lessons Learned

The experience of developing the FDRS for Southeast Asia provided insight that can benefit the implementation of similar projects in other regions. The most critical aspect of developing new FDRS is identification and analysis of the local fire problem. Understanding of local fire climate, vegetation as fuel, fire regime, fire management capabilities and policy, and culture is required to understand the fire problem. The next important step is to determine what information the FDRS need to provide so that fire managers can address the fire problem. By developing the FDRS around fire management, the FDRS are designed to serve as a decision-making tool with practical application. The final step is to link the information required by fire managers to the underlying physical aspects of the fire environment that are causing the fire problem, and then connecting those factors to FDRS indicators. To illustrate the process in this project, haze was identified as the main fire problem in Southeast Asia. The underlying cause was smouldering peat fires and the best indicator of this potential is the DC, which is a relative measure of the dryness of deep organic layers in the forest floor.

Sustainability of new FDRS also requires local capacity strengthening. Technology transfer through workshops, training courses, and information sessions is important for the successful implementation of FDRS. Fire science education is

Chapter 20: Developing a global early warning system for wildland fire 365

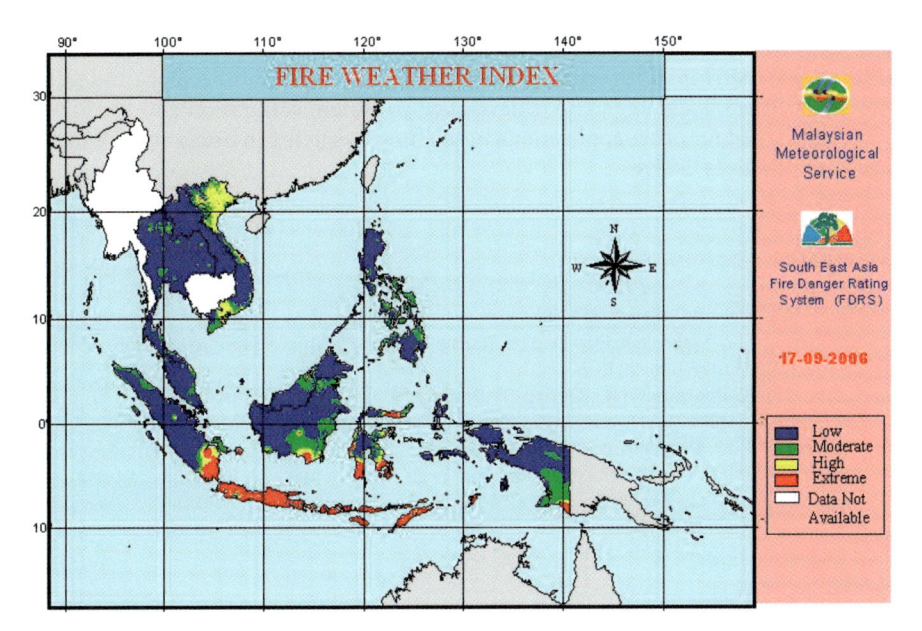

Fig. 20.3. Daily Fire Weather Index map from South East Asia Fire Danger Rating System which has been operated since September 2003 by the Malaysian Meteorological Service.

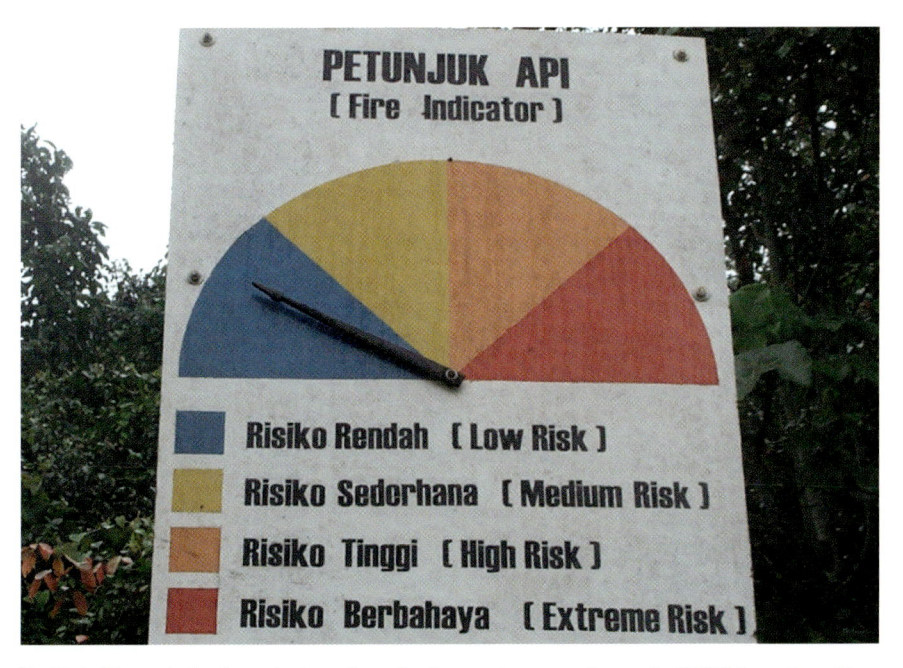

Fig. 20.4. Through the formulation of standard operating procedures, the FDRS is used to adjust daily fire danger levels on sign boards located in fire prone areas.

also important to understanding FDRS principles, and this aspect can be achieved through partnerships with universities. By training people in the science and technology of FDRS, local expertise is developed in system calibration and the FDRS can be designed for other applications in the future as new fire issues and fire management practices evolve.

References

de Groot, WJ, Field RD, Brady MA, Roswintiarti O, Mohamad M (2006) Development of the Indonesian and Malaysian Fire Danger Rating Systems. Mitig. Adapt. Strat. Glob. Change doi: 10.1007/s11027-006-9043-8

de Groot, WJ, Goldammer JG, Keenan T, Brady MA, Lynham TJ, Justice CO, Csiszar IA, O'Loughlin K (2007) . Developing a global early warning system for wildland fire. In D. X. Viegas (Ed.), Proceedings of the V International Conference on Forest Fire Research, November 24th to December 1st, 2006, Figueira da Foz, Portugal (In Press).

CHAPTER 21

Scientific and Economic Rationale for Weather Risk Insurance for Agriculture

Peter Höppe

21.1 Introduction

Munich Re is one of the largest global reinsurers and has a long tradition in both the assessment of weather-related hazards and their impact on crop production and the development of appropriate risk management tools and crop insurance schemes. Munich Re's underwriting of agricultural risks throughout the world is concentrated in its agricultural underwriting department. This department has long been a world-renowned centre of competence. It develops Munich Re's strategy and underwriting guidelines and is responsible for the underwriting of agricultural insurance within the Munich Re Group. This involves a wide range of segments like crops, crop hail, multi-peril and named perils, livestock including aquaculture and greenhouses. With a premium income of about €400m in 2006, Munich Re is the world's largest agricultural reinsurer.

Munich Re also has a long tradition of analysing natural disasters and their trends. The company recruited its first two geoscientists in 1974. Now the department has a staff of 27. The team consists of experts in geophysics, meteorology, geography, hydrology and many other fields, all contributing to the analyses of natural disasters.

21.2 Natural Disasters and Losses

Munich Re has one of the world's largest global databases on natural catastrophes, the Munich Re NatCatSERVICE. Currently, it contains entries for more than 23,000 individual natural events which have caused human suffering and/or property losses. Figure 21.1 shows the development in the number of great natural disasters (causing billion dollar losses and/or thousands of fatalities) since 1950, broken down into the different perils: floods, windstorms, geophysical disasters (earthquakes, tsunamis, volcanic eruptions) and other weather-related events (heatwaves, forest fires, droughts). The figure clearly shows a steep increase in the number of such events. While in the 1950s there were about two of them per year, the expected value has now risen to about seven. Most of the trend is driven by weather-related disasters, whereas there has been only a small trend upwards as far as geophysical events are concerned.

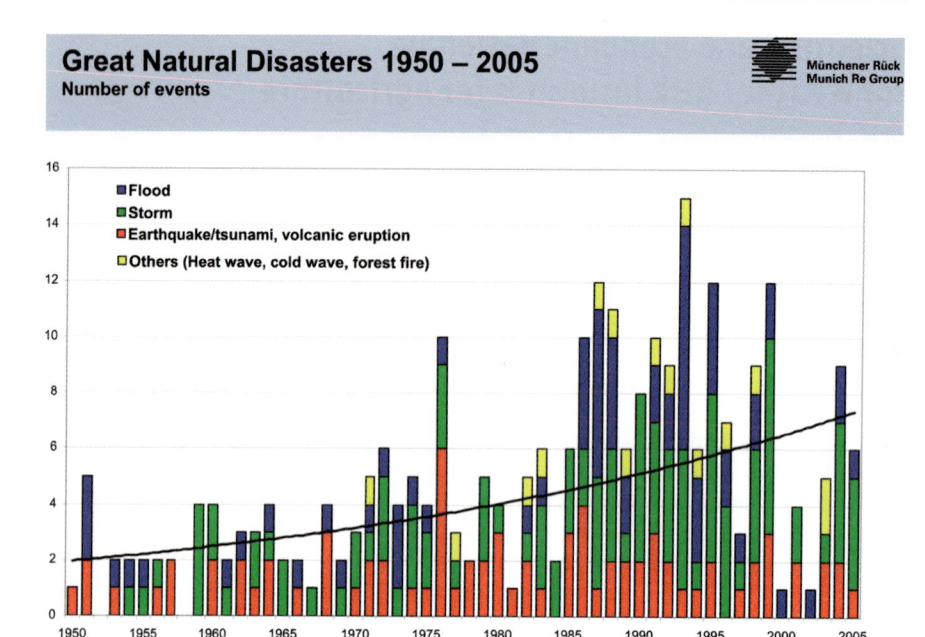

Great Natural Disasters 1950 – 2005
Number of events

Münchener Rück
Munich Re Group

- Flood
- Storm
- Earthquake/tsunami, volcanic eruption
- Others (Heat wave, cold wave, forest fire)

© 2006 NatCatSERVICE, Geo Risks Research, Munich Re

Fig. 21.1. Development of the number of Great Natural Disasters between 1950 and 2005 (Data from Munich Re NatCatSERVICE)

As a result of the growing number of weather-related disasters, but also as a result of population growth and increasing standards of living with higher values at risk, the losses from such disasters have increased dramatically. As Figure 21.2 shows on a global level, there has been an exponential increase in both overall economic and insured losses (both adjusted for inflation) since the 1950s, reaching a record level in 2004, which was topped again by new loss records in 2005.

Similar to the global trends, there has been an increase in the frequency of weather-related disasters in India, too. Figure 21.3 shows the development of weather-related disasters in India since 1980. Their number has risen from about five per year at the beginning of the 1980s to more than 20.

This increase in the disaster frequency in India is reflected in the corresponding losses (Figure 21.4). Here too we see a general trend towards higher losses, although the losses caused by the Gujarat flood in 1993 still mark the record. Figure 21.4 also clearly shows that, in India, insured losses account for a very small proportion of the overall economic losses, in most years far below 10%. This means that the people affected by these disasters and eventually the state as a financial source of last resort have to cope with these losses. In a country where the majority of the population still relies on agriculture for their livelihood, the effects of extreme weather

Great Weather Disasters 1950 – 2005
Economic and insured losses
(as at March 28, 2006)

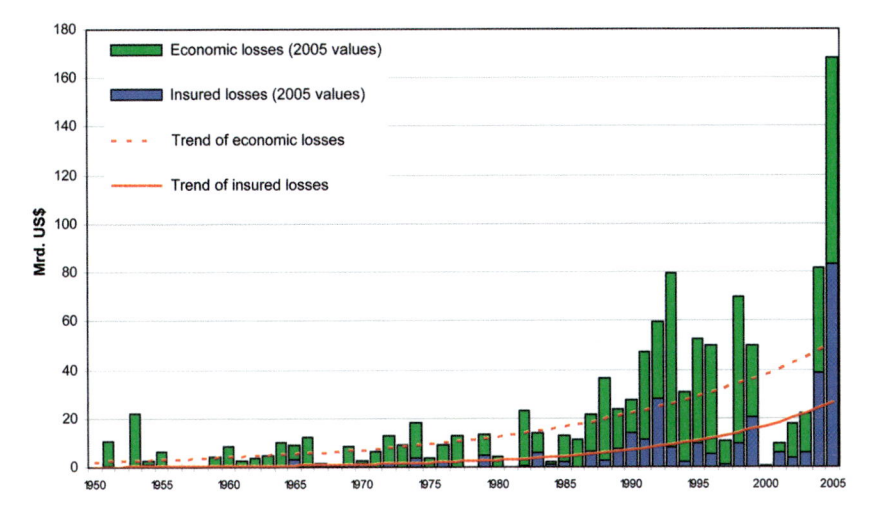

Münchener Rück — Munich Re Group

© 2006 NatCat*SERVICE*, Geo Risk Research, Munich Re

Fig. 21.2. Development of economic and insured losses from Great Weather Disasters between 1950 and 2005 (Data from Munich Re NatCatSERVICE)

events are severe because in many cases they mean a loss of the entire harvest and, as a consequence, the people's sole source of revenue.

Figure 21.5 maps the regional distribution of natural disasters in India between 1980 and 2005. It shows that hardly any state in India has not been affected by such disasters, and that windstorms (cyclones) mostly affect the states along the east coast, while floods and disasters caused by extreme temperatures are distributed more evenly over almost all states.

The last few years have brought records in the frequencies and intensities of weather-related catastrophes in many regions of the globe. In 2002, a hundred-year flood occurred in eastern Germany, causing economic losses of about €16bn, of which €3.4bn had been insured. Just one year later, the extremely hot summer in Europe killed more than 35,000 people. It was the largest human catastrophe in Europe for centuries. New records were then set in 2004 for hurricane losses in the USA. Four hurricanes making landfall in Florida in one season set a new record for that state with overall economic losses of US$ 62bn and insured losses of about half this amount. In March the same year, the first hurricane occurred in a region that had previously been thought to be hurricane-free, the South Atlantic. Just one year later, in 2005, we saw the largest loss from a single event in history, caused by Hurricane Katrina with overall economic losses of US$ 125bn and insured losses of US$ 60bn. The 2005 hurricane season was the most active since records began in

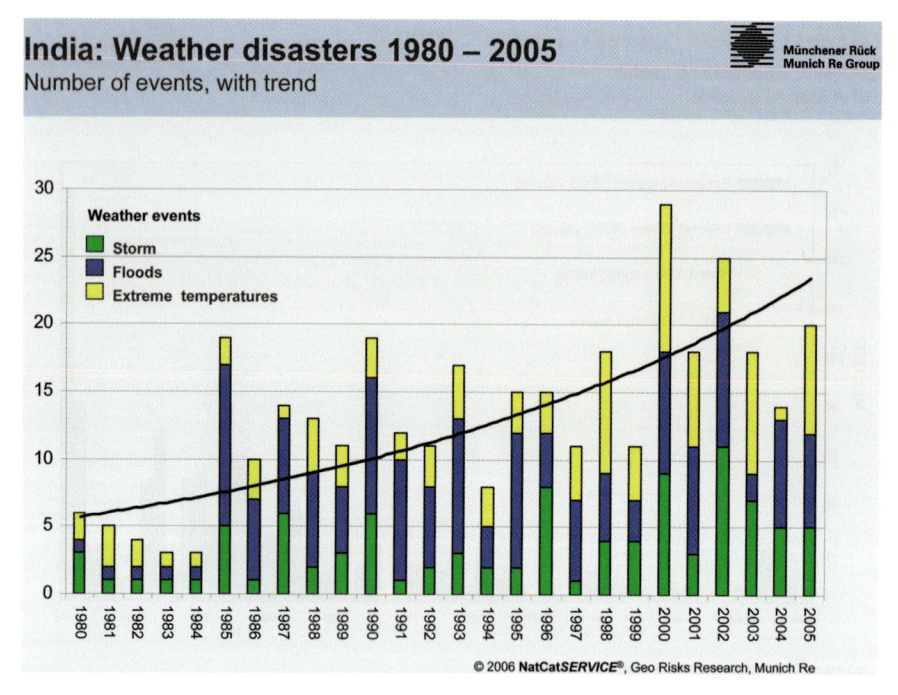

Fig. 21.3. Development of the number of Weather Disasters in India between 1980 and 2005 (Data from Munich Re NatCatSERVICE)

1851, with 27 named storms (old record: 21), including the strongest (Wilma), the fourth strongest (Rita) and the sixth strongest (Katrina). Then there was the first real hurricane system to approach Europe: Hurricane Vince, which formed close to Madeira and made its way into Spain and Portugal. The year 2005 also brought extremes in terms of floods. For Switzerland, the August 2005 floods in the northern Alps were the most expensive natural catastrophe ever (overall economic loss: US$ 2.1bn). In India, the highest ever 24-hour precipitation was measured in the Mumbai area: 944 mm. This extreme flood killed more than 1,000 people and caused economic losses of US$ 5bn and insured losses of US$ 770m. In 2006, there were again major floods in several Indian states, especially Gujarat.

21.3
Climate Change and Natural Disasters

Scientific research is now producing mounting evidence of a causal link between the increasing frequencies and intensities of weather-related disasters and anthropogenic climate change. Global warming is a fact and will further accelerate in the coming decades. A British study (Stott et al. 2004), for example, shows that human influence has already at least doubled the risk of a heatwave exceeding the magnitude of the European heatwave in 2003. Another study modelling the effects of climate change

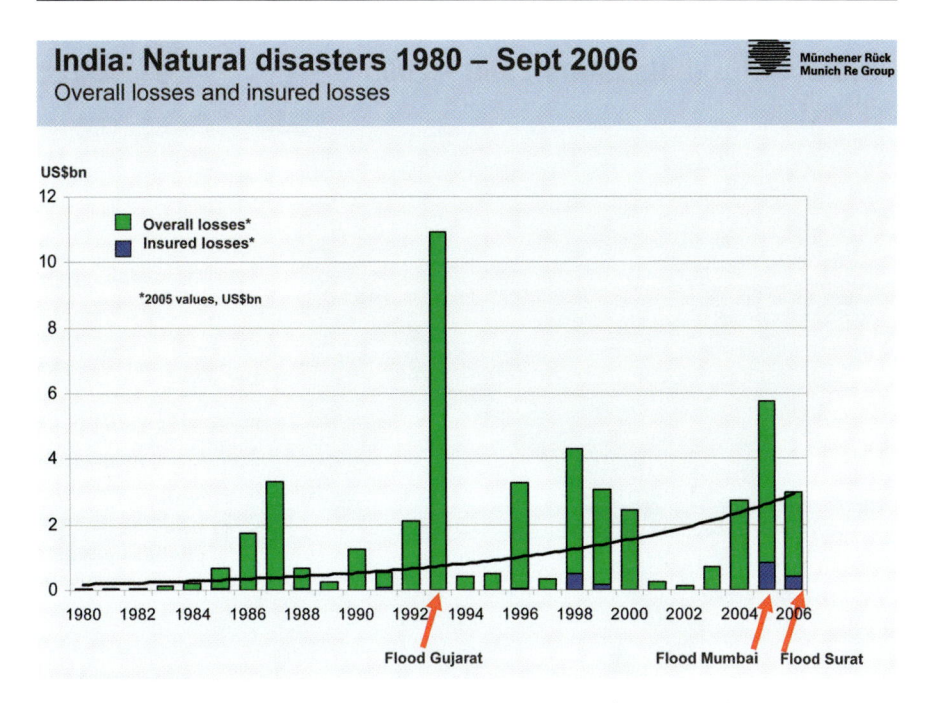

Fig. 21.4. Development of economic and insured losses in India from Natural Disasters between 1980 and 2005 (Data from Munich Re NatCatSERVICE)

on hurricanes has found that, due to global warming, the maximum wind speed of hurricanes and the associated precipitation will increase (Emanuel 2005). The number of major tropical storms has already increased in all ocean basins (Webster et al. 2005). There is now evidence that, evidence that, again due to global warming, sea surface temperatures have already increased by about 0.5°C (Barnett et al. 2005). Hoyos et al. (2006) show that of all the factors that drive a major tropical storm, only the steady increase in sea surface temperatures over the last 35 years can account for the rising strength of storms in six ocean basins around the world. So the logical link between the fact that global warming has increased sea surface temperatures and the fact that only this increase can explain the greater intensities of tropical storms results in the suggestion that anthropogenic climate change has already increased the intensity of these storms. This implies that if sea surface temperatures continue to increase due to global warming, we have to expect even more loss and damage.

21.4
Agricultural Risk Insurance

Agriculture is definitely the industry which is most vulnerable in respect of extreme weather and will therefore be the first to be affected by global warming. According to Kumar et al. (2007), two-thirds of the total sown area in India is already drought-

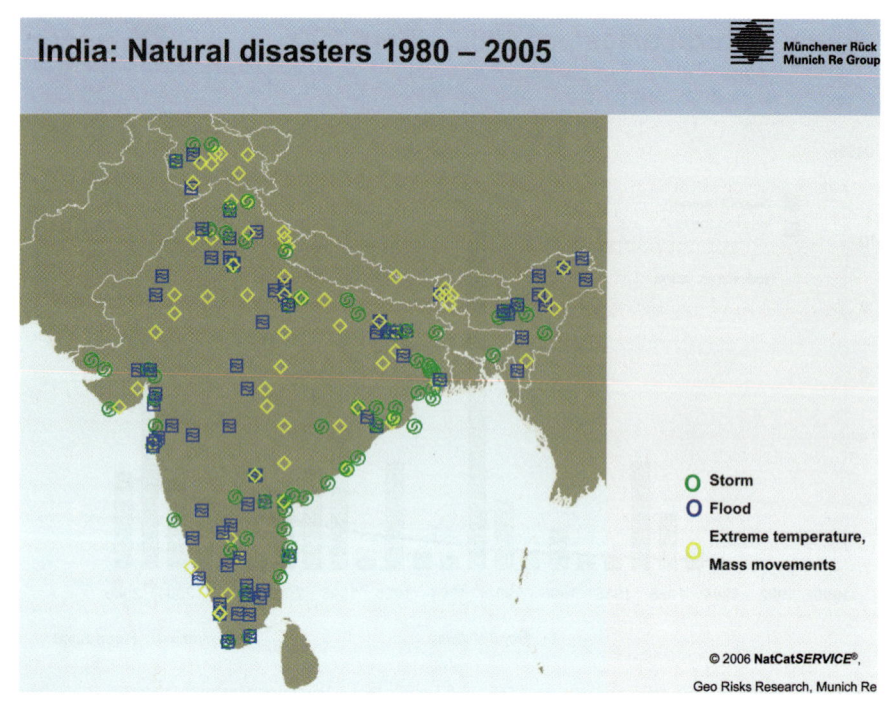

Fig. 21.5. Regional distribution of Natural Disasters in India between 1980 and 2005 (Data from Munich Re NatCatSERVICE)

MPCI- and crop hail markets 2005 (premium in Mio. Euro)

Country	MPCI	Hail	Total
USA	3.267	360	3.627
Canada	575	126	701
Spain	319	-	319
Italy	42	215	257
France	55	200	255
India	95	-	95
Austria	53	-	53
Portugal	28	-	28

Fig. 21.6. Multi Peril Crop Insurance (MPCI) and Crop/Hail markets in 2005 (premium in million Euros)

prone today, 40 million ha are liable to be flooded and the coastlines are vulnerable to tropical cyclones. The prospects for India in the future: further increases in air temperature, a decrease in the number of rainfall days and a rise in rainfall intensity, an increase in the severity of droughts and the intensity of floods, and rising sea levels may lead to a loss of settlements, property, and agricultural infrastructure.

Agricultural insurance can help farmers to cope with the increasing risks to their business. The requirements for successful crop insurance programmes are that the government introduces a Private-Public-Partnership Insurance Scheme by law, that insurance carriers specialise in agricultural insurance, that there is competition in services, that special insurance products are developed and that insurers have loss adjustment expertise.

21.4.1
Crop Insurance Products

There are basically four different groups of crop insurance products:
1. Loss insurance (hail and named peril) with fixed sums insured (e.g. hail insurance in Europe) or adjustable sums insured (e.g. cotton insurance in Australia)
2. Yield guarantee insurance (MPCI) with coverage of regional average yield (e.g. new MPCI programs) or individual historic yields (APH) (e.g. MPCI in USA, Spain)
3. Index insurance with meteorological triggers (single parameters or indices), area yield triggers (e.g. Group Risk Plan in USA), vegetation indices (increasing use of satellite data) and modelling of yields (a current example being the grassland programme in Spain)
4. Revenue insurance with covers comprising yield and price elements, which are only feasible for crops traded in existing commodity markets and boards of trade.

The requirements for successful crop insurance programmes are the control of anti-selection and the moral hazard, risk-adequate rates (exposure, crop type, fluctuation) facilitated by premium subsidies, regional differentiation of rates, sufficient market penetration as well as coverage levels and deductibles geared to the specific exposure. Figure 21.6 shows the 2005 premium levels of multi-peril crop insurance (MPCI) and crop hail insurance in different markets.

21.4.2
Crop Insurance in Developing Countries

In 2005, gross premium in India amounted to €95m. Crop insurance started there in 1972 and then developed in different phases. The National Agriculture Insurance Scheme (NAIS by AIC) has provided insurance products in all states and to all farmers since 1999. The main covers are yield guarantee and area yield, and a pilot project is under way for a product with a meteorological trigger. Thirty different crops can be insured during *Kharif* (SW monsoon, July-October), 25 crops during *Rabi* (winter months, only in the irrigated areas). In 2004/05, nearly 18 million farmers had bought insurance cover, 4 million farmers received indemnification.

Emerging markets are characterised by special circumstances with regard to agricultural insurance. Many of them provide dual agricultural structures, i.e. small-scale farms with limited access to technology and markets and modern farms. Governments have limited financial capacity to support agricultural insurance programmes and the lack of an insurance tradition makes the marketing of insurance products difficult. Small-scale farms hardly have any chance to have access to the traditional insurance market. A solution for them may be microinsurance systems especially designed for agriculture. These are based on low premiums and correspondingly low indemnification in the event of loss or damage, but they enable farmers to restore their basic business, i.e. to buy seeds for the next growing season.

Munich Re and the Munich Re Foundation, along with the traditional agricultural insurance sector, have started several initiatives especially for developing countries. The Munich Re Foundation has organised several conferences on microinsurance over the last two years and recently published a 600-page book on this. In 2007, another international microinsurance conference is to be held in Mumbai. In 2005, Munich Re launched the Munich Climate Insurance Initiative (MCII): a group of experts from the World Bank, NGOs, science and the insurance industry working on new insurance solutions to cover increasing loss and damage from weather-related disasters in developing countries. They are working in close contact with the negotiations of a Post-Kyoto Protocol under the umbrella of the UN Framework Convention on Climate Change.

21.5
Conclusions

Weather-related catastrophes like windstorms, floods and droughts have been increasing worldwide and will do so even more in the future. There is mounting evidence that global warming is contributing to the hazard situation. Agriculture is especially vulnerable to the changing weather patterns if no adaptation measures are taken (new seeds, production techniques). Proper insurance systems can help farmers to cope with the increasing volatility of their losses. Munich Re is the leading reinsurer worldwide for agro risks and offers its expertise to promote agro insurance systems in developing countries. The Munich Re Foundation has been – and will continue to be – very active in promoting microinsurance systems.

The insurance industry is quite powerful in supporting climate protection. It has a great potential to help raise awareness among decision-makers of the problems associated with global warming. By directing investments into companies and projects that promote renewable energies and sustainable processes, the insurance industry can help to reduce greenhouse gas emissions.

References

Barnett TP, Pierce DW, AchutaRao KM, Gleckler PJ, Santer BD, Gregory JM, Washington WM (2005) A Warning from Warmer Oceans. Science 309:284–287.
Emanuel K (2005) Increasing Destructiveness of Tropical Cyclones over the past 30 Years. Nature 436:686-688.
Hoyos CD, Agudelo PA, Webster PJ, Curry JA (2006) Deconvolution of the Factors Contributing to the Increase in Global Hurricane Intensity. Science 312:94-97.
Kumar, Ritu et al. (2007) The Indian insurance industry and climate change: exposure, opportunities, and strategies ahead. Climate Policy *(In Press)*
Stott PA, Stone DA, Allen MR (2004) Human Contribution to the European Heat Wave of 2003, Nature 432: 610-614.
Webster PJ, Holland GJ, Curry JA, Chang HR (2005) Changes in Tropical Cyclone Number, Duration, and Intensity in a Warming Environment. Science 309:1844–1846.

CHAPTER 22

Weather index insurance for coping with risks in agricultural production

Ulrich Hess

22.1 Introduction

This article presents innovations in agricultural risk management for natural disasters, focusing on the role of weather derivatives and weather index insurance in developing agricultural risk management strategies. The success story of weather risk insurance in India demonstrates to the world, that organized markets for risk can emerge and finance agricultural losses. Currently, many developing countries are particularly exposed to natural disaster risk without the benefit of ex-ante structures to finance losses. Instead, following each major drought event or other natural disaster, those affected must appeal for financial support and are left vulnerable to the mercy of ad-hoc responses from donor governments. Livelihoods are rarely insured by international insurance or reinsurance providers, capital markets, or even government budgets in developing countries where natural disasters and agricultural price risk impede development of both formal and informal banking. Trapped into this cycle of institutional underdevelopment, poor, risk-averse farmers are locked in poverty, burdened with old technology and faced with an inefficient allocation of resources.

22.1.1 Are there any effective precedents for agricultural insurance mechanisms in developing countries?

While financial innovations are just taking hold in some countries and others are continually being tested, progress has been made with weather insurance for farmers in India, Ukraine, Nicaragua, Malawi, Ethiopia, and Mexico. Several other experiments are also documented in this work.

Weather insured farmers in India say they either have a good crop in which case it does not matter if they do not recoup the insurance premium, or they have a monsoon failure in which case they receive an insurance payout. The payout at least covers the farmers' cash outlay and absorbs part of the agricultural income deficit resulting from shock, so that families may keep their children in school and preserve household assets that they would otherwise be forced to liquidate, often at greatly reduced prices. Given the right incentives, farmers can invest a little more in the right seeds and fertilizer and optimize their returns. Quantifying the impact of insurance is a delicate task requiring robust monitoring and evaluation systems, several years of household data including baseline data from both control and experimental groups as well as controls for the numerous other shocks that impact vulnerable populations. A large impact assessment is underway, led by the World Bank's Research Department and Professor Townsend, which will soon provide more information on the effectiveness of these risk management systems. What is already clear, however, is that when offered the choice, many farmers in India are willing to pay for fully-priced weather insurance. Even farmers with access to the government-subsidized crop insurance product choose to buy the market-priced weather insurance product, claiming preference for the objective nature of the weather index whose values they may personally verify with easily accessible weather station measurements. Farmers also prefer the timeliness of payout, guaranteed by index insurance to be made shortly after the close of the agricultural season, as opposed to payout made under the national crop insurance product, which may be paid out as much as eighteen months following the season.

22.1.2
Is this kind of insurance only suitable for large-scale commercial farmers?

Another advantage of weather insurance is that it can be sold to small farmers, as no monitoring is needed to verify farm-level losses. The Indian experience demonstrates that small farmers find value in weather insurance. BASIX , India's largest microfinance organization based in Andhra Pradesh, estimates that all of the 427 farmers who bought weather insurance policies in 2003 have small to medium-sized farms ranging from two to ten acres, producing average annual incomes of 15,000 to 30,000 Rupees, that is an average income of US$1 to US$2 per day. Currently, many farmers buying weather insurance in India are repeat customers whose low-income status does not limit them from buying the product. Early survey results demonstrate that more than half of those purchasing insurance products list managing risk as their primary reason. Some opt for insurance over the prospect of paying high interest to moneylenders when cash is urgently needed following harvest failure.

22.1.3
Is India's insurance program sustainable?

The weather insurance pilot programs launched in India in 2003 were mainstreamed into insurance programs offered by the major insurance companies sold to around half a million farmers. The biggest private insurer offering the product has broken even after two years. BASIX, the MFI that started the product has also mainstreamed the weather insurance product and automated delivery to

---

See Giné, Xavier, Robert Townsend and James Vickery, 2007, Patterns of Rainfall Insurance Participation in Rural India, working paper, 2007.

more than 10,000 clients for the 2006 season. Countries in sub-Saharan Africa, Latin America as well as other countries in Asia are setting up their own weather insurance projects at micro- and macro-levels. Malawi is now in its third year of smallholder based weather insurance as part of a credit package for quality seeds, however the package program experiences problems due smallholders unwillingness to repay despite good harvests. Local traders finally offered higher prices and therefore some farmers defaulted on their obligations to the programme. These problems are not related to the weather insurance which works well, yet they hinder scale-up and might even to an early termination. In 2006, the Government of Ethiopia successfully established contingency funding for emergency drought response in the form of weather index insurance. Thailand has a pilot for flood index insurance, Nicaragua launched a pilot in 2005, and Vietnam is setting up a large scale weather index insurance program. Markets seem to indicate that weather insurance is a sound and sustainable business, especially considering that the India experience was spurred without the support of government subsidies.

This article begins with an overview of risk in agriculture, focusing on how decision-makers currently cope with and manage risk in developing countries and on obstacles that impede development of effective risk transfer markets. Section 22.3 reviews the experiences of some developed countries with agricultural risk transfer. Section 22.4 explores alternative solutions based on the concept of weather index insurance, highlighting the advantages of such systems for developing countries. Section 22.5 describes the role of government in these markets. Section 22.6 provides an overview of a number of ongoing agricultural risk pilot programs and case studies in various countries.

22.2
Risk and Risk Management in Agriculture

To set the stage for the discussion on how to deal with risk in agriculture, we classify the different sources of risk. Agriculture is often characterized by high variability of production outcomes, that is, by *production risk*. Unlike most other entrepreneurs, agricultural producers cannot predict with certainty the amount of output their production process will yield, due to external factors such as weather, pests, and diseases. Agricultural producers can also be hindered by adverse events during harvesting or collecting that may result in production losses. In discussing how to design appropriate risk management policies, it is useful to understand strategies and mechanisms employed by producers to deal with risk, including the distinction between informal and formal risk management mechanisms and between ex-ante and ex-post strategies.

As explained in the 2000/2001 World Development Report (World Bank 2001), informal strategies are identified as "arrangements that involve individuals or households or such groups as communities or villages," while formal arrangements are "market-based activities and publicly provided mechanisms." The ex ante or ex-post classification identifies the time in which the response to risk takes place: ex-ante responses take place before the potential harming event; ex post responses take thereafter. Ex-ante informal strategies are characterized by diversification of income sources and choice of agricultural production strategy. One strategy producers employ is risk avoidance. Extreme poverty, in many cases, makes producers very risk-averse, pushing them to avoid high-risk activities, even though the income gains to be generated might be far greater than those gotten through less risky choices. This inability to accept and manage risk respectively reflected in the inability to accumulate and retain wealth is sometimes referred to as the "the poverty trap" (World Bank 2001).

22.2.1
Informal risk management mechanisms

The production strategy selected becomes an important means of mitigating the risk of crop failure. Traditional cropping systems in many places rely on crop and plot diversification. Crop diversification and intercropping systems signify common strategies of reducing the risk of crop failure due to adverse weather events, crop pests, or insect attacks. Morduch (1995) presents evidence that households whose consumption levels are close to subsistence (and which are therefore highly vulnerable to income shocks) devote a larger share of land to safer, traditional varieties such as rice and castor than to riskier, high-yielding varieties. Morduch also finds that near-subsistence households diversify their plots spatially to reduce the impact of weather shocks that vary by location. Apart from altering agricultural production strategies, households also smooth income by diversifying income sources, thus minimizing the effect of a negative shock to any one of them. According to Walker and Ryan (1990), most rural households in villages of semi-arid India surveyed by the International Crops Research Institute for the Semi-Arid Tropics (ICRISAT) generate income from at least two different sources; typically, crop income is accompanied by some livestock or dairy income. Off-farm seasonal labour, trade, and sale of handicrafts are also common income sources. The importance to risk management of income source diversification is emphasized by Rosenzweig and Stark (1989), who find that households with high farm-profit volatility are more likely to have a household member engaged in steady wage employment. Accumulating a buffer stock of crops or liquid assets at the expense of credit expenditure presents obvious means by which households can smooth consumption. Lim and Townsend (1998) show that currency and crop inventories function as buffers or precautionary savings. Crop-sharing arrangements in renting land and hiring labour can also provide an effective means of sharing risk among individuals, thus reducing producer risk exposure (Hazell 1992). Other risk sharing mechanisms, such as community-level risk pooling, occur in specific communities or extended households where group members transfer resources among themselves to rebalance marginal utilities (World Bank 2001). These arrangements, however, while effective for counterbalancing the consequences of events affecting only some members of the community, do not work well in the cases of covariate income shocks (Hazell 1992). That is, in the event of widespread shock where most members of the same community are affected, as is often the case with natural disasters, small-scale risk pooling offers little in the way of absorbing the impact of shock.

Typical ex-post informal income-smoothing mechanisms include the sale of assets, such as land or livestock (Rosenzweig and Wolpin 1993), or the reallocation of labour resources to off-farm labour activities. Gadgil et al. (2002) argue that southern Indian farmers who expect poor monsoon rains can quickly shift from 100 percent on-farm labour activities to predominantly off-farm activities. Fafchamps (1993), in his analysis of rain-fed agriculture among West African farmers, emphasizes the importance of building labour flexibility into the production strategy. In contrast, Rosenzweig and Binswanger (1993), Morduch (1995), and Kurosaki and Fafchamps (2002) all find considerable efficiency losses associated with risk mitigation, typically due to lack of specialization and inability to reach economies of scale. In effect, farmers stabilize income flows at the detriment of maximizing profits; this tendency to smooth consumption not only against idiosyncratic shocks but also against correlated shocks comes at a serious cost in terms of production efficiency and reduced profits, thus lowering overall levels of household consumption and prospects for asset accumulation.

A major consideration for innovation would be to shift correlated risk away from rural households (Skees 2003). One obvious solution would be for rural households to share risk with households or institutions from areas largely uncorrelated with the local risk conditions. Examples of such extra-regional risk sharing systems are found in the literature, including, credit and transfers between distant relatives (Rosenzweig 1988; Miller and Paulson 2000); migration and marriages (Rosenzweig and Stark 1989); or ethnic networks (Deaton and Grimard 1992). Although the examples above convey some degree of risk sharing and thus of informal insurance measures against weather, such systems cannot be scaled up to offer wider coverage nor do come even close to providing a fully efficient insurance mechanism. The world's most vulnerable households are therefore left largely unprotected against correlated risks, the main source of which is weather.

22.2.2.1
Formal risk management mechanisms

Formal risk management mechanisms can be classified as publicly provided or market-based. Government action plays an important role in agricultural risk management, both ex-ante and ex-post. Ex-ante education and services provided by agricultural extension help familiarize producers with the consequences of risk and help them adopt related coping strategies. Governments also help to reduce the impact of risk by developing relevant infrastructure and by adopting social schemes and cash transfers for relief after shocks have occurred. As mentioned under the explanation of informal mechanisms, production and market risks probably inflict the largest impact on agricultural producers. Various market-based, risk management solutions have been developed to address these sources of risk. Insurance is another formal mechanism used in many countries to share production risks. However, insurance does not as efficiently manage production risk as efficiently as do derivative merkets. Price risk is highly spatially correlated, and futures and options are tailor-made derivative instruments appropriate for dealing with spatially correlated risks. In contrast, insurance is most appropriate for managing independent risks that are spatially uncorrelated.

22.2.2.2
Challenges for traditional crop insurance

Agricultural production risks typically lack sufficient spatial correlation to be effectively hedged using only exchange-traded futures or options. At the same time, agricultural production risks are generally not perfectly spatially independent; therefore, insurance markets do not work at their best. Skees and Barnett (1999) refer to these risks as "in-between" risks. According to Ahsan et al. (1982), "good or bad weather may have similar effects on all farmers in adjoining areas," and, consequently, "the law of large numbers, on which premium and indemnity calculations are based, breaks down." In fact, positive spatial correlation in losses limits the risk reduction capacity obtained by pooling risks from different geographical areas, thus increasing the variance in indemnities paid by insurers. In general, the greater the positive correlation in losses the less efficient traditional insurance is as a risk-transfer mechanism.

The lack of statistical independence is not the only problem with providing insurance in agriculture. Another set of problems relates to asymmetric information, the situation in which the insured has more knowledge about his or her risk profile than does the insurer. Asymmetric information causes two problems: *adverse selection* and *moral hazard*. In the case of adverse selection, farmers have better knowledge than do the insurers about the probability distribution of losses. The farmers thus occupy the privileged situation of knowing whether or not the insurance premium accurately reflects the risk they face. Consequently, only farmers bearing greater risks will purchase the coverage, generating an imbalance between indemnities paid and premiums collected. Moral hazard similarly affects the incentive structure of the relationship between insurer and insured. After entering the contract, the farmer's incentive to take proper care of the crop diminishes, while the insurer has limited effective means to monitor what may prove hazardous behaviour by the farmer. Insurers may thus incur greater than anticipated losses. Agricultural insurance is often characterized by high administrative costs, due, in part, to the risk classification and monitoring systems that insurers must put in place to forestall asymmetric information problems. Other costs include acquiring the data needed to establish accurate premium rates and conducting claims adjustments. As a percentage of the premium, the smaller the policy, typically, the larger the administrative costs. Spatially correlated risk, moral hazard, adverse selection, and high administrative costs are all important reasons why agricultural insurance markets may fail.

Cognitive failure among potential insurance purchasers and ambiguity loading on the part of insurance suppliers are other possible causes of agricultural insurance market failure. If consumers fail to recognize and plan for low-frequency, high-consequence events, the likelihood that an insurance market will emerge diminishes. When considering an insurance purchase, the consumer may have difficulty determining the value of the contract or, more specifically, the probability and magnitude of loss relative to the premium (Kunreuther and Pauly 2001). Many decision makers tend to underestimate their exposure to low-frequency, high-consequence losses and thus are unwilling to pay the full costs of an insurance product that protects them against these losses. Low-frequency events, even when severe, are frequently discounted or ignored altogether by producers trying to determine the value of an insurance contract. This happens because the evaluation of probability assessments regarding future events is complex and often entails high search costs. Many people resort to various simplifying heuristics, but probability estimates based on these heuristics may differ greatly from the true probability distribution (Schade et al. 2002; Morgan and Henrion 1990).

Evidence indicates that agricultural producers forget extreme low-yield events. The general finding regarding subjective crop-yield distributions is that agricultural producers tend to overestimate the mean yield and underestimate the variance (Buzby et al. 1994; Pease et al. 1993; Dismukes et al. 1989). On the other side, insurers will typically load premium rates heavily for low-frequency, high-consequence events where considerable ambiguity surrounds the actual likelihood of the event (Schade et al. 2002; Kunreuther et al. 1995). Ambiguity is especially serious when considering highly skewed probability distributions with long tails, as is typical of crop yields. Uncertainty is further compounded when the historical data used to estimate probability distributions are incomplete or of poor quality, a very common problem in developing countries.

Small sample size creates large measurement error, especially when the underlying probability distribution is heavily skewed. Kunreuther et al. (1993) demonstrate via experimental economics that when risk estimates are ambiguous, loads on insurance premiums can be 1.8 times higher than when insuring events with well specified probability and loss estimates. Together, these effects create a wedge between the prices that farmers are willing to pay for catastrophic agricultural insurance and the prices that insurers are willing to accept. Thus, functioning private-sector markets may fail to materialize or, if they do materialize, they may cover only a small portion of the overall risk exposure (Pomareda 1986).

To better understand agricultural risk management markets and government policies to facilitate access to risk management instruments, it is worthwhile to analyze critically the experiences of some developed countries. The experiences of the United States, Canada, and Spain are thus described for reference, but it is important to consider that these systems may not be replicable in or suitable for most developing countries. In addition, many developed countries have involved market support and income transfer programs that extend well beyond crop insurance. To the extent they are based on farm income, these programs involve levels of protection against severe crop failures. The European community has extensive policies focusing on income protection.

22.3
Crop Insurance Programs in Developed Countries

This section provides an overview of agricultural risk management programs in three developed countries: the United States, Canada, and Spain. Substantial progress in developing commercial crop insurance markets has taken hold in these countries through which agricultural producers may reduce yield and revenue risk.

While these programs offer a variety of risk management products for farmers, the programs require levels of government support unfeasible for most, especially developing, countries.

22.3.1
The United States

In the United States, multiple peril yield and revenue insurance products are offered through the Federal Crop Insurance Program (FCIP), a public/private partnership between the federal government and various private sector insurance companies. The program seeks to address both social welfare and economic efficiency objectives. With regard to social welfare, private companies selling federal crop insurance policies may not refuse to sell to any eligible farmer, regardless of past loss history. At the same time, the program aims to be actuarially sound. Policies are available for over one hundred commodities but in 2004 just four crops–corn, soybeans, wheat, and cotton–accounted for approximately 79 percent of the US$4 billion in total premiums. Excluding pasture, rangeland, and forage, approximately 72 percent of the national crop acreage is currently insured under the FCIP. About 73 percent of total premiums are for revenue insurance policies, while 25 percent are for yield insurance policies.

Most FCIP policies trigger indemnities at the farm, or even sub-farm, level. Yield insurance offers are based on a rolling four-to-ten-year average yield, known as the actual production history (APH) yield. The federal government provides farmers with a base catastrophic yield insurance policy, free of any premium costs. Farmers may then choose to purchase, at federally subsidized prices, additional insurance coverage beyond the catastrophic level. This additional coverage, often called "buy-up" coverage, may be either yield or revenue insurance. Farm-level revenue insurance offers are based on the product of the APH yield and a price index that reflects national price movements for the particular commodity. For some crops and regions, defined along county barriers, area yield and/or area revenue buy-up insurance policies are offered through FCIP. On a per acre insured basis, area-level insurance products tend to be less expensive than farm-level insurance products. Thus, in 2004, area yield and area revenue policies accounted for 7.4 percent of total acreage insured but less than 3 percent of total premiums.

The U.S. government also provides a reinsurance mechanism that allows insurance companies to determine (within certain bounds) which policies they will retain and which they will cede to the government. This arrangement is referred to as the standard reinsurance agreement (SRA). The SRA is quite complex, with both quota share reinsurance and stop losses by state and insurance pools; however, in essence, it allows the private insurance companies to adversely select against the government. This is considered necessary since the companies do not establish premium rates or underwriting guidelines but are required to sell policies to all eligible farmers. The federal costs associated with the U.S. program have four components.

1) Federal premium subsidies range from 100 percent of total premium for catastrophic (CAT) policies to 38 percent of premium for buy-up policies at the highest coverage levels. Across all FCIP products and coverage levels, the average premium subsidy in 2004 was 59 percent of total premiums.
2) The federal government reimburses administrative and operating expenses for private insurance companies that sell and service FCIP policies. This reimbursement is approximately 22 percent of total premiums.
3) The SRA has an embedded federal subsidy with an expected value of about 14 percent of total premiums.
4) The program, by law, can be considered actuarially sound at a loss ratio of 1.075. This implies an additional federal subsidy of 7.5 percent of total premiums. On average, the federal government pays approximately 70 percent of the total cost for the FCIP - farmer-paid premiums accounting for the remaining 30 percent. While the direct farmer subsidy varies by coverage level, the United States has consistently passed legislation increasing the subsidy level to farmers for crop and revenue insurance products. The rate of subsidy is one component that has influenced the growth in overall premium. The growth in premium subsidy is greater than the growth in farmer-paid premiums. The rate of subsidy increased in 1995 and 2001.

22.3.2
Canada

In 2003, Canada revised its agricultural risk management programs. The "Business Risk Management" element of the new Agricultural Policy Framework (APF) is composed of two main schemes: Production Insurance and Income Stabilization. The Production Insurance (PI) scheme offers producers a variety of multiple-peril production or production value loss products similar to many of those sold in the United States. One major distinction, however, is that the Canadian program is marketed, delivered, and serviced entirely and jointly by federal and provincial government entities, although it is the provincial authorities who are ultimately responsible for insurance provision. This allows provinces some leeway to tailor products to fit their regions and to offer additional products. Production insurance plans are offered for over one hundred different crops, and provisions have been made to include plans covering livestock losses as well. Crop insurance plans are available based on either individual yields (or production value in the case of certain items, such as stone-fruits) or area-based yields. Unlike the U.S. program, Canadian producers are not allowed to separately insure different parcels but rather must insure together all parcels of a given crop type. This means that low yields on one parcel may be offset by high yields on another parcel when determining whether or not an overall production loss has occurred. Insurance can also be purchased for loss of quality, unseeded acreage, replanting, spot loss, and emergency works. The latter coverage is a loss mitigation benefit meant to encourage producers to take actions that reduce the magnitude of crop damage caused by an insured peril. Cost sharing between the federal government and each province for the entire insurance program was fixed at 60:40, respectively, in 2006. Federal subsidies as a percentage of premium costs vary, however, from 60 percent for catastrophic loss policies to 20 percent for low deductible production coverage. Combined, the federal and provincial governments cover approximately 66 percent of program costs, including administrative costs. This is roughly equivalent to the percentage of total program costs borne by the federal government in the U.S. program. Provincial authorities are responsible for the solvency of their insurance portfolio. In Canada, the federal government competes with private reinsurance firms in offering deficit financing agreements to provincial authorities.

Beginning in 2004, the Canadian Agricultural Income Stabilization (CAIS) scheme replaced and integrated former income stabilization programs. CAIS is based on the producer production margin, where a margin is defined as "allowable farm income," including proceeds from production insurance minus "allowable (direct production) expenses." The program generates a payment when a producer's current year production margin falls below that producer's reference margin, which is based on an average of the program's previous five-year margins, less the highest and lowest. One important feature of CAIS is that producers must participate in the program with their own resources. In particular, a producer is required to open a CAIS account at a participating financial institution and deposit an amount based on the level of protection chosen (coverage levels range from 70 percent to 100 percent of the "reference margin"). Once producers file their income tax returns, the CAIS program administration uses the tax information to calculate the producer's program year production margin. If the program year margin has declined below the reference margin, some of the funds from the producers' CAIS accounts will be available for withdrawal. Governments match the producers' withdrawals in different proportions for different coverage levels. The total investment by federal and provincial governments for the "business risk management" programs is CAN$1.8 billion per year. In 2004, approximately CAN$600 million was provided by governments as insurance premium subsidies.

22.3.3
Spain

The Spanish agricultural insurance system is structured around an established public/private partnership. On the public side is the National Agricultural Insurance Agency (ENESA), which coordinates the system and manages resources for subsidizing insurance premiums, and the Insurance Compensation Agency *(Consorcio de Compensación de Seguros)* that, together with private reinsurers, provides reinsurance for the agricultural insurance market. Local governments are involved only to the extent that they are allowed to augment premium subsidies offered at the national level. On the private side, insurance contracts are sold by Agroseguro, a coinsurance pool of companies that aggregates all insurance companies active in agriculture. Farmers, insurers, and institutional representatives are all part of a general commission hosted by ENESA that functions as the managing board of the Spanish agricultural insurance system.

Similar to programs in the United States and Canada, Spain's combined program offers insurance policies covering multiple perils. Policies are available for crops, livestock, and aquaculture activities, with risks being pooled across the country by Agroseguro. Compared to the United States and Canada, however, farmers' associations are more actively involved in implementation and development of agricultural insurance. The government has reserves to cover extreme losses, and, as a final resort, the government treasury covers losses that occur beyond these reserves. Total premiums for agriculture insurance policies purchased reached around US$550 million (€490 million) in 2003, of which approximately US$225 million (€200 million) or 41 percent of total cost, have been provided by the government (Burgaz 2004). The rationale behind subsidizing agricultural insurance is that this outlay serves as a disincentive for the government to also provide free ad-hoc disaster assistance. To reinforce the point, Spanish producers are ineligible for disaster payments for perils for which insurance is offered. For non-covered perils, ad hoc disaster payments are available, but only if the producer had already purchased agricultural insurance for covered perils.

22.3.4
Experiences of developed countries provide inadequate models for developing countries

For various reasons, developing countries should avoid adopting approaches to risk management mirroring those adopted in developed countries. Clearly, developing countries are more limited by fiscal resources than are developed countries. Even more importantly, the opportunity costs of employing those limited fiscal resources are significantly greater. Thus, it is critical for developing country governments to consider carefully how much risk management support is appropriate and how to leverage limited government dollars to spur insurance markets. In developed countries, government risk management programs are as much about income transfers between economic sectors as they are about risk management. Developing countries cannot afford to facilitate similar income transfers, given the high proportion of the population engaging in agriculture. On the other hand, because large segments of developing country populations depend on agriculture or agriculture-related industries for their livelihoods, catastrophic agricultural losses will have much greater relative impact on overall GDP as well on affected populations.

Policymakers should also carefully consider the varying structural characteristics of agriculture in different countries. In general, farms in developing countries are significantly smaller than are farms in countries like the United States and Canada. For traditional crop insurance products, smaller farms typically imply higher administrative costs as a percentage of total premiums. A portion of these costs is related to marketing and servicing (loss adjustment) insurance policies. Another portion is related to the lack of farm-level data and cost-effective mechanisms for controlling moral hazard.

Developing countries also have far less access to global crop reinsurance markets than do developed countries. Reinsurance contracts typically involve high transaction costs related to due diligence. Reinsurers must understand every aspect of the insurance product development and transaction process, including contract design, pricing, underwriting and establishing controls against adverse selection and moral hazard. Some minimum volume of business, or the prospect for strong future business, must be present to rationalize incurring these largely fixed transaction costs. For a global reinsurer to be willing to enter a market, the enabling environment must foster confidence in contract enforcement and institutional regulations. An enabling environment is, in fact, a prerequisite for effective and efficient insurance markets, and these components are largely missing in developing countries. Setting rules and then precedents assuring that premiums will be collected and indemnities paid is not a trivial undertaking. The alternative risk management products discussed in Sections 22.4 and 22.5 are structured to overcome many of these problems.

22.4
Weather index insurance alternatives

Given the problems with some traditional crop insurance programs in developed countries, finding new solutions to help mitigate several aspects of the problems outlined above has become critical. Index insurance products, designed to overcome many of the problems plaguing multiple-peril farm-level crop insurance products, offer some potential in this regard (Skees et al. 1999). These contingent claims contracts base payments on an independent measure, such as rainfall or temperature that is highly correlated with farm-level yield or revenue outcomes. Unlike traditional crop insurance that attempts to measure individual farm yields or revenues, index insurance makes use of variables exogenous to the individual policyholder but have a strong correlation to farm-level losses. For most insurance products, a precondition for insurability is that the loss for each exposure unit be uncorrelated (Rejda 2001). For index insurance, a precondition is that risk be spatially correlated. When yield losses are spatially correlated, index insurance contracts can be an effective alternative to traditional farm-level crop insurance. Index products also facilitate transfer of risk into financial markets where investors acquire index contracts as another investment in a diversified portfolio. In fact, index contracts may offer significant diversification benefits, since the returns generally should be uncorrelated with returns from traditional debt and equity markets.

22.4.1
Basic characteristics of an index

The underlying index used for index insurance products must be correlated with yield or revenue outcomes for farms across a large geographic area. In addition, the index must satisfy a number of additional properties affecting the degree of confidence or trust that market participants have that the index is believable, reliable, and void of human manipulation; that is, the measurement risk for the index must

be low (Ruck 1999). A suitable index required that the random variable measured meet the following criteria:
- observable and easily measured;
- objective;
- transparent;
- independently verifiable;
- reportable in a timely manner (Turvey 2002; Ramamurtie 1999); and
- stable and sustainable over time.

Publicly available measures of weather variables generally satisfy these properties.

For weather indexes, the units of measurement should convey meaningful information about the state of the weather variable during the contract period, and they are often shaped by the needs and conventions of market participants. Indexes are frequently cumulative measures of precipitation or temperature during a specified time. In some applications, average precipitation or temperature measures are used instead of cumulative measures. New innovations in technology, including sophisticated satellite imagery from which high resolution weather data may be extracted and low-cost weather monitoring stations that can be placed in many locations, will expand the number of areas in which weather variables can be measured as well as of the types of measurable variables. Measurement redundancy and automated instrument calibration further increase the credibility of an index.

22.4.2
Structure of index insurance contracts

The terminology used to describe features of index insurance contracts resembles that used for futures and options contracts rather than for other insurance contracts. Rather than referring to the point at which payments begin as a trigger, for example, index contracts typically refer to it as a strike. They also pay in increments called ticks. Consider a contract being written to protect against deficient cumulative rainfall during a cropping season. The writer of the contract may choose to make a fixed payment for every one millimetre (mm) of rainfall below the strike. If an individual purchases a contract where the strike is one hundred millimetres of rain and the limit is fifty mm, the amount of payment for each tick would be a function of how much liability is purchased. There are fifty ticks between the one hundred mm strike and fifty mm limit. Thus, if $50,000 of liability were purchased, the payment for each one mm below one hundred mm would be equal to $50,000/(100 – 50), or $1,000. Once the tick and the payment for each tick are known, the indemnity payments are easy to calculate. A realized rainfall of ninety mm, for example, results in ten payment ticks of $1,000 each, for an indemnity payment of $10,000. Figure 22.1 maps the payout structure for a hypothetical $50,000 rainfall contract with a strike of one hundred mm and a limit of fifty mm.

In developed countries, index contracts that protect against unfavourable weather events are now sufficiently well developed that some standardized contracts are

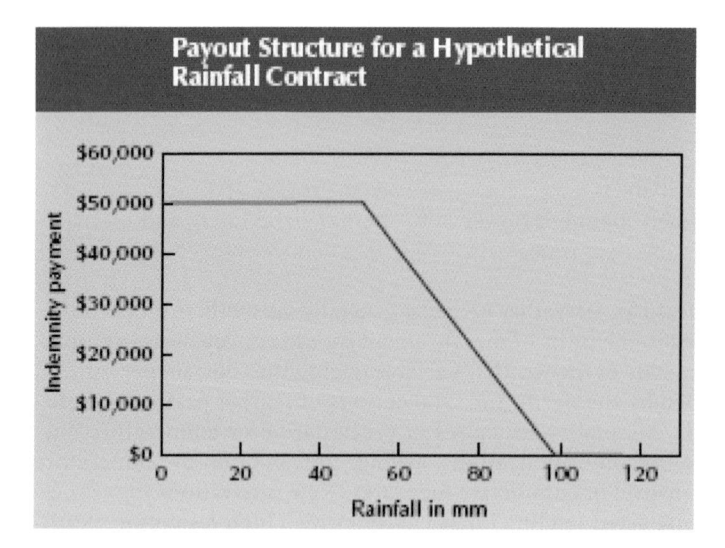

Fig. 22.1. Payout structure for a hypothetical rainfall contract

traded in exchange markets. These exchange-traded contracts are used primarily by firms in the energy sector, although the range of weather phenomena that might potentially be insured using index contracts appears to be limited only by imagination and the ability to parameterize the event. A few examples include excess or deficient precipitation during different times of the year, insufficient or damaging wind, tropical weather events such as typhoons, various measures of air temperature, measures of sea surface temperature, the El Niño Southern Oscillation (ENSO) tied to El Niño and La Niña, and even celestial weather events such as disruptive geomagnetic radiation from solar flare activity. Contracts are also designed for combinations of weather events, such as snow and temperature (Dischel 2002; Ruck 1999).

The potential for the use of index insurance products in agriculture is thus significant (Skees 2003). A major challenge in designing an index insurance product is minimizing basis risk. Basis risk refers to the potential mismatch between index triggered payouts and actual losses. It occurs when an insured has a loss and does not receive an insurance payment sufficient to cover the loss (minus any deductible) or when an insured has a loss and receives a payment that exceeds the amount of loss. Since index-insurance indemnities are triggered by exogenous random variables, such as area yields or weather events, an index-insurance policyholder can experience a yield or revenue loss and not receive an indemnity. The policyholder may also experience no yield or revenue loss and still receive an indemnity. The effectiveness of index insurance as a risk management tool depends on how positively correlated farm yield losses are with the underlying index. In general, the more homogeneous the area, the lower the basis risk and the more effective area-yield insurance will be as a farm-level risk management tool. Similarly, the more closely a given weather index actually represents weather events on the farm, the more effective the index will be as a farm-level risk management tool.

22.4.3
Relative advantages and disadvantages of index insurance

Index insurance can sometimes offer superior risk protection compared to traditional farm-level, multiple- peril crop insurance. Deductibles, co-payments, or other partial payments for loss are commonly used by farm-level, multiple-peril insurance providers to mitigate asymmetric information problems such as adverse selection and moral hazard. Asymmetric information problems are much lower with index insurance because, first, a producer has little more information than the insurer regarding the index value, and second, individual producers are generally unable to influence the index value. This characteristic of index insurance means that there is less need for deductibles and co-payments. Similarly, unlike traditional insurance, few restrictions need be placed on the amount of coverage an individual purchases. As long as the individual farmer cannot influence the realized value of the index, liability need not be restricted. An exception occurs when governments offer premium subsidies as a percentage of total premiums. In this case, the government may want to restrict liability (and thus, premium) to limit the amount of subsidy paid to a given policyholder. As more sophisticated systems, such as satellite imagery, are developed to measure events causing widespread losses, indexing major events should become more straightforward and quite acceptable to international capital markets. Under these conditions, traditional reinsurers and primary providers may begin offering insurance in countries they would have never previously considered. New risk management opportunities can develop if relevant, reliable, and trustworthy indexes can be constructed.

22.4.4
The trade-off between basis risk and transaction costs

Among the most significant issues for any insurance product is the question of how much monitoring and administration is needed to keep moral hazard and adverse selection to a minimum. To accomplish this goal, coinsurance and deductibles are used so that the insured shares the risk and any mistakes in offering too generous coverage are mitigated. Considerable information is needed to tailor insurance products and to minimize the basis risk even for individual insurance contracts. Increased information gathering and monitoring involve higher transaction costs, which convert directly into the higher premiums needed to cover them. Index insurance significantly reduces these transaction costs and can be written with lower deductibles and without introducing coinsurance. When farm yields are highly correlated with the index being used to provide insurance, offering higher levels of protection can result in risk transfer superior even to individual multiple-peril crop insurance (Barnett et al. 2005). The direct trade-off between basis risk and transaction costs has implications for achieving sustainable product designs and for outlining the role of governments and markets. Section 21.5 will introduce the idea of layering risk – an approach that also greatly depends on understanding the trade-off between basis risk and transaction costs. Under the risk-layering exercise, at least one party is assumed to accept a certain degree of basis risk at each layer of the risk transfer if the product is to be both sustainable and affordable. Otherwise, extremely high, and most likely, unaffordable transaction costs must be paid for products specifically designed to minimize or nullify basis risk. In effect, the social cost of having products with some basis risk may be significantly lower than the social cost of having products with no basis risk but high transaction costs.

22.4.5
Where index insurance is inappropriate

As with traditional crop insurance, index insurance contracts are not suitable for all agricultural producers. Many agricultural commodities are grown in microclimates. For example, coffee grows on mountainsides in countries with varied climates, and fruit such as apples and cherries also commonly grow in areas with very large differences in weather patterns within only a few miles. In highly spatially heterogeneous production areas, basis risk will likely be so high as to make index insurance problematic. Under these conditions, index insurance will work only if it is highly localized and/or can be written to protect only against the most extreme loss events. Even in these cases, it may be critical to tie index insurance to lending, since loans are one method of mitigating basis risk.

Over fitting the data is another concern with index insurance. If one has a limited amount of crop yield data, fitting the statistical relationship between the index and those limited data can become problematic. Small sample sizes and fitting regressions within the sample can lead to complex contract designs that may or may not be effective hedging mechanisms for individual farmers. Standard procedures that assume linear relationships between the index and realized farm-level losses may be inappropriate. While scientists are tempted to fit complex relationships to crop patterns, interviews with farmers may reveal more about the types of weather events of most concern. When designing a weather index contract, one may be tempted to focus on the relationship between weather events and a single crop. When it fails to rain for an extended period of time, however, many crops will be adversely affected. Likewise, when it rains for an extended period of time, resulting in significant cloud cover during critical photosynthesis periods, a number of crops may suffer.

Finally, when designing index insurance contracts, significant care must be taken to assure that the insured has no better information about the likelihood and magnitude of loss than does the insurer. Farmers' weather forecasts are often highly accurate. Potato farmers in Peru, using celestial observations and other indicators in nature, are able to forecast El Niño at least as well as many climate experts (Orlove et al. 2002). In 1988, an insurer offered drought insurance in the U.S. Midwest. As the sales closing date neared, the company noted that farmers were significantly increasing their purchases of these contracts. Rather than recognize that these farmers had already made a conditional forecast that the summer was going to be very dry, the company extended the sales closing date and sold even more rainfall insurance contracts. The company experienced very high losses and was unable to meet the full commitment of the contracts. Rainfall insurance for agriculture in the United States suffered a significant setback. The lesson learned is that when writing insurance based on weather events, it is crucial to be diligent in following and understanding weather forecasts and any relevant information available to farmers. Farmers have a vested interest in understanding the weather and climate. Insurance providers who venture into weather index insurance must know at least as much as farmers do about conditional weather forecasts. If not, intertemporal adverse selection will render the index insurance product unsustainable. These issues can be addressed; typically, the sales closing date must be established in advance of any potential forecasting information that would change the probability of a loss beyond the norm. But beyond simply setting a sales closing date, the insurance provider must have the discipline and the systems in place to ensure that no policies are sold beyond that date.

22.5
Application of weather index insurance in developing countries: The role of government

Should the lack of effective private-sector agricultural insurance markets in developing countries be addressed through government intervention? High transactions costs preclude emergence of many markets, but this does not necessarily justify government intervention. In the case of high-frequency, low-consequence losses, government intervention is likely to distort incentives and create rent-seeking opportunities, possibly to an extent that actually reduces net social welfare. Instead, farmers can employ other risk management mechanisms to cover these losses. In fact, insurance products for *high-frequency, low-consequence* losses are seldom offered because the transaction costs associated with loss adjustment renders the insurance cost prohibitive for most potential purchasers. Governments may have no inherent advantage over markets in trying to facilitate the provision of individual farm-level yield or revenue insurance products. The private sector typically does not provide these insurance products in part because of information asymmetries that cause moral hazard and adverse selection problems (Miranda and Glauber 1997); it is difficult to see how a government provider would have any advantage in addressing this problem.

22.5.1
Premise: The concept of risk layering

Segmenting risk into different "layers" is a key risk management principle. Consider, for example, Figure 22.2, which shows the probability distribution for average April to October rainfall at thirteen weather stations in Malawi (Rejda 2001). Suppose that farmers start incurring production losses whenever rainfall is less than one thousand millimetres. The domain of losses might be segregated into three risk layers, with different entities holding each layer:
- For rainfall in excess of 700 mm, farmers would retain the loss risk, either individually or with financial service providers: the risk retention layer.

- For rainfall between 500 and 700 mm, the risk would be transferred to an insurance company via a weather index insurance product: the market insurance layer.
- For rainfall levels below 500 mm, the risk in this example would not be insured due to cognitive failure and ambiguity loading: the market failure layer.

Farmers would absorb losses in the risk retention layer using self-insurance strategies such as those described in Chapter 22.2. Strategies for effectively transferring the other risk layers are described below.

Referring again to Figure 22.2, suppose that an insurance provider writes a rainfall index insurance contract with a strike of 700 mm and a limit of 500 mm. Limits are commonly used by weather index insurance writers to avoid open-ended exposure to catastrophic weather events. The insured would select the amount of insurance (the liability) and the payment per tick would be calculated using this formula.

$$\text{Payment per Tick} = \frac{\text{Liability}}{\text{Limit} - \text{Strike}}$$

Assume that a farmer has a crop with an expected value of $15,000. At only 500 mm of rainfall, the farmer is estimated to lose two-thirds of the value of the crop. Thus, the farmer purchases $10,000 of liability, with a payment for each tick (each mm of rainfall) of fifty ($10,000 divided by (700 × 500)). If the realized value for the rainfall index is 600 mm, for example, the indemnity will be $5,000 ((700 × 600) $50) (Barnett et al. 2005). The limit of 500 mm caps the insurance provider's loss exposure on the index insurance product. Without the limit, the contract would be extremely expensive, since it would protect against losses in the extreme lower tail of the probability distribution. Buyers would exhibit cognitive failure regarding

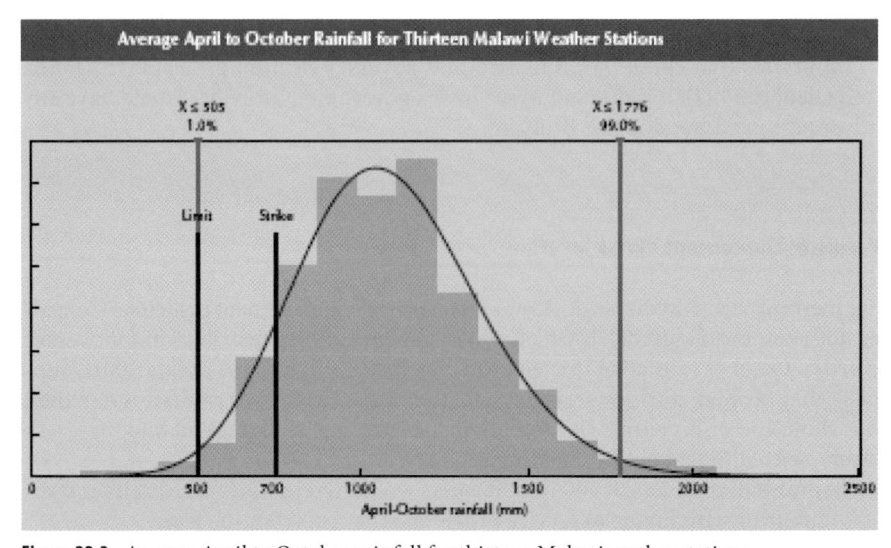

Figure 22.2. Average April to October rainfall for thirteen Malawi weather stations

the probability of events with less than 500 mm of rainfall, while insurance providers would load the premium for ambiguity regarding these same events. Thus, even if insurance was available to protect against rainfall events of less than 500 mm, few transactions would be likely, since the premium would exceed most buyer's willingness to pay.

Market failure layer: At the catastrophic loss layer represented by market failure, private decision makers will likely not purchase adequate insurance due to cognitive failure, ambiguity loading of premiums rates, and perhaps, expectations of government or donor disaster relief. Some form of government intervention may be required to facilitate adequate transfer of the risk.

22.5.2
Policy instruments

Risk layering provides an helpful conceptual framework for thinking about government intervention in risk transfer markets. The discussion of the market insurance layer described situations in which government packaging or pooling of risk could potentially reduce the transaction costs associated with risk transfer and thus the premiums paid by end users. This section explores other possible government interventions, including government facilitation of risk transfer in the market failure layer, the role of government subsidies in risk transfer markets, and potential uses of index insurance instruments to finance government disaster relief and safety net policies.

22.5.2.1
Government Disaster Option for CAT Risk: A Policy for the Market Failure Layer

Cognitive failure and ambiguity loading occur primarily with events in the extreme tail of the loss distribution, the area previously termed the market failure layer. For this reason, and as a substitute for ad hoc disaster relief payments, governments may decide to co-finance risk transfer mechanisms for these events. A government, for example, could design Disaster Option for CAT risk (DOC) index reinsurance contracts for catastrophic risks. Returning to the example in Figure 22.2, a DOC could insure against rainfall less than 500 mm with a payment per tick of say, US$50. Primary insurers could then offer coverage beyond the earlier imposed limit of 500 mm and transfer the catastrophic tail risk to the government using the DOC. Even if primary insurers are selling traditional crop insurance, they could use a DOC to transfer part of the catastrophic tail risk in their portfolio of crop insurance policies (Schade et al. 2002). DOCs could be offered for a variety of strikes and settlement weather stations, as long as the coverage is for catastrophic risk layers and can be offset in international weather risk markets. The government could even offer other DOC indexes (for example, excess rainfall or wind speed) to reinsure other lines of insurance, such as property and casualty. The government would reinsure DOCs in international reinsurance or capital markets using any of the three risk transfer strategies described earlier (Skees and Barnett 1999). Since

DOCs would address only extreme catastrophic loss events, reinsurance premium rates would likely contain an ambiguity load. Premiums could be subsidized to offset part of this ambiguity load so that DOC purchasers would pay something closer to a pure premium rate (Skees 2003). DOCs could be tailor-made to individual insurers' needs; for example, DOCs could be based on individual weather stations or written as regional weighted average baskets of weather stations. Strikes should be set so that the DOC covers only infrequent events (for example, events with an expected frequency of once every thirty years or less). This is the domain of the probability distribution over which potential insurance purchasers tend to experience cognitive failure and insurance providers engage in ambiguity loading. Primary insurers and ultimately insured parties would pay a premium for this catastrophic protection, but it would be significantly less than what the market would charge. Those who reinsure DOC contracts will insist on verifying the credibility of the underlying indexes. The premium required to transfer the risk to international markets would provide a baseline for setting DOC premium rates. The risk-layering approach proposed here would institutionalize the social role of government in subsidizing extreme risk events at the local level. Premium rates could be subsidized to offset ambiguity loading. Furthermore, by organizing DOC contracts at the local level, victims of isolated severe events that fail to capture national policymakers' attention could still receive some structured assistance. The following list summarizes the major advantages of offering index-based DOCs:

- DOC contract provisions established *ex ante* allow for better planning than do ad hoc disaster payments.
- DOCs provide a structure that provides more spatial and temporal equity in government disaster assistance.
- DOCs facilitate commercial insurance product development by providing a means by which catastrophic risk layers can be effectively transferred into international markets.
- DOCs can be subsidized to address the market failure associated with ambiguity loading and cognitive failure.
- Governments can estimate their own DOC subsidy cost exposure based on actuarial estimates of the risk inherent in the index. Reinsurance coverage adds a market check on the credibility of the index and the adequacy of DOC premium rates.
- While DOCs may be partially subsidized, end users still pay part of the cost to transfer the risk into international markets. This reduces the potential for perverse incentives that could encourage excessive risk taking.

22.5.2.2
Justification for government involvement

In the case of *low-frequency, high-consequence* loss events, however, government intervention may be justified. As explained in the Section 21.2 on production/weather risk management, research suggests that many decision makers tend to underestimate their exposure to low-frequency, high-consequence losses - a tendency that is reinforced when the decision-maker believes the government will provide assistance in the event of a disaster. Thus, producers thinking in this way will be unwilling to pay the full costs of insurance products that protect against these losses. Those who do buy insurance against low-frequency, high-consequence losses often cancel the policy if they do not receive an indemnity for an extended period. Thus, it seems that successful agricultural insurance products must be constructed so that they make indemnity payments with reasonable frequency, for example, once every seven or ten years. On the supply side, insurers will typically load premium rates heavily for low-frequency, high-consequence loss events where considerable ambiguity surrounds the actual likelihood of the event. Together, these effects create a gap between the prices farmers will pay for catastrophic agricultural insurance and the prices insurers will accept. Thus functioning private sector markets fail to materialize, or, if they do materialize, they cover only a small portion of the overall risk exposure. This type of market failure is commonly cited as justification for government interventions to facilitate provision of products or services not otherwise provided (or provided in sufficient quantity) by private markets.

Subsidies for catastrophic reinsurance are a type of government intervention that can facilitate the provision of insurance for low-frequency, high-consequence loss events. Hardaker et al. (2004), provide the following arguments for such an approach:
- Governments already provide disaster relief, even though providing assistance through reinsurance might be more efficient;
- The financial involvement of a government may address the moral hazard problem: many catastrophes can either be prevented or magnified by government policies or lack thereof. For example, governments that are financially responsible for some losses might provide incentive for putting appropriate hazard management and mitigation measures in place;
- A government's financial involvement in reinsurance may reduce political pressure to provide distorting and often capricious ad-hoc disaster relief;
- Governments can potentially provide reinsurance more economically than can commercial reinsurers. A government's advantages, including its deep credit capacity and unique position as the country's largest entity, enable it to spread risks more broadly. If governments are to intervene in agricultural insurance markets, the social benefits of reducing the inefficiencies brought on by risk must outweigh the social cost of making agricultural insurance work.

22.5.2.3
Policy objectives

Governments that seek to spur growth and eradicate poverty almost inevitably mix economic policies meant to enhance efficiency and growth with social policies meant to address poverty and vulnerability. Governments also often pursue equity or income redistribution objectives. Thus, government policies related to agriculture and rural areas tend to pursue the following objectives:
- *Growth*. Economic growth in rural areas–particularly, higher agricultural yields and value-added processing as well as development of off-farm activities–is perceived to be the best way out of poverty in the medium-term. While better in-

centives for market players and an enabling infrastructure are key drivers, better management of agricultural production risk is also critical for growth, as it enhances access to credit and adoption of new technologies.
- *Reduction of poverty and vulnerability in rural areas.* To achieve social and equity goals, governments directly intervene in a targeted manner, because free markets do not necessarily alleviate poverty for those in society who cannot effectively participate in them. Safety nets provide one tool for such government intervention.

22.5.2.4
Implementing policy objectives

Given the limited resources in developing countries and the existence of other sectors requiring government attention, these objectives are typically pursued within an environment of binding fiscal constraints. The two objectives target different segments of the rural population and different risk profiles. Growth objectives focus on increasing profitability so that less poor farmers can continue adopting production technologies even when high-frequency, low-consequence loss events occur. Poverty reduction policies seek to increase the average income of poor farmers, thus decreasing the volatility of their income and the likelihood that a risk event will wipe out hard-won asset gains.

A precondition for achieving sustainable growth and poverty reduction is an *ex-ante* system for disaster risk management. Disaster risk management covers severe and very infrequent events affecting mostly the poor, because the poor are more vulnerable and tend to live in marginal and more risk exposed areas. Susceptibility to and the experience of major natural disasters tend to trap people in poverty, due to the lack of efficient risk management at the household level. Government disaster challenge is to deliver timely and predictable aid in disaster situations. This requires *ex-ante* planning rather than just *ex-post* disaster responses. This also implies efforts to forestall political demands for *ex-post*, ad hoc government disaster assistance. Indeed, a credible and reliable disaster risk management system can put farmers and countries on a higher growth path by making people more comfortable with taking calculated and protected risks.

Naturally growth and poverty-reduction objectives overlap, but this makes it even more important to identify clear objectives and to design effective and cost-efficient ways to achieve them. Mixing objectives can lead to suboptimal outcomes. Many government-facilitated crop insurance programs, for example, attempt to accomplish social welfare and economic efficiency objectives simultaneously.

22.5.2.5
Subsidies

Governments frequently subsidize agricultural insurance products. These subsidies take a variety of forms. The government may co-finance insurance purchasing with direct premium subsidies, reimburse primary insurers for administrative or product development costs, or provide reinsurance at below-market premium rates. Regardless of the form, government subsidies are generally designed to increase insurance purchasing by lowering the premiums charged to agricultural insurance buyers. Such subsidies are extremely controversial, and for some, are seen to benefit operators of larger farms more than those of smaller farms. A wide range of stakeholders can and will engage in rent-seeking once subsidies are introduced. Subsidies are costly to maintain and are subject to close scrutiny regarding social costs versus social benefits. Many times, subsidies are provided based on the rationalization that agricultural insurance markets are missing or incomplete, without careful consideration of the core reasons why such market limitations exist. This document has carefully considered why agricultural insurance is missing or incomplete in many settings: adverse selection and moral hazard, high transaction costs, cognitive failure and ambiguity loading, and exposure to highly correlated loss events. Any government subsidies should be carefully targeted to address one of these specific sources of market failure. Even then, however, the costs of addressing that market failure may simply be too high to justify use of limited government resources to that end. The rents resulting from even the most carefully targeted subsidies can still be captured by politically powerful elites. Government insurance subsidies may crowd out demand for private sector risk transfer instruments. Any efforts to facilitate the provision of risk transfer instruments should be based on careful consideration of whether subsidies or grants can be provided without distorting or inhibiting the growth of private sector financial markets.

Some types of subsidies are likely to be less distorting than others. Subsidies and grants for supporting financial intermediaries and financial infrastructure, such as technical assistance and data systems needed to develop effective index insurance products, generally create little distortion. Beyond distortions in the markets, legitimate reasons exist for supporting infrastructure to improve market access among the rural poor. Finally, some public support for product development may be justifiable because of the free rider problem. Innovative insurance products are costly to develop, yet it is difficult to recoup these costs in a competitive market. Any firm can simply copy and compete with the new product without the expense of recovering product development costs. Unfamiliarity with index insurance products can heighten these problems in many developing countries. Examples of subsidies for financial intermediaries and infrastructure include:
- Providing technical assistance to financial intermediaries to improve systems that enhance efficiency, such as management information systems;
- Developing and introducing demand-driven products on a pilot basis;
- Helping to develop or improve service delivery mechanisms that enable greater outreach into rural areas;
- Covering a portion of the cost of establishing new branches in areas lacking financial intermediaries to serve the poor;
- Creating capacity within regulatory and supervisory bodies;
- Supporting the creation of industry associations;
- Developing training institutes and insurance information agencies;
- Supporting data for weather stations or other data to be used to develop effective indexes; and

- Providing technical assistance to develop new products in an emerging market in developing countries.

22.5.2.6
Premium Subsidies

While it is common for developed countries to co-finance premiums for farmers with direct premium subsidies, these types of subsidies are particularly problematic. Generally, direct premium subsidies reflect income enhancement objectives as much or more than they do risk management objectives. Such subsidies are typically provided on a percentage basis. This clearly benefits higher risk areas relatively more than lower risk areas. Even attempts to subsidize to levels that represent a pure premium or expected loss basis may favour higher risk areas relatively more than lower risk areas, since in a commercial market, premium rates for higher risk areas would likely contain higher catastrophic loads. Thus, any attempt to introduce premium subsidies will likely be distorting. In principle, if subsidies are targeted to the market failure layer, as described above, market distortions should be minimal. Given the ambiguity loading and cognitive failure that occur in this layer, carefully targeted subsidies may even be welfare enhancing. For the market insurance layer, however, subsidies should, in general, be avoided. Any subsidies in the market insurance layer should be targeted to reducing uncertainty loads in premium rates. Commercial insurers will tend to load premium rates based on the quantity and quality of data used to generate pure premium rates. The better (worse) the data used to generate the pure premium rates, the lower (higher) the premium load. These loads could be offset with co financing from donors. Here again, however, governments should be very clear about the level of these subsidies and the intent behind them.

22.6
Overview of ongoing agricultural risk pilot programs

22.6.1
India

In 1991, a household survey in India addressing rural access to finance revealed that barely one-sixth of rural households had loans from formal rural finance institutions and that only 35 to 37 percent of the actual credit needs of the rural poor were being met through these formal channels (Hess 2003). These findings implied that over a half of all rural household debt was to informal sources, such as moneylenders charging annual interest rates ranging from 40 to 120 percent. A survey based on the Economic Census of 1998 (Hess 2003) showed that India's formal financial intermediaries reportedly met only 2.5 percent of the credit needs of the unorganized sector through commercial lending programs. In this context, the CRMG, in collaboration with the Hyderabad-based microfinance institution BASIX and the Indian insurance company ICICI Lombard, a subsidiary of ICICI Bank, initiated a project to explore the feasibility of weather insurance for Indian farmers and to determine if, by reducing exposure to weather risk, it would be possible to extend the reach of financial services to the rural sector.

22.6.2
Malawi

Malawi and SADC: weather risk transfer to strengthen livelihoods and food security

22.6.2.1
Country context and risk profile

Malawi is dominated by smallholder agriculture, cultivating mostly maize – the staple food. Maize is very weather sensitive and requires a series of inputs. The economy and livelihoods are affected by rainfall risk (and resulting food insecurity), soil depletion, lack of credit, and limited access to inputs. Malawi suffers serious capacity constraints because it is ravaged by poverty and AIDS. Very few people have the energy and skills to build financial service programs.

22.6.2.2
Current response

Malawi used to have a paternalistic state culture. The role of the state in agricultural marketing (mainly tobacco, and also maize) is still strong and therefore, prices are not free and smallholder incentives are distorted due to food aid and the state marketing board sale of subsidized maize. The state and donors respond to recurrent drought-induced food crises by ad hoc disaster relief programs.

22.6.2.3
Proposed agricultural risk management structures

Micro-level: At the farm level, weather-based index insurance allows for more stable income streams and could thus be a way to protect peoples' livelihoods and improve their access to finance. The Government of Malawi has supported the implementation of an index-based weather insurance program at the smallholder level. In 2005, 900 groundnut farmers in Malawi bought weather insurance to increase their ability to manage drought risk and in turn access credit for better inputs. National Smallholder Farmer Association of Malawi, in conjunction with the Insurance Association of Malawi and the CRMG of the World Bank, designed an index-based weather insurance contract that would payout if the rainfall needed for groundnut production in four pilot areas was insufficient for groundnut production. Because these weather contracts could mitigate the weather risk associated with lending to farmers, Opportunity International Bank of Malawi and Malawi Rural Finance Corporation agreed to lend farmers the money necessary to purchase higher-yielding certified groundnut seed if the farmers bought weather insurance as part of the loan package. Given the success of 2005 pilot, 2500 farmers secured weather insurance-linked crop production loans for groundnut and hybrid maize in 2006, in a second pilot involving the same stakeholders. It is recognized that more weathers stations are required so that farmers in more locations can access insurance. To address this issue the Government of Malawi installed two new automatic weather stations and together with CRMG invested in creation of synthetic historical data at these sites. For the 2006/2007 pilot, one of these two new stations allowed previously excluded farmers to access weather insurance and input financing and the products were sold in five areas in 2006. There are plans to expand the pilot in 2007. CRMG and DECRG of the World Bank conducted a baseline survey partnering with the International Crops Research Institute for the Semi-Arid Tropics (ICRISAT) in 2006 to begin monitoring and evaluating the program (World Bank 2001).

Macro-level: A specific nationwide maize production index for the entire country could form the basis of an index-based insurance policy or an objective trigger to a contingent credit line for the government in the event of food emergencies that put pressure on government budgets. Applying the approach outlined above to the macro-level situation, we can define a Malawi Maize Production Index (MMPI) as the weighted average of farmer maize indexes measured at weather stations located throughout the country, with each station's contribution weighted by the corresponding average or expected maize production in that location. Given the objective nature of the MMPI, and the quality of weather data from the Malawi Meteorological Office, such a structure could be placed in the weather risk market. Analysis shows that Malawi could need up to US$70 million per year to financially compensate the government in case of an extreme food emergency (World Bank 2001). The Government of Malawi has recently expressed interest in engaging in such a drought risk management program and with assistance from the World Bank[1]. The proposal is to pilot the use of an index-based insurance contract to transfer the financial risk of severe and catastrophic national drought that adversely impacts the Government's budget to the international risk markets. The aim of such a contract would be to secure timely and reliable funds for the Government if a contractually specified severe and catastrophic shortfall in precipitation occurs during the agricultural season, as measured by weather stations throughout the country. Access to such contingency funds in a time of crisis would generate a supplemental source of emergency financing in May to complement existing budget resources, giving the Government more flexibility in its drought response and enhancing the Government's ability to launch an efficient and cost-effective drought response.

The weather index/drought risk management approach suggested for Malawi is one that could be extended to a regional level to include all members of SADC at some point in the future. Weather risk can be retained and managed internally if the areas under management are significantly diverse in their weather risk char-

---

[1] CRMG, January 2007.

acteristics. This immediately suggests that the weather sensitivity of neighboring countries, the SADC members, must be taken into account when considering Malawi's weather risk profile and its need for outside insurance. Analysis of the SADC region shows that on average, two countries suffer a drought each year (Hess and Syroka 2005). However, the distribution of drought events in SADC is extremely long-tailed, with the possibility of widespread drought events that could potentially devastate the region. This indicates that the most efficient way to layer and thus manage the risk is as follows:

SADC Fund: If the average financial impact of four average droughts in the region is approximately US$80 million for example, this could be the size of the SADC fund, with each member contributing its share determined by an actuarially fair assessment of the expected claim of each country.

Reinsurance and/or contingent credit lines: SADC-wide events incurring a financial loss of say US$80-350million could be transferred to the weather-risk reinsurance/professional investor market. Alternatively, the SADC members could have access to a World Bank contingent credit line in such situations.

Securitization: The final and extreme layer of risk, such as drought in 10 countries, occurring 1 percent of the time, could be securitized and issued as a CAT bond (investors lose the principal if the event occurs in exchange for a higher coupon) in the capital markets. The advantage of capital markets for this risk transfer is the immense financial capacity of these markets and also the longer tenure of CAT bonds – up to three years, possibly longer.

A more efficient means of transferring risk implies that costs could be greatly reduced for the member countries by transferring risk as part of a regional strategy rather than by transferring that risk one country at a time. For example, the SADC fund approach above would reduce insurance costs by 22 percent for Malawi due to risk pooling effects (Hess and Syroka 2005).

22.7
Conclusions

Agricultural producers and other rural residents are often exposed to a variety of biological, geological, and climatic factors that can negatively affect household income and/or wealth, as well as tremendous variability in output and/or input prices. Given this environment, risk-averse individuals often make investment decisions that reduce risk exposure but also reduce the potential for income gains and wealth accumulation. Thus, risk contributes to the "poverty trap" experienced by rural people in many developing countries. For a variety of reasons, markets for transferring these risks are typically either very limited or nonexistent. This "market failure" has stimulated a number of policy responses. Many developed countries have highly subsidized, farm-level agricultural insurance programs. Critics argue that, in addition to being very expensive, these programs stimulate rent-seeking activity, are highly inefficient, and may actually increase risk exposure by encouraging agricultural production in high-risk environments. Given fiscal constraints in most developing countries, highly subsidized, farm-level agricultural insurance programs are not a realistic policy option. Index-based insurance products have been proposed as an alternative risk-transfer mechanism for rural areas in developing countries. While not a panacea for all risk problems, index-based insurance products may prove to be valuable instruments for transferring the financial impacts of low-frequency, high-consequence systemic risks out of rural areas. For a variety of reasons, however, government intervention may be required to generate socially optimal quantities of risk transfer. Governments must carefully consider the extent and nature of any intervention in markets for index-based insurance products. These efforts can be facilitated by international organizations policy advice, lending instruments, technical support (WMO and FAO in particular) and monitoring and evaluation systems.

References

Barnett BJ, Roy Black J, Hu Y, Jerry RS (2005). Is Area-Yield Insurance Competitive with Farm-Yield Insurance? J Agric Res Econ 30 (2): 285–301.

Burgaz FJ (2004) Il sistema delle assicurazioni agricole combinate in Spagna. In: La Gestione Del Rischio in Agricoltura:Strumenti E Politiche, ed. F. Filippis. Rome, Italy: Coldiretti.

Buzby JC, Kenkel PL, Skees JR, Pease JW, Benson, FJ (1994) A Comparison of Subjective and Historical Yield Distributions with Implications for Multiple-Peril Crop Insurance. Agric Fin Rev 54: 15–23.

Deaton A, Grimard F (1992) Demand Analysis and Tax Reform in Pakistan. The World Bank Living Standards and Measurement Survey Working Paper, No. 85, The World Bank, Washington, D.C.

Dischel R (ed.) (2002) Climate Risk and the Weather Market. London: Risk Books.

Dismukes R, Allen G, Morzuch BJ (1989) Participation in Multiple-Peril Crop Insurance: Risk Assessments and Risk Preferences of Cranberry Growers. Northeast J Agric Res Econ 18:109–17.

Doherty NA (1997) Financial Innovation in the Management of Catastrophe Risk. J Appl Corp Fin 10:84-95

Fafchamps M (1993) Sequential Labor Decisions under Uncertainty: An Estimable Household Model of West African Farmers. Econometr 61: 1173–97.

Gadgil S, Seshagiri Rao PR, Narahari Raom K (2002) Use of Climate Information for Farm-Level Decision Making: Rainfed Groundnut in Southern India. Agric Syst 74: 431–57.

Hardaker JB, Huirne RBM, Anderson JR, Lien G (2004) Coping with Risk in Agriculture, 2nd ed. Wallingford, Oxon, U.K.: CABI Publishing.

Hazell PBR (1992) The Appropriate Role of Agricultural Insurance in Developing Countries. J Int Dev 4:567–81.

Hess U (2003) Innovative Financial Services for Rural India: Monsoon-Indexed Lending and Insurance for Smallholders. Agriculture and Rural Development Working Paper 9, The World Bank, Washington, D.C. See http://microfinancegateway.org/content/article/detail/11435

Hess U, Syroka J (2005) Weather-based Insurance in Southern Africa: The Case of Malawi,. World Bank ARD Working Paper http://www.microfinancegateway.org/content/article/detail/37405?PHPSESSID=109a87990

Hess U, Bryla E, Dana J, Varangis P (2004) Risk Management: Pricing, Insurance: the use of price and weather risk management instruments. Guarantees Paving the way forward for rural Finance – An international conference on best practices – Case Study.

Kunreuther H, Hogarth RM, Meszaros J (1993) Insurer Ambiguity and Market Failure. J Risk Uncert 7: 71–87.

Kunreuther H, Meszaros J, Hogarth RM, Sprance M (1995) Ambiguity and Underwriter Decision Processes. J Econ Beh Org 26:337–52.

Kunreuther H, Pauly M (2001) Ignoring Disaster, Don't Sweat Big Stuff. Wharton Risk Center Working Paper 01-16-HK, Wharton Business School, University of Pennsylvania.

Lim Y, Townsend R (1998) General Equilibrium Models of Financial Systems: Theory and Measurement in Village Economies. Rev Econ Dyn 1:59–118.

Miller D, Paulson A (2000) Informal Insurance and Moral Hazard: Gambling and Remittances in Thailand. Contributed paper No. 1463, Econometric Society World Congress. http://econpapers.repec.org/paper/ecmwc2000/1463.htm.

Miranda MJ, Glauber JW (1997) Systemic Risk, Reinsurance, and the Failure of Crop Insurance Markets. Amer J Agric Econ 79:206–15.

Morduch J (1995) Income Smoothing and Consumption Smoothing. J Econ Persp 3:103–14.

Morgan M, Henrion M (1990) Uncertainty: A Guide to Dealing with Uncertainty in Quantitative Risk and Policy Analysis. Cambridge: Cambridge University Press.

Pease J, Wade E, Skees JR, Shrestha CM (1993) Comparisons between Subjective and Statistical Forecasts of Crop Yields. Rev Agric Econ 15: 339–50.

Pomareda C (1986) An Evaluation of the Impact of Credit Insurance on Bank Performance in Panama? In Crop Insurance for Agricultural Development: Issues and Experience, eds.. Baltimore: Johns Hopkins University Press.

Ramamurtie, S. 1999. "Weather Derivatives and Hedging Weather Risks." In: (Hazell PBR, Pomareda C, Valdes A, ed.) Insurance and Weather Derivatives: From Exotic Options to Exotic Underlyings. H. Geman. London: Risk Books.

Rejda GE (2001) Principles of Risk Management and Insurance, 7th ed. Boston: Addison Wesley Longman.

Rosenzweig M (1988) Risk, Implicit Contracts and the Family in Rural Areas of Low-Income Countries. Econ J 98:1148–70.

Rosenzweig M, Stark O (1989) Consumption Smoothing, Migration, and Marriage: Evidence from Rural India. J Pol Econ 97: 905–26.

Rosenzweig M, Wolpin K (1993) Credit Market Constraints, Consumption Smoothing and the Accumulation of Durable Production Assets in Low-Income Countries: Investments in Bullocks in India. J Pol Econ 2: 223–44.

Ruck T (1999) Hedging Precipitation Risk. In: Insurance and Weather Derivatives: From Exotic Options to Exotic Underlyings. H. Geman. London: Risk Books.

Schade C, Kunreuther H, Kaas K (2002) Low-Probability Insurance Decisions: The Role of Concern. Wharton Risk Center Working Paper No. 02-10-HK, Wharton Business School, University of Pennsylvania.

Skees JR, Barnett BJ (1999) Conceptual and Practical Considerations for Sharing Catastrophic/Systemic Risks. Rev Agric Econ 21: 424–41.

Skees JR (2003) Risk Management Challenges in Rural Financial Markets: Blending Risk Management Innovations with Rural Finance. Thematic papers presented at the USAID Conference: Paving the Way Forward for Rural Finance: An International Conference on Best Practices, June 2–4, Washington, D.C.

Turvey CG (2002) Insuring Heat Related Risks in Agriculture with Degree-Day Weather Derivatives. Paper presented at the AAEA Annual Conference, July 28–31, Long Beach, CA.

Walker TS, Ryan JG (1990) Village and Household Economies in India's Semi-Arid Tropics. Baltimore: The Johns Hopkins University Press.

World Bank (2001) World Development Report 2000/2001: Attacking Poverty. Washington, D.C.: The World Bank.

CHAPTER 23

Weather Risk Insurance for Coping with Risks to Agricultural Production

Pranav Prashad

23.1
Weather and Indian Agriculture

Indian agriculture has high dependence on weather, especially monsoons. A causal analysis of agricultural losses as compiled by General Insurance Corporation of India's crop insurance cell (Table 23.1) shows that a major reason of crop losses can be attributed to weather vagaries.

23.2
Introduction to Weather Insurance

As the name suggests, Weather Insurance is an insurance coverage against the vagaries of weather. It is an insurance product based on a weather index, hence, provides financial protection based on the performance of specified index in relation to a specified trigger. Detailed correlation analysis is carried out to ascertain the way weather impacts yields of the crops to arrive at compensation levels. The weather indices could be deficit/excess rainfall, extreme fluctuations of temperature, relative humidity and/or a combination of above.

23.2.1
Process of making an index based product

The steps involved in the development and implementing an index based insurance programme are:

Table 23.1. Major causes of agricultural losses in India (Source: Varsha Bonds and Options – Rajas Parchure)

Cause	Proportion of Loss
Drought / Low Rainfall	0.7
Floods / Excess Rainfall	0.2
Others*	0.1

*(Storms, Pests, Negligence, Earthquakes)

23.2.1.1
Peril Identification

Peril identification involves appreciation of agronomic properties of the crops or nature of the economic activity. Detailed correlation analysis is carried out to ascertain the way weather impacts yields of the crops/ output of other economic activities.

23.2.1.2
Index Setting

The index is created by assigning weights to critical time periods of crop growth. The past weather data are mapped on to this index to arrive at a normal threshold index. The actual weather data are then mapped to the index to arrive at the actual index level. In case there is a material deviation between the normal index and the actual index, compensation is paid out to the insured on the basis of a pre-agreed formula.

23.2.1.3
Back testing for payouts

In order to ensure the robustness of the structure, the normal index is extensively tested based on historical data to ascertain if the payouts made on the basis of the chosen indices would have adequately indemnified the loss in the past or not.

23.2.1.4
Pricing

Pricing is determined based on components of expected loss, volatility of historical losses and management expenses.

23.2.1.5
Monitoring

This entails collection of weather data during the policy period and concurrent assessment of the ground conditions.

23.2.1.6
Claims Settlement

The claim settlement is a hassle-free process, as the beneficiary is not required to file a claim for loss to receive a payout. Instead ICICI Lombard compensates the beneficiary at the end of the crop season for any deviations from the normal conditions on the basis of the data collected from an independent source accessible to all, like a local weather station, thus removing the need for carrying out field surveys.

23.3
Advantages of Index based Insurance products like Weather Insurance

Index based insurance products like weather insurance carry the following advantages:
- A long term sustainable solution
- A market-based alternative to traditional crop insurance, which overcomes challenges of
  - High monitoring and administrative cost
  - Moral hazard and adverse selection
- Transparency – replaces human subjective assessment with objective weather parameters
- Scientific way of designing product
- Simple terms of insurance delivery
- Speedy claims settlement process

Weather insurance has multiplier effect on the economy as it enables access to factors of production. Adequate protection offered through the weather insurance product enhances the risk taking capacity of the farmers, banks, micro-finance lenders and agro-based industries. This in turn would result in boosting the entire rural economy.

Further, as the product is developed on the foundation of universally acceptable parameters, it is easier to transfer the risk to international financial markets through reinsurance. This allows for global pooling of risk and thereby more competitive "portfolio adjusted" pricing for the insurer and ultimately for the farmers.

23.4
Initiatives in Weather Insurance

ICICI Lombard has been a pioneer in bringing weather insurance solutions to India's farming community. Beginning with a small pilot for 230 groundnut and castor farmers in Mahbubnagar, as on date close to 80 weather insurance deals were executed across the country which have provided weather insurance solutions to 150,000 farmers covering an area of 225,000 acres.

These 80 deals represent experience in wide-ranging crops such as groundnut, castor, cotton, black gram, soybean, grapes, paddy and oranges. The deals were executed across 9 states: Andhra Pradesh, Madhya Pradesh, Uttar Pradesh, Rajasthan,

Punjab, Karnataka, Gujarat, Maharashtra and Tamil Nadu, with a wide variety of intermediaries such as micro finance organizations, agri-input corporates, non governmental organizations, banks, governments, and Internet kiosks.

ICICI Lombard has a Memorandum of Understanding with Commodity Risk Management Group of the World Bank wherein the two parties have committed themselves to jointly work on devising innovative weather index based insurance solutions.

23.5
Innovative ways to reach to the hinterland – reduction of basis risk

A common issue faced at the field level while providing Weather Insurance is Basis risk. Most of ICICI's current weather stations owned and maintained by Indian Meteorology Department (IMD) are located at the district headquarters, but most of the agricultural activities are carried out in much interior locations. As a result, most of the time, ICICI was unable to measure weather data at precisely the customer location. To build up the network of weather stations in the interiors, ICICI Lombard has a tie up with National Collateral Management Services Limited for installing Automated Weather Stations (AWS) at the block level. These data supplement India Meteorological Department's district level weather stations and ICICI gets sub-district level data which help in better monitoring of the policies. A major advantage of these AWS is that they provide real time daily data through automated calling process. Currently, through this network 91 locations are covered to reduce basis risk (up from 64 locations in 2005).

23.6
Designing Crop and situation specific products

ICICI attempts to cover the entire crop cycle which is divided into phases, so that complete protection can be provided. Different phases may involve different weather parameters which are illustrated as follows:

23.6.1
Wheat

In the early 20th century, Howard, a British Scientist at the Pusa Institute observed, 'Wheat cultivation in India is a gamble with temperature'.

After analyzing the impacts that temperature has on wheat cultivation, ICICI Lombard had designed a Weather Insurance Product for wheat cultivators which addresses the dual risks of extreme temperature fluctuations and unseasonal rainfall (Table 23.2).

Table 23.2. Weather Insurance Product for wheat

Time Period	Jan - Mar	Mar - Apr
Stage	Grain Filling Stage	Harvesting phase
Risk	Extreme temperature fluctuations	Unseasonal rainfall
Weather Index	Deviation in fortnightly average Tmin and Tmax on higher side from benchmark	Max. rainfall on any single day

Table 23.3. Weather Insurance Product for Apple

Time Period	Dec to Mar	Mar to May	April to Aug
Stage	Dormancy (Rest)	Flowering	Fruit set and development
Risk	Non availability of sufficient chilling and moisture	Extreme temperature fluctuations	Non availability of sufficient water
Weather Index	Chilling units as per Utah model Aggregate precipitation	GDD, a Tmin and Tmax based index that captures deviations on higher and lower side	Aggregate rainfall during subphase of the crop phase

Table 23.4 Weather Insurance Product for salt manufacturers

Time Period	Mar - Sep
Stage	Production period, when normally rainfall is not expected
Risk	Unseasonal rainfall
Weather Index	A Non Production Day (NPD) has been defined as a day with rainfall during the production period.
	NPDs have further been categorised as type A (1 - 9 mm), type B (10 - 24 mm) and type C (> = 25 mm)
	Compensation paid if NPDs observed is higher than a pre-determined level. Compensation is highest for type C NPDs and lowest for type A NPDs.

23.6.2
Apples

Weather Insurance Product for apples is shown in Table 23.3.

23.6.3
Salt manufacturing

Expanding the market to get benefit of diversification for agricultural risks, the first Weather Insurance deal for salt manufacturers was designed (Table 23.4), Weather Insurance was provided to an industry which was non-agricultural in nature in India.

23.7
Snapshot of 2005-2006

Till the financial year 2005-06, ICICI Lombard had covered 9 states through 11 major products, covering approximately 150,000 farmers on 180,000 acres of land. Table 23.5 provides an overview of the program run in 2005-06.

23.8
Distribution: a key challenge

The major challenge faced not only by Weather Insurance product, but by all rural financial products are that sheer spread and diversity of target customers makes

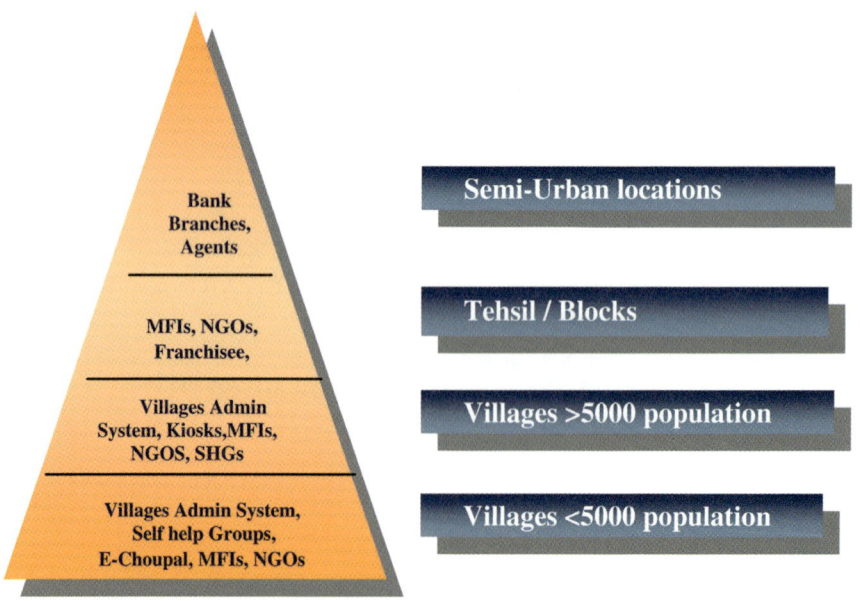

Locally available channels are effective since trust is the cornerstone of relationships

Fig 23.1. The challenge of distribution of weather insurance products

Table 23.5. Overview of ICICI Weather Insurance during 2005-06

Crop	Risk Details	States	Number of farmers covered	Area covered (in acres)	Sum Insured (Rs mn)
Soybean	Deficit rainfall	RJ, MP	4,112	16,418	66
Oranges	– Deficit rainfall – Prolonged dry spell	RJ	453	1,223	6
Generic product for all field crops	– Deficit & Excess rainfall	AP, MP, MH, Jharkhand, KK, Orissa, RJ and TN	19,100	22,000	66
Grapes	– Deficit & Excess rainfall, Temp	MH, AP	365	395	20
Paddy	– Prolonged dry spell – Excessive rainfall	Punjab	1,625	7,643	30
Cumin	– High relative humidity	RJ	686	688	6
Coriander	– Frost like temperature – Unseasonal rainfall	RJ	2,075	2,200	6
Fenugreek	– Excessively high temperature during days with high RH	RJ	70	260	2
Kinnu	– Excessively high temperature – Deficit rainfall	RJ	62	80	4
Wheat	– High temperature – Unseasonal rainfall	Punjab, Haryana	874	875	4
Cotton	– Deficit rainfall	MH	100,018	100,084	160
Total			150,000	225,000	

cost effective distribution a big challenge. In India, 37% of the urban population lives in 23 cities whereas 37% of the rural population lives in 100,000 villages.

To overcome this challenge, a three pronged strategy is important to implement wherein ICICI Lombard takes help of all contact points to reach farmers and sell the product concept (Fig. 23.1). ICICI Lombard takes help of various aggregators to sell the policies and is in touch with various State Governments and the Central Government to endorse the product and also has its own dedicated distribution channel to market the product. It has been realized that locally available channels are not only cost effective but also trustworthy for the end customer.

Technology based solutions like smart cards offer a cost effective distribution and also quicker service delivery.

23.9
Conclusions

In conclusion, insurance backed development is an effective way forward since:
- Rural demand for products and services is no different from urban requirements provided
- A fairly priced and relevant product is made available
- Cost effective distribution systems are established
- Effective administration is ensured
- Easy accessibility and quality service is ensured
- Focused approach along with appropriate regulation will help build a model which is viable, sustainable and scalable
- Availability of financial services and insurance would change the rural landscape in future

CHAPTER 24

Contingency planning for drought – a case study in coping with agrometeorological risks and uncertainties

Roger C Stone, Holger Meinke

24.1
Introduction

"Contingency Planning – a plan designed to take account of a possible future event or circumstances" (Australian Oxford Dictionary).

Contingency planning provides an important component in improving measures to reduce the impacts of climate variability on crop production and other agricultural production systems. In particular, ways must be found to avoid, reduce, or cope with climatic risk. In some of the drought-prone areas around the world, contingency planning is used by governments as an effective strategy to cope with risks.

Past attempts to manage droughts and its impacts through a reactive, crisis management approach, have been ineffective, poorly coordinated, and untimely. The so-called 'hydro-illogical cycle' illustrates this issue (Wilhite 2005). Because of the ineffectiveness of the crisis management approach there has been increasing interest in the adoption of a more proactive risk-based management approach. An interesting aspect is that these actions are partly due to the apparently more recent occurrence of drought episodes or of more severe droughts in some instances. Additionally, there may be little respite to allow recovery plans to take effect between more frequent drought episodes. The potential for increased frequency of drought in the future, possibly as a result of long-term climate change, has also caused greater anxiety about the absence of drought preparation plans (Wilhite 2005).

However, contingency planning for agrometeorological purposes is not necessarily receiving high attention in scientific literature, despite the fact that there is a trend towards the need to improve drought preparedness and policy development worldwide. This is needed to alleviate the increasing costs or impacts associated with drought, to manage the complexity of impacts on sectors well beyond agriculture, to alleviate increasing social and environmental effects, and to better manage increasing water conflicts between users. Another factor that may be responsible for constraint in drought preparedness has been the dearth of methodologies available to planners to guide them through the planning process. Drought differs in its characteristics between climate regimes and drought impacts are more locally defined by unique economic, social, and environmental characteristics. The aim of this paper is to provide examples of recent case studies and approaches to contingency planning with a particular emphasis on drought contingency planning as a means of furthering its use in government and agricultural industry world-wide.

24.2
The basis of drought contingency planning

Drought planning and water crisis management needs to be proactive. This is largely because overall policy, legislation, and specific mitigation strategies should be in place before a drought or water crisis affects the use of the country's water resources. Bruins (2001) provided the basic elements involved in the development of proactive drought contingency planning and their respective relationships. These basic elements (Figure 24.1) involve drought risk analysis, drought impact assessment, and drought scenarios. Assessments have to be made of the impact of drought on the various water resources, economic sectors, population centres, and the environment. Different types of drought should be considered in the impact assessment studies. Drought scenarios have to be calculated on the basis of available information, including development of a frequency and severity index. From this, drought risk assessment can be investigated, primarily on the basis of meteorological data but may also include paleoclimatic information and other historical data in relation to climatic variation. The latter may be important as the time scale of recurrence of severe drought may well exceed the average human lifespan. Finally, proactive drought contingency planning needs to be developed (Bruins 2001).

To aid in drought assessment remarkably comprehensive meteorological and climatological information and advice is now starting to appear as output prod-

Fig. 24.1. Basic elements involved in proactive drought contingency planning (Bruins, 2001).

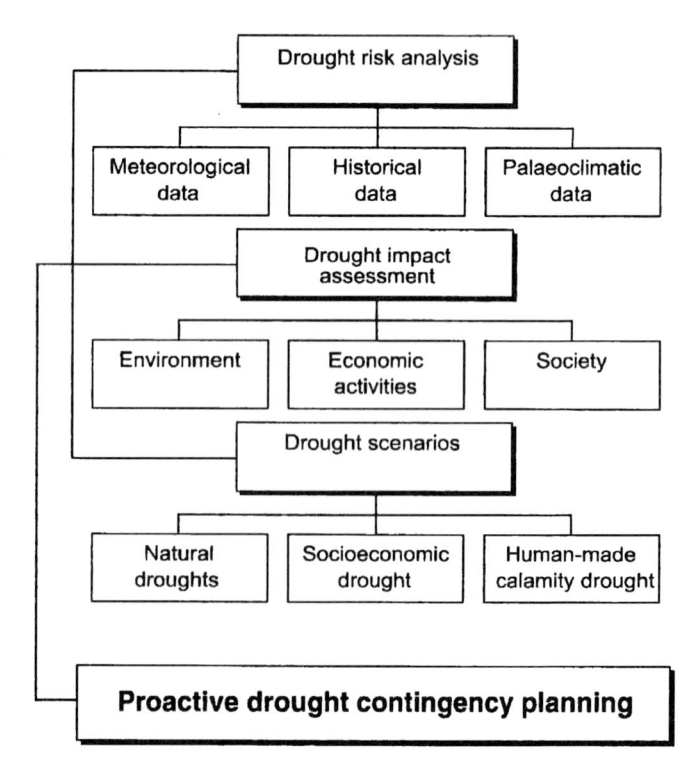

ucts emanating from meteorological and agricultural departments of governments worldwide. These can have fundamental value to those organizations that need to assess information related to the potential likelihood for drought and exceptional drought conditions in a region. Examples include those under The World Agrometeorological Information Service (WAMIS) of WMO in which products, (e.g. output from Germany) include monthly agro-weather bulletins, soil temperature information, actual and potential evaporation, grain dampness, forest and grass fire indices, and animal thermal load (http://www.wamis.org).

Policy makers, analysts, and farmers in regions prone to drought requiring access to the most recent climate, drought status, and crop prospects are able to use information on specialised drought exceptional circumstance information on sites such as the Australian National Agricultural Monitoring System (NAMS) (http://www.nams.gov.au). However, major stakeholder groups involved in preparedness planning (policy makers, regulators and large agribusinesses, including financial institutions) and those involved directly in crop production (farmers, farm managers, rural businesses and consultants) may require widely varying different information. This means that information sources need to be continually re-evaluated and assessed as to their appropriateness in provision of most appropriate information for key users. Tactical as well as strategic preparedness decisions need to be made continuously and some climate forecast-related information outputs might only be relevant for some of these decisions.

Another approach in drought assessment is to apply crop or pasture simulation models that contain over 100 years of historical daily weather data to construct time series of modeled crop yield or pasture growth and so provide frequency data on occurrence of the likelihood of extremely low yields. In an Australian example for assessing exceptionally poor wheat yields the APSIM model (Keating et al. 2003) has been applied to over 100 years of historical data to provide simulated wheat yields for each of the 100 years in the study. From these data an assessment of the relative severity of the current or recent drought was made (Fig. 24.2) and the results provided to authorities responsible for exceptional drought circumstance payments (Keating and Meinke 1998; deJager et al. 1998). For grazing industry needs the Aussie GRASS modelling framework (Carter et al. 2000) has been adapted to provide alerts of impending drought for Australian shires. This approach combines calculations of pasture utilization by livestock and other animals. Additionally, the Aussie GRASS system has been modified to provide pasture condition alerts. Figure 24.3 provides an illustration of the structure of this drought alert system. Figure 24.4 provides an example of routine forecast of simulated pasture growth values using known soil moisture levels, recent and forecast temperature and rainfall values. The pasture condition alert takes into account the fact that high pasture utilization under conditions of relatively high pasture growth can have different impacts (e.g. woody weed increase) than that under low pasture growth (e.g. loss of perennial grasses and pasture cover). The development of an associated 'drought alert' system in 1991 produced interpolated percentile maps of pasture growth (as well as rainfall) and similar products that were particularly instrumental in convincing policy agencies as to the severity and spatial extent of drought. At the end of the drought period, indices such as pasture growth can be used in the drought assessment process to provide additional evidence that, although rainfall

Simulated Wheat Yield

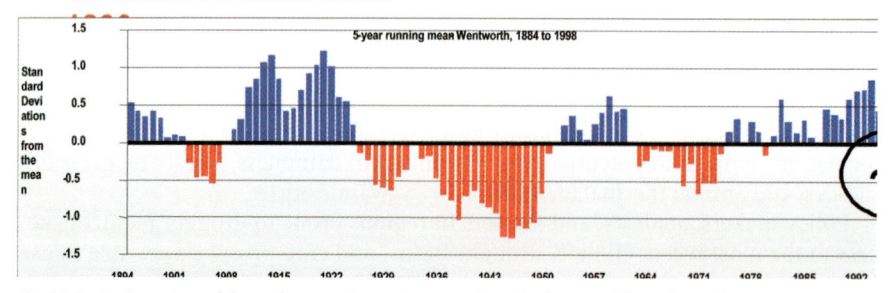

Fig. 24.2. Estimation of drought severity using simulated wheat yield analysis from 1890 to 1997 at Wentworth, Australia (Keating and Meinke 1998).

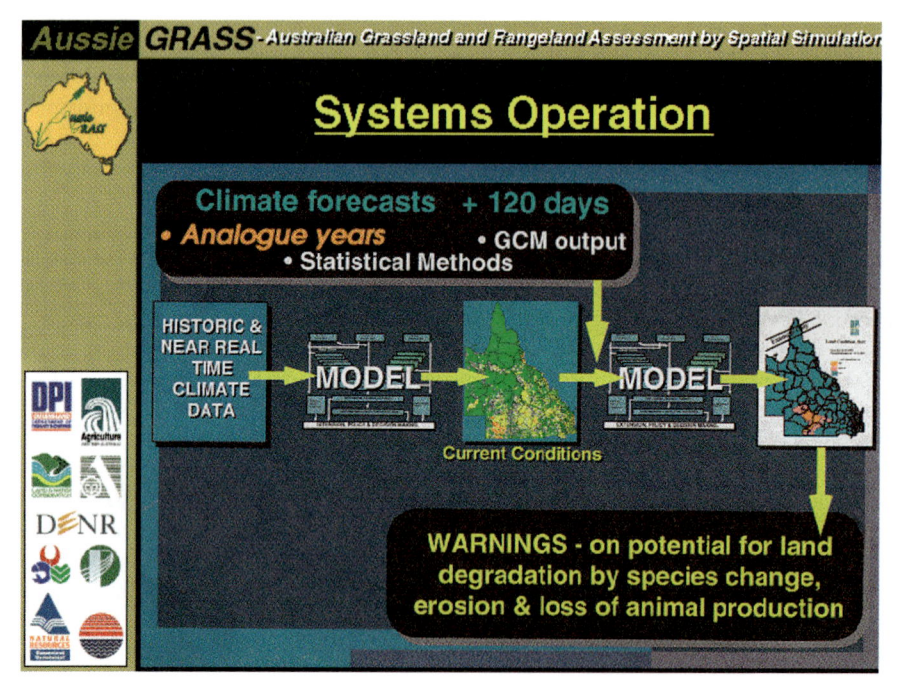

Fig. 24.3. Flow diagram depicting Aussie GRASS system for drought and land degradation warning (Day et al, 2003).

patterns had returned to 'normal', the outputs from the pasture growth model indicated drought conditions were continuing as the rainfall received was still insufficient to produce growth in major grass species in Queensland (Day et al. 2003).

It has been suggested that more recent interest in, and development of, contingency planning programs to aid agrometeorological coping strategies may be due to important shifts in underlying climatic systems that have had a large influence

Chance of Exceeding Median Growth
September to November 2006

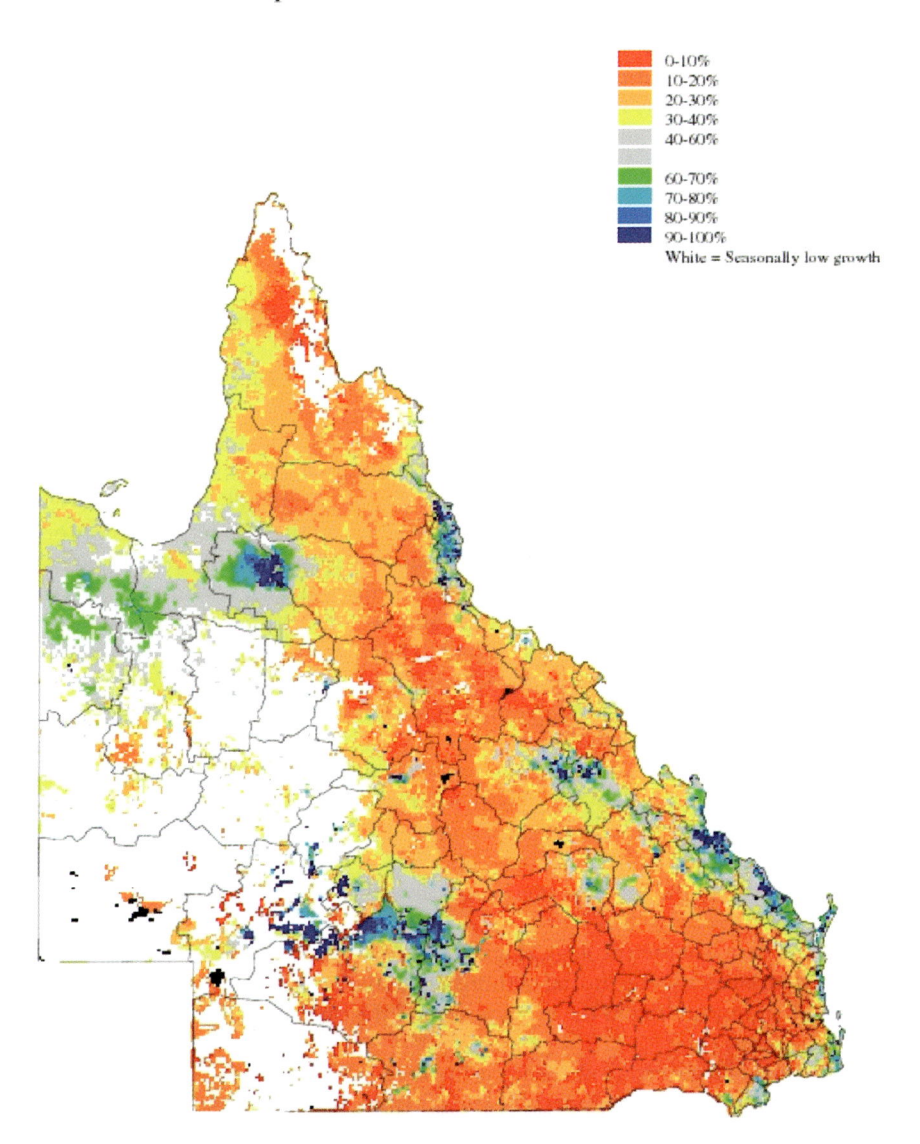

- 0–10%
- 10–20%
- 20–30%
- 30–40%
- 40–60%
- 60–70%
- 70–80%
- 80–90%
- 90–100%

White = Seasonally low growth

www.LongPaddock.qld.gov.au

Fig. 24.4. Example of forecast output from Aussie GRASS depicting pasture growth likelihood for subsequent three month period (September to November 2006) using climate forecast system integrated into pasture growth simulation model for 5 degree pixel regions.

on agricultural production and sustainability over the past 30 years. For instance, the long period of dominance of La Niña patterns from 1950 to 1975 brought improved conditions for agriculture in many countries through generally increased rainfall, with flow-on benefits for pastoral and cropping industries. The beginning of this period coincided with rapid agricultural development in many regions encouraged by high world prices for many commodities. Provision of drought relief during these periods would have constituted a relatively simple process in the relatively few El Niño-induced drought years (such as 1957, 1965, or 1972) (Haylock and Ericksen 2001). However, as La Nina's dominance faded in the mid-1970s so did the standard of living of many rural economies. Unfortunately, instigation of government assistance schemes encouraged farming in marginal country which also helped to intensify farming in some areas that were more susceptible to drought impacts. The response to losses from drought was for governments to intervene with various forms of assistance (Haylock and Ericksen 2001).

Reformed drought policy has been associated with the development of contingency plans for more severe droughts in New Zealand. In New Zealand an overarching Contingency Plan refers to and reflects the central government policies for adverse climatic events assistance. In this example this overarching Contingency Plan makes a distinction between local events and national events. Local events are those which have impacts on the rural sector but which are not considered severe enough to warrant central government involvement. In these events, producer organizations representative of the affected groups are responsible for managing the local response with the assistance of local government agencies. While the Ministry of Agriculture and Fisheries (MAF) has no direct operational role it liaises with local organizations, and may be contracted to help such organizations develop disaster response plans (Haylock and Ericksen 2001). Nevertheless, as is common in a number of countries, central government may have a role in responding to national events, particularly those which are beyond the coping capacity of the local community. In this case, a series of actions and responsibilities operates. Local farmer organizations, in association with local government agencies in the affected areas, make a request for assistance to the central government through the relevant Ministry (Haylock and Ericksen 2001).

In New Zealand the general criteria for a drought event to be of national significance include:
- An exceedence of meteorological norms,
- An extension across territorial authority boundaries,
- Proof that responding to the event is beyond the capacity of local communities and agencies,
- More than one government agency is involved in the recovery operation, and
- Considerable stock and property is at risk (Haylock and Ericksen 2001).

The Contingency Plans extend responsibility for initial response to extreme drought events from the central government to local communities and agencies. Remarkably, anticipated changes to New Zealand drought policy include increasing the meteorological criteria for an event of national significance from a one in twenty year event (5% annual probability of exceedence) to a one in fifty year event (2% annual probability of exceedence). It is unknown whether attention to likely changes in drought frequency or intensity under long-term climate change has been assessed in formulating these changes (see later sections in this chapter for further discussion). Further, a new category of event has been created where in addition to local and national events are regional significant events which may attract limited government assistance such as social welfare grants and labour assistance (Haylock and Ericksen 2001).

In terms of water supply issues drought contingency planning can take on the following processes:
- The need to forecast the supply situation in relation to demand,
- The need to assess drought mitigation options such as supply enhancement and demand reduction and timing necessary to obtain results,
- The need to establish triggering levels for mitigation actions by working backwards from a worst case scenario,
- The need to develop phased demand reduction programs,
- The need to adopt a drought plan with a key component to work with users and also the press and associated media and with special interest ('lobby') groups affected by drought,
- The need to adopt a formal adoption process,
- The need to monitor results and adjust responses as necessary,
- The need to provide evaluation of the contingency plan in any post-drought period (Macy 1993).

Engagement of the community is especially important in the development of contingency plans for drought. The community is then more likely to implement a water conservation program to combat likely drought rather than pay for other options. Including community needs in contingency planning such as this is difficult as the community may well provide widely differing responses depending on when the community consultation takes place. For instance, there are likely to be widely differing responses in a drought or dry years compared to a wet year (Macy 1993).

Additionally, it is important to include contingency planning within the context of an overall drought preparedness program. In particular, it is important to improve preparedness and triggering mechanisms to produce timely and appropriate responses. While contingency planning may remain at the level of theory, Davies (2001) argues more concerted risk management can be regarded as being at the core of drought preparedness. Nevertheless, preparedness and contingency planning can form the very important first stages of the risk management process (Davies 2001). Davies (2001) also notes that those in drought-prone areas have highly developed information systems which they use to predict likely disruptions in rainfall and to then diagnose and produce responses. Notably, in India even in 'normal years', Indian farmers remain aware of the possibility of drought and undertake water management, crop management, and contingency planning. Then, it is through their informed decisions that they make that they produce their relief and mitigation strategies. All these local strategies (at individual, household, and community levels) can then become part of an effective overall contingency planning process (Subbiah 2001).

Many drought-prone countries have some form of drought contingency plan on paper but many have not been translated into effective policy covering the range of activities required to address short and long-term consequences. In many cases, information is not linked to response and the use of information that is available is partial and unsystematic. Additionally, coordination between the different sectors, agencies, or individuals involved in the response planning may be poor or impossible. The process from policy development to effective drought preparedness may not be effective or well thought out with the result that little effective drought preparedness actually takes place by the farmer or industry. As Davies (2001) points out 'there is more to (drought) contingency planning than simply drawing up a document'. Additionally, contingency planning must be strong enough to withstand the pressures of crisis and relief operations. Contingency planning must continually be reinforced, must involve local people, personnel must be trained, and flexible responses must be field tested in advance of emergency situations (Davies 2001).

A final point is that, as a means towards proactive drought contingency planning, effective and interactive management systems need to be set in place. This approach helps ensure that future adjustments can be made through actual realizations during drought events and updated as may be required (Bruins and Lithwick 1998; Bruins 2001).

24.3
Preparedness strategies

Few countries have actually implemented risk-based drought policies and preparedness programs or strategies. However, Australia is an exception to this and some components associated with effectiveness in program development can be provided.

Wilhite (2003; 2005) identified four key components that have been identified for effective drought risk reduction strategy. These are:
- The availability of timely and reliable information on which to base decisions,
- Policies and institutional arrangements that encourage assessment, communication, and application of that information,
- A suite of appropriate risk management measures for decision makers,
- Actions by decision makers that are effective and consistent.

Thus, the Australian national drought policy has developed in the context of deregulation of the agricultural sector with the aim of creating efficient and internationally competitive farming systems. The policy has multiple objectives that include (adapted from Drought Policy Task Force 1997 and cited in Botterill and Wilhite 2005):
- Encouraging self-reliant approaches to managing climate variability;
- Protecting the natural resource base in times of extreme climate stress;
- Ensuring adequate welfare support for farm families commensurate with that available to other Australians;
- Ensuring that the policy does not impede structural adjustment in the farm sector; and

- A high level of awareness and understanding of drought and drought policy (also Wilhite 2003, 2005).

Wilhite (1996) argued that drought policy should be informed about the impact of climate variability on key policy relevant outcomes such as farm incomes and profitability. Thompson and Powell (1998) showed how this could be done using whole-farm models. They concluded that a preoccupation with climate and environmental definitions of drought was not consistent with a need to holistically manage all sources of risk on farms. The relevance gap that has emerged between the information necessary to support drought policy, and that being supplied has been highlighted through the work of Nelson et al. (2005). Nelson et al. (2005) showed that understanding the capacity of rural households to cope with risk requires broad measures of the human, social, natural, physical and especially financial assets from which rural livelihoods are derived (Figure 24.5a). However, Meinke et al. (2006) suggest that the information used to inform the implementation of climate policy in Australia agriculture is mostly related to measures of variability in rainfall, soil water analysis and plant growth (see Bureau of Rural Sciences 2006). However, policy may be somewhat powerless to influence the impact of climate variability on these scientific measures of exposure to climate risk and may also bear little relation to more holistic measures of the key policy outcome: the coping capacity of rural communities (Figure 24.5b) (Meinke et al. 2006).

It is also suggested that beyond engaging with public policy makers, public-private partnership models need to be further explored in order to 'mainstream'

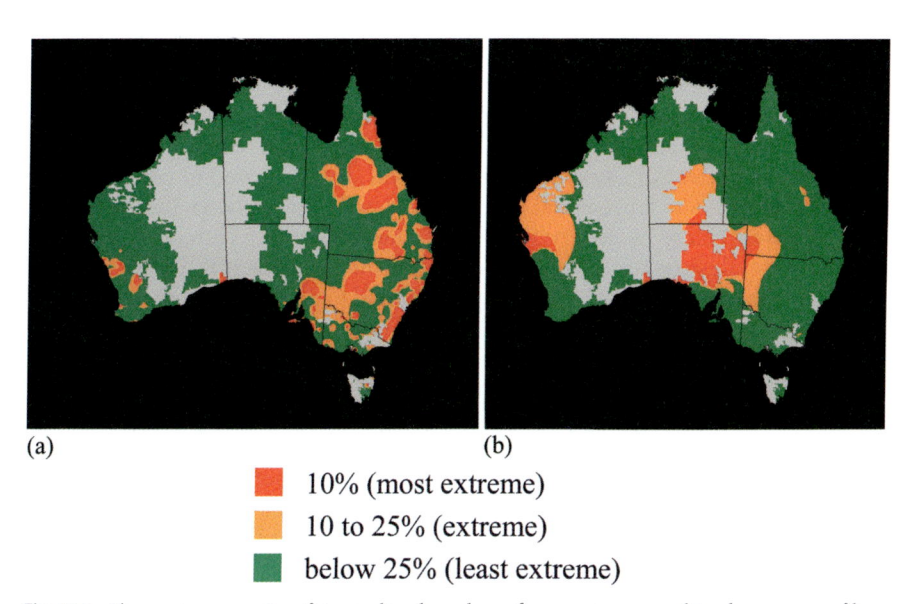

Fig. 24.5. The coping capacity of Australian broadacre farms using a combined measures of human, social, natural, physical and financial capital (**a**, left) and exposure of Australian broadacres farms to climate risk using a measure of extreme pasture growth conditions (**b**, right; after Nelson et al. 2005).

drought risk management. Mainstreaming involves developing risk management tools and approaches within the context of overall rural livelihood strategies, integrating risk arising from markets, management skill and threats to the natural resource base. It also involves communicating risk management knowledge through functional, existing communication networks of farmers and other landholders, rather than pursuing specific communication programs. Although some small steps have been taken in this direction, this has in most cases not yet advanced beyond the proof-of-concept stage (Nelson and Kokic 2006).

The degree to which central government assistance is provided can depend to an extent on the effectiveness of local farming lobby groups or local government bodies in persuading central government of the need for assistance. In this respect, it is difficult to ascertain the extent to which central government's resolve to restrict drought relief to the exclusively severe events will hold in the future, especially in respect to the potential for changing climatic patterns under greenhouse induced climate change. In fact, in Australia and New Zealand central government policy goals for a reduction in drought relief have occasionally been undermined as a result of various economic factors and political processes (Stehlik et al. 1999; Haylock and Ericksen 2001).

In southern Africa, Wilhite (2005) notes that one of the common problems with drought planning is maintaining interest beyond the relatively short window of opportunity that follows the event, 'given the on-again, off-again nature of drought'. Indeed we have noted that government interest in developing drought research and development programs, drought planning programs, and other associated issues can wane almost immediately when it starts raining, irrespective of whether the rains have provided the type of moisture levels to ensure a return to improved grazing, cropping, or water resource conditions. This may be due to what Wilhite (2005) considers to be a problem related to governmental agencies trying to cope with the 'mysterious processes' of effective drought planning that in our view are made more complex because of all the associated complexities of dealing with emotional and stressful situations of the farmers, issues related to compiling accurate economic data for regional or industry losses, issues related to dealing with stress situations faced by departmental staff who are working with drought affected families, and issues related to needing to continually work effectively with rural industry lobby groups who are also being pressured to provide immediate response to their constituents (also Stehlik et al. 1999). It is no surprise, therefore, that interest in undertaking further planning beyond the critical core drought period and its immediate aftermath can quickly wane. This appears to be especially the case in those countries where drought is regarded as a completely unforeseen type of event that could be prepared for. Remarkably, while farmers may treat both normal and reduced rainfall due to drought as an integral part of climatic variation and manage the situation in a holistic manner, public policy implementers in many countries perceive drought as a discrete and unexpected event (Bruins 2001; Haylock and Ericksen 2001; Subbiah 2001; Wilhite et al. 2005).

Meinke et al. (2006) point out that drought should be regarded as a social construct and represents the risk that agricultural activity will be severely disrupted given spatial and temporal variations in rainfall. The basic philosophy of the Australian Government's drought policy is to encourage primary producers to adopt self-reliant approaches in managing the risks associated with climate variability. Australian drought policy recognises that producers are responsible for managing the commercial performance of their enterprise and for ensuring that agricultural activity is carried out in an economically and environmentally sustainable manner (Botterill 2003). The concept also recognises that the Australian Government should not inhibit the natural course of structural adjustment due to other pressures such as declining commodity prices (White and Karssies 1999; Botterill 2003).

In arid, semi-arid, and marginal areas with a probability of drought incidence it is important for those responsible for land-use planning and agricultural programs to seek expert climatological advice regarding rainfall expectations. Drought can also be regarded as being the result of the interaction of human patterns of land use and the rainfall regimes (Das 2005). There is therefore an urgent need for detailed examination of rainfall records (including trends) related to these regions. Additionally, climate and weather forecasting systems can play a very important role in providing advance warning of rainfall likelihood or deficits (Das 2005). More sophisticated integrated crop production/climate forecasting systems can also provide scenario analysis of likely crop production under varying soil moisture and likely climatic states for the forthcoming season (Stone and Meinke 2005). While at this stage there may be limited uptake of these more recent advances in forecasting systems, there are a few examples where they form a necessary component of drought contingency plans within overall drought policy (e.g. Queensland Government 1992).

As a preparedness strategy, food reserves are needed to meet emergency requirements of up to two consecutive droughts. Also a variety of policy decisions on farming, human migration, population dynamics, livestock survival, and ecology must be formulated. Furthermore, sustainable strategies must be developed to alleviate the impact of drought on crop productivity. This may be achieved through varietal manipulation, that can also be facilitated through judicious use of seasonal climate forecasts, through which drought effects can be minimized by adopting varieties that are more drought resistant at different growth stages. Alternatively, strategies that include changing planting times and nitrogen fertilizer application to better suit the likely drought conditions can be employed (Das 1999, 2005; Stone and Meinke 2005, 2006).

However, Podesta et al. (2002), in their case study of farmer's use of climate forecasts in Argentina as a means of drought preparedness, found a reluctance to use seasonal forecasts for drought preparedness because the temporal and spatial resolution of the forecasts was perceived as not relevant to local conditions (Buizer et al. 2000). These types of issues must be taken into account in order to improve the relevance and potential adoption of seasonal climate or crop forecasts in drought preparedness. For example, for effective management systems to be put into place, integrated climate-crop modeling systems need to be developed at the appropriate farm or regional scale suitable for the decision-makers needs (Meinke and Stone 2005).

Challinor et al. (2003) also make the point that reliable forecast output for drought preparedness will not result from simply linking climate and crop models. In this respect, they suggest consideration should be given to the spatial and temporal scales on which the models operate, the relative strengths and weaknesses of the individual models, and the nature and accuracy of the model predictions. A key aspect of this approach is that on longer timescales, process-based forecasting has the potential to provide skilful forecasts for possible future climates where empirical methods would not necessarily be expected to perform well.

Another preparedness strategy may include identifying the potential to harvest other types of crops in different months of the year which, in turn, may require the assessment of the suitability of land for cultivation of other crops, the availability of water, and the orientation and training of farmers to harvest new crops. This preparedness approach may help mitigate the effects of failure of the main crop on farming families (Das 2005).

Continuing issues are related to determining more precisely what the precise trigger points should be that would indicate a need for government assistance in some countries. For example, it is well known that certain meteorological criteria usually need to be met although these have generally been made more stringent in recent years. However, some countries have introduced criteria related to a region's economy. Yet, while such criteria, such as the need to establish 'significant' financial loss during a drought is a stated aim of national drought relief policies, in practice it has proved difficult to determine the degree and type of regional economic impact that would then be used as a threshold. In addition, in practice it has proved difficult for agencies to provide lengthy economic data to support their claims for assistance as a one in twenty or one in fifty year event (Stehlik et al. 1999; Haylock and Ericksen 2001).

In the case of establishing triggering levels there is a need to know drought demands, demands with mitigation actions in place, the drought inflows into storage systems (includes knowledge of worst historical drought, use of drought modelling systems to develop likely return intervals), the timing necessary to achieve the objective of drought mitigation plans (especially lead times), and the need to know the worst case scenario of supply such as reaching the 'dead storage capacity' of a reservoir. For multiple storages it is necessary to set up multiple trigger levels for each of the storages. There is an additional need to establish least-cost planning where it is necessary to undertake a detailed assessment of the cost of various measures and their impacts on the environment. It is necessary to ask users what their demands will be and which of these do the water authorities need to meet with 100% reliability or those with less reliability. The user then determines what they are willing to pay for based on the costs involved. This type of contingency planning may be a detailed and lengthy process (Macy 1993).

24.4
Risk management systems and tools

Risk management tools suitable for drought management are now becoming more common world-wide. These tools appear to have high benefit at the broader industry level where the additional aspect of use of seasonal climate forecasting can have considerable benefit. We would argue that use of integrated agricultural management, crop simulation models, and climate forecast systems offers the highest benefit. This is especially the case when applied at the whole farming system scale and across industry value chains. This approach has the potential to benefit industries in many areas. For example these systems can produce strategies that include:

- Improved on-farm profitability by better using scarce water resources, increasing water use efficiency and enabling higher production with consequent minimal movement of nutrients and pesticides off-farm,
- Improved planning for early season supply and better scheduling of milling operations leading to more effective use of resources (e.g. milling capacity, haulage capacity, haulage equipment, shipping, together with enhanced on-farm profitability),
- Enhanced industry competitiveness through more effective forward selling of the commodity based on enhanced knowledge of the amount of supply and improved efficiency of commodity shipments (Everingham et al. 2002).

The value of integrated climate/crop modelling efforts in strategic management and contingency planning can also be seen when probability distributions of a large number of simulated yields and gross margins can be produced and incorporated into risk assessment tools. Furthermore, the large number of simulations using the modelling approach allows the exploration of climate influences such as ENSO on extreme climate outcomes such as drought, a difficult approach with purely historical series that are typically short in duration (Sivakumar 2002; Podesta et al. 2002; Meinke and Stone 2005).

Strategic management decisions that could benefit from more integrated and targeted forecasts can be made at a range of temporal and spatial scales. These range from more tactical decisions regarding the scheduling of planting or harvest operations to broader policy decisions regarding land use allocation (eg. grazing systems vs cropping systems).

Hammer et al. (2001) stress that the most useful lessons for strategic management lie in the value of an interdisciplinary *systems* approach in connecting knowledge from particular disciplines in a manner most suited to decision-makers. The RES AGRICOLA project is an example of evolution of the 'end-to-end' concept proposed by Manton et al. (2000). Importantly, it distinguishes three discipline groups that need to interact closely if farmers and others impacted by drought or similar agrometeorological events are going to benefit. These fundamentally important discipline groups are: (i) climate sciences, (ii) agricultural systems science (including economics) and (iii) rural sociology (Meinke and Stone 2005).

Improved pay-offs for those impacted by drought are facilitated when such an integrated systems approach is employed that includes farmers, government planners, and scientists across the various disciplines which ensures that the issues that are addressed are relevant to the farmer (Meinke et al. 2001).

Hansen (2002) stressed that the sustained use of such a framework requires institutional commitment and favourable government policies. An example where links could be strengthened is in the area of connecting agricultural simulation with both whole farm economic analyses and broader government policy analyses. Using a case study, Ruben et al. (2000) reviewed the available options for adapting land use systems and labour allocation for typical households in a region in Mali. They showed that compensatory strategic policy devices could, at least, partially offset consequences of climatic patterns, largely through better-informed price policies which would enable welfare-enhancing adjustments for better-endowed farm households, while poor farmers would benefit from reductions in transaction costs.

Decision-support systems as risk management tools have often been cited as an effective means of providing output of integrated climate-agronomic information in the form of scenario analyses relating to impending drought or within the drought period that can be valuable to users. Such examples include the rainfall analysis computer package: *'Rainman'* (Clewett et al. 1994); crop management planning systems: *'Wheatman'* (Woodruff 1992), and *'Whopper Cropper'* (Nelson et al. 2002); grazing management systems: *Aussie GRASS* (Day et al. 2003); irrigation management systems such as *'Flowcast'* (Abawi et al. 2001; Ritchie et al. 2004); and agricultural management systems: CLIMPACTS (Campbell et al. 1999); CropSyst (Stoeckle et al. 2003), DSSAT (Jones et al. 2003) that provide sophisticated crop simulation platforms useful for integrating and simulating future climate systems scenarios, and 'UAS' (UAS 1999). Yet, despite the amount of effort developing these systems there is also a perception that these systems have been less than ideal in their overall effectiveness (Stone and Meinke 2005). However, when used in the manner of *'discussion-support'* users can engage in discussions with advisors and government agencies regarding drought patterns and crop and grazing management scenarios while maintaining ownership of the overall processes and final decision-making. In this way discussion-support systems move beyond traditional notions of supply-driven decision-support systems and can compliment participative action research. The critical role of interaction and dialogue among the key participants impacted by drought – strategic planners, policy agencies, farmers, advisors, crop modellers, and climate or meteorological scientists - is paramount (Plant 2000; Podesta et al. 2002; Nelson et al. 2002).

24.5
Issues associated with contingency planning for drought under climate change

Contingency planning for future risks associated with long-term climate change could be regarded as managing low frequency components of climate variability with the important point that the same quantitative approaches currently being developed for strategic planning issues associated with climate variability could also be applied to the complex issues likely to be encountered under climate change (Meinke and Stone 2005). Moreover, despite there being likely complex issues associated with climate change such as changes in land cover and changes in runoff associated with altered precipitation and temperature patterns, some suggest that farmers are likely to cope and adapt to climate change. Some studies assessing future economic impacts of climate change on agriculture suggest that farmers will continue to produce the same commodities on the same land using the same management tools (Rogers and McCarty 2000; Abler et al. 2002).

Yet, extreme weather events, including extreme droughts, over the past 30 years have already caused severe crop damage and induced significant economic toll for

United States' farmers alone. This has occurred against a back-drop of greater variability in crop yields, price and farmer income, part of which can be related to long-term climate change (Rosenzweig et al. 2000). Additionally, it is the authors' experience that farmer needs and requests in eastern Australia for climate-related information (while engaging in participative workshops) have shifted markedly over recent years from needs for more tactical issues to be addressed that are only associated with three month to two year patterns of climate variability to more high-level strategic issues associated with climate change. Issues include the need to find ways of coping with perceived more extreme weather events that are likely to be linked to more complex whole-farm economic issues associated with 2-20 year planning horizons. Typical management examples often cited relate to long-term reduction of cattle stocking rates on available land because of perceived long-term decline in rainfall and pasture availability (with some expectation these conditions will, more or less, continue) or to, otherwise, make the high-risk decision to purchase an adjoining property in order to maintain constant stocking rates under potentially increasing drying and warming conditions.

Many strategic impact assessments regarding climate change on agriculture are typically based on smoothly varying climatic change trends, whereas Schneider et al. (2000) note 'farmers in the real world will need to adapt to climate change trends embedded in a very noisy background of natural climatic variability'. They argue this variability can mask slow trends and delay necessary adaptive responses by government agencies. To add to this complexity, strategic climate change information is needed to anticipate and plan in a dynamic world in which many factors are changing both simultaneously, and not necessarily independently. Contingency plans developed in response to anticipated climate change will need information on shifting market and social conditions which may render adaptive behaviour for climate change much more multi-faceted that may be assumed (Risbey et al. 1999; Schneider et al. 2000). While, in some developed countries, research and development agencies continually monitor environmental trends, potentially leading to contingency planning and subsequent adaptive strategies, in developing countries problems with agricultural pests, extreme weather events and lack of capital to invest in adaptive strategies and infrastructure may be a serious impediment to reducing climatic impacts for agriculture (Schneider et al. 2000).

It has also been suggested that information dissemination networks, such as agricultural extension services, should now start to carefully provide data on trends and observed weather in local regions (Fankhauser et al. 1999). It is suggested that a more focused and urgent effort be made world-wide to provide enhanced and targeted climate trend and scenario information that is of direct relevance and value to a government's contingency planning policy. This may especially be the case in developing countries where climate change may shift farming regions into increasingly more vulnerable farming zones (Rosenzweig and Parry 1994).

24.6
Conclusions

It is clear that drought contingency planning provides an important component in improving measures to reduce the impacts of drought on agricultural production and natural resource systems. Droughts have mostly been managed through a reactive, crisis management approach but have largely been ineffective. Nevertheless, it is clear that investments in preparedness and mitigation will pay large dividends in reducing the impacts of drought and a growing number of countries are realizing the potential advantages of drought contingency planning. Improved preparedness programmes that involve suitable drought impact assessments and drought risk analyses (including use of agricultural simulation models) will help, not only current drought impacts but enable improved capability to respond to longer-term changes in climate. A common and key requirement for contingency planning to be successful is to develop a formal adoption process and to engage the likely affected community. Otherwise, as Susanna Davies points out, the plan may remain merely 'a piece of paper'.

A worthwhile development has been the initiation of decision-support systems that can aid in drought management preparedness practices. However, more recent assessment of these systems suggests they could be used more as discussion support tools that are then able to be more effective in providing planning tools for drought preparedness.

Aspects associated with global warming and long-term climate change appears to have received little attention in relation to future drought contingency planning. Additionally, current or recently developed policies regarding exceptional circumstance provisions for exceptional droughts (rare and exceptional droughts with a current frequency of one in twenty years or similar criterion) may not be relevant in the future as the frequency of severe droughts may potentially change either in severity or in frequency in some regions. This means that the impacts associated with a current one in twenty year drought may be encountered more often in some regions in the future (e.g. one in five years). It is suggested any changes in severe drought frequency due to climate change will greatly impact on drought contingency planning policies with a possible result that governments will revert to again taking a more reactive role. (Indeed, this may already be happening). It is suggested aspects of drought contingency planning, drought preparedness, and drought impact assistance policies need to be urgently considered as to their future effectiveness under long-term climate change.

References

Abawi Y, Dutta S, Ritchie J, Harris T, McClymont D, Crane A, Keogh D, and Rattray D (2001) A decision support system for improving water use efficiency in the northern Murray Darling Basin. Natural Resources Management Strategy Project 17403, Final Report to the Murray Darling Basin Commission. Toowoomba: Queensland Centre for Climate Applications, Department of Natural Resources and Mines. 58pp.

Abler DG, Shortle J, Carmichael J, and Horan R (2002) Climate change, agriculture, and water quality in the Chesapeake Bay region. Clim Chang 55:339-359.

Botterill LC, and Wilhite DA (2005) From Disaster Response to Risk Management: Australia's National Drought Policy. Springer, Dordrecht

Botterill LC (2003) Uncertain climate: the recent history of drought policy in Australia. Austr J Pol Hist 49:61-74

Bruins HJ, and Lithwick H (1998) Proactive planning and interactive management in arid frontier development. In: Bruins, H.J, and Lithwick H (eds) The Arid Frontier: Interactive Management of Environment and Development, Dordrecht: Kluwer Academic Publishers, 3-29

Bruins HJ (2001) Drought hazards in Israel and Jordan: policy recommendations for disaster mitigation. In: Wilhite D A (ed.) Drought: A global assessment Volume II. Routledge, London, pp.178-193

Buizer KF, Foster J, Lund, D (2000) Global impacts and regional actions: preparing for the 1997-98 El Niño. Bull Amer Met Soc 81: 2121-2139.

Bureau of Rural Sciences (2006) Scientific evaluation of drought events. Australian Government Bureau of Rural Sciences: http://www.affa.gov.au/content/output.cfm?ObjectID=BB52664E-530A-411C-A4CCA49107C70F93

Campbell B, Ogle G, McCall D, Lambert G, Jamieson P, Warwick R (1999) Methods for linking forecasts with decision-making. In: The 1997/98 El Nino: Lessons and opportunities. The Royal Society of New Zealand, Wellington, New Zealand.

Carter J, Hall W, Brook K, McKeon G, Day K, Paull C (2000) Aussie GRASS: Australian grassland and rangeland assessment by spatial simulation. In: Hammer G.L, Nicholls N, Mitchell C (eds) Applications of Seasonal Climate Forecasting in Agricultural and Natural Ecosystems: The Australian Experience. Kluwer Academic Publishers, Dordrecht, The Netherlands, pp. 329-351

Challinor AJ, Slingo JM, Wheeler TR, Craufurd PQ, Grimes DIF (2003) Toward a combined seasonal weather and crop productivity forecasting system: Determination of the working spatial scale. J Appl Meteorol 42: 175-192

Clewett JF, Clarkson NM, Owens DT, Arbrecht DG (eds) (1994) Australian Rainman: Rainfall information for better management. Brisbane. Queensland Department of Primary Industries

Das HP (1999) Management and mitigation of adverse effects of drought phenomenon. In: Natural Disasters – some issues and concerns. Natural Disasters management cell, Visva Bharati, Shantiniketan, Calcutta, India, pp. 87-97

Das HP (2005) Agrometeorological impact assessment of natural disasters and extreme events and agricultural strategies adopted in areas with high weather risks. In: Sivakumar M V K, Motha R P, Das H P (eds) Natural Disasters and Extreme Events in Agriculture: Impacts and Mitigation. Springer. Berlin. pp.93-117

Davies S (2001) Effective drought mitigation: linking micro and macro levels. In: Wilhite D A (ed) Drought: A global assessment Volume II. Routledge, London, pp.3-17

Day K, Ahrens D, McKeon G (2003) Simulating historical droughts: some lessons for drought policy. In: Stone RC, Partridge I (eds) Science for Drought: Proceedings of the National Drought Forum, Brisbane, April, 2003, pp. 141-151

deJager JM, Potgieter AB, van den Berg WJ (1998) Framework for forecasting the extent and severity of drought in maize in the Free State Province of South Africa. Agric Syst 57:351-365

Everingham YL, Muchow RC, Stone RC, Inman-Bamber G, Singels A, Bezuidenhout CN (2002) Enhanced risk management and decision-making capability across the sugarcane industry value chain based on seasonal climate forecasts. Agric Syst 74:459-477

Fankhauser S, Smith JB, Tol RS J (1999) Weathering climate change: some simple rules to guide adaptation decisions. Ecol Econ 30:67-78

Hammer GL, Hansen JW, Phillips JG, Mjelde JW, Hill H, Love A, Potgieter A (2001) Advances in application of climate prediction in agriculture. Agric Syst 70:515-553

Hansen JW (2002) Applying seasonal climate prediction to agricultural production. Agric Syst 74:305-307

Haylock HJK, Ericksen NJ (2001) From State dependency to self-reliance: agricultural drought policies and practices in New Zealand. In: Wilhite D A (ed) Drought: A global assessment Volume II. Routledge, London, pp. 178-193

Jones JW, Hoogenboom G, Porter CH, Boote KJ, Batchela WD, Hunt LA, Wilkens PW, Singh U, Gersman AJ, Ritchie IT (2003) The DSSAT cropping system model. Europ J Agron 18:235-265

Keating BA, Meinke H (1998) Assessing exceptional drought with a cropping systems simulator: a case study for grain production in north-east Australia. Agric Syst 57:315-332

Keating BA, Asseng S, Brown S, Carberry PS, Chapman S, Dimes JP, Freebairn DM, Hammer GL, Hargreaves JNG, Hochman Z, Holzworth D, Huth NI, Meinke H, McCown RL, Mclean G, Probert ME, Robertson MJ, Silburn M, Smith CJ, Snow V, Verburg K, Wang E (2003) The Agricultural Production Systems Simulator (APSIM): its history and current capability. Europ J Agron 18:267-288.

Macy P (1993) Presentation on drought planning. In: Water Resources Management Committee Occasional Paper WRMC No. 4. National Workshop on Drought Planning and Management for Water Supply Systems, Agriculture and Resource Management Council of Australia and New Zealand. Water Forum. 40-45

Manton MJ, Brindaban PS, Gadgil S, Hammer GL, Hansen J, Jones JW (2000) Implementation of a CLIMAG demonstration project. In: Sivakumar MVK (ed.) Climate prediction and agriculture. Proceedings of the START/WMO International Workshop, Geneva, Switzerland, 27-29 September 1999. Washington DC, USA, International START Secretariat, pp. 287-293

Meinke H, Baethgen WE, Carberry PS, Donatelli M, Hammer GL, Selvaraju R, Stockle CO (2001) Increasing profits and reducing risks in crop production using participatory systems simulation approaches. Agric Syst 70:493-513

Meinke H, and Stone RC (2005) Seasonal and inter-annual climate forecasting: the new tool for increasing preparedness to climate variability and change in agricultural planning and operations. Clim Chang 70:221-253

Meinke H, Nelson R, Kokic P, Stone RC, Selvaraju R, Baethgen W (2006) Actionable climate knowledge – from analysis to synthesis. Clim Res 33:101-110.

Nelson RA, Holzworth DP, Hammer GL, Hayman PT (2002) Infusing the use of seasonal climate forecasting into crop management practice in north east Australia using discussion support software. Agric Syst 74:393-414

Nelson R, Kokic P, Elliston L, King J (2005) Structural adjustment: a vulnerability index for Australian broadacre agriculture. Austr Commod 12:171-179

Nelson R, Kokic P (2006) Part II: Forecasting the incomes of Australian mixed crop-livestock farms at a regional scale. Aus J Agric Res (submitted)

Plant S (2000) The relevance of seasonal climate forecasting to a rural producer. In: Hammer GL, Nicholls N, Mitchell C (eds) Applications of Seasonal Climate Forecasting in Agricultural and Natural Ecosystems: The Australian Experience. Kluwer Academic Publishers, Dordrecht, Netherlands, pp. 23-29

Queensland Government (1992) Drought: Managing for Self Reliance: A Policy Paper. Queensland Department of Primary Industries, Brisbane, Queensland, Australia. 22pp

Risby J, Kandlikar M, Dowlatabadi H, Graetz D (1999) Scale, context, and decision-making in agricultural adaptation to climate variability and change. Mitig Adapt Strat Glob Chang 4:137-165

Ritchie JW, Abawi GY, Dutta S, Harris TR, Bange M (2004) Risk management strategies using seasonal climate forecasting in irrigated cotton production: a tale of stochastic dominance. Aus J Agric Res Econ 48:65-93

Rogers CE, McCarty JP (2000) Climate change and ecosystems of the Mid-Atlantic region. Clim Res 14:235-244

Rosenzweig C, Parry ML (1994) Potential impact of climate change on world food supply. Nature 367:133-138

Rosenzweig C, Iglesias A, Yang XB, Epstein PR, Chivian E (2000) Climate change and U.S. agriculture: The impacts of warming and extreme weather events on productivity, plant diseases, and pests. Center for Health and the Global Environment, Harvard Medical School, Boston. 47 pp

Ruben R, Kruseman G, Kuyvenhoven A, Brons J (2000) Climate Variability, risk-coping and agrarian policies: Farm households' supply response under variable rainfall conditions. Report for NOP Project 'Impact of Climate change on drylands (ICCD), Wageningen, The Netherlands, October, 2000

Schneider SH, Easterling WE, Mearns LO (2000) Adaptation: sensitivity to natural variability, agent assumptions and dynamic climate changes. Clim Chang 45:203-221

Sivakumar MVK (2002) Opening Remarks. In: Sivakumar MVK (ed) Improving Agrometeorological Bulletins. Proceeding of the Inter-Regional Workshop October 2001, Bridgetown, Barbados. pp. 15-19

Stehlik D, Gray I, Lawrence G (1999) Drought in the 1990s: Australian Farm Families' Experiences. Rural Industries Research and Development Corporation Publication No. 99/14 Project No. UCQ-5A. Barton, ACT, Australia. 119pp

Stoeckle CO, Donatelli M, Nelson R (2003) CropSyst, a cropping systems simulation model. Europ J Agron 18:289-307

Stone R C, Meinke H (2005) Operational seasonal forecasting of crop performance. Phil Trans Roy Soc B, 360:2109-2124

Stone RC, Meinke H (2006) Weather, Climate, and Farmers: An Overview. Met Applns 13 (Supplement 1):7-20

Subbiah AR (2001) Response strategies of local farmers in India. In: Wilhite D A (ed) Drought: A global assessment Volume II. Routledge, London, pp. 29-34

Thompson D, Powell R (1998) Exceptional circumstances provisions in Australia—is there too much emphasis on drought? Agric Syst 57:469-488

UAS (1999) Package of practices for high yield for central dry zone. Publication of the University of Agricultural Sciences, G.K.V.K., Bangalore, India.

White DH, Karssies L (1999) Australia's national drought policy. Aims, analyses, and implementation. Water Int 24:2-9

Wilhite DA (1996) A methodology for drought preparedness. Nat Haz 13(3):229-252

Wilhite DA (2003) Drought policy and preparedness: The Australian experience in an international context. In: Botterill L C, Fisher M (eds) Beyond Drought: People, Policy, and Perspectives. Melbourne, CSIRO Publishing, pp. 175-199

Wilhite DA (2005) Drought policy and preparedness: The Australian Experience in an International Context. In: Botterill LC, Wilhite DA (eds) From Disaster Response to Risk Management: Australia's National Drought Policy. Springer Publishers, pp. 157-177

Wilhite DA, Hayes MJ, Knutson CL (2005) Drought preparedness planning: building institutional capacity. In: Wilhite DA (ed) Drought and water crises: Science, technology, and management issues. Taylor & Francis. Boca Raton, Florida

Woodruff DR (1992) 'WHEATMAN' a decision support system for wheat management in subtropical Australia. Aus J Agric Res 43:1483-1499.

CHAPTER 25

Agrometeorological services to cope with risks and uncertainties

Raymond P. Motha, V.R.K. Murthy

25.1.
Introduction

Agriculture is defined as "the production and processing of plant and animal life for the use of human beings" and also "a system for harvesting or exploiting the solar radiation." Meteorology is "the science of atmosphere." "The study of those aspects of meteorology, which have direct relevance to agriculture," is defined as agricultural meteorology for which the abbreviated form is "Agrometeorology" (Murthy 1996). The primary aim of agricultural meteorology is to extend and fully utilize the knowledge of atmospheric and other related processes to optimize sustainable agricultural production with maximum use of weather resources and with little or no damage to the environment. This science also provides the necessary information to help reduce the negative impact of risks and uncertainties of adverse weather through agrometeorological services (Motha et al. 2006). According to Murthy and Stigter (2003), agrometeorological services are "all agrometeorological and climatological information that can be directly applied to improve or protect agricultural production (yield quality, quantity, and income obtained from yields) while protecting the agricultural production base from degradation." Sivakumar et al. (2004) observed that the current status of agricultural production and increasing concerns with related environmental issues call for improved agrometeorological services for enhancing and sustaining agricultural productivity and food security around the world.

25.2
Weather, Natural Disasters, and Agriculture

The risks and uncertainties in agriculture are directly related to the nature of weather and climatic hazards and their associated damages, and their potential impacts on loss of crops, animals, land, produce, etc. Agrometeorological services play a valuable part in making daily and seasonal farm management decisions and in management of risks and uncertainties pertaining to agriculture, forestry, rangelands, environment, and livestock.

25.2.1
Fundamental importance of weather in agriculture

Agriculture is one of the most weather dependent industries. It is the world's single largest employer and one of the main sources of export earnings, thereby significant from foreign exchange point of view (Sivakumar et al. 2004). Weather is defined as "the day-to-day condition of the atmosphere at a particular place and given instant of time," whereas, climate is "the summation of weather conditions over a given region during a comparatively longer period." The knowledge of weather is essential in the daily management of crops and animals, and the science of climate aids in the selection of crops and animals. Of all the weather elements, solar radiation controls organic life by heating the earth and atmosphere and also provides the energy required in photosynthesis for the conversion of carbon dioxide and water into primary source of food (carbohydrates). Air temperature influences rates of biochemical reactions in crops (approximately double with each 10°C rise in temperature), leaf production, expansion, and flowering. Soil temperature influences crop production. Murthy et al. (2002) mentioned that both cold and heat waves and abnormal soil temperatures are risks to crop growth and development. Humidity is an important factor in crop production and is closely related to rainfall and temperature. It is of great importance in determining the vegetation of a region and affects the internal water potential of plants, which in turn, determines the water requirements of crops. The significance of wind on crops and animals was reported by several authors, among which Murthy (2003) reported that the moderate wind turbulence promotes the consumption of carbon dioxide in photosynthesis and prevents frost by disrupting the temperature inversion. The economy of many nations, more so the under-developed and developing countries, depends largely on the amount and distribution of rainfall received each year. This is a significant agricultural input because rainfall is the major source of water, which is essential for seedbed preparation and plant growth and development. Plant physiological processes like cell division and enlargement depend largely on water (rainfall), which in turn influences plant growth and development.

25.2.2
Impact of natural disasters in agriculture, rangeland, forestry, and environment

Agriculture, rangelands, and forestry are highly dependent on weather and climate. Climate change is inevitable and the impacts are visible, which are adversely effecting the environment. Anderson (1990) defines natural disasters as "the temporary events triggered by natural hazards that overwhelm local response capacity and seriously affect the social and economic development of a region." Increasing climate variability and anthropogenic climate change lead to floods, cyclones, earthquakes, hurricanes, tornados, snow storms, avalanches, tidal waves, heat and cold waves, land slides, forest fires, droughts, blasting mildew, frost, droughts, sand and dust storms, etc. These are the natural disasters that principally impact agriculture, rangelands, forestry, and worldwide environmental degradation. Information on natural disasters and trends is basically available from global databases that provide essential information on the occurrence, recurrence, and location of disasters and disaster trends over time (World Disaster Report 2003). The natural disasters, risks, and uncertainties of weather and climate have shown an increase from 1993 to 2002. There is evidence of this from different parts of the world. Of a grand total of 2,654 disasters during this period, floods and windstorms account for about 70% and the remaining 30% of the disasters are accounted for by droughts, landslides, forest fires, and heat and cold waves. At the regional level, in South East Asia and Bangladesh over the last century, 700 disasters have occurred of which 23% were between 1900 and 1979, and 77% between 1972 and 1996. For the Latin American and Caribbean region, Charveriat (2000) showed a noticeable trend of increase in the frequency of natural disasters. Loss in crop and animal production and land degradation from droughts that occur twice a year cost Australia billions of dollars. The economic cost associated with all natural disasters has increased 14-fold since the 1950s (World Disasters Report 2001). Worldwide, annual economic costs related to natural disasters have been estimated at about USD $50 to $100 billion. The world land use data (FAO 1999) show that 70% of the global land use is for agriculture, rangeland, and forestry with 12% of the land use for arable and permanent crops, 31% for forest and woodlands, and 27% for permanent pasture. By the year 2050, it is predicted that the global cost to natural disasters could top USD $300 billion annually. In India, according to Roy et al. (2002), the Orissa state has been disaster-affected for 90 years (floods have occurred for 49 years, droughts for 30 years, and cyclones have hit the state for 11 years). The United States experienced more disasters between 1970 and 1999 than any other region, but the impact on national development was not as severe as in some of the developing countries. For example, Hurricane Andrew in 1992 caused a total damage of USD $26.5 billions in the United States, but it was a mere 0.4% of GDP.

25.2.2.1
Impact of natural disasters on agriculture

Agriculture is one of the sectors most heavily impacted by natural disasters. As Das (2003) explained, the impact of natural disasters on agriculture can be direct or indirect in their effect. Direct impacts arise from the direct physical damage on crops, animals, and horticultural orchards caused by extreme hydro-meteorological events. Disasters also cause indirect damage through increased costs of crop production. Floods occur due to overflow of rivers. Sudden hazards such as storms usually have fewer long lasting effects than droughts, which are often described as creeping in nature because of the slow rate at which they develop. The most prolonged and widespread droughts occurred in 1973 and 1984, when almost all the African countries were affected, and in 1992 when all southern African countries experienced extreme food shortages. In 1973 alone, drought killed 100,000 people in the Sahel (Gommes and Petrassi 1996). The Committee of Agricultural Organisation in the Union (COPA) and the General Committee for Agricultural Co-operation in the European Union (COGECA) estimate that drought and fires during 2003 have cost European Union farmers $13.5 billion – with Austria, France,

Germany, Italy, Portugal, and Spain identified as countries most affected. One of the major impacts of droughts is regressive distributional effects across communities and across households within communities. Scott and Litchfield (1994) provided evidence of these impacts in the rural regions of Coquimbo in Chile where inequality and poverty at the community level increased significantly in Las Tazas due to the cumulative effects of very low rainfall over the years. The projected climate change and the attendant impacts on agriculture in the arid and semi-arid tropical regions add additional layers of risk and uncertainties to agricultural systems that are already impacted by land degradation due to growing population pressures.

25.2.2.2
Impact of natural disasters on rangeland

Rangelands include natural grasslands, savannas, shrub lands, most deserts, tundra, alpine communities, coastal marshes, and wet meadows in certain parts of the globe. These expanse lands are suitable for livestock to wander and graze on. Due to the impact of natural disasters on rangelands, it was estimated that hundreds of thousands of people died and nearly half of the entire livestock herds and two million herds of wild animals were killed due to the severe droughts and land desertification at the southern edge of the Sahara desert. Droughts cause inappropriate herding practices, forcing liquidation of livestock at depressed prices (Sivakumar et al. 2004). Cyclones cause destruction of vegetation and livestock in rangelands. Similarly, floods are the greatest natural disasters that cause extensive damage to rangelands because of loss of pasture use, permanent damage to perennial crops, trees, livestock, etc.

25.2.2.3
Impact of natural disasters on forestry

The forests cover nearly 30% (3,500 Mha) of the world's land area. The response of forest ecosystems to climate change, risks, and uncertainties can be expressed in terms of boundaries shift, changes in productivity, and risk of damages (e.g. forest fires). In Europe, most climate change scenarios suggest a possible overall enlargement of the climatic zone suitable for Boreal forests. During 1982/83 and 1994/95 El Niño events, South East Asia experienced severe smoke and haze episodes associated with the forest and bush fires due to reduced rainfall and drought conditions. According to the National Interagency Fire Center (NIFC) published material covering the period January 2003 to early November 2003 inclusive, around 56,000 United States' wild fires have affected 3.8 million or so acres. Forest fires are another natural disaster that causes large scale damage to plants, animals, and human lives and property. Forest fires seriously affect human health, economy, and environment. The extent of damage to human health from smoke inhalation depends on the constituents of smoke, concentration, and exposure time. The environmental impacts from forest/bush fires range from local to global. Local impacts include increased soil degradation, increased risks of wet-season flooding and dry-season drought, reduced number of animals and plants, and increased risk of recurrent fires. The global effects of these fires include the release of large amounts of various amounts of greenhouse gases, reduced rainfall, and increased day lighting and reduction of biodiversity through extinction of populations and species. Emissions from forest fires are lofted high into the atmosphere and significantly impact air quality on local, regional, and even global scales. The forest fire plumes are known to contain a highly variable mix of gases and particulate pollution.

25.2.2.4
Impact of natural disasters on environment

Droughts, floods, heat waves, frost, and extreme weather periodically wreak havoc on crops, pastures, livestock, and contribute to pollution both downstream and off-farm. Environmental degradation is one of the major factors contributing to the vulnerability of agriculture, forestry, and rangelands to natural disasters because it directly magnifies the risk of natural disasters, or by destroying natural barriers leaving agriculture, forestry and rangelands more vulnerable to their effects. Deforestation, land clearing, weed invasion, and loss of wetlands could lead to ecosystem alteration, including changes to vegetation cover and composition and the incidence of diseases and pests on plants and animals. Water erosion, wind erosion, siltation and sedimentation, and coastal erosion could result in transport of soil and deposition elsewhere. Soil salinity, degradation of soil structure, soil fertility decline, soil acidification, water logging, and soil pollution could lead to soil degradation, involving the alteration of soil properties in situ. Clearing of vegetation, rapid abandonment of exhausted cropland, and expansion of cropping into new and marginal land all set up a vicious pollution cycle that is hard to break. Poverty and environmental degradation are closely linked, often in a self-perpetuating negative spiral in which poverty accelerates environmental degradation and degradation results in or exacerbates poverty. While poverty is not the only cause of environmental degradation, it does pose the most serious environmental threat in many low income countries. Expanding aquaculture could also increase pollution and the use of scarce water and land resources, threatening the environment in under-developed and developing countries.

25.2.3
The role of Indigenous Technical Knowledge (ITK) in agrometeorological services

Human beings *(Homo sapiens)* have been on the earth for approximately two million years. They have been hunters and gatherers for 99.5% of their existence. They started domesticating plants and animals only 10,000 years ago. Of all the human endeavors, agriculture is the first sector in which there is evidence to show dependence and strong relationship with weather and climate. However, humans were often mystified and frustrated by the variability of the weather, climate, and seasons. Therefore, they developed some agrometeorological services and managed their environments for generations by following environmentally-friendly agricultural practices (Reddy 2002) and without significantly damaging local ecologies (Singh et al. 2004). This is strong proof that indigenous knowledge has immense potential to manage the disasters, risks and uncertainties like climate change and variability, floods, cyclones, droughts, and pests and diseases on crops and animals through agrometeorological services.

These services were developed by humans based on their understanding of weather and climatic patterns. Such services have been used over the centuries and continue to be used right up to the present day. The ITK products on weather in Asia are strong knowledge pools developed by different communities through keen observations, natural selection, and centuries of trial and error. All Asian cultures and civilizations have developed a form of astrology to understand seasons and weather in relation to movements of planets. As the position of planets was predetermined, the rainfall was forecast for any time in future. The first agrometeorological service developed and being used in the present day modern technology era in India is the classical Hindu almanac known as "Panchanga" (Murthy 2003). It is a book or record of astronomical phenomena containing calendar days, weeks, and months of the year. Weather prognostications, predicted rainfall distribution, based on cropping patterns and area were decided for a state or country.

The other ITK based agrometeorological services in India include the use of calotropis to control thrips and mealy bugs, the use of cow dung cake gas as burrow fumigant, and use of bow traps to control field rats, use of leaf powder of Margosa (Azadiracha indica), Nicotine (nicotiana tobaccum) and extract of custard apple (anona squamosa) for pest control on crops and animals and storing the grains in wooden bins. In addition, different species and varieties of crop seeds are mixed and sown to delay the onset and spread of pests and diseases, which also modify the micro climate for better harvest of all crops. The Himalayan farmers maintain their own rangelands plus a share in village managed community rangelands. They rare the animals and use cow and buffalo milk and milk products during failure of crops in drought situations. According to Dafu et al. (1990), the farmers of China in mountainous regions follow irrigated (level), dry (bench), and complex (bench + level) terracing to manage water scarcity for crops and rangelands. Similarly, farmers in Pakistan follow the same practices to derive the same benefits in addition to sustaining the productivity of mountain soil against landslides (Bhatt 1992). It was stated by Singh et al. (2004) that "three north" system of shelterbelts was evolved a few thousand years ago and is still being practiced in China. This system protects the existing forests and rangelands against flood damages.

Alteri (1991) described an indigenous system called *waru-warus* followed in Peru for over 3000 years. This system consists of a platform of soil surrounded by ditches filled with water. During droughts, the moisture from canals slowly ascends through the roots in capillary action, and during floods, the furrows drain away excess run off. This system produces bumper crops even in the events of both droughts and floods. In Mexico, a low-cost and self-sufficient farming system called *chinampa* has been practiced for centuries. In this system, small plots of land are prepared, which are separated by water channels. The water in the channels helps to grow fish and is also useful for irrigation (when in excess) and when scanty (drought), plants grown in the sides of *chinampas* give income to the farmers. The "Sami" are the indigenous people in the northern Scandinavia. They live in Sweden, Norway, Finland, and Kola Peninsula of Russia. They are a 70,000 strong population of which 16% of 17,000 Swedish Sami are still reindeer herders who live in close contact with nature. They are not nomadic because the reindeer herding is their culture. However, they follow the path of reindeer, an agrometeorological service, between summer grazing lands in mountain regions and winter grazing lands in forests.

25.2.4
The role of contemporary technological advances in agrometeorological services

Weather and climate data systems for agricultural activities are necessary to expedite the generation of products, analyses, and forecasts (Sivakumar et al. 2004) to develop preparedness measures against natural disasters, management strategies for risks and uncertainties in agriculture, forestry, rangelands, and environmental protection (Rao et al. 2004). The following products, tools, and services of contemporary science and information technology have been providing newer dimensions to effectively monitor and manage the weather and climate related disasters, risks, and uncertainties.

25.2.4.1
Satellites and remote sensing

According to Maracchi et al. (2000), one of the most important sources of agrometeorological data that compliments traditional methods of data collection is remote sensing. The satellite remote sensing is a new technique which provides spatial coverage of earth's surface and the surrounding atmosphere. Use of remote sensing for rainfall estimation, monitoring crop condition progress, etc., is state-of-the-art technology for food security. Recent advances in satellite and computer technology have led to significant progress in remote sensing and proved its potential to meet the requirements of farmers at the operational level. Towards this direction, a new numerical method was developed for drought detection and impact assessment in any part of the world from NOAA operational environmental satellites (Kogan 2000). In the MARS project, a software system (SPACE) was implemented for crop growth monitoring. In order to quantify vegetation stress and to monitor drought, the Portuguese Meteorological Institute calculates the NDVI based on NOAA-AVHRR data. The wildland fire assessment system (WFWAS), developed by the USDA Forest Service, became operational in the middle 1990s and has been providing useful information such as a "greenness" map using AVHRR-NDVI (Burgan and Klamer 1998). In India, Landsat TM optical bands data were used for computation of regional surface albedo. An albedo image was generated for the snow and forest covered Himalayan Mountains of India using

NOAA-AVHRR ch1 and ch2 data following an empirical relationship for broad band albedo and narrow band albedo. Thapliyal et al. (2004) used microwave radiometer data for assessment of drought with 6.6 GHz brightness temperature. They are developing a suitable index to indicate the severity of drought conditions over India. According to Susman et al. (1983), a methodology was developed to estimate flood damage to rice production using temporal synthetic aperture radar data. It was concluded that the method could easily be adopted to estimate flood damage to the crop by just acquiring a single date data coinciding with the flood event. It is observed that the methodology is cost-effective in Indian context where rice is grown over 30 mha, often in a large contiguous area as a single most dominant crop during rainy season. Madhavi et al. (2004) analyzed IRS WIFS data at 10-day intervals over the Raisen Forest division in India and assessed the spread of fire damaged areas in different forest segments. The studies provided information on progression and recovery of fire burnt areas, which are quite useful in planning for control operations.

25.2.4.2
Geographical Information Systems (GIS) and Geographical Positioning Systems (GPS)

The review of Jayasheelan and Chandrasekhar (2002) indicates that GIS refers to a description of characteristics and tools used in the organization and management of geographical data. The term GIS is currently applied to computerized storage, processing and retrieval systems (Murthy 2006) that have hardware and software specially designed to cope with geographically referenced spatial data and corresponding informative attribute. GIS enables management of large datasets such as traditional digital maps, databases, and models. The quantitative data handling capability offered by GIS would assist the users to overlay numerous spatial data sets and statistically analyze the same. Through this procedure it is possible to develop quantitative relationships which are not achievable through the use of simple map drawing or graphics display programmes. GIS technology is a powerful agrometeorological tool for combining, or overlaying, various map and satellite information sources in meteorological and climatological models that simulate interactions of complex natural systems. Therefore, it is possible to prepare decision support systems based on GIS to manage weather related disasters, risks and uncertainties. These tools are state-of-the-art technology for appropriate planning, coordination, and monitoring of these events. The expected cyclone characteristics, rainfall patterns, water levels in rivers, environmental impact assessments and vulnerabilities, can assist in modeling disaster consequences accurately. They can also help to evolve effective decision-making on logistic and infrastructure requirement and their development in an area.

The ultimate use of GIS lies in its modeling capability, using real world data to represent natural behaviour and to simulate the effect of specific processes. With this scientific background the frost risk map and the desertification climatic index were developed using spatial modeling with GIS, at the Portuguese Institute of Meteorology. On the same lines several information products (soils, crops, and meteorology) were integrated in ISOP project in France, which is in operational use and it assesses real-time forage production over France. The prediction and management maps of chilling injury of banana and litchi trees were developed using GIS in Guangdong province, China (Wang et al. 2003), which are highly successful and are being used by the farmers. A new major programme LANDFIRE uses GIS technology to map all wildfire fuels across the United States at 30m spatial resolution. This is intended to be the safety net for land management agencies. A few studies have focused on the collection of historical data and habitat conditions with the dynamics of locust development stages, and synthesis of data using GIS and evolving decision support systems (Hodell et al. 1995). This system integrates remotely sensed and farm soil texture, soil moisture and vegetation density with the daily weather data to forecast the suitable breeding sites and time of onset of locust upsurge in and around study area. Reliable drought interpretation requires a GIS based approach, since the topography, soil type, spatial rainfall variability, crop type and variety, and irrigation support and management practices are relevant parameters. The conventional methods of surveying and navigation require tedious field and astronomical observations for deriving positional and directional information. The GPS service consists of three components: space, control, and user. Rapid advancement in higher frequency signal transmission and precise clock signals along with advanced satellite technology have led to the development of GPS (Ramesh et al. 2004). The outcome of a typical GPS survey includes geocentric positions accurate to 10 m and relative positions between receiver locations to centimeter level or better. The capabilities of surveying, mapping, and locating geophysical positioning of GIS immensely help in combating the natural disasters in agriculture, forestry, rangelands, and effective control of environmental pollution.

25.2.4.3
Information technologies and communications

It is important that information is disseminated to the user by agrometeorological services, which are easily understandable and in time through a rapid communication system. For effective communication, information related to disasters, risks, and uncertainties is a major challenge in the developing and under developed countries. Reliable communication networks connect the scientific and technological advances of the developed countries with these nations. The Internet, digital satellite technology, wind-up machines, computers, etc., are new possibilities for rural areas in the under developed and developing world. Mobile phones, facsimiles, e-mail, wireless technologies, etc., which are available in the developed countries, offer the greatest potential and must be recommended for the developing countries and under developed countries. The Internet can accomplish accurate, timely, useful, and cost-effective information to the rural areas. The RANET (Radio and Internet) system is an innovative system that brings new communications and technologies together and delivers operational agrometeorological services on risks and uncertainties of weather and climate over a distributed network in Africa. This is managed by the local communities and is rated as the most successful communication network for efficient communication of agrometeorological services.

In a developing country like India, NCMRWF's weather forecast bulletin for the subsequent three days is disseminated biweekly to AAS units every Tuesday and Friday over the telephone, telefax, or satellite based very small aperture terminal (VSAT) communication system. The VSAT system has the capabilities for reliable interactive data communication and picture transmissions. Also, dissemination of short- and medium-range forecasts to the farmers is operational through radio and television. The long-range forecast regarding onset of monsoon seasonal rain is disseminated through newspapers. The newly developed agrometeorological techniques are being communicated through the extension workers who reach the end users directly (Kashyapi 1998). The introduction of the satellite-based cyclone warning dissemination system in the 1980s in Andhra Pradesh state of India was the single most important step to improve the speed and credibility of transmission of warnings for operational use by the farmers. The farming communities in a developed country like Germany are interested in receiving the longest-term and most exact weather forecasts possible, as well as information about the expected conditions at the production sites. Therefore, telefax, T-online, and agrometeorological online services are in operation (Rudolph 1998).

25.3. Operational Agrometeorological Services to Cope with Risks and Uncertainties of Natural Disasters

25.3.1
United States (U.S.A.)

The U.S. National Weather Service (NWS) was created in 1870 and was officially transferred to the Department of Agriculture (USDA) in 1891 and then to the Department of Commerce (DOC) on June 30, 1940. As early as 1941, operational agrometeorological services were provided for orchardists under DOC and USDA weather, climate, and agricultural activities. The farmers were advised on conditions suitable for spray on fruit trees. Pioneering research at USDA resulted in the publication of "Atlas of Climatic Types in the United States 1900-1939" (Motha et al. 2006), categorizing climate by moisture regimes, providing definitions of effective precipitation, the use of vegetation as climatic indicators, and discussions of climate variations. While the meteorological requirements of USDA are numerous, they can be categorized into four basic areas. They include: current measurement and observational data and services; climate services including the summarization of historical weather data, the analyses of climatological data to characterize climate conditions or regimes for different geographical areas or time periods, and the development of normals, freeze probabilities, and drought indices; forecasting services including the prediction of future weather events or climatic conditions and their associated probabilities that impact agriculture, forestry and rangelands; and other services such as consultation, analyses of particular weather events, interpretation of forecast materials, monitoring and summarizing recent weather events, weather briefings and summaries, special studies and analyses, and user education (Motha et al. 1997; Rippey et al. 2000). USDA has a vast need for weather and climate data to assist agriculturalists with their management operations, which include strategic decisions such as what to plant, or tactical decisions including when to irrigate. As a result, USDA agrometeorological services that assist farmers directly or indirectly in their decision-making process require a detailed set of weather information. Based on this information, certain operational agrometeorological services have been developed by USDA of which the most important are as follows:

25.3.1.1
Weekly Weather and Crop Bulletin (WWCB)

In 1872, the then "Division of telegrams and reports for the benefit of commerce" in the War Department began publishing the "Weekly Weather Chronicle" for the benefit of commerce and agriculture. It evolved into the present Weekly Weather and Crop Bulletin (WWCB). It contains a global summary of weather for each week and also provides information pertinent to regional, national, and international agricultural weather. Detailed maps and tables of agrometeorological information for appropriate seasons along with a summary of weather and crop information are provided. In the bulletin, the report usually discusses crop weather conditions suitable for field work and crop development, pests and disease outbreak, soil moisture conditions, crop progress, and livestock conditions. The bulletin is an effective means of distributing weather information to farmers and the public. The WWCB has been cooperatively produced by USDA since 1978. The users of this agrometeorological service range from farmers to marketing agencies of agricultural products.

25.3.1.2
Joint Agricultural Weather Facility (JAWF)

This is a global agrometeorological service which was established in 1978 and serves as USDA's overall focal point for weather/climate information and agricultural impact assessments. The JAWF was created as a world agricultural weather information center, located in USDA, and is jointly staffed and operated by the Departments of Commerce and Agriculture. The primary mission of this facility is to monitor global weather and determine the potential impacts on agriculture. The JAWF provides information on weather related developments and their effect on crops and livestock. This information in turn helps the decision-makers in formulating crop production forecasts and trade policy. In addition, JAWF developed a Data Base Management System (DBMS), which effectively manages a global weather, climate, and agriculture data base for analyses. GIS techniques are utilized in the products of DBMS, which enhance the analytical capability of the agrometeorologists to produce crop-weather assessments. In May 1996, a JAWF field office (a weather/GIS data center) was established at the Mississippi State University (MSU) to meet the local demands for agricultural weather information required for research and production agriculture in the Delta growing region. The main mission of this facility is to ensure collection and archival of vital agricultural weather data in the Mississippi Delta. This center provides weather and climate data, crop progress information, tailored products using GIS, and weekly weather briefings to researchers, producers, county extension agents, and agricultural industries in the Delta. A product called "Node above white flower five rule" is an unique agrometeorological product developed from the research at the center. The cotton planting recommendation is based on soil temperature and other agrometeorological products from this center.

25.3.1.3
U.S. Drought Monitor

Comprehensive weather, water, soil moisture, mountain snow amount, and climate observations are the foundations of monitoring and assessment activity that alerts the nation to impending drought. An operational drought product called the U.S. Drought Monitor was established in 1999 and since then has been considered a major agrometeorological service to the nation. JAWF contributes to the Drought Monitor. It monitors drought conditions including aerial extent, severity, and type around the country. Over the last 6 years, this product has become a highly successful agrometeorological tool for assessing the development and duration of drought conditions. The significant reason for the outstanding success of this agrometeorological service is the process of information by many experts located across the country. The Drought Monitor is a dynamic product, the uses of which range from farmers to government policy-makers.

25.3.1.4
National Resources Conservation Service's (NRCS)
Water and Climate Center (WCC)

NRCS's cooperative Snow Survey and Water Supply Forecast (SS/WSF) program provides farmers and other water management groups in the western states with water supply forecasts to enable them to plan for efficient water use management. The program also provides the public and the scientific community with a database that can be used to accurately determine the extent and amount of seasonal snow resources. The SS/WSF also provides nationwide climate services to the NRCS and USDA in partnership with other Federal agencies and universities. The SS/WSF operates a 715-station SNOTEL (SNOw TELemetry) network in the western U.S. and a 111-station SCAN (Soil Climate Analysis Network) in 39 states throughout the U.S. and Caribbean. The SS/WSF program provides essential products from these data networks necessary to monitor and mitigate drought and floods. The SS/WSF program has also sponsored the development of the Applied Climate Information System (ACIS) for information dissemination, and the PRISM climate mapping technology used for spatial analysis worldwide.

25.3.2
India

The India Meteorological Department (IMD) was established in 1875. It is the national meteorological service of the country and the principal Government agency in all matters pertaining to agricultural meteorology. One of the major objectives of this department is to provide meteorological statistics required for agriculture (Rao 2004). India has diverse agroclimatic regions with large variations of rainfall ranging from 150 mm in the northwest/western part of the country to over 10,000 mm in the northeast region. Two thirds of the country comes under arid and semi-arid region. This area is prone to recurrent droughts. The vast Indian coastal belt is frequently affected by cyclonic storms. Over 40 million ha of land area in the country is vulnerable to floods, out of which, about 8 million ha is severely affected by floods each year. The country's hilly region is prone to land slides due to heavy rains and the Himalayan region to avalanches. These risks and uncertainties cause major setbacks to agriculture and the economy. Therefore, a number of initiatives are taken by the Government of India to improve various types of agrometeorological services for managing the weather related disasters, risks, and uncertainties as detailed below.

25.3.2.1
Flood meteorological offices and cyclone forecasting and warning systems

The IMD established 10 flood meteorological offices in areas prone to floods. These offices issue operational agrometeorological services like prevailing synoptic situation, heavy rainfall warnings and quantitative precipitation forecasts which are highly useful to the all farmers in general and the poorest of the poor in particular across the nation. The IMD also has developed a well established organizational setup for observing, detecting, tracking, and forecasting cyclones and issuing cyclone warnings. The frequency of occurrence of cyclones in eastern India is presented in Figure 25.1. The cyclone warning bulletins are issued to state owned radio stations and televisions and disseminated through landline telegrams, police stations, telex, telephones, fax, cell phones, bulletins to the press, internet, satellites, etc.

25.3.2.2
Agrometeorological advisories

The Agrometeorological Advisory Service in the country was started first in the India Meteorological Department in 1976. Agrometeorological advisories are bul-

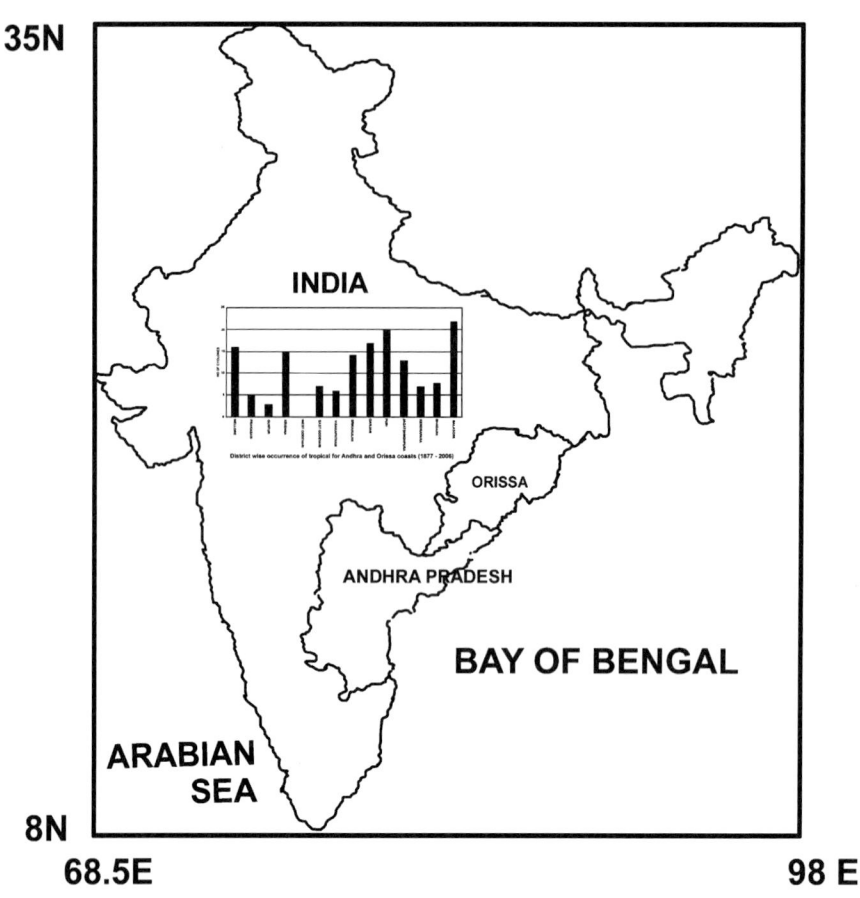

Fig. 25.1. District wise occurrence of tropical cyclones for Andhra and Orissa coasts (1877 – 2000) (source: Chittibabu et al., 2004)

letins prepared for farmers, taking into account prevailing weather, soil, crop condition, and weather prediction. In these bulletins, the suggestions on measures practiced are provided to minimize the losses and optimize inputs in the form of irrigation, fertilizers and pesticides, drought mitigation strategies like water management, minor and micro irrigation, water conservation, crop management, crop rotation, inter cropping, planting date, crop variety, alternate land use systems, fuel and fodder plantation, Silvi pasture, Agro horticulture, Agro forestry, and livestock management. The IMD issues Agrometeorological Advisory Service bulletins twice a week in an operational mode in consultation with the Director of Agriculture of the respective States and disseminated through All India Radio, Doordarshan, newspaper, Internet, etc. On the basis of long-range forecast, an important agrometeorological service, various pre-emptive and relief measures like procurement, transport, storage, and distribution of food grains are taken by the Government.

25.3.2.3
National Center for Medium-Range Weather Forecasting (NCMRWF)

The NCMRWF was established in 1988 by the Government of India as a scientific mission to develop operational numerical weather prediction (NWP) models for forecasting weather in the medium-range (3-10 days in advance) scale and setting up of agrometeorology advisory service units in the 127 agroclimatic zones of the country. The Agrometeorology Advisory Service farmers (AAS Farmers) receive the medium-range weather forecast based on the agrometeorological advisories, including optimum use of inputs for different farm operations. Due to judicious and timely use of inputs, the cost of production of AAS farmers reduced approximately by 3-6%. At the same time, the yield level of the AAS farmers also increased. The increased yield level and reduced cost of production led to increased net returns.

25.3.2.4
Other services

Some other activities tailored to meet the user communities are on the dry farming meteorology, which helps the dry land farmers to minimize the uncertainty of rainfed agriculture. Similarly, the forecasts on desert locust meteorology help in issuing the warnings about the migrations of locust in relation to low-level flow and to support the plant protection and quarantine and storage directorate in their locust eradication programmes. The satellite instructional television experiment (SITE) started functioning in 1975, catering to mass communications by giving weather information, advising on sowing operations, application of insecticides, cautioning on floods, etc. The crop weather calendars (Figure 25.2) prepared by the IMD on cereals, pulses, oilseeds crops, etc., improve the knowledge of farmers on the weather elements influencing the crop and help in planning agricultural operations and taking up precautionary measures. The drought research unit, started at Pune in 1967, prepares aridity anomaly maps for the country during kharif season (June to September) in order to monitor droughts over the country on a real-time basis. These are prepared on a fortnightly basis and also provide information about the crop stress conditions and its intensity experienced by the plants, and thus help to monitor the drought situation in the country.

25.4
Strategies to Improve the Agrometeorological Services to Cope with Risks and Uncertainties

Sivakumar et al. (1998 and 2000) emphasized that the agrometeorological information plays a valuable part not only in making daily and seasonal farm management decisions but also in the management of disasters, risks, and uncertainties. Earlier, Das (1999) expressed that it may not be possible to prevent the occurrence of natural disasters, but agreed with the observations of Sivakumar et al. (2004) that the

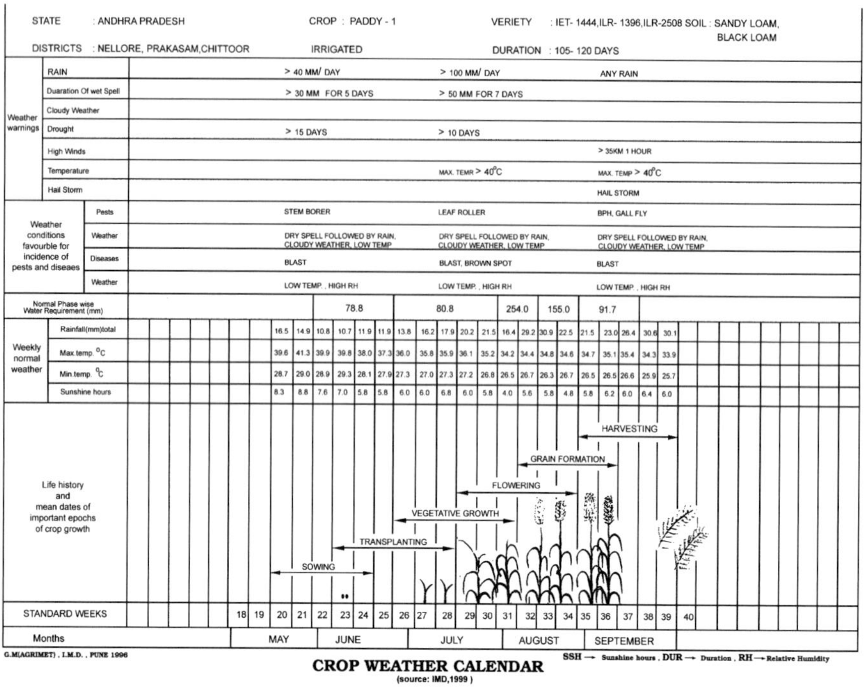

Fig. 25.2. Crop Weather Calendar

resultant negative and disastrous effects can be reduced considerably through the agrometeorological services. Therefore, the role of agrometeorological services is crucial for advance planning in the management of disasters, risks, and uncertainties because of the significance of their impact and influence on the overall well-being of humanity and livestock. Murthy and Stigter (2003) stated that the agrometeorological services play a key role in strategic and tactical planning and efficient monitoring of crops and is a growing recognition of the importance of operational agrometeorological services in all the sectors mentioned above around the world.

25.4.1
Improving the agrometeorological services

To cope with the risks and uncertainties pertaining to agriculture, forestry, rangeland, environment, and livestock, agrometeorological services must develop strategies to improve management and operational decisions in an efficient manner. Some of these strategies have been examined by CAgM since 1999.

25.4.1.1
Agrometeorological characterization, using different methodologies

The purpose of agroclimatic characterization is to identify those aspects of climate which distinguish a region from the nearby regions and to draw inferences on the influence of climatic factors on crop production. The hypothesis is that under given climatic conditions there are similarities in crop growth and development, there by the yield in that homogenous region. According to Reddy (2002), four decades ago, Chang observed that the failures or disappointing results of agricultural development projects in various parts of the world including projects to produce pineapples in Philippines, sugar in Puerto Rico, peanuts in East Africa, and rubber in the Amazon basin may have been largely due to failures in proper agroclimatic classification. It was also stressed that adequate assessment of agroclimatic resources is an essential prerequisite for proper planning of agricultural development. The homogenous crop zone boundaries with relevant crops and cropping patterns could be delineated with different methodologies through appropriate and scientific agrometeorological characterization. The resultant agroecological zoning offers the potential for developing strategies for efficient and sustainable natural resource management, including sustainable management of agriculture, forestry, rangeland, environment, livestock, etc. Therefore, it is imperative to give top priority to agroecological zoning in agrometeorological services. In addition, agricultural risk zoning is an essential component of natural disaster mitigation and preparedness strategies. Given the complex nature of databases, GIS and remote sensing should be employed in the future in any studies to facilitate strategic and tactical applications at the farm and policy levels.

25.4.1.2
Advice on microclimate management

The climate of a region determines the extent of adaptability of a crop (or animal) species and weather influences its day to day growth. In turn, the crops not only modify their own microclimate and weather within their canopies but also the soil underneath them due to emission of long wave radiant energy. According to Robert (2000), any modification in soil agricultural practices or ground cover vegetation may have consequences for the global carbon cycle via their impact on the dynamics of soil organic matter. Conversely, changes in composition of atmosphere may bring about changes in certain soil characteristics. Stigter et al. (2004) give the examples of parkland agroforestry and other stabilizing intensive management of scattered or clumped or allayed trees for microclimate management and manipulation to cope with temperature changes in the northern China Plain. There should be more research at the micro level into the physical behaviour of crop growth like profiles of solar radiation, temperature, wind speed, vapour pressure, carbon dioxide demand, and moisture regimes to develop better agricultural mitigation strategies at the micro level against risks and uncertainties. Micrometeorological knowledge about energy exchange and transports at the surface has useful applications in agriculture. The changes on a small scale are relatively easy to initiate and control. Therefore, studies also on increasing surface absorptive power, exposure through site selection, and artificial or natural shading for reduction of day length need to be encouraged.

25.4.1.3
Advice on crop phenology using recent climatic variability data and agrometeorological information

Response farming is defined as a method of identifying and quantifying the seasonal rainfall variability and predictability to address the problem of the farmers at field level. However, Stigter et al. (2005) suggested that response farming should not only be considered with respect to fitting cropping seasons to variable rainfall patterns but also for temperature patterns. A case study from Vietnam shows that either a planting date or a combination of planting date and variety could be varied to make sure that the rice flowers optimally with the detailed knowledge of temperature. Response farming has become a promising technology in the past two decades to alter cropping systems/patterns in relation to fluctuations in seasonal weather. Therefore, it is suggested that response farming be considered with indigenous technical knowledge to form new solutions to farming problems by improved use of available forecasting in the cropping season(s).

25.4.1.4
Establishing measures to reduce the impacts and to mitigate the consequences of weather and climate related natural disasters for agricultural production

The plan of implementation of the world summit on sustainable development (WSSD) held in Johannesberg in 2002 highlighted the need to mitigate the effects of disasters, risks, and uncertainties. Measures were suggested such as improved use of climate and weather information and forecasts, early warning systems, land and natural resource management, agricultural practices, and ecosystem conservation. The WSSD noted that these measures would reverse the current trends and minimize the degradation of land and water resources, which are the basic needs for agricultural production. There is a need to promote the access and transfer of technology related to early warning systems, and disaster mitigation programmes to the developing countries which are seriously affected by agrometeorological risks and uncertainties. A comprehensive documentation of risks and uncertainties related to agriculture and allied fields at national, regional, and international levels is very important. This process helps to develop mechanisms for more efficient assessment of the impacts of the risks and uncertainties in all fields in general and pertaining to agriculture in particular. Collaboration with other international and regional agencies is essential to develop an integrated coastal management approach in reducing the impacts of natural disaster on agriculture, forestry, rangelands, environment, and livestock given the importance of storm surges to coastal lowlands.

25.4.1.5
Monitoring and early warning exercises directly connected to already established measures

The role of early warning and advance planning for natural disaster management and the mitigation of extreme weather/climate events is crucial for agriculture, forestry, rangelands, environment, livestock, etc. The application of weather and climate information to improve the effectiveness and efficiency of emergency preparedness and response activities is essential. Critical thresholds must be monitored that should trigger early warnings. So, it is essential to survey the status of trends in land degradation and to report on appropriate criteria to conserve and manage material and environmental resources for the benefit of these sectors. Rapid advances in information technology need to be rapidly transferred to operational applications to more effectively disseminate agrometeorological information to the user community. All users of the information as well as the providers of the information must be involved to ensure that the right information is delivered to the right user at the right time for early warnings. Information gaps must be identified and guidelines and procedures must be established to improve the flow of timely and accurate information to farmers, including both monitoring and early warning systems. Current natural disaster management is largely crisis driven. There is also an urgent need for a more risk-based management approach to natural disaster planning in agriculture, rangelands, forestry, environment, livestock, etc. The concept of the drought monitor map product should be promoted as a tool for all drought prone countries to better understand drought severity using multiple indicators. The feasibility of organizing joint training workshops on national and regional drought monitor products under the auspices of WMO and the NDMC should be examined. Indices used in China in their agrometeorological bulletin could be effective training tools.

25.4.1.6
Climate predictions and forecasts and meteorological forecasts for agriculture and related activities

Weather forecasts play an important role in agriculture. The study of climatic fluctuations in the rainfall and their impact of agriculture has become an important area of climatology in recent decades. The average distribution of weather and climate phenomena along with the average variations in frequency and extent give a better insight into agronomic importance. One of the persistent demands of agriculturists is for more reliable forecasts of seasonal climate information to make appropriate crop decisions. These decisions include which crops and cropping patterns are chosen well ahead of the growing season in order to avoid undue risks and uncertainties. Hence, there is also an urgent need to assess the forecasting skills for natural disasters to determine those where greater research is needed. Lack of good forecast skill in drought, for example, is a constraint to improved adaptation, management, and mitigation. This has to be pursued uncompromisingly. For the identification of climatic fluctuations in rainfall data, long and continuous records are needed. Such data series are available only at a few stations in developing countries. Different techniques like relative rainfall probabilities, moving average, and iterative auto-regression may be adopted for better defining the climatic fluctuations, cause and effect of such fluctuations, and expected fluctuations to food-producing ecosystems.

25.4.1.7
Development and validation of adaptation strategies to increasing climate variability and climate change in the physical, social, and economic environments of farmers

Scientific assessments have shown that over the past several decades, human activities, especially burning of fossil fuels for energy production and transportation, are changing the natural composition of the atmosphere. Providing an adequate standard of living (adequate food, water, energy, health, environment, etc.) for the current and future generations is a major challenge. So, there is a need in many agricultural areas around the world to enhance the understanding of climate variability in order to assess the impact of causal factors (natural and human). A better understanding of the climate of the major ecosystems of the world where agriculture and related sectors are at risk could help develop effective in situ coping strategies (Salinger et al. 2005). There is a need for thorough understanding of the effects of changes in regional climate on crop production, forestry, rangelands, environment, livestock, etc. Given the growing incidence of dust and sand storms around the world, it is essential to include measurement of Aeolian sedimentation loads in the standard agrometeorological stations of NMHSs. It is also essential to include a routine and comprehensive analysis of wind speed and direction data and disseminate this information to the users. These data should be applied to analyze the impact of sand storms on agriculture. The issue of distinguishing long-term climate variability (e.g., IPO) and long-term climate change is important, as is the need to consider the impacts of both on agriculture, water resource management, and disasters such as bushfires. This is important because there are implications for long-term sustainability of certain types of activities, especially agriculture. There is modeling work at more overseas institutions (e.g., the Hadley Centre) that would be of relevance here. These issues need to be drawn to the attention of national policy-makers.

25.4.1.8
Specific weather forecasts for agriculture

In natural ecosystems and also in cultivated or forest ecosystems, climate change is capable of disturbing the balance between the species, whether they are plant or animal, in terms of individual and population (Lorean et al. 2001). The effect of climate changes on development of pests and diseases could manifest a direct effect on the biological cycle of parasites and host parasite interaction. Weather influences the degree to which plants and animals are attacked by pests and diseases or harbour them. It also affects the biology of insects and disease organisms, and determines the nature, numbers, and activity of pests (and of predators on pests) and extent and influence of diseases. In crop and livestock protection, the spread and aerial transport of pests and diseases and the effectiveness of applied control or eradication methods depend upon atmospheric agencies. Agrometeorologists need accurate and reliable climate forecasts to assist the agricultural community in planning and operations.

25.4.1.9
Advice on measures to reduce the contributions of agricultural production to global warming

Agriculture in the 21st century not only will have to make its contribution to the reduction of GHG emissions (particularly CO_2, CH_4, N_2O) to satisfy the vital needs of populations in food, energy, fiber, and other products. The adoption of agriculture to global warming triggers new requirements from major contemporary research efforts. This research shall aim both to increase the forecasting capacity and to anticipate the design of new cropping and forestry systems. Global warming in all sectors of agricultural production necessitates the adoption measures that must be economically feasible. Such measures could improve resilience of agricultural production systems to global warming, but, do not necessarily reduce emissions from the agricultural sector. To serve the agricultural sector, there is a need to thoroughly review the interactions between greenhouse gas emissions and agricultural activities. Also, there exist needs to document both positive and negative influences of agriculture on weather and climate systems and develop guidelines for increasing awareness within the farming communities of the related adaptation/mitigation strategies to address global warming and poverty issues. More attention should be given to the impacts of potentially increasing frequency and severity of extreme events associated with global warming and appropriate mitigation strategies.

25.4.1.10
Measures for sustainable agricultural development with strong agrometeorological components

While scientific and technological advances have resulted in higher quality information and increased capabilities in providing agrometeorological information, major difficulties remain. Technological advances in remote sensing (hand-held and space-based) such as soil moisture detection and evapotranspiration estimations and GIS constitute new sources of data for many agrometeorological applications and should help reduce some of these problems in the future. The RS and GIS data sets not only complement ground observations but also offer new types of data and also provide greater global coverage and improve aerial averaging. Regional and global cooperation can hold the countries that lack the financial and technical resources to acquire such data. As indicated above, recent technological advances in GIS offer significant improvements in spatial analyses of meteorological and agricultural databases. Sustainable research and development has occurred in the application of crop models ranging from the field level to country level and even larger scale modeling. Various modeling techniques range from statistics-based regression analysis to more complex process-oriented approaches. Models are also used in global change impact studies. The problem is how to develop an integrated information management system with technology and standardized analytical techniques that can be applied operationally for validation of selected models in agriculture, rangelands, forestry, environment, etc., at the eco-regional level. Therefore, it is recommended that an integration of GIS, remote sensing, simulation models, and other computational techniques be used to develop more effective early warning alerts of natural disasters risks and uncertainties. Also, there is a need and opportunity for the agrometeorologists to supply design requirements for new satellite sensors. This applies in particular to droughts, rangeland management, and to combat forest fires from a disaster mitigation point of view.

25.4.2
Improving the support systems of agrometeorological services

To cope with the risks and uncertainties pertaining to agriculture, forestry, rangeland, and environment agrometeorological support systems must have a solid foundation of data, analytical support, educating and training extension service for application and dissemination of operational results, and policy support.

25.4.2.1
Data

At present, collection, management, and analysis of atmospheric and surface data are being done with both manual and automated station networks. The new techniques like remote sensing and GIS cover the data from near ground to outer space. However, these data systems have grown only in the developed countries. To overcome the difficulties faced by the developing and under developed countries, the cooperation among international, regional, national, and where possible local specialized bureaus and organizations is essential. Assessment of the impact of natural disasters on agriculture, rangelands, forestry, and environment requires the design of a comprehensive data base in accordance with the users needs. There is a need for an integrated data management system, from adequate collection to quality control, analysis, presentation, and also metadata not just meteorological data. Presentation should make use of the best available technology, e.g., GIS and Internet. Effective management of, and preparedness for, natural disasters, risks, and uncertainties require free and unlimited access to relevant databases that will allow monitoring, assessment, and prediction. It is recommended that all agencies responsible for these databases develop good collaborating links for the exchange of information included in these databases.

25.4.2.2
Research

Agrometeorological services need regional, national, and local coordination. The priority items identified at Accra (1999) are agrometeorological aspects of the efficient use and management of resources in the full production environment; reduction of impacts on the resource base, yields, and income from natural disasters, risks, and uncertainties; validations and applications of databases and models for well specified systems and users; and, ways to ensure that research results are adopted in farming. There is an urgent need to assess the forecasting skills for natural disasters to determine where greater research is needed.

25. 4.2.3
Education/training/extension

With increasing incidence of natural disasters, risks, and uncertainties around the world, a comprehensive assessment of their impacts on agriculture, forestry, rangelands, environment, and livestock and strategies for mitigation of natural disasters is critical for sustainable development, especially in the developing countries. Education and training is an important component in these sectors. It is recommended that strategies for education and training address the needs at national, regional, and international levels in order to exploit the synergies and share experiences. Community involvement and education is essential in preparedness and mitigation.

25.4.2.4
Policies

An appropriate policy environment based on social concerns and environmental considerations can help develop the right mix of strategies for preparedness and problem solving practices against natural disasters, risks, and uncertainties. It is recommended that countries develop policies aimed at effective natural disaster management. Such policies should emphasize incentives over insurance, insurance over relief, and relief over regulation. The growing frequency of natural disasters requires effective use of the media to better inform and educate the general public and policy-makers about the potential impact of natural disasters and the need to adopt the preparedness strategies. Given the regional and global nature of natural disasters, it is essential to promote and foster collaboration between agencies and between international and regional programs and build partnerships.

25.4.3
A comprehensive agrometeorological service strategy to cope with risks and uncertainties

With the increasing incidence of events such as natural disasters, risks, and uncertainties around the world, a comprehensive assessment of their impacts on agriculture, forestry, rangelands, environment, and livestock needs to be addressed by country. Strategies for mitigation are critical for sustainable development, especially in the developing countries. Agrometeorological services and information must increasingly be made available to assist farmers in characterization of agroclimate, microclimate management and manipulation, advisories on response farming, monitoring of and early warning on natural disasters. Measures need to be established to reduce their consequences. Climate prediction and forecasting along with forecasts for pests and diseases and management of natural resources must be improved. Remote sensing, GIS applications, simulation models, and other computational techniques hold a lot of promise for improving operational agrometeorological services. These technologies must be used along with indigenous technical knowledge as Blended Technologies (BTs) for more effective early warning alerts of these events. More attention needs to be paid to enhancing such applications. There should be more research into the physical behaviour of crop growth and moisture regimes to develop better agricultural mitigation strategies and standardization of services and products. The research shall also be aimed at improved agrometeorological services, not just for enhancing agricultural productivity, but also for protecting the environment and biodiversity, coping with climate change, and drought and desertification for ensuring sustainable development. Training programmes and education at regional levels, which include aspects related to climate modeling and integration of satellite imagery oriented to agricultural GIS analytical tools, need to become a standard feature. Regional exchange programmes that will consider the transfer of methodologies and knowledge of professionals of different services, by means of seminars, workshops, and/or hands on training, must be given top priority. The regional and global nature of the natural disasters, risks, and uncertainties and the complexity of issues involved demands promoting and fostering collaboration between agencies and between international and regional programs and builds partnerships. This process helps in increasing the network of agrometeorological stations, maintaining the existing ones, and developing competitive agrometeorological products. New initiatives such as the World AgroMeteorological Information Service (WAMIS) could help strengthen operational agrometeorological services through the provision of agrometeorological products on a near real time basis on the Internet and through training modules to enhance the quality of agrometeorological products. Priority shall also be given and enough funds allocated for dissemination of meteorological tools applied to agriculture and oriented towards small and medium farmers.

25.5
Conclusions

Agrometeorological services help the farmers in reducing the impact of risks and uncertainties, in addition to efficient management of pests and diseases on their crops, there by helping to increase their agricultural production. Agrometeorological services are useful in crop management systems that extension services provide to the agricultural community and help the extension personnel in performing their functions more efficiently at the end-user. The research scientists work in collaboration with different departments as a multi-disciplinary team with the expert knowledge to evolve agrometeorological products and services for farmers in different farming systems. They include forecasts of weather and climate, monitoring and early warning products for drought, floods, or other calamities, general agrometeorological advisories, etc. These products and services would increase the preparedness of the farmers, well in advance, to cope with risks and uncertainties. In education, the successes, failures, and experiences of researchers and extension specialists are taught in the curricula of agrometeorology departments in the universities, training centers, and other capacity building institutions. Such exercises will enlighten the classical training and strengthen the usefulness of the services for the farmers and other user communities. These agrometeorological services help the government decision-makers to involve the private sector on priority issues like crop insurances, providing infrastructure facilities, enhanced cooperation between the institutions providing information and relevant advisories, and those responsible for their transfer to the farming community to solve the risks and uncertainties associated with production agriculture, forestry, rangelands, environment, and livestock.

Agrometeorological services are the end products of agrometeorological research. Improved management of climatic and weather related events such as disasters, risks and uncertainties is central to the profitability of our rural industries and the ecological sustainability of its resource base. Agriculture, forestry, rangelands, environment, livestock, etc., are the most important sectors heavily impacted by these events. The current status of agricultural production and increasing concerns with related environmental issues calls for improved agrometeorological services for enhancing and sustaining agricultural productivity around the world. The requirements for the agrometeorological services were described in the light of emerging issues related to environment, climate change, biodiversity, drought and desertification, food security, and sustainable development. The Agenda 21, International Conventions including the United Nations Framework Convention on Climate Change (UNFCCC), the Convention on Biological Diversity (CBD), and the United Nations Convention to Combat Desertification (UNCCD), the World Food Summit Plan of Action and the World Summit on Sustainable Development (WFSPAWSSD) include the elements that have implications for strengthening operational agrometeorological services. The weather related disasters around the world have been increasing in the recent past and operational agrometeorological services could help the farming community with better preparedness and mitigation strategies.

References

Altieri, M. (1991) Traditional Farming in Latin America. The Ecologist 11:93-96.

Anderson, M. (1990) Analysing the costs and benefits of natural disaster responses in the context of development. Environment Working Paper 29, World Bank, Washington, D.C., USA.

Bhatt, A.K., Shakya, L.M., Bhatta, B.R. & Mohammd, N. (1992) Seabuckthorn in the Hindu Kush-Himalayas, Nepal, India, Pakistan: Status of resource potential and traditional uses. ICIMOD, Mountain Farming Systems Commissioned Study Report (Mimeo).

Burgan, K. and K. Klaver (1998) Fuel Models and Fire Potential from Satellite and Surface Observations. Int J Wildland Fire 8:159-170.

Charveriat C (2000) Natural disasters in Latin America and the Caribbean: an overview of risk. Research Department Working Paper No.434, Inter-American Development Bank, Washington DC, USA.Chittibabu, P., Dube, S. K., Mohanty, U. C., Murthy, C. S., Rao, A. D. and P. C. Sinha. (2004). Mitigation of flooding and cyclone hazard in Anndhra. Pradesh and Orissa, Inia. Proceeding of the International Sympsium on Natural Hazards. (Intromet–2004., Hyderabad 24–27 February, 2004.

Dafu Yu, Guojie Wang Fei, Yu Side, Fan Hong (1990) Farmers strategies and sustainability of mountain agriculture in West Sichuan, China. Report of the ICIMOD working paper No.10. ICIMOD, Kathmandu, Nepal.

Das HP (2003) Incidence, prediction, monitoring and mitigation measures of tropical cyclones and storm surges. In agrometeorology related to Extreme Events. WMO No. 943, World Meteorological Organization, Geneva, pp.

FAO (1999) New concepts and approaches to land management in the tropics with emphasis on steeplands, FAO Soils Bulletin No.75, Food and Agriculture Organisation, Rome.

Gommes R, Petrassi F (1996) Rainfall variability and drought in Sub-Saharan Africa since 1960. FAO Agrometeorology Series Working Paper No.9, Food and Agriculture Organisation, Rome.

Hodell DA, Curtis JH, Brenner M (1995) Possible role of climate in the collapse of classic maya civilization, Nature 375:391-393

Jeyaseelan AT, Chandrasekar K (2002) Satellite based identification for updation of Drought prone area in India. ISPRS-TC-VII. International Symposium on Resource and Environmental Monitoring, Hyderabad.

Kashyapi A (1998) Agrometeorological information for user community – Experience of India. Proceedings of the International Workshop on user requirements for agrometeorological services. 10-14 November, 1997, Pune, India. pp 202-216.

Kogan F (2000) Contribution of remote sensing to drought early warning. In: Wilhite DA, Sivakumar MVK, Wood DA (Eds) Proceedings of an Expert Group Meeting, 5-7 September 2000, Lisbon, Portugal, WMO/TD No.1037, World Meteorological Organisation, Geneva.

Loreau M, Naeem S, Inchausti P, Bengtsson J, Grime JP, Hector A., Hooper FU, Huston MA, Raffaelli D, Schmid B, Tilman D, Wardle DA (2001)

Madhavi K, Kiran Chand TR, Gherai B, Badhrinath KVS, Murthy MSR (2004) Studies on forest fires using satellite data and ground based measurements. Intromet-2004, pp 59-60.

Maracchi G, Perarnand V, Kleschenko AD (2000) Applications of geographical information systems and remote sensing in agrometeorology. Agric For Meteorol 103:119-136.

Motha RP, Sivakumar MVK, Bernardi M (2006) Strengthening operational agrometeorlogical services at the national level. Proceedings of the Inter-regional workshop, March 22-26, 2004, Manila, Philippines.

Motha R, Peterlin A, Puterbaugh T, Stefanski R, Brusberg M (1997) The Definition of Climate Services for Agriculture, American Meteorological Society, 10th Conference of Applied Climatology, Reno, NV, October 1997, pp. 329-332.

Murthy VRK (1996). Terminology on Agricultural Meteorology. Sri Venkateswara Publishers. Ashok Nagar, Hyderabad, AP, India. 225 pp.

Murthy VRK (2003) The role of crop growth models in Agricultural production. Training Workshop on Satellite Remote Sensing and GIS applications in Agricultural Meteorology. Dehradun. July 7-11, 2003.

Murthy VRK, Stigter CJ (2003) Stigter's diagnostic conceptual framework for generation and transfer of agricultural meteorological services and information for end users. Paper in: Agrometeorology in the new millennium – perspectives and challenges. Proceedings of the Second National Seminar of the Association of Agrometeorologists in India, Ludhiana, October 26-28, 2003.

Murthy VRK (2006). Risk management through Blended Technologies of climate change, variability, sustainable agriculture and food security field schools. Project submitted to START, UNITAR under, ACCCA.

Murthy VRK, Mohammed SK, Prasad PVV, Satyanarayana V (2002) Resource capture mechanisms – an aid to promote nursery growth in paddy for higher yields in winter. Symposium of Association of Agrometeorologists, Anand. October 26-28, 2002.

Ramesh KJ, Bhadram CVV, Ramachandran M, Pattabhi Ramarao E, Akhilesh Gupta (2004) Pre-Symposium proceeding of International Symposium on natural hazards (Intromet-2004). Hyderabad 24-27 February 2004.

Rao AD, Sujatha D, Babu SV (2004) Ocean response to the passage of 1999 Orissa Super cyclone. Intromet 2004. pp 26-28.

Rao GS (2004) Role of India Meteorological Department (IMD) in mitigating natural disasters. Lecture in model training course on Agricultural disaster management for the officers of department of agriculture of all states. 4-11 March, 2005. EEI, Rajendranagar, Hyderabad, Andhra Pradesh, India.

Reddy J (2002) Dryland agriculture. BS Publication, 4-4-309, Giriraj lane, Sultan Bazar, Hyderabad.

Rippey B, Peterlin A, Deprey D (2000) The U.S. Department of Agriculture AWIPS Link to NOAAPORT. Preprints, 10th International Conference Interactive Information and Processing Systems for Meteorology, Oceanography, and Hydrology, Long Beach, California, Amer Meteor Soc 348-350.

Robert M (2000) 'Effets potentials des changements climatiques sur les sols: Impact potentiel du changement climatique en France au XXIe siecle. Ministere de l'Amenagement du territoire et de *l'Environment*.

Roy BC, Mruthyunjaya Selvarajan S (2002) Vulnerability of climate induced natural disasters with special emphasis on coping strategies of the rural poor in Coastal Orissa, India. Paper presented at the UNFCC COP 8 Conference organized by the Government of India, UNEP and FICCI, 23 October to 1 November 2002, New Delhi, India.

Rudolph B (1998) Agrometeorological planning aids for the farmer. Proceedings of the International Workshop on user requirements for agrometeorological services. 10-14 November, 1997, Pune, India. pp 199-201.

Salinger J, Sivakumar MVK, Motha RP (2005) Increasing climate variability and change, reducing the vulnerability of agriculture and forestry. Springer, P.O. Box 322, 3300 AH Dordrecht, The Netherlands ISBN 1-4020-3354-0.

Scott CD, Litchfield JA (1994) Inequality, mobility and the determinants of income among the rural poor in Chile, 1968-1986. London School of Economics, London.

Shaik Mohammed, Murthy VRK (2001) Simple and indirect method of leaf area measurement in brinjal (solonum melongene L.). Research on crops 2: 51-53

Singh RP, Jhamtani A, Kumar GAK (2004) Revival of indigenous technical knowhow in Agriculture. Jain Brothers, 16/873, East Park Road, Karolbagh, New Delhi.

Sivakumar MVK, Roy PS, Harmsen K, Saha SK (2004) Satellite remote sensing and GIS applications in agricultural meteorology. AGM-8 WMO/TD No.1182, WMO 7 bis, Avenue de la Paix 1211 Geneva 2 Switzerland.

Sivakumar MVK, Motha RP, Das HP (2000) Natural disasters and extreme events in agriculture. Impacts and Mitigation. Salinger, J., Verlag Berlin Heidelberg 2005 ISBN-10 3-540- 90-4.

Sivakumar MVK, De US, Sinha Ray K, Rajeevan M (1998) User requirements for agrometeorological services. Proceedings of an international workshop, 10-14 November 1997, Pune, India.

Stigter CJ (2003) The establishment of needs for climate forecasts and other agrometeorological information for agriculture by local, national and regional decision makers and users' communities. Proceedings of the WMO (CAgM/CLIPS) RA I (Africa) Expert Group Meeting on Applications of Climate Forecasts for Agriculture, Banjul, the Gambia, 15 p.

Stigter CJ, Zheng Dawei., Xurong M, Onyewoto LOZ (2005) Using traditional methods and indigenous technologies for coping with climate variability. Clim Chang 70:255-271.

Susman P, O'Keefe P, Wisner B (1983) Global disasters, a radical interpretation. In: Hewitt K (ed) Interpretations of calamity from the view point of human ecology. Allen and Unwin, London.

Thapliyal PK, Pal PK, Gupta A (2004) Use of passive microwave radiometer data for monitoring drought conditions. Intromet-2004. pp 47-49.

Wang Chunlin, Liu Jinluan, Zhou Guoyi (2003) Research on real-time cold-disaster watching and prediction in Guangdong Province based on GIS technology. Quart J Appl Meteorol 14: 487-495

WMO (1997) Extreme agrometeorological events. CAgM Report No.73, TD No.836, World Meteorological Organization, Geneva.

World Disasters Report (2001) International Federation of Red Cross and Red Crescent Societies, Geneva.

World Disasters Report (2003) International Federation of Red Cross and Red Crescent Societies, Geneva.

CHAPTER 26

Using Simulation Modelling as a Policy Option in Coping with Agrometeorological Risks and Uncertainties

Simone Orlandini, A. Dalla Marta, L. Martinelli

26.1
Introduction

Agricultural systems are largely dependent on weather and climate, then management and planning decisions are made in condition of risk or uncertainty due to the high level of complexity of the agricultural systems. Despite the important advances in technology over the last decades, many production factors are not well defined and they are outside of the farmer control (Orlandini and Cappugi 2001). The lack of precise information increases the level of uncertainty in farm management. To overcome these problems, farmers increased the level of energy and chemical inputs above the necessary requirements with the aim of decreasing the impacts of the variability of agricultural systems. Unfortunately, the consequence of this strategy was the increasing of environmental impact and production costs without obtaining the expected goal (Travis et al. 1992). A solution to interrupt this negative trend is to substitute expensive and pollutant chemical and energy inputs with elaborated information of high quality. In this way it is possible to decrease the risk of the uncertainties of decision making and thus to minimise the application of excessive inputs and increase the potential income (Maracchi 2001).

Therefore, the monitoring of environmental variables and the elaboration of information represent a necessary support for decision making both for long-term and short-term management of agricultural activities. Information can hardly be used as a raw datum, but it needs to be analysed, processed and organised according to the final operational use. A new approach to agriculture seeks to increase the application of agrometeorological information for the development of models for the assessment of the quality of agricultural products, estimation and monitoring of yields, environmental protection and cultural rural heritage conservation. Agriculture needs agrometeorological models to minimise environmental costs of its activity and to determine short and long term consequences (reduction of soil fertility).

Agrometeorological models are basically formal expressions of biological, physical and chemical functions fed with environmental and climatic forcing variables (Table 26.1). Models are often the only tools available to study the behaviour of complex systems, and they offer unique insights to understand the frequent non-linear interactions among processes in soil-plant systems.

In the last few years an increasing interest in this subject was observed and a high number of computer applications for agrometeorological purposes was developed.

Table 26.1. Examples of agrometeorological forcing variables and their effect in epidemiological models.

Variable	Effect
Temperature	Phenological development
Solar radiation	Biomass assimilation and growth
High temperature	Rate of infection Higher threshold of development and survival
Low temperature	Spore and insect conservation Lower threshold of development and survival
Leaf wetness	Inoculation Survival of organism
Precipitation	Dispersion of spore and insect Survival of organism
Relative humidity	Presence of saturation conditions Survival of organism
Wind	Dispersion of spore and insect Modification of temperature and humidity

Modelling can find useful applications in many fields: plant growth and development, crop yield quality and quantity estimation, water balance, plant protection against pests, diseases, weed and weather hazards, climatic changes, generation of weather data, spatial and temporal interpolation or extrapolation, soil erosion and conservation, etc. (Orlandini 1996).

Moreover, concerning a more complex and articulated context, models can be included for setting up Decision Support Systems (DSS) and Early Warning Systems (EWS). In this case, models are usually integrated with other technologies such as remote sensing, geographic information systems, and numerical weather models. Information elaborated is used not only for regulating the agricultural and land management activities but also, in more critical cases such as in developing countries or in particularly vulnerable areas, to manage food security concerns.

26.2
Conditions of model implementation and application

Agrometeorological models can be set up with different methodologies. From the simplest to most complex: tables for manual calculations (Goidanich and Mills tables), electronic plant stations, computer and integrated systems, which combine models, monitoring networks and GIS for the production of information spatially distributed on the territory. The quality of the information increases in the same direction, but obviously also technological requirements have the same trend. So, particularly in developing countries, simple methodologies seem to be preferable.

The required inputs are meteorological (temperature, rainfall, relative humidity, leaf wetness, solar radiation, wind direction and speed), physical (CO_2 concentration, soil structure) and biological (observed symptoms, crop monitoring, plant parameters) data. Meteorological data are generally required with hourly time step for epidemiological models, while daily data are required for the other kinds of simulations; soil erosion models require a shorter time step (minutes). Sometimes, also historical data are needed to define the climatic characteristics of the agricultural environment. The availability of meteorological information can be improved by further developing the spatial interpolation methods and by a more effective use of weather radar and satellite information in addition to traditional meteorological ground data. Automation of weather observing stations may have impacts on the availability of some meteorological parameters (Mestre 2006). The use of atmospheric models as a source of meteorological data is also an alternative that is worth considering (Dalla Marta et al. 2003).

Model outputs represent the basis for implementing support systems based on information technologies to disseminate advices and early warnings to the potential users: policy-makers, extension services, farmers, plant breeders. Insurance companies are also interested in the results of these analyses and they will be involved in the evaluation of agricultural risk insurance. Risk maps and other methods (graphics, tables, etc.) can be used to provide the end users with a detailed description of agricultural system conditions (Friesland and Orlandini 2006).

Particularly in case of applications oriented to support farmers, local or territory alternatives can be chosen. In the first case, the model is applied directly by farmers, with evident benefits in the assessment of real epidemiological condition and microclimate assessment. On the other hand, the management of the simulations and the updating of the systems represent big obstacles. The second (territory) is probably preferable because it allows a better management and updating of the system. This solution requires the application of suitable methods for the information dissemination among the users (personal contact, newspaper and magazines, radio and television, videotel, televideo, telefax, mail, phone, Internet, and SMS). The use of mobile phones to acquire information is interesting because it enables access from the field and does not require the use of computer. PlanteInfo (Jensen and Thysen 2003) contains weather forecast and plant protection warnings developed in two ways: "push-type" sent regularly when criteria are met, as specified by the user, while "pull-type" are sent on the user's request by SMS.

Also in future, difficulties in validation will remain with some models due to compatibility of simulation results and field assessments, but nevertheless validation is necessary and often is possible. Uncertainty of models and their output can be minimised only within limits, by improving biometeorological understanding, extending results into an area, or better weather forecasts. Validated models still grow in practical importance, and an increasing number of them is and will be used in routine procedures for the benefit of agricultural users (Friesland and Orlandini 2006).

However starting from knowledge and technologies developed since the last decades, the operational application to the agricultural practice of this knowledge

is limited by the following constraints, practically everywhere and in any case, in Europe:
- Meteorological data at local scale are not available or not accurate enough.
- Models are not accurate enough in term of information compared with the empirical rules adopted by the farmers.
- Links between the farmers and the agricultural extension services are too weak and in many cases the activity of extension services are more devoted to help the farmers in burocratic duties instead of supporting them in the technical choices.
- The time taken in delivering information is often too long to help the farmers to take decisions in time.
- No agency did a reasonable study on the improvements of agricultural practice and the consequent economic benefits that can be derived through the utilisation of agrometeorological information.

26.3
Examples of Using Agrometeorological Models

26.3.1
Models for soil erosion

Soil erosion is a natural phenomenon: it has occurred over the millennia as part of geological processes and climate change. However, erosion is more severe nowadays: soil degradation affects almost 2 billion ha of arable and grazing land (Table 26.2). More than 55% of this damage is caused by water erosion and nearly 33% by wind erosion.

Every year soil erosion and other forms of land degradation rob the world of 5-7 m ha of farming land and 2.5 billion tonnes of topsoil are washed away. The United States lost about one-third of its topsoil since settled agriculture began. Worldwide, soil erosion puts the livelihoods of nearly one billion people at risk. The effects of

Table 26.2. Soil degradation by area and type (million ha) (www.fao.org).

	Water erosion	Wind erosion	Chemical degradation	Physical degradation	Total
Asia	440	222	73	12	747
Africa	227	187	61	19	494
South-America	123	42	70	8	243
Europe	115	42	26	36	219
North and Central America	106	39	7	6	158
Southwest Pacific	83	17	1	2	103

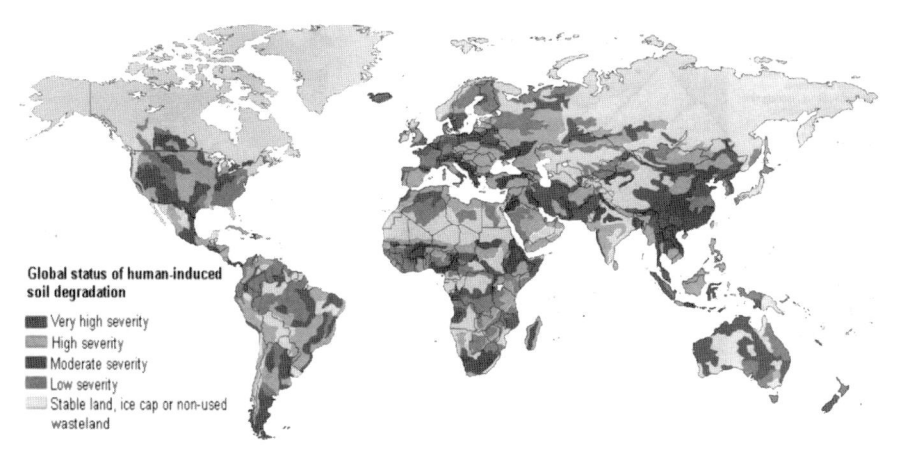

Fig. 26.1. Global status of human-induced soil degradation (www.fao.org)

erosion are legion (Fig. 26.1). Soil washed off bare hillsides, ruins aquatic habitats and clogs waterways. Riverbeds rise, increasing the risk of floods. However, erosion can be reduced and eroded land can be restored.

Soil erosion is among the major environmental threats related to agricultural land use (Helming et al. 2005). Important European policies and directives, such as the Water Framework Directive, the European Commission Strategy for Soil Protection as well as agro-environmental measures address the issues of soil erosion. During recent decades, international research has greatly contributed to an improved understanding of soil erosion processes at various scales from single plots to complex watersheds. Research focus evolved from descriptive approaches over process analyses of soil-hydrological dynamics to in depth studies of the temporal interactions of rainfall and soil erosion. The analysis of spatial dynamics of soil surface characteristics, runoff and erosion patterns is a recent topic of research, which is a crucial piece of the puzzle when analysing connectivity issues and when linking upland area processes of sediment production with channel processes of sediment transport.

Patterns of runoff and soil erosion represent the two-dimensional response of the landscape to rainstorm events. These patterns illustrate in a complex and yet incompletely understood way the spatial variability of important soil, land use and landscape characteristics. There has been increasing recognition of the significance of such patterns for understanding and predicting erosion and its environmental impacts. For example, improved insight into the way in which patterns of runoff and soil loss evolve over time and vary with spatial scale may help us to better comprehend the value and limitations of short-term plot studies with respect to erosion in a wider, real-landscape context. A better appreciation of the role of dynamic connectivity in runoff and sediment delivery may assist in improving our estimates and surface functions of landscape response to rainfall events.

At present, numerous models for the estimation of soil erosion have been set up and some of them are available on-line (Table 26.3).

Table 26.3. Examples of soil erosion simulation models.

AGNPS (Agricultural Non-Point Source pollution model)
AGNPS-UM (Agricultural Non-Point Source pollution model, modified)
ANSWERS (Areal Nonpoint Source Watershed Environment Response Simulation)
CREAMS (Chemicals, Runoff and Erosion from Agricultural Management Systems)
EPIC (Erosion-Productivity Impact Calculator)
EROSION-3D
EUROSEM (European Soil Erosion Model)
GLEAMS (Groundwater Loading Effects of Agricultural Management Systems)
KINEROS2
LISEM (Limburg Soil Erosion Model)
MEDRUSH
MOSES (Modular Soil Erosion System) project
MWISED (Modelling Within-Storm Sediment Dynamics) project (link down)
RillGrow 1 and 2
RUSLE (Revised Universal Soil Loss Equation)
SWAT (Soil and Water Assessment Tool)
USLE (Universal Soil Loss Equation)   – APSIM (Agricultural Production Simulator)   – TMDL (Total Maximum Daily Load)   – USLE-2D (Universal Soil Loss Equation 2D)   – USLE (MS Excel version)   – USLE-M (Universal Soil Loss Equation Modification)   – USPED (Unit Stream Power-based Erosion Deposition)
WATEM (Water and Tillage Erosion Model)
WEPP (Water Erosion Prediction Project)
GeoWEPP (Geo-spatial interface for WEPP)
WEPP interfaces (US Forest Service)

26.4
Water balance and irrigation

Water is a finite resource: there are some 1.4 billion km^3 on earth and circulating through the hydrological cycle. Nearly all of this is salt water and most of the rest is frozen or under ground. Only one-hundredth of 1% of the world's water is readily available for human use. In many countries, the amount of water available to each person is falling, as populations rise. Of the three main ways in which people use water:

- Municipal (drinking water and sewage treatment);
- Industrial;
- Agricultural (mostly irrigation);

farming accounts for the largest part, some 65% globally in 1990. Irrigation systems have existed for almost as long as settled agriculture and they are essential to feed the world. Although only 17% of the world's cropland is irrigated, it produces over 33% of our food, making it two and a half times as productive as rain-fed agriculture. Nevertheless, in some cases up to 60% of the water withdrawn for use in irrigation never reaches the crops (Fig. 26.2). In addition, waterlogging and salinization have sapped the productivity of nearly 50% of the world's irrigated lands. Other problems include the accumulation of pollutants and sediments in large dams and reservoirs, and the fact that irrigation systems provide an ideal habitat for the vectors of waterborne diseases. The key to improve irrigation lies in recycling waste water, proper drainage and especially in more efficient use of water.

Water balance models play a central role in most of the agricultural system models describing soil-crop-atmosphere interactions. Water is not only the dominant factor affecting crop growth, it also determines soil processes like transport of chemicals, biological activity and matter transformations, surface runoff, groundwater recharge and pest and disease development on soil and crop surface.

Agricultural production growth and stabilization, under conditions of the extreme climatic events occurrence, such as severe droughts, can be ensured by using many methods, the most important one being the irrigation, provided attention is paid to environmental preservation and protection. Soil moisture information is of interest to a wide range of users and is crucial for making informed decisions on land and water resources management. Many existing water balance models can be used to evaluate the soil moisture dynamics and soil water deficits at the rooting depth of the different agricultural crops, in order to provide information necessary in taking decisions on irrigation planning and management.

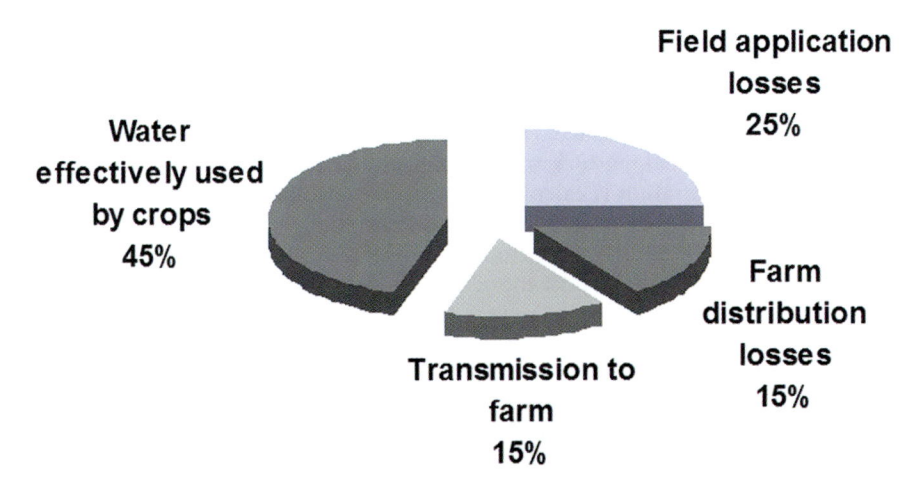

Fig. 26.2. Irrigation losses in agriculture (www.fao.org)

In the last years, the importance of water balance models to assess different decisions regarding the irrigation planning and management was considerably enhanced. Many models exist for calculations on irrigation management. In particular, four models were given as an example in COST Action 718 "Meteorology application for agriculture" of European Union final report: AMBAV, CROPWAT, IRRFIB, and SWAP (Kroes et al. 2006).

AMBAV (Agrarmeteorologisches Modell zur Berechnung der aktuellen Verdunstung-agrometeorological model for calculating the actual evapotranspiration) is part of the complex agrometeorological model toolbox AMBER (Löpmeier 1994) developed by the Agrometeorological Research Braunschweig (German Weather Service, Deutscher Wetterdienst - DWD). The model calculates the potential and real evapotranspiration and the soil water balance for different crop covers. It is used for producing irrigation recommendations which are disseminated by the DWD via fax service for different soil types using hourly data from the meteorological station network, including weather forecast up to 5 days. The model is designed to be used by local meteorological advisory services. It considers 13 different crops: winter wheat, spring wheat, winter barley, rye, oats, maize, sugar beets, potatoes, oilseed rape, grassland, fruit trees, coniferous and deciduous forest.

The CROPWAT was developed by FAO (Allen et al. 1998). Its main functions are to calculate reference evapotranspiration, crop water requirements, irrigation requirements, scheme water supply, develop irrigation schedules under various management conditions, evaluate rainfed production and drought effects and evaluate the efficiency of irrigation practices. CROPWAT is meant as a practical tool to help agrometeorologists and irrigation engineers to carry out standard calculations for evapotranspiration, crop water-use studies and more specifically, the design and management of irrigation schemes. It allows the development of recommendations for improved irrigation practices, the planning of irrigation schedules under varying water supply conditions, and the assessment of production under rainfed conditions or deficit irrigation.

IRRFIB model calculates reference daily water balance for different regions and represents agricultural decision support tool in the frame of agrometeorological information system. Its recent development enabled quick and accurate transfer of information to end users. An open code solution developed in Linux platform is based on the PostgreSQL database. In the SAgMIS meteorological, soil, crop and agrotechnical data are integrated. Daily meteorological data from regional stations in the frame of national network are automatically delivered into the system. Reference evapotranspiration is calculated by Penman-Monteith equation using air temperature, wind speed, air humidity and net radiation.

The model SWAP (Soil-Water-Atmosphere-Plant) is the successor of the agrohydrological model SWATR and some of its numerous derivatives. It has a long history with the first publications in the year 1978. The latest version was published as SWAP3.0 by Van Dam (2000) and Kroes and Van Dam (2003). Top soils show the largest concentration of biological activity on earth. Water movement in the upper soil determines the rate of plant transpiration, soil evaporation, runoff and recharge to the groundwater. In this way, unsaturated soil water flow is a key factor in the hydrological cycle, transporting large amounts of solutes, ranging from nutrients to all kind of contaminations. Therefore the model SWAP aims at an accurate description of unsaturated soil water movement to derive proper management conditions for vegetation growth, irrigation conditions and environmental protection in agricultural and natural systems.

26.5
Crop protection

Pesticides use multiplied by a factor of 32 between 1950 and 1986, with developing countries now accounting for a quarter of the world's pesticide use. Inappropriate and excessive use can cause contamination of both food and environment and, in some cases, damage the health of farmers and consumers. Pesticides also kill the natural predators of pests, allowing them to multiply; meanwhile the number of

Fig. 26.3. Number of insects (bold), pathogens (dotted) and weeds (thin) resistant to pesticides (www.fao.org)

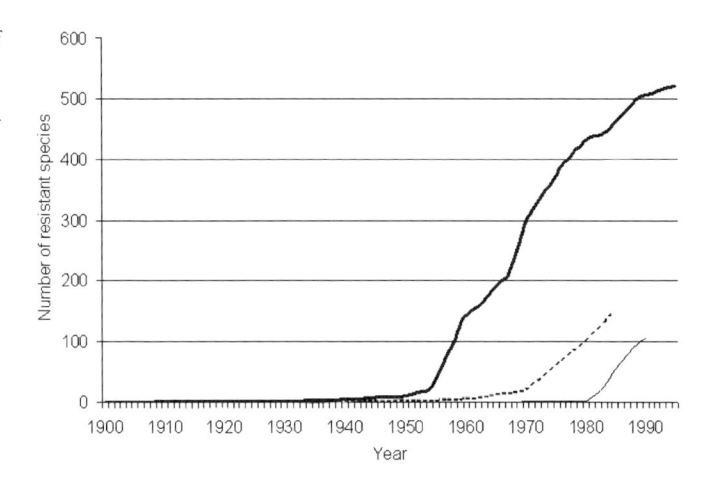

Fig. 26.4. Geographical repartition of simulation models developed since '80

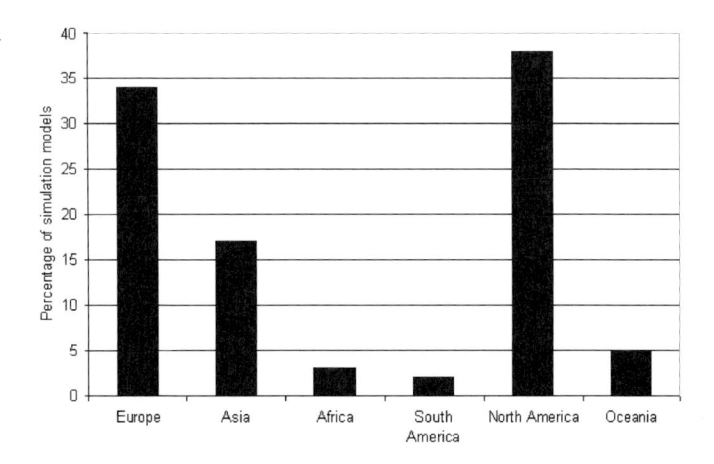

Table 26.4. Crops considered for the application of epidemiological models (Friesland and Orlandini 2006).

Category	Crop
Cereals	wheat, barley, rye, oats, maize, sorghum, rice
Row crops and other	potatoes, sugar beet, oilseed rape, soybean, sunflower, hop, tobacco, alfalfa
Vegetables	cabbage, onions, leek, tomato, carrot, celery, bean, paprika, lettuce, pea, turnip, aubergine
Fruit	apple, grapevine, citrus, plum, pear, cherry, melon, olive, strawberry, watermelon
Other	elm, mustrad, pinewood, rose, chestnut, almond, poplar, oak

pest species with resistance to pesticides has increased from a handful 50 years ago to over 700 now (Fig. 26.3).

A large number of simulation models has been formulated in the last decades, starting during the '80. The higher contribution is from Europe and North America, while Asia (mainly oriented to rice application), Africa, Oceania and South America show a smaller activity (Fig. 26.4).

The main crops have been studied, from field annual crops, to forestry, trees and flowers (Table 26.4).

P.Rada is a project funded by the European Community in the frame of the initiative "Interreg IIIA Italia-Slovenia 2000-2006" created to promote an operational collaboration between the two neighbouring countries.

In particular, the interested area includes Friuli Venezia Giulia region (North East part of Italy) and the eastern part of Slovenia. This system is applied on the entire regional area and represents a fundamental DSS for most agricultural workers both in public administrations and in farms. It is the result of the integrated use of epidemiological simulation model, remote sensing and the GIS. It is composed by three subroutines: a module for meteorological data spatialisation, a leaf wetness and a disease development simulation model.

The necessary agrometeorological data are collected by ground stations scattered on the two countries, spatialised and integrated with rainfall data collected by the meteorological radar of Fossalon di Grado (Italy). The territorial information obtained is then used by the system to feed two agrometeorological models: the first for the estimation of leaf wetness and the second for the simulation of grapevine downy mildew. The main output is represented by daily maps containing operational indications about the current meteorological situation, the presence and the stage of downy mildew development and the evaluation of the potential risk (Fig. 26.5). In the current system great attention is paid to grapevine downy mildew, nevertheless its modular structure allows to consider other biological processes thanks to new algorithms and subroutines.

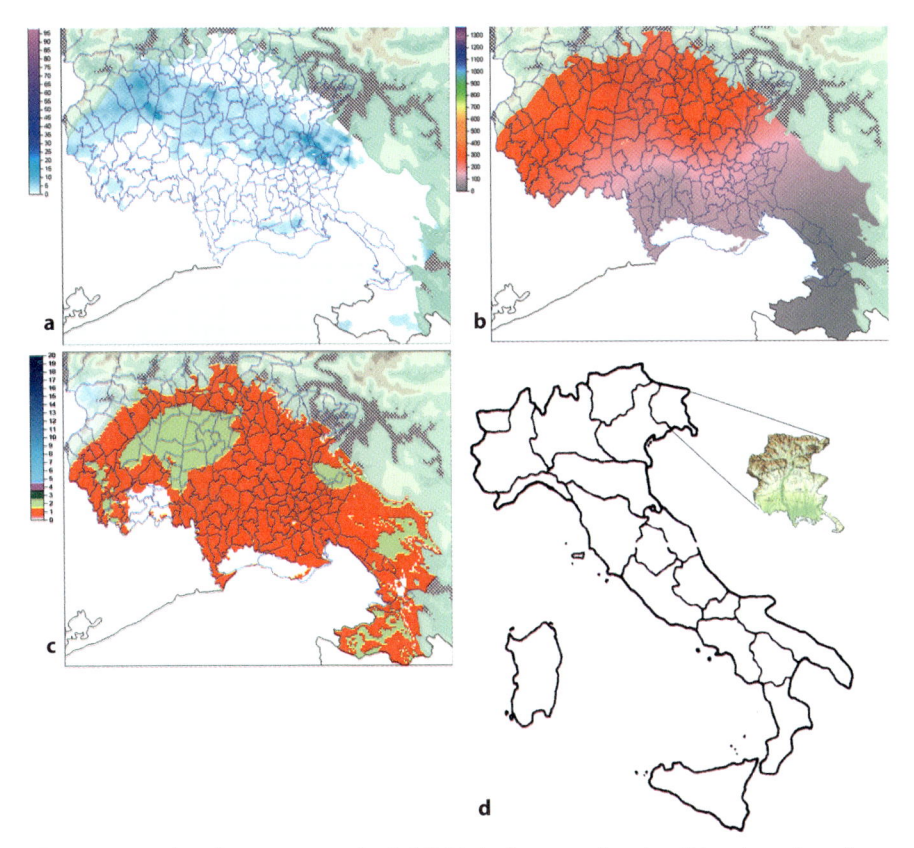

Fig. 26.5 Examples of output maps of rainfall (**a**), leaf wetness duration (**b**) and number of current downy mildew infections (**c**) in Friuli Venezia Giulia region (Italy)

As far as remote sensing is concerned, it is important to emphasize the fact that it is a very helpful tool and research efforts should be supported, particularly with regard to operational applications. Indeed, the use of these techniques allows to obtain accurate data estimations, reducing both expenses and installation/maintenance works.

Nevertheless, it is acknowledged that both radar and, particularly, satellite data frequently provide discordant rain estimates as compared to traditional rain gauges. As a matter of fact, remote sensing provides indirect phenomenon estimations and the measures are affected by a low spatial and temporal accuracy. On the other hand, rain is an extremely variable unit both in spatial and temporal terms. These differences in measurement principles greatly amplify the disagreement due to instrument characteristic errors. As a result, satellite or radar estimates cannot exactly reply rain gauge measurements. Nevertheless, an integration between these methodologies is desirable, given that remote sensing data are characterized by high temporal resolution and spatial continuity.

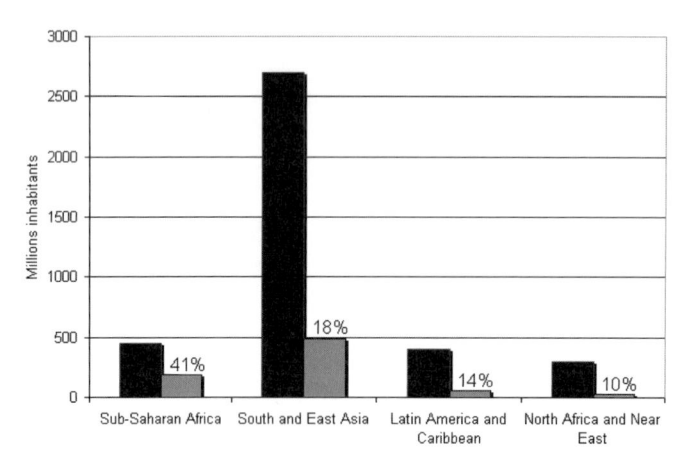

Fig. 26.6. Number and percentage of chronically undernourished (grey) on total population (black) in developing regions 1990-1992 (www.fao.org)

26.6. Early Warning Systems (EWS)

Over 800 million people, mostly in the developing world, are chronically undernourished, eating too little to meet minimal energy requirements (Fig. 26.6). Millions more suffer acute malnutrition during transitory or seasonal food insecurity. Over 200 million children suffer from protein-energy malnutrition and each year nearly 13 million under the age of five die as a direct or indirect result of hunger and malnutrition.

The essential purpose of EWS is to give decision makers sufficient time to take action to avoid the worst effects of impending drought or poor harvests, for example, in an effort to protect the most susceptible areas.

EWS are based on a very extensive multidisciplinary analysis. The utilisation of satellite remote sensing for the provision of meteorological information, also integrated by ground data, vegetation and land cover maps is the common denominator. As reported in "Proceedings of Early Warning Systems and Desertification" (AA. VV. 1999) a list of EWS used operationally in developing countries is given (Table 26.5).

The socio-economic aspect is often predominant, however, a more statistical, agricultural and food approach is adopted, for example, by GIEWS (Global Information and Early Warning System), while AP3A (Alerte Précoce et Prévision des Production Agricoles) extends its range of action from agrometeorology to livestock, also integrating them with socio-economic baseline information.

In particular the AP3A methodology, instead of being centred on the most classic economic aspects (prices, markets, etc.), pays greater attention to the agrometeorological and agro-pastoral analyses. Agricultural production is the factor that mostly determines food availability: in the Sahel region this factor is based on rainfed crops and is mostly destined to self-consumption.

Therefore, the so-called food risk zones are those where the rainfed cereal production is insufficient. The agrometeorological aspect is, anyway, integrated with the socio-economic aspects, represented by basic information on the agricultural and pastoral production, data on population, etc. GIEWS monitors the condition

Table 26.5 Early Warning Systems (EWS) used in developing countries.

Agrhymet Alerte Précoce et Prévision des Production Agricoles (AP3A) project	http://www.ibimet.cnr.it/Case/ap3a/
USAID's Famine Early Warning System (FEWS)	http://www.fews.net/
SADC Food Security Programme (/REWU)	http://www.sadc.int/english/fanr/food_security/food_earlywarning.php
FAO Global Information and Early Warning System (GIEWS) on Food and Agriculture	http://www.fao.org/giews/english/index.htm

of food crops in all regions and countries of the world. Information is gathered on all factors that might influence planted area and yields. In many drought prone countries, particularly in sub-Saharan Africa, there is a lack of continuous, reliable information on weather and crop conditions. For this reason, GIEWS, in collaboration with FAO's Africa Real Time Environmental Monitoring Information System (ARTEMIS) established a crop monitoring system using near real-time satellite images. Data from satellite systems are used for monitoring the various crop seasons. Data received directly at FAO ARTEMIS from the European METEOSAT satellite are used to produce cold cloud duration (CCD) images for Africa every 10 days. These provide a proxy estimate for rainfall, as cold clouds are often responsible for rain, and high cold cloud duration over an area is indicative of significant rainfall. In addition to rainfall monitoring, the System makes extensive use of Normalised Difference Vegetation Index (NDVI) images that provide an indication of the vigour and extent of vegetation cover. These allow GIEWS analysts to monitor crop conditions throughout the season.

26.7
Conclusions

Climate plays a fundamental role in agriculture due to its direct and indirect influence on production. Each physical, chemical and biological process determining the agricultural activity is regulated by specific climatic requirements and any deviation from these patterns may exert a negative influence. Agriculture of developed countries, mainly oriented towards the production of high quality food, is prone to being subjected to meteorological hazard impacts because it is based on highly developed farming techniques. On the other hand, agriculture in developing countries can be strongly affected by weather conditions, responsible for dramatic reduction in yield and could lead to famines.

Finally, current and future trends of model outputs can be analysed to evaluate the hazard levels for agriculture and the possible consequences for natural resources due to climate change. Consequences in terms of production quality and quantity, biological and physical damages and seasonal changes can be mainly considered. Risk assessment can be carried out considering the sensitivity to climatic hazards of different agricultural systems, defining specific critical thresholds according to farming characteristics in agricultural areas. Based on this, possible modification of crop protection methods, irrigation programs, cultivation techniques, harvesting, storage and commercialisation strategies can be evaluated in conjunction with economic aspects. Climate change impact assessment is the first step for implementing support systems based on information technologies to disseminate advices and early warnings to the potential end-users.

References

AA. VV. (1999) Proceedings of early warning systems and desertification regional workshop. Agrymet Regional Center Niamey, Niger, 25-29 Oct. 1999, 40 pp
Allen RG, Pereira LS, Raes D, Smith M (1998) Crop evapotranspiration. Guidelines for computing crop water requirements. In: Irrigation and Drainage Paper 56, FAO, Rome, Italy, 300 pp
Dalla Marta A, Gozzini B, Grifoni D, Orlandini S (2003) Applications of RAMS models for the creation of agrometeorological maps at territorial level. In: Act of the Sixth European Conference on Applications of Meteorology (CD ROM). Rome, 15-19 Sept. 2003
Friesland H, Orlandini S (2006) Simulation models and plant pests and diseases. COST Action 718, Meteorology application for Agriculture. Office for Official Publications of the European Communities, Luxembourg, pp 81-98
Helming K, Auzet, AV, Favis-Mortlock D (2005) Soil erosion patterns: evolution, spatio-temporal dynamics and connectivity. Earth Surf Proc and Landforms 30:131-132. Published online in Wiley Inter Science (www.interscience.wiley.com). DOI 10.1002/esp.1179
Jensen AL, Thysen I (2003) Agricultural information and decision support by SMS. In: EFITA 2003 Conference, Debrecen – Budapest (Hungary), pp 286-292
Kroes JG, Van Dam JC (2003) Reference Manual SWAP version 3.0.3. Wageningen, Alterra, Green World Research. Alterra-report 773. 211 pp
Kroes JG, Kersebaum KC, Marica A, Susnik A (2006) Irrigation modelling related to agrometeorology. COST Action 718, Meteorology application for Agriculture. Office for Official Publications of the European Communities, Luxembourg, pp 99-130
Löpmeier FJ (1994) The calculation of soil moisture and evapotranspiration with agrometeorological models (in German). Zeitschrift f. Bewaesserungswirtschaft 29:157-167
Maracchi G (2001) Meteorologia e climatologia applicate. Istituto Geografico Militare, Firenze, Italy
Mestre A (2006) Meteorological information as input in agrometeorological models: analysis of the potential use of data from numerical weather models in agrometeorology. COST Action 718, Meteorology application for Agriculture. Office for Official Publications of the European Communities, Luxembourg, pp 5-46
Orlandini S (1996) Agrometeorological models for crop protection. In: (Dalezios NR ed.), Proceedings of the International Symposium on Applied Agrometeorology and Agroclimatology Volos, Greece, 24-26 April 1996. European Comission.
Orlandini S, Cappugi A (2001) Sistemi di supporto alle decisioni e servizi agrometeorologici. In: Meteorologia e climatologia applicate. Istituto Geografico Militare, Firenze, Italia, pp 279-295
Travis JW, Rajotte E, Bankert R, Hickey KD, Hull LA, Eby V, Heinemann PH, Crassweller R, McClure J, Bowser T, Laughland D (1992) A working description of the Penn State apple orchard consultant, an expert system. Plant Dis 76:545-554
Van Dam JC (2000) Field scale water flow and solute transport. SWAP model concepts, parameter estimation and case studies. PhD thesis, Wageningen University, 167 pp

CHAPTER 27

Managing Weather and Climate Risks in Agriculture Summary and Recommendations

Mannava V.K. Sivakumar, Raymond P. Motha

27.1
Introduction

Agriculture is a complex system, within which changes are driven by the joint effects of economic, environmental, political and social forces (Olmstead 1970; Bryant and Johnston 1992). It is very well known that agriculture is inherently sensitive to climate conditions and is among the sectors most vulnerable to weather and climate risks. Of the total annual crop losses in world agriculture, many are due to direct weather and climatic effects such as droughts, flash floods, untimely rains, frost, hail, and severe storms (Hay 2007). Chattopadhyay and Lal (2007) estimated that around 28% of the land in India is vulnerable to droughts, 12% to floods and 8% to cyclones. But in the year 1918, which was ranked as the worst drought year of the last century in India, about 68.7% of the total area of the country was affected by drought (Chowdhury et al. 1989).

Farm decision-making is seen as an on-going process, whereby producers are continually making short-term and long-term decisions to manage risks emanating from a variety of climatic and non-climatic sources (Ilbery 1985). The decisions farmers make have a significant impact on the returns to their investments and on their overall family welfare. The climate-based decisions that farmers make are mainly strategic in nature eg., choice of a crop/cropping system, allocation of acreage, purchase of inputs such as seed and fertilizer ahead of the cropping season, etc. In contrast, the weather-based decisions are tactical in nature and affect the operational activities such as sowing, fertilizer application, irrigation, weeding, harvesting, etc. Farm-level risk management strategies have to deal with both the changing and variable climatic conditions as well as the weather conditions.

27.2
Risk and Risk Management in Agriculture

Basically, risk is the chance of something happening that will impact on the objectives of farmers. Chance implies uncertainty and hence risk management in agriculture basically involves managing uncertainty.

As Hay (2007) explained, risk levels can change, including as a result of potentially detrimental changes in the climate (e.g. warming, decreasing rainfall). Changes in levels of exposure, due to altering levels of investment, also influence risk levels. Risk combines both the likelihood of a harm occurring and the consequences of it doing so. Thus according to Hay (2007), in risk terms, an unlikely hazard or condition causing considerable harm (e.g. a category 5 hurricane, such the cyclone in the state of Orissa that devastated parts of India in 1999), may be compared to a hazard or condition which causes less harm but has a higher probability of occurrence (e.g. a seasonal drought).

Risk management research recognizes that decisions in agriculture involve both risk assessment and specific actions taken to reduce, hedge, transfer or mitigate risk (Wandel and Smit 2000). Risk management strategies in agriculture could involve:
- Avoiding the dangers
- Preventing/reducing the frequency of impacts
- Controlling/reducing the consequences (adaptation measures)
- Transferring the risk (e.g. insurance)
- Responding appropriately to incidents/accidents (e.g. disaster management)
- Recovering or rehabilitating as soon as possible (e.g. media response)

As Hay (2007) described, increasingly, farm managers and other practitioners are seeking more rational and quantitative guidance for decision making, including cost benefit analyses. A risk-based approach to managing the adverse consequences of weather extremes and climate anomalies for agriculture can indeed provide a direct functional link between, on the one hand, assessing exposure to the adverse consequences of extreme weather and anomalous climatic conditions and, on the other, the identification, prioritization and retrospective evaluation of management interventions designed to reduce anticipated consequences to tolerable levels.

Coping with agrometeorological risk and uncertainties is the process of assessing agrometeorological risks and uncertainties and then developing strategies to cope with these risks. High preparedness, prior knowledge of the timing and magnitude of weather events and climatic anomalies and effective recovery plans will do much to reduce their impact on production levels, on land resources and on other assets such as structures and infrastructure and natural ecosystems that are integral to agricultural operations (Hay 2007). When user-focused weather and climate information are readily available, and used wisely by farmers and others in the agriculture sector, losses resulting from adverse weather and climatic conditions can be minimized, thereby improving the yield and quality of agricultural products (Hay 2007).

27.3
Addressing Agrometeorological Risk Management during the Workshop

During the International Workshop on Agrometeorological Risk Management: Challenges and Opportunities, the issue of agrometeorological risk management was addressed in six technical sessions:
- Weather and Climate Risks, Preparedness and Coping Strategies: Overview

- Challenges to Coping Strategies with Agrometeorological Risks and Uncertainties – Regional Perspectives
- Agrometeorological Risks and Uncertainties – Perspectives for Farm Applications
- Coping Strategies with Agrometeorological Risks and Uncertainties
- Weather Risk Insurance for Agriculture – A Special Symposium
- Coping with Agrometeorological Risks and Uncertainties – Policies and Services

27.4 Workshop Summary

The presentations and the discussions during the six technical sessions of the workshop could be summarized under the following headings:
- Risk in agriculture
- Risk and risk characterization
- Approaches for dealing with risk
- Risk coping strategies
- Perspectives for farm applications
- Challenges to coping strategies

27.4.1 Risk in Agriculture

The global food and fiber system -- from the producer to the final consumer -- is subject to a wide range of risks and uncertainties (Menzie 2007). For example, in the South Pacific islands there is little forestry, and traditionally agriculture has been based on subsistence and cash crops for survival and economic development. Subsistence agriculture has existed for several hundreds of years and is subject to weather and climate risks.

Risk in agriculture can be broadly defined into several categories (USDA 2006). These include: yield risk, production risk, price or market risk, institutional risk, human or personal risk, and financial risk. The relationships between weather, climate and production risk are well recognised (George et al. 2005) and Hay (2007) provided excellent examples to illustrate the strength and importance of these relationships.

Risk due to weather and climate extremes is larger in some regions of the world than others. According to Mukhala and Chavula (2007), in sub-Saharan Africa, 90% of agricultural production is rainfed which makes agriculture susceptible to inter-annual rainfall variability. The main climatic characteristics in South America, during the El Niño event, are frequent anomalies (Velazco 2007). Together with an increase in the temperatures in the western coasts of the Pacific Ocean, it modifies the atmospheric circulation patterns, pressure, precipitation, river discharges and the water level of the lakes. El Niño causes above normal rains, and droughts in several places. The El Niño-Southern Oscillation (ENSO) provides a large source of seasonal to interannual variability across the southwest Pacific region, with significant influences on weather and climate extremes, including floods and droughts and warmer and cooler seasons at higher latitudes (Trenberth and Caron 2000). According to Salinger (2007), ENSO events can bring drought and widespread decreased precipitation anomalies over the Philippines, Indonesia, northern and eastern Australia, the subtropical Southwest Pacific and the north east of New Zealand. Increased precipitation occurs in the equatorial Pacific from Kiribati (west of the Date Line) through to the Galapagos Islands. El Niño events produce widespread impacts on communities across the Southwest Pacific, as documented by the 1997-98 event (Shea et al. 2001). Drought severely affected Fiji, Papua-New Guinea, the Solomon Islands, Tonga and the Marquesas Islands of French Polynesia. The reverse anomalies occur in La Niña episodes. The Interdecadal Pacific Oscillation (IPO) (Trenberth and Hurrell 1994; Deser et al. 2004) is also an important source of multidecadal climate fluctuations in southwest Pacific and causes shifts in climate across the region. Brunini et al. (2007) discussed the complex nature of drought in Brazil. The high frequency of drought occurrence in Brazil is associated with the frequency of the El Niño/La Niña phenomena in the east-central Pacific Ocean and the Atlantic Ocean dipole.

While yield risk, and its impact on the producer, is the first and most recognized source of uncertainty, myriad related impacts affect economic conditions throughout the marketing chain. All factors must be considered when estimating global supply and demand (Menzie 2007). Price and income effects stemming from a particular set of meteorological conditions in a given crop cycle can also influence cropping patterns in subsequent crop cycles. These influences must be factored into the global supply and demand estimates. In order to highlight the scope of the potential impacts of meteorological events on global supply and demand, Menzie (2007) presented an excellent summary of the implications of a hypothetical drought affecting the soybean crop in the western U.S. Corn Belt.

27.4.2
Risk and Risk Characterization

Risk considers not only the potential level of harm arising from an event or condition, but also the likelihood that such harm will occur. Climate anomalies and extreme climatic events both dominate the challenges for coping with agrometeorological risks and uncertainties.

Risk has both natural and social components. The risk associated with weather and climate for any region is a product of both the region's exposure to the event (i.e., probability of occurrence at various severity levels) and the vulnerability of society to the event. This aspect was elaborated by Wilhite (2007) in his excellent analysis of the drought hazard and societal vulnerability. While drought hazard is a result of the occurrence of persistent large-scale disruptions in the global circulation pattern of the atmosphere, vulnerability to drought is determined by social factors such as population changes, population shifts (regional and rural to urban), demographic characteristics, land use, environmental degradation, environmental awareness, water use trends, technology, policy, and social behavior. For example, Velazco (2007) explains that in South America, vulnerability is aggravated because of the location of human activities in some places of great risk, natural resources subject to excessive pressure of poverty, lack of environmental management policies, excessive centralization, little agricultural technology, and lack of education of the population to prevent and face risks.

Much of the agricultural activity in the poorest and marginal countries takes place in high risk environments and the extreme poverty makes people very risk averse. Ravallion (1988) explains that two of the most widely accepted stylized facts about agriculture in Sub-Saharan Africa, South Asia and elsewhere are that income is highly uncertain from one year to another and that deep and widespread poverty exists. Under such circumstances, producers often avoid activities that entail significant risk, even though the income gains might be larger than for less risky choices. This inability to accept and manage risk and accumulate and retain wealth is sometimes referred to as the "the poverty trap" which fosters hopelessness and insecurity.

To facilitate exit from poverty traps, it is important to reduce exposure to risk through improved tools for managing risk. The *ex ante* approaches here involve diversification opportunities, information systems, preventive health care, mobility and stabilization while *ex post* approaches are primarily through safety nets. In most developing countries, livelihoods are not insured by international insurance/reinsurance providers, capital markets, or even government budgets. Without access to credit, risk-averse poor farmers are locked in poverty, burdened with old technology, and faced with an inefficient allocation of resources.

Agrometeorological risk and uncertainty permeate the entire marketing system with far-reaching consequences. Menzie (2007) explained that in order to optimize business decisions relative to these risks and uncertainties for every economic agent within the global agricultural production and distribution system, accurate, timely, consistent, and widely available information is essential. This information requirement can be met in part through periodic review and estimation of global supply and demand for agricultural commodities.

27.4.3
Approaches to Dealing with Risks

As Hay (2007) explained, there is a well established approach to characterizing and managing risks. This includes risk scoping, risk characterization and evaluation, risk management and monitoring and review.

In risk scoping, risk reduction targets and criteria are established through a consultative process, involving stakeholders as well as relevant experts, as required (Hay 2007). In risk characterization, scenarios are developed in order to provide a basis for estimating the likelihood of each risk event, for present conditions and into the future if change is anticipated, for example as a consequence of climate change (Hay 2007). The consequences of a given risk event are quantified in terms of individual and annualized costs. The overall findings are compiled into a risk profile.

The efficient management and planning of agricultural activities requires policies and tools that allow communities to face agro-meteorological risks and uncertainties. Velazco (2007) described a number of such policies and tools. Farmers have many options for managing the risks they face, and most use a combination of strategies and tools. Some strategies deal with only one kind of risk, while others address multiple risks. Most producers use a mix of tools and strategies to manage risks. Since the willingness and ability to bear risks differ from farm to farm, there is usually variation in the risk management strategies used by producers (Hay 2007). Preparedness planning, risk assessments, and improved early warning systems can greatly lessen societal vulnerability to weather and climate risks.

The goal of effective risk management is to impose management and policy changes between hazard events such that the risk associated with the next event is reduced through the implementation of well-formulated policies, plans, and mitigation actions that have been embraced by stakeholders. An important point to remember is that food security and weather risk management are inextricably linked: weather risk management, or the lack of it, determines the level of systemic risk in the food security system (Mukhala and Chavula 2007). The exposure to weather risk drives overall food insecurity.

Maracchi et al. (2007) discussed the potential for the use of seasonal forecasts in water and crop management as a fundamental tool to avoid serious health problems. The importance of sea-surface temperatures to force the long-term atmospheric anomalies at the regional scale has led to the development of a large number of model simulations, i.e., Global (GCM) or Regional (RCM) circulation models. As a result, weather predictions today are established on solid theoretical and practical bases, and their reliability and accuracy are steadily increasing. These forecasts of future trends in precipitation three months or more in advance could be extremely important to agriculture, forestry, and land management by potentially forecasting drought or heat waves, for example. These outlooks have strategic relevance to national policy with respect to planning to help alleviate food shortages, lessen the impact of droughts, and provide distribution of energy.

Enterprise diversification, vertical integration, contracting, hedging, liquidity, crop yield insurance, crop revenue insurance and household off-farm employment or investment are some of the useful approaches in dealing with risks. Over many years, crop insurance had been one of the approaches to dealing with risks. In India crop insurance was considered by the central government as early in 1947-48. The National Agricultural Insurance Scheme (NAIS) had been in operation since 1999-2000, and at present, is implemented by 19 States and 2 union territories (Chattopadhyay and Lal 2007).

Weather derivatives and weather index insurance play a role in developing agricultural risk management strategies. Weather based index insurance is slowly gaining recognition as one of the methodologies that can be used sustain livelihoods and reduce poverty as part of the Millennium Development Goals (MDGs). Examples from Malawi and Ethiopia described by Mukhala and Chavula (2007) illustrate the potential for this approach.

Menzie (2007) points out that accurate, timely, consistent, and widely-available information is essential to optimize decisions relative to these risks and uncertainties within the global agricultural production and distribution system. He presented several examples of the meteorological tools and methods of crop assessments typically used in developing monthly world agricultural supply and demand estimates at the United States Department of Agriculture.

27.4.4
Risk Coping Strategies

Velazco (2007) recommends that in order to better cope with weather and climate risks, the agro-economic planning at a short and long term and at a local, regional or national scale, should be formulated more rationally including among its variables, the agro-meteorological and agro-climatic information. For the developed countries (e.g Australia, New Zealand) coping strategies are more sophisticated and involve both structural and non-structural measures to reduce the impacts of change on crop and livestock production (Salinger 2007). In these countries it will be the rate of change that will pose the risks.

Huda et al. (2007) outlined approaches to cope with integrated pest management risks, based on the Australian experiences with wheat and canola. Collaborative activity is essential between scientists, risk managers, government, and local farmers to determine best practice approaches for addressing pest management, in order to achieve economically sound and ecologically sustainable outcomes. A major focus of Australian research is the optimization of natural controls relating to informed planting strategies, and the minimization of pesticide application through the prediction of climatic influences, which can in turn lead to the optimal effectiveness in the control of disease agents. The relationship between meso- and microclimate, and the effects on the cycles of disease agents needs special attention if quantity of applied pesticide is to be minimized, while optimizing disease amelioration outcomes.

Brunini et al. (2007) explained that the mitigation measures for drought must take into account the cultural aspects of the population, the climate regime, and the agricultural development. Various indices have proven adequate for monitoring and mitigating drought effects; however, adjustments are necessary for each of these indices for each region and by crop. Hence, Brunini et al. (2007) emphasized that it is important for researchers and specialists from all disciplines to work together to cope with the drought phenomena.

Wang et al. (2007) reviewed strategies to cope with desertification, illustrating numerous cases in China. Some structural measures include biological, agronomic and engineering measures. Non-structural measures for combating desertification involve desertification monitoring of meteorological conditions, research on the relationship between climate and drought occurrence, development of desertification as well as combating countermeasures, and agrometeorological information services in decision-making to cope with desertification.

Tibig and Lansigan (2007) reviewed seven types of management strategies to cope with risks and uncertainties in agrometeorology. These include: optional use of resources (crop diversification); use of appropriate cultivars (varietal diversification); improved cultural/farming practices (organic farming and flexible calendar to fit weather/climate, i.e., farm afforestation, land topography change); local indigenous knowledge (coping mechanisms of farmers to various environmental and natural challenges); technological innovations (direct seeded rice (DSR) cropping system to increase net income); and, farmers can opt to reduce their production area if conditions warrant.

Doraiswamy et al. (2007) discussed intervention management techniques for the farming systems to help minimize soil erosion and soil loss from excessive surface water runoff. Coping strategies to reduce risks and uncertainties in crop production related to soil management practices include tillage, the crops cultivated, sequence of cropping with cover crops planted where possible, the use of crop residue, and development of mechanical barriers to slow down the runoff over sloping landscapes. The specific strategy suitable for a particular site would depend on many factors such as landscape characteristics, soil properties, rainfall patterns and intensity, and adaptability of soil and crop management practices in the region. Developing a proper combination of these strategies based on a good assessment of the problem is critical for a successful implementation of coping with risks and uncertainties in crop production. These methods need to be adaptable for worldwide application.

27.4.5
Perspectives for Farm Applications

One of the major constraints for farm applications of agrometeorological risks such as droughts is that in general there are no operational procedures to forecast the impending drought conditions with respect to area of impact, extent and duration. Stigter (2004) showed that the main bottlenecks here were insufficient considerations of the actual conditions of the livelihood of farmers and therefore the development of inappropriate support systems.

Eitzinger et al. (2007) reviewed the key points to optimization of farm technologies for both high input agricultural systems and low input farming systems. Proper management of resources is essential to sustain agricultural production within specific agro-ecosystems. Production variability depends directly upon weather and climate variability, water availability, soil nutrients, crop management and microclimatic conditions. The combination of locally adapted traditional farming technologies, seasonal weather forecasts and warning/forecasting methods may help farmers improve productivity, food production, and income.

Stigter et al. (2007) discussed information needs and demands for four different rural income groups in China. They reviewed the socio-economic levels of farmers in China and India, the implications for information approaches and technologies, and their needs for capacity building for agrometeorological services, and presented some brief examples of these services in Cuba, Nigeria, Sudan, and Vietnam. The authors emphasized the utmost need for "middle level" intermediaries who work as two-way guidance between the providers and users of agrometeorological data and services. While reports on using new technologies and pilot projects offer some promising results, lack of resources and skills have prevented significant technological progress in most rural areas of many developing countries.

Lee (2007) proposed an Emergence Response System (ERS) in agricultural management to be considered as an on-farm application for decision-making support system (DMSS) against agricultural hazards.

27.4.6
Challenges to Coping Strategies

In the south west Pacific, climate and extreme climatic events dominate in providing challenges for coping with agrometeorological risks and uncertainties (Salinger 2007). Tropical cyclones are one of the most devastating risks for agrometeorology on the small island developing states in the region. These generally cause large scale destruction to crops and infrastructure through high intensity rainfall and severe winds.

Governments often undertake emergency operations as part of their coping strategies with natural disasters. The primary objectives of the emergency relief response mechanism is to undertake immediate rescue and relief operations. As Rathore and Stigter (2007) explained, the mechanism requires planners to identify disasters and their probability, evolve signal/warning mechanisms, identify the activities and sub-activities, define the level of response, specify authorities, determine the response kind, work out individual activity plans, have quicker response teams, undergo preparedness drills, provide appropriate delegations and have alternative plans.

Uncertainties in agrometeorology are part of everyday farmer conditions and Stigter et al. (2005) have, for example, extensively dealt with traditional methods and indigenous technologies to cope with such consequences of climate variability. That such variability is increasing makes it more important to improve and extend the mitigating practices involved and pay attention to farmer innovations and to products from NMHSs, research institutes and universities that can be absorbed by farmers to better cope with increasing uncertainties and disasters.

Lee (2007) explained that there are numerous challenges to an operational ERS, including: establishing and maintaining observing systems and data management systems; maintaining archives, including quality control and digitization of historical data; obtaining systematic environmental data for vulnerability analysis; and, securing institutional mandates for collection and analysis of vulnerability data. Other issues for consideration include communication systems, early-warning and dissemination to the farm communities.

According to Wang et al. (2007), while progress has been made, there are still serious challenges in coping with desertification. Global climate warming, frequent and severe drought, and future climate uncertainties will continue to influence desertification. At the same time, human-driven factors leading to deteriorating vegetation in sandy areas are exacerbating desertification. Combating desertification requires vigilant monitoring research between climate and desertification occurrence, and, measures for ecological restoration.

27.5 Recommendations

Workshop recommendations covered the broad areas of risk management, risk management tools, research needs, policy issues, emphasis on user needs, communication and marketing. Salient recommendations under each of these areas are given below:

27.5.1 Risk Management

- Develop a pro-active risk-based management approach to deal with the adverse consequences of weather extremes and climate anomalies which includes risk scoping, risk characterization and evaluation, risk management and monitoring and review.
- Emphasize preparedness planning and improved early warning systems to lessen societal vulnerability to weather and climate risks.
- Provide accurate, timely, consistent, and widely-available information to optimize decisions relative to the risks and uncertainties within the global agricultural production and distribution system.

27.5.2 Risk Management Tools

- Use of decision-support systems as risk management tools should be promoted as an effective means of providing output of integrated climate-agronomic information in the form of scenario analyses relating to impending risks that can be valuable to users
- For medium and low input systems in the developing countries, crop or agro-ecosystem modeling should be used to guide general decision-making on a higher institutional or farm advising level.
- Current and future trends of simulation model outputs should be analysed for sensitivity to climatic hazards of different agricultural systems and defining specific critical thresholds according to farming characteristics in agricultural areas. Based on this, possible modification of crop protection methods, irrigation programs, cultivation techniques, harvesting, storage and commercialisation strategies can be evaluated in conjunction with economic aspects.
- Risk assessment and risk management models supporting coping strategies for integrated pest management could be used in a prototype conceptual framework that can be utilized in other agricultural-related risk approaches.
- Statistical forecasting tools to link observed weather data to crop yields in major crop-producing regions should be developed.
- Emergency response system (ERS) based on advanced Information Technology (IT) such as information network, simulation models, tools for GIS and remote sensing could be developed to address agricultural hazards and early warning.

- Climatic risk zoning could be used for quantifying climate-plant relationships and the risk of meteorological extremes in agricultural financing programs.

27.5.3
Research Needs

- Local indigenous knowledge has been blended with specific and important weather patterns in a cultural tradition in many poor, rural areas. Introducing new scientific-based weather/climate forecast services, which provide accurate and reliable outlooks into this cultural system may help farmers improve yields and cope with risks.
- There is an essential need for the development of standards, protocols, and procedures for the international exchange of data, bulletins, and alerts for some types of agricultural hazards. World Agrometeorolgical Information Service (WAMIS) offers the potential to assist with this technology transfer.
- The application of seasonal forecasts for crop management strategies, risk management planning, and national policy implications needs to be considered, as these outlooks become more accurate and reliable.
- Developing methods for screening satellite imagery to identify crop-specific impacts of weather in crop regions around the world should be research priority.
- For effective management systems to be put into place, integrated climate-crop modeling systems need to be developed at the appropriate farm or regional scale suitable for the decision-makers needs.
- In many developing countries, the inability of poor in rural areas to gain access to support mechanisms in terms of technical expertise or technological innovations, including formal sources of credit or crop insurance, requires urgent attention.
- Agrometeorological services and support systems for agrometeorological services should be strengthened for effective management of weather and climate risks.
- Aspects of drought contingency planning, drought preparedness, and drought impact assistance policies need to be urgently considered as to their future effectiveness under long-term climate change.
- Drought contingency plans on paper should be translated into an effective policy covering the range of activities required to address short and long-term consequences. Effective and interactive management systems need to be set in place.
- Public-private partnership models need to be further explored in order to 'mainstream' drought risk management. Involving the development of risk management tools and approaches within the context of overall rural livelihood strategies, and integrating risk arising from markets and threats to the natural resource base is important. It also involves communicating risk management knowledge through functional, existing communication networks of farmers and other landholders, rather than pursuing specific communication programs.

- The concept of a drought mitigation and monitoring center, coordinated by both meteorological and agricultural agencies at national and state levels, to define standards and policy for monitoring and mitigation of drought at both state and national levels should be promoted.
- A scientific desertification monitoring and evaluation system involving all appropriate sectors including agriculture, forestry, water conservation, environmental protection, meteorological and natural resource conservation should be established.
- Measures to combat desertification must be vigorously pursued. These include: shelterbelts, windbreaks, converting cropland to forests, grazing prohibition, grassland construction, water-saving irrigation project, and integrated ecological agro-forest measures or integrated ecological agro-economic measures.

27.5.4
Emphasis on User Needs

- Develop clear and useful guidelines on the exact nature of agrometeorological products needed for local user communities
- Strengthen the use of intermediaries in training farmers and the use of information technologies fit for target groups.
- Implement an effective user-driven delivery system comprising of decision support tools and the training of users on their application at critical decision points in farming.

27.5.5
Communication

- Communication and dissemination are critical links to the transfer of early warning information to the right decision-makers. For disseminating warnings, Internet is a useful medium for expanding coverage and reducing time lags and its active use should be promoted.
- Enhancements in communication channels for the improved dissemination of agricultural meteorological information must take into account the literacy levels of users, socio-economic conditions, level of technological development, and accessibility to improved technology and farming systems.
- For effective inter-sectoral and multi-stages communication of risk, appropriate involvement of communication pathways and common dialogue between scientists, managers, and communities should be promoted.
- A documentation of the many and varied types of management strategies to cope with agrometeorological risks and uncertainties should be posted on WAMIS web server.
- Methodologies and tools to assess precipitation anomalies and drought should be posted on the WAMIS web server for potential application elsewhere.
- Efficient irrigation water management plays a key role in agricultural productivity and also protects the soil health. Proper and timely agro-advisories related

to irrigation scheduling, fertilizer management helps the farmers in better planning of agricultural operations.

27.5.6
Marketing

- The entire global agricultural economy encounters price, income and other forms of risk related to weather uncertainty. Increased interdisciplinary collaboration between meteorologists, agronomists, and economists can improve the quality of information upon which agriculture-related businesses and agricultural policy-makers around the world depend.

References

Bryant CR, Johnston TRR (1992) Agriculture in the City's Countryside. Toronto: University of Toronto Press.

Brunini O, da Anunciação YMT, Fortes LTG, Abramides PL, Blain GC, Brunini APC, de Carvalho JP (2007). Coping strategies with agrometeorological risks and uncertainties for drought examples in Brazil. In: Sivakumar MVK, Motha R (Eds.) Managing Weather and Climate Risks in Agriculture. Springer, Berlin Heidelberg, pp. 191-207

Chattopadhyay N, Lal B (2007) Agrometeorological Risk and Coping Strategies - Perspective from Indian Subcontinent. In: Sivakumar MVK, Motha R (Eds.) Managing Weather and Climate Risks in Agriculture. Springer, Berlin Heidelberg, pp. 83-98

Chowdhury A, Dandekar MM, Raut PS (1989) Variability of drought incidence over India: A Statistical Approach, Mausam 40:207-214

Deser C, Phillips AS, Hurrell JW (2004) Pacific interdecadal climate variability: Linkages between the tropics and the north Pacific during boreal winter since 1900. J. Climate 17: 3109–3124

Dessai S, Hulme M (2004) Does climate adaptation policy need probabilities? Climate Policy 4:107–128

Doraiswamy PC, Hunt ER, Murthy VRK (2007) Coping strategies with agrometeorological risks and uncertainties for water erosion, runoff and soil loss In: Sivakumar MVK, Motha R (Eds.) Managing Weather and Climate Risks in Agriculture. Springer, Berlin Heidelberg, pp. 343-353

Eitzinger J, Utset A, Trnka M, Zalud Z, Nikolaev M, Uskov I (2007) Weather and climate and optimization of farm technologies at different input levels. In: Sivakumar MVK, Motha R (Eds.) Managing Weather and Climate Risks in Agriculture. Springer, Berlin Heidelberg, pp. 141-170

George DA, Birch C, Buckley D, Partridge IJ, Clewett JF (2005) Assessing climate risk to improve farm business management. Extension Farming Systems Journal vol 1(1).

Hay J (2007) Extreme Weather and Climate Events, and Farming Risks. In: Sivakumar MVK, Motha R (Eds.) Managing Weather and Climate Risks in Agriculture. Springer, Berlin Heidelberg, pp. 1-19.

Huda AKS, Hind-Lanoiselet T, Derry C, Murray G, Spooner-Hart RN (2007) Examples of coping strategies with agrometeorological risks and uncertainties for Integrated Pest Management. In: Sivakumar MVK, Motha R (Eds.) Managing Weather and Climate Risks in Agriculture. Springer, Berlin Heidelberg, pp. 265-280

Lee B-L (2007) Information technology and decision support system for on-farm applications to cope effectively with agrometeorological risks and uncertainties. In: Sivakumar MVK,

Motha R (Eds.) Managing Weather and Climate Risks in Agriculture. Springer, Berlin Heidelberg, pp. 191-207

Ilbery B (1985) Agricultural Geography: A Social and Economic Analysis. London, Clarendon Press.

Maracchi G, Pasqui M, Piani F (2007) Water management in a semi-arid region: an analogue algorithm approach for rainfall seasonal forecasting. In: Sivakumar MVK, Motha R (Eds.) Managing Weather and Climate Risks in Agriculture. Springer, Berlin Heidelberg, pp. 237-244

Menzie K (2007) Methods of Evaluating Agrometeorological Risks and Uncertainties for Estimating Global Agricultural Supply and Demand. In: Sivakumar MVK, Motha R (Eds.) Managing Weather and Climate Risks in Agriculture. Springer, Berlin Heidelberg, pp. 125-140

Mukhala and Chavula (2007) Challenges to coping strategies with Agrometeorological risks and uncertainties in Africa. In: Sivakumar MVK, Motha R (Eds.) Managing Weather and Climate Risks in Agriculture. Springer, Berlin Heidelberg, pp. 39-51

Olmstead CW (1970) The phenomena, functioning units and systems of agriculture. Geographica Polonica 19, 31–41.

Rathore LS, Stigter CJ (2007) Challenges to coping strategies with agrometeorological risks and uncertainties in Asian regions. In: Sivakumar MVK, Motha R (Eds.) Managing Weather and Climate Risks in Agriculture. Springer, Berlin Heidelberg, pp. 53-69

Ravallion M (1988) Expected Poverty Under Risk-Induced Welfare Variability. The Economic Journal, 98: 1171-1182.

Salinger J (2007) Challenges to coping strategies in Agrometeorology: The Southwest Pacific. In: Sivakumar MVK, Motha R (Eds.) Managing Weather and Climate Risks in Agriculture. Springer, Berlin Heidelberg, pp. 99-112

Shea EL, Dolcemascolo G, Anderson CL, Barnston A, Guard CP, Hamnett MP, Kubota ST, Lewis N, Loschnigg J, Meehl G (2001) Preparing for a changing climate: The potential consequences of climate variability and change. A report of the Pacific Islands Regional Assessment Team for the U.S Global Change Research Program. East-West Center, 102 pp.

Stigter CJ (2004) The establishment of needs for climate forecasts and other agromet information for agriculture by local, national and regional decision-makers and users communities. In: Applications of Climate Forecasts for Agriculture. Proceedings of the RA I (Africa) Expert Group Meeting in Banjul, the Gambia (December 2002). AGM-7/WCAC-1, WMO/TD-No 1223, WMO, Geneva, pp 73–86

Stigter CJ, Ying T, Das HP, Dawei Z, Rivero Vega RE, Van Viet N, Bakheit NI and Abdullahi YM (2007) Complying with farmers' conditions and needs using new weather and climate information approaches and technologies. In: Sivakumar MVK, Motha R (Eds.) Managing Weather and Climate Risks in Agriculture. Springer, Berlin Heidelberg, pp. 171-190

Stigter K (Ed), with contributions from Barrie I, Chan A, Gommes R, Lomas J, Milford J, Ravelo A, Stigter K, Walker S, Wang S, Weiss A (2005) Support systems in policy making for agrometeorological services: the role of intermediaries. Policy paper for a CAgM/MG meeting in Guaruja, Brazil. WMO, Geneva, Management Group meeting of 30 March - 2 April, document 7.1, 6pp + 1 App. Also available at the INSAM website (www.agrometeorology.org) under "Needs for Agrometeorological Solutions of Farming Problems"

Tibig L, Lansigan F (2007) Coping Strategies with Agrometeorological Risks and ncertainties for Crop Yield. In: Sivakumar MVK, Motha R (Eds.) Managing Weather and Climate Risks in Agriculture. Springer, Berlin Heidelberg, pp. 209-235

Trenberth KE, Caron JM, (2000) The Southern Oscillation revisited: Sea level pressures, surface temperatures and precipitation. J. Climate 13: 4358–4365.

Trenberth KE, Hurrell JW (1994) Decadal atmosphere–ocean variations in the Pacific. Climate Dyn 9: 303–319.

USDA (2006) Farm Risk Management: Risk in Agriculture, United States Department of Agriculture. http://www.ers.usda.gov/Briefing/RiskManagement/RiskinAgriculture.htm

Velazco CA (2007) Challenges and Strategies to face Agrometeorological Risks and Uncertainties - Regional Perspective in South America. In: Sivakumar MVK, Motha R (Eds.) Managing Weather and Climate Risks in Agriculture. Springer, Berlin Heidelberg, pp. 71-82

Wandel J, Smit B (2000) Agricultural risk management in light of climate variability and change. In: H. Milward H, Beesley K, Ilbery B, Harrington L (eds.). Agricultural and Environmental Sustainability in the New Countryside, Winnipeg, Hignell Printing Limited, pp. 30–39.

Wang S, Yuping M, Wang Yinshun H (2007) Coping Strategies with Desertification in China. In: Sivakumar MVK, Motha R (Eds.) Managing Weather and Climate Risks in Agriculture. Springer, Berlin Heidelberg, pp. 317-341

Wilhite DA (2007) Preparedness and Coping Strategies for Agricultural Drought Risk Management: Recent Progress and Trends. In: Sivakumar MVK, Motha R (Eds.) Managing Weather and Climate Risks in Agriculture. Springer, Berlin Heidelberg, pp. 21-38

Subject Index

A

ACIS, *see* applied climate information system
actual
- evapotranspiration standardized index (IPER)   292
- production history (APH)   382
afforestation   89
- belt with irrigation   331
- program   59
African meteorological service   43
agriculture/agricultural
- Africa   219
- Australia   273
- Canada   216
- China   175
- climate change   12
- drought   24
- Europe   113
- impacts   23
- India   83
- innovations   182
- insurance program   401
- – Spain   384
- management   192
- market information   175
- natural disasters   435
- optimization strategies   144, 156
- prices   126
- production   39, 199
- – role of climate   2
- risk   192
- – insurance   372
- – management   375
- – pilot programs   398
- technological information   175
- user-focused information   1
- yield risk   125
agroclimatic
- information   80
- zone   88
agroecosystem   141, 142

agroforestry   55
- complex ecosystem   324
- system   163
agrometeorological
- adaptation strategies   54
- advisory service (AAS)   63, 65
- characterization   449
- coping strategies   169, 343
- index   290
- information   42, 80, 433
- models   462, 464
- monitoring   121
- research needs   121
- risk   77, 79, 191, 193, 206, 265, 476
- – India   83
- – management   115, 194
- – South America   80
- services   61, 183, 185, 433, 455
- uncertainties   58
agropastoral system in the Andes   219
agro-resource environment   211
analogue year   239
animal husbandry in steppe   337
annual rainfall variability   39
anthropogenic climate change   117
anticyclonic circulation   116
AP3A (Alerte Précoce et Prévision des Production Agricoles)   472
APH, *see* actual production history
applied climate information system (ACIS)   445
APSIM model   153
Atlantic cyclonic activity   116
atmospheric
- carbon dioxide   273
- circulation   103
automated weather station (AWS)   408
awareness-raising program   198
AWS, *see* automated weather station

B

BASIX   376
biofuel   122
biological hazard   193
bio-pharmaceuticals   122
blended technologies   456
bush farming   258
bushfire   57, 61

C

CAgM   200
camellones   162
carbohydrate   434
carbon dioxide   434
CCD, *see* cold cloud duration
CCIS, *see* comprehensive crop insurance scheme
cereal production   8
Chacras hundidas   148
change rates   238
chinampa   438
CIIAGRO   291
climate
– anomalies   1, 11
– anthropogenic change   117
– Australia   105
– change   54, 74, 115, 120, 143, 193, 257, 371, 426
– – coping strategies   173
– – effects on desertification   335
– – impact on agriculture   12
– – in Europe   114
– data system   65, 439
– erosivity   344
– extreme events   108
– forecast   44, 214
– information   45, 169, 184, 433
– inter-annual variability   113
– modelling   2, 18
– New Zealand   106
– prediction   456
– prognosis   303
– risk
– – evaluation   310
– – South America   81
– – zoning   309
– Southern hemisphere   103
– Southwest Pacific   99
– stress   39
– user-focused information   1

– variability   16, 54, 173, 421
– – South America   72
clubroot of Brassica spp.   270
CMI, *see* crop moisture index
CO_2
– concentration   114
– gas emission   14
cold
– cloud duration (CCD)   472
– injury   86
– waves   84
combating desertification   339
– of grassland   336
common agricultural policy (CAP)   122
communication
– channel   186, 196, 486
– network   441
comprehensive crop insurance scheme (CCIS)   95
conservation
– farming (CF)   41, 42
– tillage (CT)   41, 120, 347
contingency planning   53, 58, 181, 182, 195, 413, 426
contour
– tillage   347
– trenching   256
conventional tillage   346
converting cropland to forest   324
coping strategies   53, 169, 216, 276, 379, 477
corn production   125

cost-benefit
– analyses   2
– ratio   185
CPC MORPHing technique (CMORPH)   133
– analysis   138
– data   138
crop
– agrometeorological system   310
– cycles   478
– development   293
– disease   267, 268
– diversification   41, 210, 211, 226, 481
– – efficient water use   251
– evapotranspiration   308
– global assessment methods   130
– heat resistance   121
– horticultural   90
– hygiene   268
– improvement   212
– – land-type diversification   213

Subject Index   495

- insurance   40, 95, 217, 266, 376, 393
- – developing countries   373
- – in developed countries   381
- – products   372
- – traditional   379
- irrigated dry (ID)   251
- losses   21
- management   152, 158
- – planning systems   426
- – strategies   349
- modelling   152
- moisture index (CMI)   293
- monitoring   160
- phenology   450
- predictions   114
- protection   469
- – against wind   92
- resources   152
- revenue insurance   17
- simulation models   424
- stress conditions   64
- water requirement index (CWRI)   309
- water stress index (CWS)   295, 296
- yield
- – distributions   381
- – insurance   17

cropland
- converting to forest shrub land   332
- shelterbelt network   330

cropping
- pattern   210
- rice-based systems   228
- system
- – direct-seeded rice (DSR)   215
- – raised bed   215

CROPWAT   468
cross validation calculation   241
cultivars   157
cumulative precipitation   136
CWRI, see water requirement index
CWS, see crop water stress index
cyclone   90, 104, 371
- early warnings   56
- forecasting   445
- South Pacific   102, 107

D

Daily Fire Weather Index map   365
daily rainfall   10
- probability   3

database management system (DBMS)   197, 443
decadal-to-interdecadal variability   103
decision(-making) support system (D(M)SS)   194, 196, 462, 484
decision-making process   192, 206
- supportive tools   271
deficit irrigation   147
deforestation   57, 73, 76
DEMETER project   115
denitrification   94
desertification
- cause analysis   319
- China   317
- – monitoring   321, 322
- combating   334
- degree   318
- development   319
- dynamic changes   319
- influence of climate change   321
- types   318
DIR, see drought impact reporter
direct seeding   229, 230
direct-seeded rice (DSR)   215, 482
disaster
- mitigation   77, 450
- preparedness   181, 182
- risk
- – mainstreaming   54
- – management   396
- standardized data collection   54
- to agriculture (Vietnam)   185
Disaster Prevention in the Anden Community (PREDECAN)   78
diversification   210, 227
DMSS, see decision-making support system
drainage technology   147
drip-irrigation system   146
drought   11, 61, 93, 145, 183
- agrometeorological aspects   304
- alert system   414
- alleviating   88
- assessment   414
- assistance programs   21
- contingency planning   428
- crop losses   21
- early warning   26
- examples in Brazil   281
- global financial impact   118
- hazard   22
- human-induced   26
- hydrological   25, 286
- impact   23, 34, 88

– – archives   31
– – reporter (DIR)   31, 32
– in India   56, 86, 184
– incidence reports   34
– index insurance   49
– information delivery   29
– mitigation   419
– modelling systems   424
– monitoring   26, 444
– – monitoring and mitigation center   306, 313
– planning   414
– – tools   36
– policy   421
– preparation plans   413
– preparedness   423
– – programs   420
– research   422
– risk   21
– – analysis   414
– – atlas   35
– social vulnerability   22
– socioeconomic   25
– South America   74
– types   23, 24
drought-specific decision support system   29
dry
– area   28
– farming meteorology   447
– spells
– – intra-season   43
– – off-season   43
dryness index   13
dry-seeding   229
dust storm   89
– in China   338

E

early warning system (EWS)   63, 78, 153, 192, 194, 219, 269, 462, 471, 486
– wildland fire   355, 357
– – development   363
– – sustainability   362
ECB, *see* European corn borer
economic empowerment   47
efficient market function   129
El Niño
– event   72, 75, 99, 105
– – forecasting systems   79
– period   281

– Southern Oscillation (ENSO)   71, 99, 105, 477
elevators enterprise   128
emergency response system (ERS)   194, 207, 484
EMS, *see* environmental management system
ENSO, *see* El Niño Southern Oscillation
enterprise diversification   16, 217

environmental
– degradation   193, 473
– management system (EMS)   271
EPIC model   344
ERS, *see* emergency response system
ETR/ETM ratio   310
European corn borer (ECB)   161
evapotranspiration   150, 292
experimental observation   272
extreme
– temperature   56, 60, 74
– weather events   1

F

fallow crop   349
farming
– coping strategies   39, 41
– diversification   224
– ecological systems   155
– management systems   191
– risks   1, 6
– technologies   142
– – optimization strategies   144
– waste   224
FDRS, *see* fire danger rating system
fermentation   346
fertilizer   157, 218
financial
– leverage   17
– risk   7, 126, 209
fire
– danger   104, 355
– – maps   359
– – rating systems (FDRS)   356
– fighting   62
– management   361, 364
– risk   108
fireproof belt   331
flood   93, 435
– control   55, 59
– frequencies   259

- hazards 55
- irrigation 148
- meteorological offices 445
- resistant construction techniques 55
- South America 73
floodplain 59
fog formation 88
food
- production 113
- risk zones 472
- security 41, 48, 343, 480
forage production 36
forecast
- accuracy 63
- anomalies 241
- improvement 46
- of duststorm 339
- precipitation 240
- seasonal 242
- submodel 337
forefront stopping sands belt 331
forest
- ecosystem 452
- fire 355
frost
- damage 163
- injury to winter crop 85
- protection 86
fuel 62, 62
fungal disease 275
fungicide 155, 268, 269, 273

G

Galerias filtrantes 148
gamma probability density function (GPDF) 287
GCM (global climate model) 12
geographic information system (GIS) 21, 81, 218, 440, 470
- information 153
- technology 356
geological hazard 193
GHG emission 453
GIEWS, *see* global information and early warning system 472
GIS, *see* geographic information system
global
- circulation model (GCM) 12, 237, 480
- climate change 265
- crop reinsurance markets 385

- earth observation system of systems (GEOSS) 357
- food and fiber system 125
- food security 343
- information and early warning system 427
- multi-hazard early warning system (GEWS) 357
- positioning system (GPS) 218, 440, 462
- reinsurer 367
- soybean production 134
- supply estimation 127
- vegetation fire 355
- warming 13, 14, 104, 142, 371, 453, 483
- - increased drought risk 119
government risk management program 385
GPS, *see* global positioning system
greenhouse gas emission 12
grid portal 202
groundnut production in India 184
groundwater
- exploitation 245, 248
- reserves 143
- utilization 250
guaranteed price 17

H

hailstorm 89, 163
hazard warning 197
hazard-relevant parameter 197
heat wave 90
heat-resistant crop 121
herbicide 230
high-input farming 164
high-intensity rainfall 59, 104
high-yielding variety (HYV) 220, 224
horticulture 269
- crop 90
host-agent-environment disease triangle 275
human/personal risk 7, 209
human-induced drought 26
humidity 434
- index 283
hydro-illogical cycle 413
hydrological
- cycle 259
- drought 25
hydrometeorological hazard 193
HYV, *see* high-yielding variety

498 Subject Index

I

impact
- assessment methodology   31
- reduction   58

in situ
- coping strategies   452
- moisture conversation   247
- water conservation   247, 250

income
- levels
- - in China   179
- risk   126

index
- insurance   372, 388, 399
- - contract   387
- - policyholder   388
- - product   386
- reinsurance
- - disaster option for CAT risk   393
- setting   406

index-based
- insurance products   407
- weather insurance   48

indigenous technical knowledge (ITK)   214, 437

information
- channels   176, 178
- communication   175
- management technique   323
- technology   169, 484

INFOSECA   291

initial
- available soil water (ISAW)   151
- spread index (ISI)   363

innovative technology   180
institutional risk   7, 127, 209

insurance   62
- contract   48
- weather-based   48

integrated pest management (IPM)   265
intensification of production   213
intercropping   212
Interdecadal Pacific Oscillation (IPO)   99
inter-tropical convergence zone (ITCZ)   72

intra-season
- dry spells (ISDS)   43
- rainfall distribution   43

irrigated dry (ID) crops   251
irrigation   21, 47, 61, 245, 466
- fertilizer   261
- management   120, 426
- Mediterranean region   143
- techniques   146

J

joint agricultural weather facility (JAWF)   443

K

kanchas   162

L

La Niña   99, 282

land
- fire   355
- topography   214

LEISA, see low external inputs sustainable agriculture

livestock
- feeder   128
- production   36

local
- fire climate   364
- knowledge   214

loss insurance   372

low external inputs sustainable agriculture (LEISA)   144

low-input
- agriculture   218
- farming   164, 482

M

macro-economic policy   174
major irrigation   249
Marcov chain model   65

market
- failure   395
- - layer   392
- information   176, 178, 179
- limitation   397
- risk   7, 209

MARS project   439
MASIPAG   222, 224
mass media   177
maximum

– annual wind gust   11, 14
– temperature   15
– – tolerance   60
meteorological
– index   286
– – decile method   288
– – quantile method   289
– information services   337
– satellite   338
– services   78
microclimate   173, 214
– management   449
– of crop stands   162
micro-environment diversification   220
microinsurance   373, 374
micro-irrigation technique   258
minimum tillage (MT)   41, 150, 160, 348
minor irrigation   249
mitigation practices   53, 181, 182
models for soil erosion   464
moisture in situ conversation   247
monitoring
– network   78
– system   80
– techniques   323
monocropping   220, 245

monsoon
– circulation   238
– depression   90
mountain ecosystem   76
MPCI, *see* multi-peril crop insurance
mulch tillage   348
mulching   350
multi-peril crop insurance (MPCI)   372
Munich Re   367
MySQL database   308

N

NatCatService   367
national weather service (NWS)   34
natural
– disaster, *see* also disaster   368   97, 371, 456
– – advance planning   451
– – coping strategies   483
– – databases   367
– – early warning   451
– – on environment   437
– – on forestry   436
– – on rangeland   436

– – South America   71
– resources   24
NDVI, *see* normalized difference vegetation index
need assessment   172, 174
net present value (NPV)   232
nitrogen   155, 158, 162
– gas   94
normalized difference vegetation index (NDVI)   472
North American Drought Monitor (NADM)   29

O

oasis agroecosystem   162
ocean-atmosphere phenomenon   99
official warning message   198
off-season dry spells (ODS)   43
on-farm application   191
over-exploitation of irrigated water   245
oxygen   62

P

Palmer Drought Severity Index (PDSI)   27, 31, 283,   285, 257, 304
pastoral farming   107
PCA, *see* principal component analysis
PDF, *see* probability distribution function
PDSI, *see* Palmer Drought Severity Index
peril identification   406
personal/human risk   7, 209
pest
– control techniques   160
– cycle   276
– management risks   481
pest-crop-climate relationship   161
pesticide   157, 158, 218, 222, 261, 266, 481
Philippine experience   220
PI, *see* production insurance
plant
– disease   270
– protoplasm   86
Plasmodiaphora brassicae   270
poikilothermic insects   160
Populus enphratica   330
PostgreSQL database   468
poverty
– alleviation   183
– reduction policies   396

– trap   377, 401, 479
prairie ecosystem   76
precipitation   116
– anomalies   100
– deficiency   24
precision farming   162, 164, 218
pre-crop disease control   265
premium subsidies   397
preparedness strategies   174
price risk   7, 126
principal component analysis (PCA)   238
probability distribution function (PDF)   115

production
– insurance (PI)   383
– management system   191
– risk   7, 125, 209, 377
PRUDENCE   116
Puccinia striiformis f.sp. *tritici*   267
pumping devices   147

Q

qochas   162

R

rain precipitation index   300
rainfall
– analysis computer package   426
– high intensity   59, 104
– intra-season distribution   43
– records   423
rainfed technology   254
rainwater
– harvesting (RWH)   43
– management
– – in rainfall areas   254
– – India   247
raised bed cropping system   215
RANET system   441
rangeland farming   107
ratooning   225
regional circulation model (RCM)   237, 480
regional warming   103
– South Pacific   107
remote sensing   456, 462, 470
renewable raw material   122
resilience   169

resource
– allocation for risks   270
– management   199
– sharing system   200
retention terraces   351
return period   10, 15
rice transplanting   229

risk
– assessment   196, 276
– characterization   4, 7
– climate-related   12, 217
– coping strategies   481
– identification   78
– layering   391, 392
– management   3, 193, 209, 266, 276, 422, 475
– – formal mechanisms   379
– – instruments   381
– – monitoring   6
– – tools   424, 484
– mitigation   47, 379
– modeling   276
– of crop failure   378
– of erosion   80
– preparedness   195
– reduction   130
– – targets   4
– resource allocation   270
– scoping   4, 479
rubber production   231
rural development   220
RustMan   273

S

SADC fund   400
salt manufacturing   409
SALT, *see* sloping agricultural land technology
sand
– barriers   327
– encroachment   317, 319
– storm
– – damage   326
– – monitoring   338
sandification prevention   317
sands
– enclosure belt   331
– stabilization   331
satellite

- instructional television experiment (SITE) 447
- remote sensing 471, 439
scenario analysis 426
Sclerotinia rot 267
- of canola 266
Sclerotinia sclerotiorum 267, 270
sea surface temperature (SST) 237, 371
- anomaly (SSTA) 237
seasonal
- forecast 242
- rainfall in India 184
- weather forecast 114
semi-arid region 150
shelterbelt
- program 326
- system
- - for railway in sandy land areas 329
- - in oasis 328, 330
- - in pasture of sandy land 329
Sis Plant technology 308
SITE, *see* satellite instructional television experiment
skill evaluation 240
slopeland management 226
sloping
- agricultural land technology (SALT) 226
- cropping fields 332
smoke potential indicator 363
social
- capital 181, 182
- development 175
- empowerment 47
society vulnerability 23
socioeconomic drought 25
soil 24
- anaerobic conditions 93
- cultivation 150
- degradation 335
- erodibility 344
- erosion 94, 108, 118, 151, 344, 351, 464
- fertility 150
- high wetness 158
- low workability 158
- management strategies 346
- moisture 293
- - monitoring 148
- productivity 343
- protection 89
- reserves 143
- resources 149
- salinization 74
- temperature 434

soil-water-atmosphere plant (SWAP) 468
South Pacific convergence zone (SPCZ) 103
Southern Oscillation Index 105, 106
sowing
- date 160
- - optimization 159
soybean
- crushers 128
- production 125
- - Brazil 134, 135
- rust 8
spatial information 64
SPI, *see* standardized precipitation index
spikelet sterility 153
sprinkler irrigation 146, 162
SRA, *see* standard reinsurance agreement
stabilizing sands technique system 326
- biological 327
- chemical 328
- engineering 327
stakeholder training 44
standard reinsurance agreement (SRA) 382
standardized precipitation index (SPI) 282, 285, 296
storage
- enterprise 128
- facilities 128
storm 60
straw barrier belt 331
strip tillage 348
stripe rust 267
- of wheat 266
strong wind 89
supplemental irrigation 21
surface
- canal system 149
- irrigation 249
- tanks 149
- temperature 101

T

Tamarix ramosissima 330
technological
- information 176, 177, 180
- innovation 215
technology transfer 269
telecommunication system 198
temperature precipitation 136
terraces 351
thunderstorm 89
tillage practices 346

timely issuance   45
tornado   60
total annual crop loss   475
Traditional Techniques of Microclimate Improvement (TTMI) project   58
transportation system   128
tropical
- cyclone, *see* cyclone
- storm, *see* storm

U

US Drought Monitor (USDM)   27, 33
- classification   28

V

Vegetation Drought Response Index (VegDRI)   35
vegetation index (VI)   64
Vegetation Outlook (VegOut)   35
vertical integration   17
volcanic eruption   73
VSAT communication system   442
vulnerability
- of agriculture   76
- of mountain ecosystem   76
- of natural forests   76
- to hazards   58

W

WAMIS   200, 202, 414
- grid portal   203, 204
- - service-oriented architecture   105
- system components   205
warning system   63, 191
waru-warus   438
water
- balance   257, 465, 467
- crisis management   414
- distribution   147
- erosion   343, 437
- harvesting   21, 47
- - storage structures   249
- in situ conservation   247
- management   146
- - rainfed regions of India   245
- - semi-arid regions   237
- - watershed program   251
- quality problems   259
- resources   145, 414
- - India   246
- scarcity   114
- shortage   145
- storage systems   93
- stress   312
- - concept   295
- - condition   311
- supply   24, 26, 429
- - in Europe   119
- transportation services   128
watershed
- development program   249, 262
- program   250
- - water management   251
weather
- data system   439
- disasters   368, 369
- extreme events   1, 16
- - return period   9
- forecast   2, 65, 214, 358, 390
- index contract   390
- index insurance   386, 480
- - government involvement   394
- - in developing countries   391
- information   169, 184
- insurance   376, 405
- - distribution of products   410
- - initiatives   407
- - weather-based   48
- numerical model   462
- numerical prediction (NWP)   447
- observing stations   463
- risk
- - insurance   375
- - management   48, 114
- seasonal forecasting   18, 114
weather-based insurance   48
weekly weather and crop bulletin (WWCB)   443
West-African semi-arid tropics (WASAT)   226
- cropping pattern   228
- production potential
wet-*seed*ing   229
wildfire   57, 61
- early warning system   355
wind
- erosion   320, 335
- protection   55, 92
- reduction   60
wind-sand damage   325

winer crop   86
– cover crop   350
– frost injury   85
WMO/CAgM   186
World Agricultural Outlook Board
   (WAOB)   129
World Agricultural Supply and Demand
   Estimates (WASDE)   130
World Weather Watch   359

Y

yield
– guarantee insurance   372
– insurance policy   382
– risk   125

Z

zone-specific forecast   66